D0152660

| Quantity | SI Equivalence | English Equivalence |
|---|---|---|
| Work or energy | $1 \text{ in.} - \text{lb}_f = 0.1129848 \text{ J}$<br>$1 \text{ ft} - \text{lb}_f = 1.355818 \text{ J}$<br>$1 \text{ Btu} = 1055.056 \text{ J}$<br>$1 \text{ kWh} = 3.6 \times 10^6 \text{ J}$ | $1 \text{ J} = 8.850744 \text{ in.} - \text{lb}_f$<br>$1 \text{ J} = 0.737562 \text{ ft} - \text{lb}_f$<br>$\quad = 0.947817 \times 10^{-3} \text{ Btu}$<br>$\quad = 0.277778 \text{ kWh}$ |
| Power | $1 \text{ in} - \text{lb}_f/\text{sec} = 0.1129848 \text{ W}$<br>$1 \text{ ft} - \text{lb}_f/\text{sec} = 1.355818 \text{ W}$<br>$\quad = 0.0018182 \text{ hp}$<br>$1 \text{ hp} = 745.7 \text{ W}$ | $1 \text{ W} = 8.850744 \text{ in.} - \text{lb}_f/\text{sec}$<br>$1 \text{ W} = 0.737562 \text{ ft} - \text{lb}_f/\text{sec}$<br>$\quad = 1.34102 \times 10^{-3} \text{ hp}$ |
| Area moment of inertia or second moment of area | $1 \text{ in}^4 = 41.6231 \times 10^{-8} \text{ m}^4$<br>$1 \text{ ft}^4 = 86.3097 \times 10^{-4} \text{ m}^4$ | $1 \text{ m}^4 = 240.251 \times 10^4 \text{ in}^4$<br>$\quad = 115.862 \text{ ft}^4$ |
| Mass moment of inertia | $1 \text{ in} - \text{lb}_f - \text{sec}^2 = 0.1129848 \text{ m}^2 \cdot \text{kg}$ | $1 \text{ m}^2 \cdot \text{kg} = 8.850744 \text{ in.} - \text{lb}_f - \text{sec}^2$ |
| Spring constant:<br>  translatory<br><br>  torsional | $1 \text{ lb}_f/\text{in.} = 175.1268 \text{ N/m}$<br>$1 \text{ lb}_f/\text{ft} = 14.5939 \text{ N/m}$<br>$1 \text{ in.} - \text{lb}_f/\text{rad} = 0.1129848 \text{ m} \cdot \text{N/rad}$ | $1 \text{ N/m} = 5.71017 \times 10^{-3} \text{ lb}_f/\text{in.}$<br>$\quad = 68.5221 \times 10^{-3} \text{ lb}_f/\text{ft}$<br>$1 \text{ m} \cdot \text{N/rad} = 8.850744 \text{ in-lb}_f/\text{rad}$<br>$\quad = 0.737562 \text{ lb}_f - \text{ft/rad}$ |
| Damping constant:<br>  translatory<br>  torsional | $1 \text{ lb}_f - \text{sec/in} = 175.1268 \text{ N} \cdot \text{s/m}$<br>$1 \text{ in} - \text{lb}_f - \text{sec/rad}$<br>$\quad = 0.1129848 \text{ m} \cdot \text{N} \cdot \text{s/rad}$ | $1 \text{ N} \cdot \text{s/m} = 5.71017 \times 10^{-3} \text{ lb}_f - \text{sec/in.}$<br>$1 \text{ m} \cdot \text{N} \cdot \text{s/rad} = 8.850744 \text{ lb}_f - \text{in} - \text{sec/rad}$ |
| Angles | $1 \text{ rad} = 57.295754 \text{ degrees};$     $1 \text{ degree} = 0.0174533 \text{ rad};$<br>$1 \text{ rpm} = 0.166667 \text{ rev/sec} = 0.104720 \text{ rad/sec};$    $1 \text{ rad/sec} = 9.54909 \text{ rpm}$ | |

# Mechanical Vibrations

**Singiresu S. Rao**

Purdue University

**Addison-Wesley Publishing Company**
Reading, Massachusetts  ·  Menlo Park, California  ·
Don Mills, Ontario  ·  Wokingham, England  ·  Amsterdam  ·
Sydney  ·  Singapore  ·  Tokyo  ·  Mexico City  ·
Bogotá  ·  Santiago  ·  San Juan

**Library of Congress Cataloging in Publication Data**

Rao, S. S.
  Mechanical vibrations.

  Includes bibliographies and index.
  1.  Vibration.    I.  Title.
TA355.A37  1984      620.3      84-18484
ISBN 0-201-06550-9

Copyright © 1986 by Addison-Wesley Publishing Company, Inc. All rights reserved. No part of this publication may be reproduced, stored in a retrieval system, or transmitted, in any form or by any means, electronic, mechanical, photocopying, recording, or otherwise, without the prior written permission of the publisher. Printed in the United States of America. Published simultaneously in Canada.

HIJ-HA-89

*To Lord Sri Venkateswara*

# Preface

The increasing application of digital computer techniques in engineering has not only shifted the emphasis among several methods of vibration analysis, but has also facilitated the development of many areas that depend heavily on the availability of a high-speed digital computer. This book is written with these trends in mind and to present the theory, computational aspects, and applications of vibrations in a manner as clear and precise as the subject permits.

The student is assumed to have an elementary knowledge of dynamics, calculus, differential equations, strength of materials, and Fortran programming.

## CONTENTS

*Mechanical Vibrations* is organized into 11 chapters and 3 appendixes. The material of the book provides flexible options for different types of vibration courses. For a one-semester senior or dual-level course, Chapters 1 through 5, portions of Chapters 6, 7, and 8, and Chapter 9 may be used. The course can be given a computer orientation by including Chapter 10 in place of Chapter 8. Alternatively, with supplementary reading for the material of Chapter 11, the text has sufficient material for a one-year sequence at the senior level. For shorter courses, the instructor can select the topics depending on the level and orientation of the course. It is hoped that the relative simplicity with which the various topics are presented makes the book useful to students as well as practicing engineers for purposes of self-study and as a reference source.

Chapter 1 starts with a brief discussion of the history and importance of vibrations. The basic concepts and terminology used in vibration analysis are introduced. The free vibration analysis of single-degree-of-freedom undamped translational and torsional systems is given in Chapter 2. The effects of viscous, Coulomb and hysteretic damping are also discussed. The harmonic response of single-degree-of-freedom systems is considered in Chapter 3. The theory of vibration measuring instruments is also presented. Chapter 4 is concerned with the response of a single-degree-of-freedom system under general forcing conditions. The roles of convolution integral, Laplace transformation, and numerical methods are discussed. The concept of response spectrum is also introduced in this chapter. The free and forced vibration of two-degree-of-freedom systems is considered in Chapter 5. The self-excited vibration and stability of the system are discussed. Chapter 6 presents the vibration analysis of multidegree-of-freedom systems. Matrix methods of analysis are used for the presentation of the theory. The modal analysis procedure is described for the solution of forced vibration problems. Several methods of determining the natural frequencies of discrete systems are outlined in Chapter 7. Dunkerley's, Rayleigh's, Holzer's, matrix iteration, and Jacobi's methods are discussed.

The vibration analysis of continuous systems including strings, bars, shafts, beams, and membranes is given in Chapter 8. The Rayleigh and Rayleigh-Ritz methods of finding the approximate natural frequencies are also described. Chapter 9 discusses the various aspects of vibration control including the problems of elimination, isolation and absorption. The balancing of rotating and reciprocating machines and whirling of shafts are also considered. Chapter 10 presents several numerical integration techniques for finding the dynamic response of discrete and continuous systems. The central difference, Runge-Kutta, Houbolt, Wilson, and Newmark methods are summarized and illustrated. An introductory treatment of nonlinear vibration, random vibration and finite element method of vibration analysis is presented in Chapter 11. Appendixes A, B, and C outline the basic relations of matrices, Laplace transforms, and SI units, respectively.

## FEATURES

Each topic in *Mechanical Vibrations* is self-contained. In derivations and developments, steps needed for continuity of understanding have been included to aid the reader at the introductory level. The computational aspects are emphasized throughout the book. Several Fortran computer programs, most of them in the form of general purpose subroutines, are given that can be adopted by most Fortran compilers, perhaps with minor modifications in format statements.

Problems, which are based on the use/development of computer programs, are given at the end of each chapter. It is recommended that the students solve these problems so that they become aware of many computational and programming problems and pitfalls.

Some subjects are presented in a somewhat unconventional manner. The topics of the last three chapters fall in this category. Most textbooks discuss the topics of isolators, absorbers, and balancing at different places. Since one of the main purposes of the study of vibrations is to control vibration response, all the topics directly related to vibration control are given in Chapter 9. Similarly, all the numerical integration methods applicable to single- and multidegree-of-freedom systems, as well as continuous systems, are unified in Chapter 10. And finally, nonlinear vibration, random vibration, and finite element analysis are all covered in Chapter 11.

Specific features include:

* Computer programs to aid the student in the numerical implementation of the methods discussed in the text.
* Illustrative examples following the presentation of most of the topics.
* Review questions to help students in reviewing and testing their understanding of the text material.
* Over 500 problems, with solutions in the instructor's manual.
* References to lead the reader to specialized and advanced literature.
* Brief biographical information about scientists and engineers who contributed to the development of theory of vibrations.

## NOTATION AND UNITS

A list of symbols, along with the associated units in SI and English systems, is given at the beginning of the book. Arrows are used over symbols to denote column vectors and square brackets are used to indicate matrices. The SI units are used in most of the examples and problems. A brief discussion of SI units as they apply to the vibration field is given in Appendix C.

## ACKNOWLEDGMENTS

Among other resources I have consulted in the preparation of this book, I wish to acknowledge the influence of *Vibration Engineering* by Professor Andrew D. Dimarogonas, West Publishing Co., ©1976, particularly in the preparation of some of the examples and problems presented in this book. I would also like to express my appreciation to the students who used the notes that led to the present text. The suggestions made by my three reviewers have been of much help. It was gratifying throughout to work with the staff of Addison-Wesley on this project. Finally I wish to thank my wife Kamala and daughters Sridevi and Shobha Rani without whose patience, encouragement, and support this book might never have been completed.

S.S.R.

# Contents

**CHAPTER 2**

# Free Vibration of Single Degree of Freedom Systems

**CHAPTER 3**

# Harmonically Excited Vibration

**CHAPTER 4**

# Vibration under General Forcing Conditions

**CHAPTER 5**

# Two Degree of Freedom Systems

**CHAPTER 6**

# Multidegree of Freedom Systems

## CHAPTER 7

# Determination of Natural Frequencies and Mode Shapes

## CHAPTER 8

# Continuous Systems

## CHAPTER 9

# Vibration Control

**CHAPTER 10**

# Numerical Integration Methods in Vibration Analysis

# LIST OF SYMBOLS

| Symbol | Meaning | English Units | SI Units |
|---|---|---|---|
| $a, a_0, a_1, a_2, \ldots$ | constants, lengths | | |
| $a_{ij}$ | flexibility coefficient | in./lb | m/N |
| $[a]$ | flexibility matrix | in./lb | m/N |
| $A$ | area | $\text{in}^2$ | $\text{m}^2$ |
| $A, A_0, A_1, \ldots$ | constants | | |
| $b, b_1, b_2, \ldots$ | constants, lengths | | |
| $B, B_1, B_2, \ldots$ | constants | | |
| $\vec{B}$ | balancing weight | lb | N |
| $c, \underline{c}$ | viscous damping coefficient | lb-sec/in. | N·s/m |
| $c, c_0, c_1, c_2, \ldots$ | constants | | |
| $c$ | wave velocity | in./sec | m/s |
| $c_c$ | critical viscous damping constant | lb-sec/in. | N·s/m |
| $c_i$ | damping constant of $i$th damper | lb-sec/in. | N·s/m |
| $c_{ij}$ | damping coefficient | lb-sec/in. | N·s/m |
| $[c]$ | damping matrix | lb-sec/in. | N·s/m |
| $C, C_1, C_2, C_1', C_2'$ | constants | | |
| $d$ | diameter, dimension | in. | m |
| $D$ | diameter | in. | m |
| $[D]$ | dynamical matrix | $\text{sec}^2$ | $\text{s}^2$ |
| $e$ | base of natural logarithms | | |
| $e$ | eccentricity | in. | m |
| $\vec{e}_x, \vec{e}_y$ | unit vectors parallel to $x$ and $y$ directions | | |
| $E$ | Young's modulus | $\text{lb/in}^2$ | Pa |
| $f$ | linear frequency | Hz | Hz |
| $f$ | force per unit length | lb/in. | N/m |
| $\underline{f}$ | unit impulse | lb-sec | N·s |
| $F, F_d$ | force | lb | N |
| $F_0$ | amplitude of force $F(t)$ | lb | N |
| $F_t, F_T$ | force transmitted | lb | N |
| $F_i$ | force acting on $i$th mass | lb | N |
| $F_x, F_y$ | forces along $x$ and $y$ directions | lb | N |
| $\vec{F}$ | force vector | lb | N |
| $\underline{F}$ | impulse | lb-sec | N·s |
| $g$ | acceleration due to gravity | $\text{in./sec}^2$ | $\text{m/s}^2$ |
| $g(t)$ | impulse response function | | |

| Symbol | Meaning | English Units | SI Units |
|---|---|---|---|
| $G$ | shear modulus | $lb/in^2$ | $N/m^2$ |
| $h$ | thickness, dimension | in. | m |
| $h$ | time increment | sec | s |
| $H(i\omega)$ | frequency response function | | |
| $i$ | $\sqrt{-1}$ | | |
| $I$ | area moment of inertia | $in^4$ | $m^4$ |
| $[I]$ | identity matrix | | |
| Im( ) | imaginary part of ( ) | | |
| $J$ | polar moment of inertia | $in^4$ | $m^4$ |
| $J_0, J_1, J_2, \ldots$ | mass moment of inertia | $lb\text{-}in./sec^2$ | $kg \cdot m^2$ |
| $k, \underline{k}$ | spring constant | $lb/in.$ | $N/m$ |
| $k_i$ | spring constant of $i$th spring | $lb/in.$ | $N/m$ |
| $k_{ij}$ | stiffness coefficient | $lb/in.$ | $N/m$ |
| $[k]$ | stiffness matrix | $lb/in.$ | $N/m$ |
| $l, l_i$ | length | in. | m |
| $l_0$ | unstretched length of spring | in. | m |
| $m, \underline{m}$ | mass | $lb\text{-}sec^2/in.$ | kg |
| $m_i$ | $i$th mass | $lb\text{-}sec^2/in.$ | kg |
| $m_{ij}$ | mass coefficient | $lb\text{-}sec^2/in.$ | kg |
| $[m]$ | mass matrix | $lb\text{-}sec^2/in.$ | kg |
| $M$ | mass | $lb\text{-}sec^2/in.$ | kg |
| $M$ | moment | $lb\text{-}in.$ | $N \cdot m$ |
| $M_t, M_{t1}, M_{t2}, \ldots$ | torque | $lb\text{-}in.$ | $N \cdot m$ |
| $M_{t0}$ | amplitude of $M_t(t)$ | $lb\text{-}in.$ | $N \cdot m$ |
| $n$ | an integer | | |
| $n$ | number of degrees of freedom | | |
| $N$ | normal force | lb | N |
| $N$ | total number of time steps | | |
| $p$ | pressure | $lb/in^2$ | $N/m^2$ |
| $P$ | force, tension | lb | N |
| $q_j$ | $j$th generalized coordinate | | |
| $\vec{q}$ | vector of generalized displacements | | |
| $\dot{\vec{q}}$ | vector of generalized velocities | | |
| $Q$ | fluid flow rate | $in^3/sec$ | $m^3/s$ |
| $Q_j$ | $j$th generalized force | | |
| $r$ | frequency ratio $= \omega/\omega_n$ | | |

# LIST OF SYMBOLS (*continued*)

| Symbol | Meaning | English Units | SI Units |
|---|---|---|---|
| $\vec{r}$ | radius vector | in. | m |
| Re( ) | real part of ( ) | | |
| $R$ | radial distance | in. | m |
| $R$ | Rayleigh's dissipation function | lb-in/sec | N·m/s |
| $R$ | Rayleigh's quotient | $1/\sec^2$ | $1/s^2$ |
| $s$ | exponential coefficient, root of equation | | |
| $S_a, S_d, S_v$ | acceleration, displacement, velocity spectrum | | |
| $t$ | time parameter | sec | s |
| $t_i$ | $i$th time station | sec | s |
| $T$ | kinetic energy | in.-lb | J |
| $T_i$ | kinetic energy of $i$th mass | in.-lb | J |
| $T_r$ | transmissibility ratio | | |
| $T(t)$ | a function of $t$ | | |
| $u$ | axial displacement | in. | m |
| $u_0$ | value of $u$ at $t=0$ | in. | m |
| $\dot{u}_0$ | value of $\dot{u}$ at $t=0$ | in./sec | m/s |
| $u_{ij}$ | an element of matrix $[U]$ | | |
| $U$ | axial displacement | in. | m |
| $U$ | potential energy | in.-lb | J |
| $\vec{U}$ | unbalanced weight | lb | N |
| $[U]$ | upper triangular matrix | | |
| $v, v_0$ | linear velocity | in./sec | m/s |
| $V$ | velocity | in./sec | m/s |
| $V$ | shear force | lb | N |
| $V$ | potential energy | in.-lb | J |
| $V_i$ | potential energy of $i$th spring | in.-lb | J |
| $w, w_1, w_2$ | transverse deflections | in. | m |
| $w_0$ | value of $w$ at $t=0$ | in. | m |
| $\dot{w}_0$ | value of $\dot{w}$ at $t=0$ | in./sec | m/s |
| $w_n$ | $n$th mode of vibration | | |
| $W$ | weight of a mass | lb | N |
| $W$ | total energy | in.-lb | J |
| $W$ | transverse deflection | in. | m |
| $W_i$ | value of $W$ at $t=t_i$ | in. | m |
| $W(x)$ | a function of $x$ | | |

| Symbol | Meaning | English Units | SI Units |
|---|---|---|---|
| $x, y, z$ | cartesian coordinates, displacements | in. | m |
| $x_0, x(0)$ | value of $x$ at $t = 0$ | in. | m |
| $\dot{x}_0, \dot{x}(0)$ | value of $\dot{x}$ at $t = 0$ | in./sec | m/s |
| $x_j$ | displacement of $j$th mass | in. | m |
| $x_j$ | value of $x$ at $t = t_j$ | in. | m |
| $\dot{x}_j$ | value of $\dot{x}$ at $t = t_j$ | in./sec | m/s |
| $x_h$ | homogeneous part of $x(t)$ | in. | m |
| $x_p$ | particular part of $x(t)$ | in. | m |
| $\vec{x}$ | vector of displacements | in. | m |
| $\vec{x}_i$ | value of $\vec{x}$ at $t = t_i$ | in. | m |
| $\dot{\vec{x}}_i$ | value of $\dot{\vec{x}}$ at $t = t_i$ | in./sec | m/s |
| $\ddot{\vec{x}}_i$ | value of $\ddot{\vec{x}}$ at $t = t_i$ | in./sec$^2$ | m/s$^2$ |
| $\vec{x}^{(i)}(t)$ | $i$th mode | | |
| $X$ | amplitude of $x(t)$ | in. | m |
| $X_j$ | amplitude of $x_j(t)$ | in. | m |
| $\vec{X}^{(i)}$ | $i$th modal vector | in. | m |
| $X_i^{(j)}$ | $i$th component of $j$th mode | in. | m |
| $[X]$ | modal matrix | in. | m |
| $\vec{X}_r$ | $r$th approximation to a mode shape | | |
| $y$ | base displacement | in. | m |
| $Y$ | amplitude of $y(t)$ | in. | m |
| $z$ | relative displacement, $x - y$ | in. | m |
| $Z$ | amplitude of $z(t)$ | in. | m |
| $Z(i\omega)$ | mechanical impedance | lb/in. | N/m |
| $\alpha$ | angle, constant | | |
| $\beta$ | angle, constant | | |
| $\beta$ | hysteresis damping constant | | |
| $\gamma$ | specific weight | lb/in$^3$ | N/m$^3$ |
| $\gamma$ | shear strain | | |
| $\delta$ | logarithmic decrement | | |
| $\delta_1, \delta_2, \ldots$ | deflections | in. | m |
| $\delta_{st}$ | static deflection | in. | m |
| $\delta_{ij}$ | Kronecker delta | | |
| $\vec{\delta}$ | column vector of Kronecker deltas | | |
| $\Delta$ | determinant | | |
| $\Delta F$ | increment in $F$ | lb | N |

## LIST OF SYMBOLS (*continued*)

| Symbol | Meaning | English Units | SI Units |
|---|---|---|---|
| $\Delta x$ | increment in $x$ | in. | m |
| $\Delta t$ | increment in time $t$ | sec | s |
| $\Delta W$ | energy dissipated in a cycle | in.-lb | J |
| $\varepsilon$ | a small quantity | | |
| $\zeta$ | damping ratio | | |
| $\theta$ | constant, angular displacement | | |
| $\theta_i$ | $i$th angular displacement | rad | rad |
| $\theta_0$ | value of $\theta$ at $t = 0$ | rad | rad |
| $\dot{\theta}_0$ | value of $\dot{\theta}$ at $t = 0$ | rad/sec | rad/s |
| $\Theta$ | amplitude of $\theta(t)$ | rad | rad |
| $\Theta_i$ | amplitude of $\theta_i(t)$ | rad | rad |
| $\lambda$ | eigenvalue $= 1/\omega^2$ | $\text{sec}^2$ | $\text{s}^2$ |
| $\mu$ | viscosity of a fluid | $\text{lb-sec}/\text{in}^2$ | $\text{kg}/\text{m}\cdot\text{s}$ |
| $\mu$ | coefficient of friction | | |
| $\rho$ | mass density | $\text{lb-sec}^2/\text{in}^4$ | $\text{kg}/\text{m}^3$ |
| $\sigma$ | stress | $\text{lb}/\text{in}^2$ | $\text{N}/\text{m}^2$ |
| $\tau$ | period of oscillation, time parameter | sec | s |
| $\tau$ | shear stress | $\text{lb}/\text{in}^2$ | $\text{N}/\text{m}^2$ |
| $\phi$ | angle, phase angle | rad | rad |
| $\phi_i$ | phase angle in $i$th mode | rad | rad |
| $\omega$ | frequency of oscillation | rad/sec | rad/s |
| $\omega_i$ | $i$th natural frequency | rad/sec | rad/s |
| $\omega_n$ | natural frequency | rad/sec | rad/s |

| Symbol | Meaning | English Units | SI Units |
|--------|---------|---------------|----------|
| $\omega_d$ | frequency of damped vibration | rad/sec | rad/s |

**SUBSCRIPTS**

| | | | |
|--------|---------|---------------|----------|
| cri | critical value | | |
| eq | equivalent value | | |
| $i$ | $i$th value | | |
| $L$ | left plane | | |
| max | maximum value | | |
| $n$ | corresponding to natural frequency | | |
| $R$ | right plane | | |
| 0 | specific or reference value | | |
| $t$ | torsional | | |

**OPERATIONS**

| Symbol | Meaning |
|--------|---------|
| $(\dot{\ })$ | $\dfrac{d(\ )}{dt}$ |
| $(\ddot{\ })$ | $\dfrac{d^2(\ )}{dt^2}$ |
| $\vec{(\ )}$ | column vector ( ) |
| [ ] | matrix |
| $[\ ]^{-1}$ | inverse of [ ] |
| $[\ ]^T$ | transpose of [ ] |
| $\Delta(\ )$ | increment in ( ) |
| $\mathscr{L}(\ )$ | Laplace transform of ( ) |
| $\mathscr{L}^{-1}(\ )$ | inverse Laplace transform of ( ) |

# CHAPTER 1

# Fundamentals
# of Vibration

Galileo Galilei (1564–1642), an Italian astronomer, philosopher, and professor of mathematics at the Universities of Pisa and Padua, in 1609 became the first man to point a telescope to the sky. He wrote the first treatise on modern dynamics in 1590. His works on the oscillations of a simple pendulum and the vibration of strings are of fundamental significance in the theory of vibrations.

Courtesy of the Granger Collection

## 1.1 PRELIMINARY REMARKS

This chapter introduces the subject of vibrations in a relatively simple manner. The chapter begins with a brief history of the subject and continues with an examination of its importance. The various steps involved in vibration analysis of an engineering system are outlined, and essential definitions and concepts of vibration are introduced. There follows a presentation of the concept of harmonic analysis, which can be used for the analysis of general periodic motions. No attempt at exhaustive treatment is made in Chapter 1; subsequent chapters will develop many of the ideas in more detail.

## 1.2 BRIEF HISTORY OF VIBRATION

People became interested in vibration when the first musical instruments, probably whistles or drums, were discovered. Since then, people have applied ingenuity and critical investigation to study the phenomenon of vibration. Galileo discovered the relationship between the length of a pendulum and its frequency and observed the resonance of two bodies that were connected by some energy transfer medium and tuned to the same natural frequency. Further, he observed the interrelationships of the density, tension, length, and frequency of a vibrating string [1.1]. Although it had long been understood that sound was related to the vibration of a mechanical system, it was not clear that pitch is determined by the frequency of vibration until Galileo found the result. At about the same time as Galileo, Hooke showed the relationship between frequency and pitch.

Among mathematicians, Taylor, Bernoulli, d'Alembert, Euler, Lagrange, and Fourier made valuable contributions to the development of vibration theory. Wallis and Sauveur observed, independently, the phenomenon of mode shapes (with stationary points, called nodes) in vibrating strings. They also established that the frequency of the second mode is twice that of the first and the frequency of the third mode three times that of the first. Sauveur is credited with coining the term *fundamental* for the lowest frequency and *harmonics* for the others. Bernoulli first proposed the principle of linear superposition of harmonics: Any general configuration of free vibration is made up of the configurations of individual harmonics, acting independently in varying strengths [1.2].

After the enunciation of Hooke's law of elasticity in 1676, Euler (1744) and Bernoulli (1751) derived the differential equation governing the lateral vibration of prismatic bars and investigated its solution for the case of small deflections. In 1784, Coulomb did both theoretical and experimental studies of the torsional oscillations of a metal cylinder suspended by a wire.

There is an interesting story related to the development of the theory of vibration of plates [1.3]. In 1802, Chladni developed the method of placing sand on a vibrating plate to find its mode shapes and observed the beauty and intricacy of the modal patterns of the vibrating plates. In 1809, the French Academy invited Chladni to give a demonstration of his experiments. Napoleon Bonaparte, who attended the meeting, was very impressed and presented a sum of 3000 francs to the Academy, to be awarded to the first person to give a satisfactory mathematical theory of the vibration of plates. By the closing date of the competition in October, 1811, only one candidate, Sophie Germain, had entered the contest. But Lagrange, who was one of the judges, noticed an error in the derivation of her differential equation of motion. The Academy opened the competition again, with a new closing date of October, 1813. Sophie Germain again entered the contest, presenting the correct form of the differential equation. However, the Academy did not award the prize to her because the judges wanted physical justification of the assumptions made in her derivation. The competition was opened once more. In her third attempt, Sophie Germain was finally awarded the prize in 1816, although the judges were not completely satisfied with her theory. In fact, it was later found that her differential equation was correct but that the boundary conditions were erroneous. The correct boundary conditions for the vibration of plates were given in 1850 by Kirchoff.

After this, vibration studies were done on a number of practical mechanical and structural systems. In 1877, Lord Rayleigh published his book on the theory of sound [1.4]; it is considered a classic on the subject of vibrations even today. Notable among the many contributions of Rayleigh is the method of finding the fundamental frequency of vibration of a conservative system by making use of the principle of conservation of energy—now known as Rayleigh's method [1.5]. In 1902, Frahm investigated the importance of torsional vibration study in the design of propeller shafts of steamships. The dynamic vibration absorber, which involves the addition of a secondary spring-mass system to eliminate the vibrations of a main system, was also proposed by Frahm in 1909. Among the modern contributors to the theory of vibrations, the names of Stodola, Timoshenko, and Mindlin are notable. Stodola's method of analyzing vibrating beams is also applicable to turbine blades. The works of Timoshenko and Mindlin resulted in improved theories of vibration of beams and plates.

It has long been recognized that many basic problems of mechanics, including those of vibrations, are nonlinear. Although the linear treatments commonly adopted are quite satisfactory for most purposes, they are not adequate in all cases. In nonlinear systems, there often occur phenomena that are theoretically impossible in linear systems. The mathematical theory of nonlinear vibrations began to develop in the works of Poincaré and Lyapunov at the end of the last century. After 1920, studies undertaken by Duffing and van der Pol brought the first definite solutions into the theory of nonlinear vibrations and drew attention to its importance in engineering. In the last 20 years, authors like Minorsky and Stoker have endeavored to collect the main results concerning nonlinear vibrations in the form of monographs [1.6, 1.7].

Random characteristics are present in diverse phenomena such as earthquakes, winds, transportation of goods on wheeled vehicles, and rocket and jet engine noise. It became necessary to devise concepts and methods of vibration analysis for these random effects. Although Einstein considered Brownian movement, a particular type of random vibration, as long ago as 1905, no applications were investigated until 1930. The introduction of the correlation function by Taylor in 1920 and of the

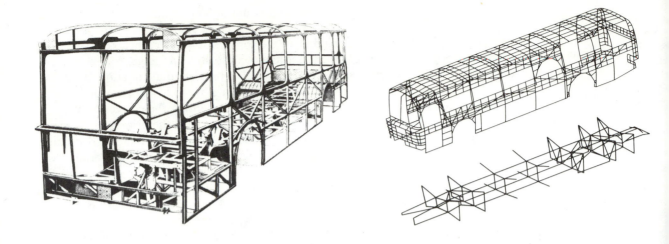

**Figure 1.1**   Finite element idealization of the body of a bus [1.12]. *(Reprinted with permission © 1974 Society of Automotive Engineers, Inc.)*

spectral density by Wiener and Khinchin in the early 1930s opened new prospects for progress in the theory of random vibrations. Papers by Lin and Rice, published between 1943 and 1945, paved the way for the application of random vibrations to practical engineering problems. The monographs of Crandall and Mark, and Robson systematized the existing knowledge in the theory of random vibrations [1.8, 1.9].

Until about 25 years ago, vibration studies, even those dealing with complex engineering systems, were done by using gross models, with only a few degrees of freedom. However, the advent of high-speed digital computers in the 1950s made it possible to treat moderately complex systems and to generate approximate solutions in semi-closed form, relying on classical solution methods but using numerical evaluation of certain terms that can not be expressed in closed form. The simultaneous development of the finite element method enabled engineers to use digital computers to conduct numerically detailed vibration analysis of complex mechanical, vehicular, and structural systems displaying thousands of degrees of freedom [1.10, 1.11]. Figure 1.1 shows the finite element idealization of the body of a bus [1.12].

## **1.3**  IMPORTANCE OF THE STUDY OF VIBRATION

Early scholars in this field concentrated their efforts on understanding the natural phenomena and developing mathematical theories to describe the vibration of physical systems. In recent times, many investigations have been motivated by the

**Figure 1.2**  Tacoma Narrows bridge during wind-induced vibration. The bridge opened on 1 July 1940 and collapsed on 7 November 1940. *(Farquharson photo, Historical Photography, Collection, University of Washington Libraries.)*

engineering applications of vibration, such as the design of machines, foundations, structures, engines, turbines, and control systems.

Most prime movers have vibrational problems due to the inherent unbalance in the engines. The unbalance may be due to faulty design or poor manufacture. Imbalance in diesel engines, for example, can cause ground waves sufficiently powerful to create a nuisance in urban areas. The wheels of some locomotives can rise more than a centimeter off the track at high speeds due to unbalance. In turbines, vibrations cause spectacular mechanical failures. Engineers have not yet been able to prevent the failures that result from blade and disk vibrations in turbines. Naturally, the structures designed to support heavy centrifugal machines, like motors and turbines, or reciprocating machines, like steam and gas engines and reciprocating pumps, are also subjected to vibration. In all these situations, the structure or machine component subjected to vibration can fail because of material fatigue resulting from the cyclic variation of the induced stress. Furthermore, the vibration causes more rapid wear of machine parts such as bearings and gears and also creates excessive noise.

Whenever the natural frequency of vibration of a machine or structure coincides with the frequency of the external excitation, there occurs a phenomenon known as *resonance*, which leads to excessive deflections and failure. The literature is full of accounts of system failures brought about by resonance and excessive vibration of components and systems (see Fig. 1.2). Because of the devastating effects that vibrations can have on machines and structures, vibration testing [1.13] has become a standard procedure in the design and development of most engineering systems (see Fig. 1.3).

**Figure 1.3**  Vibration testing of the space shuttle Enterprise. *(Courtesy of NASA.)*

In many engineering systems, a human being acts as an integral part of the system. The transmission of vibration to human beings results in discomfort and loss of efficiency [1.14]. Vibration of instrument panels can cause their malfunction or difficulty in reading the meters [1.15]. Thus one of the important purposes of vibration study is to reduce vibration through proper design of machines and their mountings. In this connection, the mechanical engineer tries to design the engine or

**Figure 1.4**  Vibratory finishing process. *(Reprinted courtesy of the Society of Manufacturing Engineers, © 1964 The Tool and Manufacturing Engineer.)*

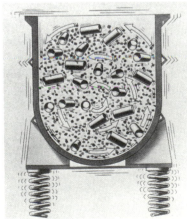

machine so as to minimize unbalance, while the structural engineer tries to design the supporting structure so as to ensure that the effect of the imbalance will not be harmful [1.16].

In spite of its detrimental effects, vibration can be utilized profitably in several industrial applications. In fact, the applications of vibratory equipment have increased considerably in recent years [1.17]. For example, vibration is put to work in vibratory conveyors, hoppers, sieves, and compactors. Vibration is also used in pile driving, vibratory testing of materials, vibratory finishing processes, and electronic circuits to filter out the unwanted frequencies (see Fig. 1.4). Vibration has been found to improve the efficiency of certain machining, casting, forging, and welding processes. It is employed to simulate earthquakes for geological research and also to conduct studies in the design of nuclear reactors.

## 1.4   BASIC CONCEPTS OF VIBRATION

### 1.4.1   Vibration

Any motion which repeats itself after an interval of time is called *vibration* or *oscillation*. The swinging of a pendulum and the motion of a plucked string are typical examples of vibration. The theory of vibration deals with the study of oscillatory motions of bodies and the forces associated with them.

### 1.4.2   Elementary Parts of Vibrating Systems

A vibratory system, in general, includes a means for storing potential energy (spring or elasticity), a means for storing kinetic energy (mass or inertia), and a means by which energy is gradually lost (damper).

The vibration of a system involves the transfer of its potential energy to kinetic energy and kinetic energy to potential energy, alternately. If the system is damped, some energy is dissipated in each cycle of vibration and must be replaced by an external source if a state of steady vibration is to be maintained.

As an example, consider the vibration of the simple pendulum shown in Fig. 1.5. Let the bob of mass $m$ be released after giving it an angular displacement $\theta$. At position 1 the velocity of the bob and hence its kinetic energy is zero. But it has a potential energy of magnitude $mgl(1 - \cos\theta)$ with respect to the datum position 2. Since the gravitational force $mg$ induces a torque $mgl\sin\theta$ about the point $O$, the bob starts swinging to the left from position 1. This imparts to the bob a certain angular acceleration in the clockwise direction, and by the time it reaches position 2, all of its potential energy will be converted into kinetic energy. Hence the bob will not stop in position 2, but will continue to swing to position 3. As it crosses the mean position 2, however, a counterclockwise torque starts acting on the bob (again due to gravity) and causes the bob to decelerate. Eventually, the velocity of the bob reduces to zero at the left extreme position. By this time, all the kinetic energy of the bob will be converted to potential energy. Again the gravity torque continues to act and gives the bob a counterclockwise velocity. Hence the bob starts swinging back with progressively increasing velocity and overshoots the mean position again. This sequence of events keeps on repeating, and the pendulum keeps on performing

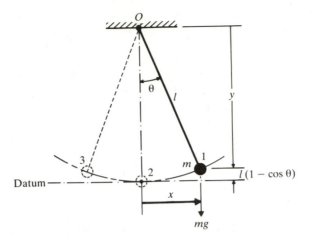

**Figure 1.5**   A simple pendulum.

oscillations. However, the magnitude of oscillation ($\theta$) gradually decreases and the pendulum ultimately stops due to the resistance (damping) offered by the surrounding medium (air). This means that some energy is dissipated in each cycle of vibration due to damping by the air.

### 1.4.3   Degree of Freedom

The minimum number of independent coordinates required to determine completely the positions of all parts of a system at any instant of time defines the degree of freedom of the system. The simple pendulum shown in Fig. 1.5, as well as each of the systems shown in Fig. 1.6, represents a single degree of freedom system. For example, the motion of the simple pendulum (Fig. 1.5) can be stated either in terms of the angle $\theta$ or in terms of the cartesian coordinates $x$ and $y$. If the coordinates $x$ and $y$ are used to describe the motion, it must be recognized that these coordinates

**Figure 1.6**   Single degree of freedom systems.

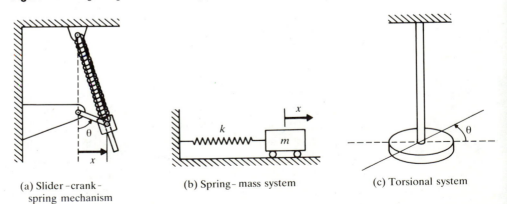

(a) Slider–crank–spring mechanism     (b) Spring–mass system     (c) Torsional system

are not independent. They are related to each other through the relation $x^2 + y^2 = l^2$ where $l$ is the constant length of the pendulum. Thus any one coordinate can describe the motion of the pendulum. In this example, we find that the choice of $\theta$ as the independent coordinate will be more convenient than the choice of $x$ or $y$. For the slider shown in Fig. 1.6(a), either the angular coordinate $\theta$ or the coordinate $x$ can be used to describe the motion. In Fig. 1.6(b), the linear coordinate $x$ can be used to specify the motion. For the torsional system (long bar with a heavy disk at the end) shown in Fig. 1.6(c), the angular coordinate $\theta$ can be used to describe the motion.

Some examples of two and three degree of freedom systems are shown in Figs. 1.7 and 1.8 respectively. Figure 1.7(a) shows a two mass-two spring system that is described by the two linear coordinates $x_1$ and $x_2$. Figure 1.7(b) denotes a two rotor system whose motion can be specified in terms of $\theta_1$ and $\theta_2$. The motion of the system shown in Fig. 1.7(c) can be described completely either by $X$ and $\theta$ or by $x$, $y$, and $X$. In the latter case, $x$ and $y$ are constrained as $x^2 + y^2 = l^2$ where $l$ is a constant. For the systems shown in Figs. 1.8(a) and 1.8(c), the coordinates $x_i$ ($i = 1, 2, 3$) and $\theta_i$ ($i = 1, 2, 3$) can be used, respectively, to describe the motion. In the case of the system shown in Fig. 1.8(b), $\theta_i$ ($i = 1, 2, 3$) specifies the positions of the masses $m_i$ ($i = 1, 2, 3$). An alternate method of describing this system is in terms of $x_i$ and $y_i$ ($i = 1, 2, 3$); but in this case the constraints $x_i^2 + y_i^2 = l_i^2$ ($i = 1, 2, 3$) have to be considered.

The coordinates necessary to describe the motion of a system constitute a set of *generalized coordinates*. The generalized coordinates are usually denoted as $q_1, q_2, \ldots$ and may represent cartesian and/or noncartesian coordinates.

### 1.4.4 Discrete and Continuous Systems

A large class of real systems can be described by a finite number of degrees of freedom, such as the simple systems shown in Figs. 1.5 to 1.8. Several other systems, such as those that include deformable members, have an infinite number of degrees of freedom. As a simple example, consider the cantilever beam shown in Fig. 1.9.

**Figure 1.7** Two degree of freedom systems.

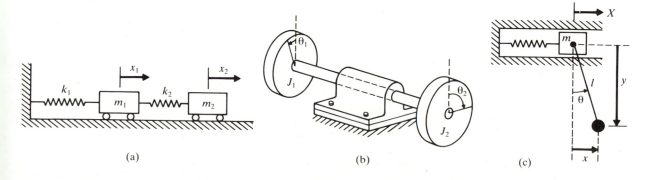

(a)

(b)

(c)

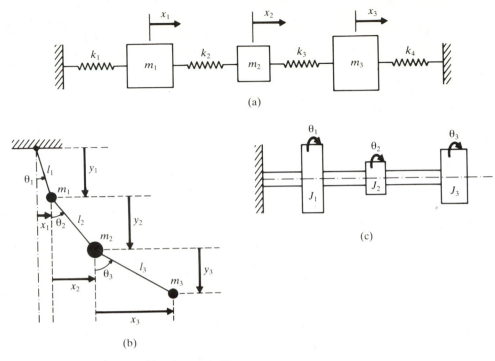

**Figure 1.8**   Three degree of freedom systems.

Since the beam has an infinite number of mass points, we need an infinite number of coordinates to specify its deflected configuration. The infinite number of coordinates gives its elastic line (deflection curve). Thus the cantilever beam has an infinite number of degrees of freedom. Most structural and machine systems have deformable (elastic) members and therefore have an infinite number of degrees of freedom.

Systems with a finite number of degrees of freedom are called *discrete* or *lumped parameter* systems, and those with an infinite number of degrees of freedom are called *continuous* or *distributed* systems.

Most of the time, continuous systems are approximated as discrete systems, and solutions are obtained in a simpler manner. Although treatment of a system as continuous gives exact results, the analysis methods available for dealing with

**Figure 1.9**   A cantilever beam (an infinite number of degrees of freedom system).

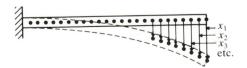

continuous systems are limited to a narrow selection of problems, such as uniform beams, slender rods, and thin plates. Hence most of the practical systems are studied by treating them as finite lumped masses, springs, and dampers. In general, more accurate results are obtained by increasing the number of masses, springs, and dampers—that is, increasing the number of degrees of freedom.

## 1.5 CLASSIFICATION OF VIBRATION

Vibration can be classified in several ways. Some of the important classifications are as follows.

### 1.5.1   Free and Forced Vibration

**Free Vibration.** If a system, after an initial disturbance, is left to vibrate on its own, the ensuing vibration is known as free vibration. No repeated external force acts on the system. The oscillation of a simple pendulum is an example of free vibration.

**Forced Vibration.** If a system is subjected to an external force (often, a repeating type of force), the resulting vibration is known as *forced vibration*. The oscillation that arises in machines such as diesel engines is an example of forced vibration.

If the frequency of the external force coincides with one of the natural frequencies of the system, a condition known as *resonance* occurs, and the system undergoes dangerously large oscillations. Failures of such structures as buildings, bridges, and airplane wings have been associated with the occurrence of resonance.

### 1.5.2   Undamped and Damped Vibration

If no energy is lost or dissipated in friction or other resistance during oscillation, the vibration is known as *undamped vibration*. If any energy is lost in this way, on the other hand, it is called *damped vibration*. In many physical systems, the amount of damping is so small that it can be disregarded for most engineering purposes. However, consideration of damping becomes extremely important in analyzing vibratory systems near resonance.

### 1.5.3   Linear and Nonlinear Vibration

If all the basic components of a vibratory system—the spring, the mass, and the damper—behave linearly, the resulting vibration is known as *linear vibration*. On the other hand, if any of the basic components behave nonlinearly, the vibration is called *nonlinear vibration*. The differential equations that govern the behavior of linear and nonlinear vibratory systems are linear and nonlinear, respectively. If the vibration is linear, the principle of superposition holds, and the mathematical techniques of analysis are well developed. For nonlinear vibration, the superposition principle is not valid, and techniques of analysis are less well known. Since all vibratory systems tend to behave nonlinearly with increasing amplitude of oscillation, some knowledge of nonlinear vibration is desirable.

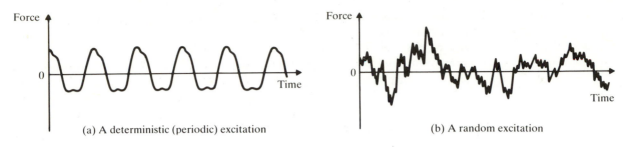

(a) A deterministic (periodic) excitation

(b) A random excitation

**Figure 1.10**

### 1.5.4   Deterministic and Random Vibration

If the value of the excitation (force or motion) acting on a vibratory system is known at any given time, the excitation is called *deterministic*. The resulting vibration is known as *deterministic vibration*.

In some cases, the excitation is *nondeterministic* or *random*; the value of the excitation at a given time cannot be predicted. In these cases, a large collection of records of the excitation may exhibit some statistical regularity. It is possible to estimate averages such as the mean and mean square values of the excitation. Examples of random excitations are wind velocity, road roughness, and ground motion during earthquakes. If the excitation is random, the resulting vibration is called *random vibration*. In the case of random vibration, the vibratory response of the system is also random; it can be described only in terms of statistical quantities. Figure 1.10 shows examples of deterministic and random excitations.

## 1.6  VIBRATION ANALYSIS PROCEDURE

The vibration analysis of an engineering system usually involves the following steps.

**Step 1: Mathematical Modeling.**  Most engineering systems are complex. It is impossible to consider every detail of the system for vibration analysis, so we must identify the elements of the system, keeping in view the purpose of the analysis [1.18]. A mathematical model that is adequate for one purpose may not be adequate for another purpose. To illustrate the procedure of mathematical modeling, consider an automobile moving over a road bump, as in Fig. 1.11(a). If the purpose of analysis is to predict the vertical vibration of the automobile, the mathematical model need not consider all the individual components such as engine, gearbox, body, frame, and passengers. Since the deflection of these components is much smaller than the deflection of the tires with respect to the road, we can lump the whole body and the components as a single mass. The tires and springs can be replaced by an equivalent spring and damper. The resulting mathematical model will look as shown in Fig. 1.11(b). A more accurate mathematical model of the automobile, including the effects of pitching along with vertical translation, is shown in Fig. 1.11(c).

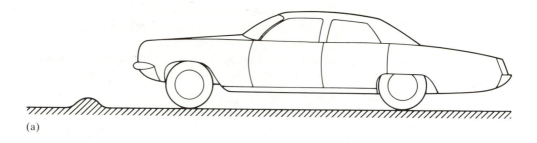

(a)

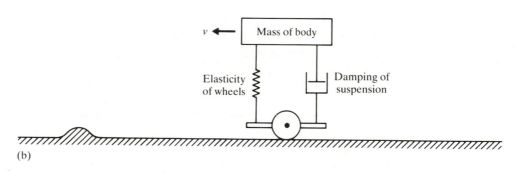

(b)

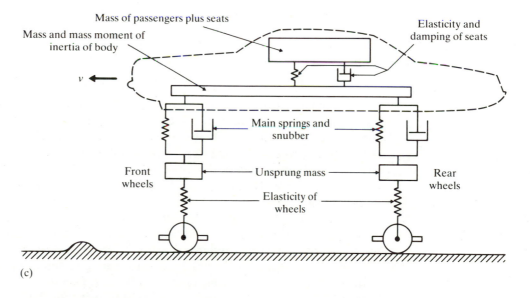

(c)

**Figure 1.11**   Modeling of an automobile.

**Step 2: Derivation of Governing Equations.** Once the mathematical model is available, we use the principles of dynamics and derive the equations that describe the vibration of the system. Several approaches are commonly used to derive the governing equations. Among them are Newton's second law of motion, d'Alembert's principle, and the principle of conservation of energy.

**Step 3: Solution of the Governing Equations.** The equations of motion must be solved to obtain the desired results. Depending on the nature of the problem, we can use one of the following techniques for finding the solution: standard methods of solving differential equations, Laplace transformation methods, matrix methods*, and numerical methods.

**Step 4: Interpretation of the Results.** The solution obtained by solving the equations of motion must be interpreted with a clear view of the purpose of the analysis and the possible design implications of the results.

Steps 1 and 4 are subjective and depend essentially on the experience and judgment of the person analyzing the problem, while steps 2 and 3 are objective and are amenable to mathematical treatment. In this book, we shall deal with steps 2 and 3 only.

## 1.7  SPRING ELEMENTS

A linear spring is a type of mechanical link which is generally assumed to have negligible mass and damping. A force is developed in the spring whenever there is relative motion between the two ends of the spring. The spring force is proportional to the amount of deformation and is given by

$$F = kx \tag{1.1}$$

where $F$ is the spring force, $x$ is the deformation (displacement of one end with respect to the other), and $k$ is the coefficient of proportionality, called the *spring stiffness* or *spring constant*. If we plot a graph between $F$ and $x$, the result is a straight line according to Eq. (1.1). The work done in deforming a spring is stored as strain or potential energy in the spring.

Actual springs are nonlinear and follow Eq. (1.1) only within a certain range of deformation. Beyond a certain value of deformation (after point $A$ in Fig. 1.12), the stress exceeds the proportionality limit of Hooke's law and the deflection is no longer proportional to the applied force. In many practical applications we assume the deflections to be small and make use of the linear relation in Eq. (1.1).

Even if the force-deflection relation of a spring is nonlinear, we often approximate it as a linear one by using a process known as *linearization*. Let the spring have a nonlinear force-deformation relation, as shown in Fig. 1.13. First apply a static force $F_0$ to the spring so that it deflects by $x_0$. There is equilibrium at this position, and the spring reaction is equal to $-F_0$. Now apply an additional force $\Delta F$, causing an additional deflection $\Delta x$ in the spring so that we have $F_0 + \Delta F = F(x_0 + \Delta x)$.

---

*The basic definitions and operations of matrix theory are given in Appendix A.

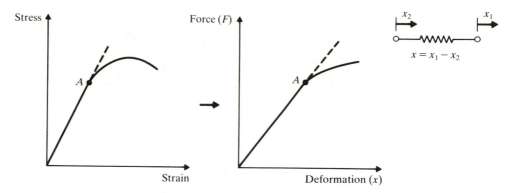

**Figure 1.12** Nonlinearity beyond proportionality limit.

The function $F$ can be expanded using Taylor's series expansion about $x_0$:

$$F_0 + \Delta F = F(x_0 + \Delta x) = F(x_0) + \left. \frac{dF}{dx} \right|_{x_0} \cdot \Delta x + \frac{1}{2!} \left. \frac{d^2 F}{dx^2} \right|_{x_0} \cdot (\Delta x)^2 + \cdots \qquad (1.2)$$

If $\Delta x$ is small, we can neglect higher order terms and write Eq. (1.2) as

$$F_0 + \Delta F \simeq F(x_0) + \left. \frac{dF}{dx} \right|_{x_0} \cdot \Delta x \qquad (1.3)$$

Since $F_0 = F(x_0)$, we obtain

$$\Delta F = \left. \frac{dF}{dx} \right|_{x_0} \cdot \Delta x = k \cdot \Delta x \qquad (1.4)$$

with

$$k = \left. \frac{dF}{dx} \right|_{x_0}$$

Equation (1.4) is the desired linear relationship and indicates that the spring constant $k$ is given by the derivative of the function $F = F(x)$, evaluated at the equilibrium point $x_0$. We may use Eq. (1.4) for simplicity, but sometimes the error involved in the approximation may be very large.

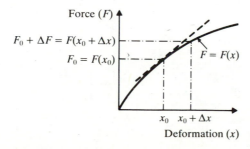

**Figure 1.13** Linearization process.

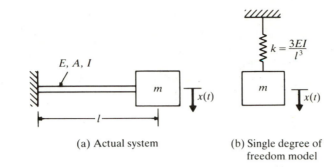

(a) Actual system                (b) Single degree of
                                     freedom model

**Figure 1.14**    Cantilever with end mass.

$$Y_{max} = \frac{Pl^3}{3EI}$$

Elastic elements like beams also behave as springs. For example, consider a cantilever beam with an end mass $m$, as shown in Fig. 1.14. We assume, for simplicity, that the mass of the beam is negligible in comparison with the mass $m$. From strength of materials [1.19], we know that the static deflection of the beam at the free end is given by

$$\delta_{st} = \frac{Wl^3}{3EI} \qquad (1.5)$$

where $W = mg$ is the weight of the mass $m$, $E$ is Young's modulus and, $I$ is the moment of inertia of the cross section of the beam. Hence the spring constant is

$$k = \frac{W}{\delta_{st}} = \frac{3EI}{l^3} \qquad (1.6)$$

Similar results can be obtained for beams with different end conditions.

### 1.7.1    Combination of Springs

In many practical applications, several linear springs are used in combination. These springs can be combined into a single equivalent spring as indicated below.

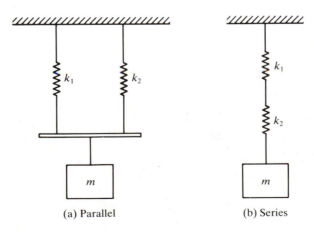

(a) Parallel                (b) Series

**Figure 1.15**    Combination of springs.

**Case (i): Springs in Parallel.** Let the springs be parallel as shown in Fig. 1.15(a). If $W$ is the weight of mass $m$, we have for equilibrium

$$W = k_1 \delta_{st} + k_2 \delta_{st} \tag{1.7}$$

where $\delta_{st}$ is the static deflection of the mass $m$. If $k_{eq}$ denotes the equivalent spring constant of the combination of the two springs, then for the same static deflection $\delta_{st}$, we have

$$W = k_{eq} \delta_{st} \tag{1.8}$$

Equations (1.7) and (1.8) give

$$k_{eq} = k_1 + k_2 \tag{1.9}$$

In general, if we have $n$ springs with spring constants $k_1, k_2, \ldots, k_n$ in parallel, then the equivalent spring constant $k_{eq}$ can be obtained:

Parallel Springs
$$k_{eq} = k_1 + k_2 + \cdots + k_n \tag{1.10}$$

**Case (ii): Springs in Series.** Next we consider two springs connected in series, as shown in Fig. 1.15(b). Since both the springs are subjected to the same force $W$, we have for equilibrium

$$\left. \begin{array}{l} W = k_1 \delta_1 \\ W = k_2 \delta_2 \end{array} \right\} \tag{1.11}$$

where $\delta_1$ and $\delta_2$ are the elongations of springs 1 and 2, respectively. As the total elongation is equal to the static deflection $\delta_{st}$,

$$\delta_1 + \delta_2 = \delta_{st} \tag{1.12}$$

If $k_{eq}$ denotes the equivalent spring constant, then for the same static deflection,

$$W = k_{eq} \delta_{st} \tag{1.13}$$

Equations (1.11) and (1.13) give

$$k_1 \delta_1 = k_2 \delta_2 = k_{eq} \delta_{st}$$

or

$$\delta_1 = \frac{k_{eq} \delta_{st}}{k_1} \qquad \text{and} \qquad \delta_2 = \frac{k_{eq} \delta_{st}}{k_2} \tag{1.14}$$

Substituting these values of $\delta_1$ and $\delta_2$ into Eq. (1.12), we obtain

$$\frac{k_{eq} \delta_{st}}{k_1} + \frac{k_{eq} \delta_{st}}{k_2} = \delta_{st}$$

that is,

$$\frac{1}{k_{eq}} = \frac{1}{k_1} + \frac{1}{k_2} \tag{1.15}$$

Equation (1.15) can be generalized to the case of $n$ springs in series:

Series Springs
$$\frac{1}{k_{eq}} = \frac{1}{k_1} + \frac{1}{k_2} + \cdots + \frac{1}{k_n} \tag{1.16}$$

In certain applications, springs are connected to levers or gears. In such cases, an equivalent spring constant can be found using energy equivalence, as indicated in Example 1.2.

---

### EXAMPLE 1.1

A weight $W$ is being lifted by an overhead traveling crane [1.20] as shown in Fig. 1.16(a). The girder is a uniform beam of length $l_1$ and flexural rigidity $EI$ and each of the two cables has a length $l_2$, diameter $d$, and Young's modulus $E$. Assuming the masses of the trolley, motor, drum, cables, and hook to be negligible, find the equivalent spring constant of the girder and the cable.

**Solution.** If we assume the girder to be a simply supported beam with central loading, its spring constant is given by

$$k_g = \frac{48EI}{l_1^3} \tag{E.1}$$

Since the cables are subjected to axial loading, the stiffness of each of the cables is

$$k_c = \frac{AE}{l_2} = \frac{\pi d^2 E}{4 l_2} \tag{E.2}$$

and the total stiffness of the two parallel cables is $2k_c$. Finally, the girder and cables can be considered as series springs (as shown in Fig. 1.16(b)) whose equivalent

**Figure 1.16** Overhead traveling crane.

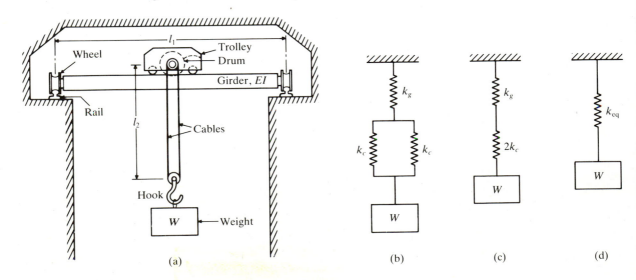

spring constant, $k_{eq}$, is given by

$$\frac{1}{k_{eq}} = \frac{1}{k_g} + \frac{1}{2k_c} \quad \text{or} \quad k_{eq} = \frac{2k_g k_c}{k_g + 2k_c} = \frac{48\pi EI d^2}{96Il_2 + \pi d^2 l_1^3} \quad \text{(E.3)}$$

## EXAMPLE 1.2

Determine the effective stiffness of the system shown in Fig. 1.17(a) in terms of the displacement $x$.

**Solution.** A displacement $x$ of point $A$ will cause the spring $k_1$ to extend by an amount $x_1 = x$ and the link $AOB$ to rotate about the pivot point $O$ by an angle $\theta$ where, for small deformations (see Fig. 1.17(b)),

$$\theta = \frac{x}{b} \quad \text{(E.1)}$$

This will cause the spring $k_2$ to compress by an amount

$$x_2 = \theta a = \frac{xa}{b} \quad \text{(E.2)}$$

and the spring $k_3$ to compress by an amount

$$x_3 = \theta c = \frac{xc}{b} \quad \text{(E.3)}$$

**Figure 1.17**

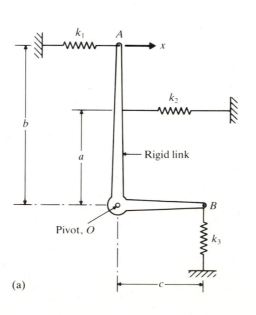
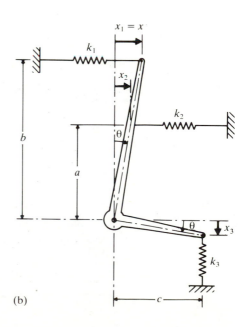

(a)                    (b)

Thus the total potential energy stored in the three springs ($U$) is given by

$$U = \frac{1}{2}k_1x_1^2 + \frac{1}{2}k_2x_2^2 + \frac{1}{2}k_3x_3^2 = \frac{1}{2}\left(k_1 + \frac{a^2}{b^2}k_2 + \frac{c^2}{b^2}k_3\right)x^2 \qquad \text{(E.4)}$$

It is implied that the equivalent spring undergoes a deformation of $x$ (i.e., it is located at the site of the spring $k_1$). Hence the potential energy of the equivalent spring ($U_{eq}$) is given by

$$U_{eq} = \frac{1}{2}k_{eq}x_{eq}^2 = \frac{1}{2}k_{eq}x^2 \qquad \text{(E.5)}$$

By setting $U = U_{eq}$, we obtain the effective stiffness of the system as

$$k_{eq} = k_1 + \frac{a^2}{b^2}k_2 + \frac{c^2}{b^2}k_3 \qquad \text{(E.6)}$$

## 1.8 MASS OR INERTIA ELEMENTS

The mass or inertia element is assumed to be a rigid body; it can gain or lose kinetic energy whenever the velocity of the body changes. From Newton's second law of motion, the product of the mass and its acceleration is equal to the force applied to the mass. Work is equal to the force multiplied by the displacement in the direction of the force and the work done on a mass is stored in the form of kinetic energy of the mass.

In most cases, we must use a mathematical model to represent the actual vibrating system, and there are often several possible models. As stated earlier, the purpose of the analysis determines which mathematical model is appropriate. Once the model is chosen, the mass or inertia elements of the system can be easily identified. For example, consider the cantilever beam with a tip mass shown in Fig. 1.14(a). For a quick and reasonably accurate analysis, the mass and damping of the beam can be disregarded; the system can be modeled as a spring-mass system, as shown in Fig. 1.14(b). The tip mass $m$ represents the mass element, and the elasticity of the beam denotes the stiffness of the spring. Next, consider a multistory building subjected to an earthquake. Assuming that the mass of the frame is negligible compared to the masses of the floors, the building can be modeled as a multidegree of freedom system, as shown in Fig. 1.18. The masses at the various floor levels represent the mass elements, and the elasticities of the vertical members denote the spring elements.

### 1.8.1 Combination of Masses

In many practical applications, several masses appear in combination. For a simple analysis, we can replace these masses by a single equivalent mass, as indicated below.

**Case (i): Translational Masses Connected by a Lever.** Let the masses be placed on a lever that is pivoted at one end, as shown in Fig. 1.19. The equivalent mass can be

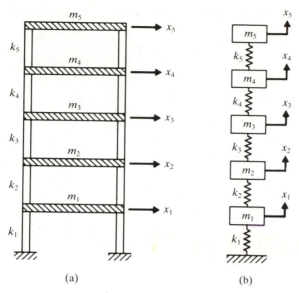

**Figure 1.18** Idealization of a multistory building as a multidegree of freedom system.

assumed to be located at any point along the lever. To be specific, we assume the location of the equivalent mass to be the site of mass $m_1$. The velocity of mass $m_2(\dot{x}_2)$ can be expressed in terms of that of mass $m_1(\dot{x}_1)$, by assuming small angular displacements for the lever, as

$$\dot{x}_2 = \frac{l_2}{l_1}\dot{x}_1 \tag{1.17}$$

and

$$\dot{x}_{eq} = \dot{x}_1 \tag{1.18}$$

**Figure 1.19** Translational masses connected by a lever.

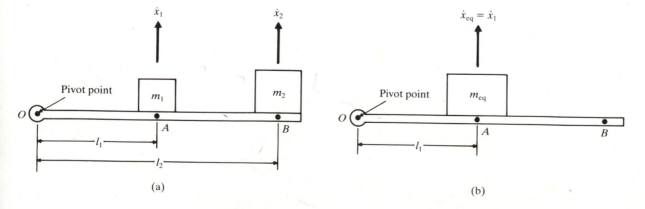

By equating the kinetic energy of the two mass system to that of the equivalent mass system, we obtain

$$\frac{1}{2}m_1\dot{x}_1^2 + \frac{1}{2}m_2\dot{x}_2^2 = \frac{1}{2}m_{eq}\dot{x}_{eq}^2 \tag{1.19}$$

This equation gives, in view of Eqs. (1.17) and (1.18),

$$m_{eq} = m_1 + \left(\frac{l_2}{l_1}\right)^2 m_2 \tag{1.20}$$

**Case (ii): Translational and Rotational Masses Coupled Together.** Let a mass $m$ having a translational velocity $\dot{x}$ be coupled to another mass (of mass moment of inertia $J_0$) having a rotational velocity $\dot{\theta}$, as shown in Fig. 1.20. These two masses can be combined to obtain either 1) a single equivalent translating mass $m_{eq}$ or 2) a single equivalent rotational mass $J_{eq}$, as shown below.

*1. Equivalent translating mass.* The kinetic energy of the two masses is given by

$$T = \frac{1}{2}m\dot{x}^2 + \frac{1}{2}J_0\dot{\theta}^2 \tag{1.21}$$

and the kinetic energy of the equivalent mass can be expressed as

$$T_{eq} = \frac{1}{2}m_{eq}\dot{x}_{eq}^2 \tag{1.22}$$

Since $\dot{x}_{eq} = \dot{x}$ and $\dot{\theta} = \dot{x}/R$, the equivalence of $T$ and $T_{eq}$ gives

$$\frac{1}{2}m_{eq}\dot{x}^2 = \frac{1}{2}m\dot{x}^2 + \frac{1}{2}J_0\left(\frac{\dot{x}}{R}\right)^2$$

that is,

$$m_{eq} = m + \frac{J_0}{R^2} \tag{1.23}$$

*2. Equivalent rotating mass.* Here $\dot{\theta}_{eq} = \dot{\theta}$   and   $\dot{x} = \dot{\theta}R$, and the equivalence of $T$ and $T_{eq}$ leads to

$$\frac{1}{2}J_{eq}\dot{\theta}^2 = \frac{1}{2}m(\dot{\theta}R)^2 + \frac{1}{2}J_0\dot{\theta}^2$$

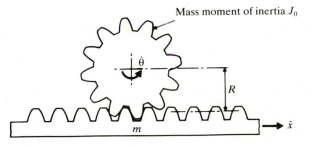

**Figure 1.20**   Translational and rotational masses coupled together.

or

$$J_{eq} = J_0 + mR^2 \qquad (1.24)$$

## EXAMPLE 1.3

A cam-follower mechanism (Fig. 1.21) is used to convert the rotary motion of a shaft into the oscillating or reciprocating motion of a valve. The follower system consists of a pushrod, a rocker arm, a valve, and a valve spring [1.21]. Find the equivalent mass ($m_{eq}$) of this cam-follower system by assuming the location of $m_{eq}$ as point $A$.

**Solution.** For simplicity, we assume the mass of the valve spring to be negligible. Due to a vertical displacement $x$ of the pushrod, the rocker arm rotates by an angle $\theta_r = x/l_1$ about the pivot point, and the valve has a downward movement of $x_v = \theta_r l_2 = x l_2 / l_1$. The kinetic energy of the system ($T$) can be expressed as

KE applies to item w/mass only  $T = \dfrac{1}{2} m_p \dot{x}_p^2 + \dfrac{1}{2} m_v \dot{x}_v^2 + \dfrac{1}{2} J_r \dot{\theta}_r^2 \qquad (E.1)$

where $\dot{x}_p$ and $\dot{x}_v$ are the linear velocities of the pushrod and the valve, respectively, and $\dot{\theta}_r$ is the angular velocity of the rocker arm. If $m_{eq}$ denotes the equivalent mass placed at point $A$, with $\dot{x}_{eq} = \dot{x}$, the kinetic energy of the equivalent mass system $T_{eq}$ is given by

$$T_{eq} = \frac{1}{2} m_{eq} \dot{x}_{eq}^2 \qquad (E.2)$$

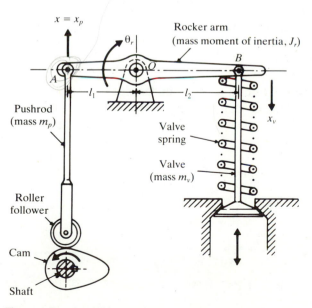

**Figure 1.21**  Cam-follower system.

By equating $T$ and $T_{eq}$, and noting that

$$\dot{x}_p = \dot{x}, \qquad \dot{x}_v = \frac{\dot{x} l_2}{l_1}, \qquad \text{and} \qquad \dot{\theta}_r = \frac{\dot{x}}{l_1}$$

we obtain

$$m_{eq} = m_p + \frac{J_r}{l_1^2} + m_v \frac{l_2^2}{l_1^2} \tag{E.3}$$

## 1.9  DAMPING ELEMENTS

A damper is assumed to have neither mass nor elasticity, and damping force exists only if there is relative velocity between the two ends of the damper. The energy or work input to a damper is converted into heat or sound; hence the damping element is nonconservative. The damping may be one or more of the following types [1.22].

**Viscous Damping.** Viscous damping is the most commonly used damping mechanism in vibration analysis. This type of damping is present whenever a viscous fluid flows through a slot, around a piston in a cylinder, or around the journal in a bearing. In viscous damping, the damping force is proportional to the velocity of the vibrating body.

**Coulomb or Dry Friction Damping.** Here the damping force is constant in magnitude but opposite in direction to that of the motion of the vibrating body. It is caused by kinetic friction between two sliding dry surfaces.

**Material or Solid or Hysteretic Damping.** When materials are deformed, energy is absorbed and dissipated by the material [1.23]. This effect is due to friction between the internal planes, which slip or slide as the deformations take place.

### 1.9.1  Construction of Viscous Dampers

A viscous damper can be constructed using two parallel plates separated by a distance $d$, with a fluid of viscosity $\mu$ between the plates (see Fig. 1.22). If one plate is fixed and the other plate moves with a velocity $v$ in its own plane, then a shearing action is produced on the fluid. This results in a horizontal force $F$ that opposes the motion. This resisting or damping force $F$ is proportional to the relative velocity $v$

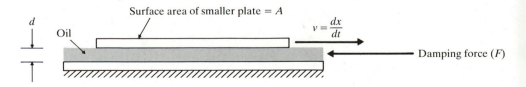

**Figure 1.22**   Parallel plates with a viscous fluid in between.

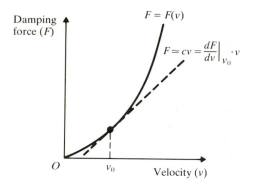

**Figure 1.23** Nonlinear damper characteristics.

and is given by

$$F = \frac{\mu A v}{d} = cv \tag{1.25}$$

where $c = \mu A / d$ is called the *damping constant* [1.24]. Although the damper shown in Fig. 1.22 is linear over a wide range of velocities $v$, it requires a large plate area $A$ or a very narrow clearance $d$ to produce substantial damping. Hence other devices having nonlinear force-velocity relationships are used, as shown in Fig. 1.23. If a damper is nonlinear, a linearization procedure is generally used, as in the case of a nonlinear spring.

### 1.9.2 Combination of Dampers

When dampers appear in combination, they can be replaced by an equivalent damper by adopting a procedure similar to the one described in Secs. 1.7 and 1.8 (see Problem 1.15).

### EXAMPLE 1.4

Develop an expression for the damping constant of the dashpot shown in Fig. 1.24.

*Solution.* As shown in Fig. 1.24, the dashpot consists of a piston of diameter $D$, and length $l$, moving with velocity $v_0$ in a cylinder filled with a liquid of viscosity $\mu$ [1.25]. Let the clearance between the piston and the cylinder wall be $d$. At a distance $y$ from the moving surface, let the velocity and shear stress be $v$ and $\tau$, and at a distance $(y + dy)$ let the velocity and shear stress be $(v - dv)$ and $(\tau + d\tau)$, respectively. The negative sign for $dv$ shows that the velocity decreases as we move towards the cylinder wall. The viscous force on this annular ring is equal to

$$F = \pi D l d\tau = \pi D l \frac{d\tau}{dy} \, dy \tag{E.1}$$

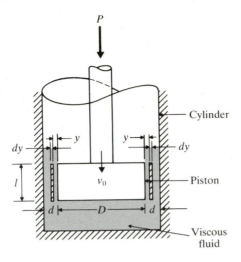

**Figure 1.24.** A dashpot.

But the shear stress is given by

$$\tau = -\mu \frac{dv}{dy} \qquad \text{(E.2)}$$

where the negative sign is consistent with a decreasing velocity gradient [1.26]. Using Eq. (E.2) in Eq. (E.1), we obtain

$$F = -\pi D l \, dy \, \mu \frac{d^2v}{dy^2} \qquad \text{(E.3)}$$

The force on the piston will cause a pressure difference on the ends of the element, given by

$$p = \frac{P}{\left(\frac{\pi D^2}{4}\right)} = \frac{4P}{\pi D^2} \qquad \text{(E.4)}$$

Thus the pressure force on the end of the element is

$$p(\pi D \, dy) = \frac{4P}{D} \cdot dy \qquad \text{(E.5)}$$

where $(\pi D \, dy)$ denotes the annular area between $y$ and $(y + dy)$. If we assume uniform mean velocity in the direction of motion of the fluid, the forces given in Eqs. (E.3) and (E.5) must be equal. Thus we get

$$\frac{4P}{D} \cdot dy = -\pi D l \, dy \, \mu \frac{d^2v}{dy^2}$$

or

$$\frac{d^2v}{dy^2} = -\frac{4P}{\pi D^2 l \mu} \qquad \text{(E.6)}$$

Integrating this equation twice and using the boundary conditions $v = -v_0$ at $y = 0$ and $v = 0$ at $y = d$, we obtain

$$v = \frac{2P}{\pi D^2 l \mu}(yd - y^2) - v_0\left(1 - \frac{y}{d}\right) \tag{E.7}$$

The rate of flow through the clearance space can be obtained by integrating the rate of flow through an element between the limits $y = 0$ and $y = d$:

$$Q = \int_0^d v \pi D \, dy = \pi D\left[\frac{2Pd^3}{6\pi D^2 l \mu} - \frac{1}{2}v_0 d\right] \tag{E.8}$$

The volume of the liquid flowing through the clearance space per second must be equal to the volume per second displaced by the piston. Hence the velocity of the piston will be equal to this rate of flow divided by the piston area. This gives

$$v_0 = \frac{Q}{\left(\frac{\pi}{4}D^2\right)} \tag{E.9}$$

Equations (E.9) and (E.8) lead to

$$P = \left[\frac{3\pi D^3 l\left(1 + \frac{2d}{D}\right)}{4d^3}\right]\mu v_0 \tag{E.10}$$

By writing the force as $P = cv_0$, the damping constant $c$ can be found as

$$c = \mu\left[\frac{3\pi D^3 l}{4d^3}\left(1 + \frac{2d}{D}\right)\right] \tag{E.11}$$

## 1.10 HARMONIC MOTION

Oscillatory motion may repeat itself regularly, as in the case of a simple pendulum, or may display considerable irregularity, as in the case of ground motion during an earthquake. If the motion is repeated after equal intervals of time, it is called *periodic motion*. The simplest type of periodic motion is *harmonic motion*. The motion of the mechanism shown in Fig. 1.25 is an example of simple harmonic motion. In this mechanism, a flywheel of radius $A$ rotates about an axis passing through its center $O$. A pin $P$ is attached to the flywheel and slides in the vertical slot of a link $Q$, which reciprocates in the horizontal guides $R$ [1.27]. When the flywheel rotates with an angular velocity $\omega$, the end point $S$ of the link $Q$ is displaced from its middle position by an amount $x$ (in time $t$) given by

$$x = A\sin\theta = A\sin\omega t \tag{1.26}$$

If the recording paper is made to move with a constant speed $V$, as shown in Fig. 1.25, the motion of the point $S$ will be recorded as a sinusoidal curve. The velocity

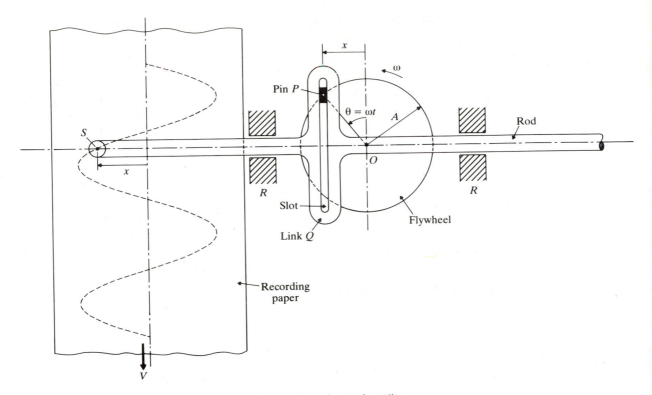

**Figure 1.25**   Mechanism producing harmonic motion.

of point $S$ at time $t$ is given by

$$\frac{dx}{dt} = \omega A \cos \omega t \tag{1.27}$$

and the acceleration by

$$\frac{d^2x}{dt^2} = -\omega^2 A \sin \omega t = -\omega^2 x \tag{1.28}$$

It can be seen that the acceleration is directly proportional to the displacement. Such a vibration, with the acceleration proportional to the displacement and directed towards the mean position, is known as *simple harmonic motion*. The motion given by $x = A \cos \omega t$ is another example of a simple harmonic motion. Figure 1.25 clearly shows the similarity between cyclic (harmonic) motion and sinusoidal motion.

### 1.10.1   Vectorial Representation of Harmonic Motion

Harmonic motion can be represented conveniently by means of a vector $\overrightarrow{OP}$ of magnitude $A$ rotating at a constant angular velocity $\omega$. In Fig. 1.26, the projection of

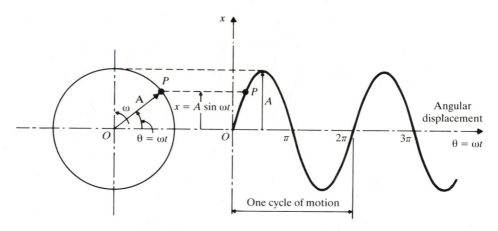

**Figure 1.26** Harmonic motion as the projection of the end of a rotating vector.

the tip of the vector $\vec{X} = \overrightarrow{OP}$ on the vertical axis is given by

$$y = A \sin \omega t \qquad (1.29)$$

and its projection on the horizontal axis by

$$x = A \cos \omega t \qquad (1.30)$$

### 1.10.2  Complex Number Representation of Harmonic Motion

Any vector $\vec{X}$ in the $xy$ plane can be represented as a complex number:

$$\vec{X} = a + ib \qquad (1.31)$$

where $i = \sqrt{-1}$ and $a$ and $b$ denote the $x$ and $y$ components of $\vec{X}$, respectively (see Fig. 1.27). Components $a$ and $b$ are also called the *real* and *imaginary* parts of the vector $\vec{X}$. If $A$ denotes the modulus or absolute value of the vector $\vec{X}$, and $\phi$ represents the argument or the angle between the vector and the $x$-axis, then $\vec{X}$ can also be expressed as

$$\vec{X} = A \cos \phi + iA \sin \phi = A e^{i\phi} \qquad (1.32)$$

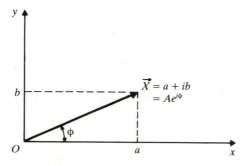

**Figure 1.27**

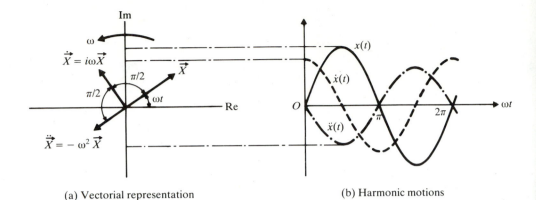

(a) Vectorial representation          (b) Harmonic motions

**Figure 1.28** Displacement, velocity, and acceleration vectors.

with

$$A = \left( a^2 + b^2 \right)^{1/2} \tag{1.33}$$

and

$$\phi = \tan^{-1} \frac{b}{a} \tag{1.34}$$

The rotating vector $\vec{X}$ of Fig. 1.26 can be represented as a complex number:

$$\vec{X} = A e^{i\omega t} \tag{1.35}$$

The differentiation of Eq. (1.35) with respect to time gives

$$\frac{d\vec{X}}{dt} = \frac{d}{dt} \left( A e^{i\omega t} \right) = i\omega A e^{i\omega t} = i\omega \vec{X} \tag{1.36}$$

$$\frac{d^2\vec{X}}{dt^2} = \frac{d}{dt} \left( i\omega A e^{i\omega t} \right) = -\omega^2 A e^{i\omega t} = -\omega^2 \vec{X} \tag{1.37}$$

Thus the displacement, velocity, and acceleration can be expressed as*

$$\text{displacement} = \quad \text{Re}[ A e^{i\omega t} ] = A \cos \omega t \tag{1.38}$$

$$\text{velocity} = \quad \text{Re}[ i\omega A e^{i\omega t} ] = -\omega A \sin \omega t$$
$$= \omega A \cos( \omega t + 90° ) \tag{1.39}$$

$$\text{acceleration} = \text{Re}[ -\omega^2 A e^{i\omega t} ] = -\omega^2 A \cos \omega t$$
$$= \omega^2 A \cos( \omega t + 180° ) \tag{1.40}$$

---

*If the harmonic displacement is originally given as $x(t) = A \sin \omega t$, then we have

$$\text{displacement} = \text{Im}[ A e^{i\omega t} ] = A \sin \omega t$$

$$\text{velocity} = \text{Im}[ i\omega A e^{i\omega t} ] = \omega A \sin( \omega t + 90° )$$

$$\text{acceleration} = \text{Im}[ -\omega^2 A e^{i\omega t} ] = \omega^2 A \sin( \omega t + 180° )$$

where Im denotes the imaginary part.

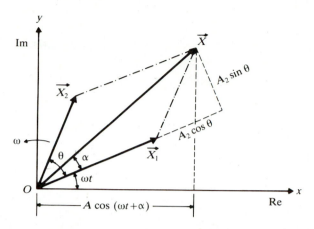

**Figure 1.29** Vectorial addition of harmonic functions.

where Re denotes the real part. These quantities are shown as rotating vectors in Fig. 1.28. It can be seen that the acceleration vector leads the velocity vector by 90°, and the latter leads the displacement vector by 90°.

Harmonic functions can be added vectorially, as shown in Fig. 1.29. If $\text{Re}(\vec{X}_1) = A_1\cos\omega t$ and $\text{Re}(\vec{X}_2) = A_2\cos(\omega t + \theta)$, then the magnitude of the resultant vector $\vec{X}$ is given by

$$A = \sqrt{(A_1 + A_2\cos\theta)^2 + (A_2\sin\theta)^2} \tag{1.41}$$

and the angle $\alpha$ by

$$\alpha = \tan^{-1}\left(\frac{A_2\sin\theta}{A_1 + A_2\cos\theta}\right) \tag{1.42}$$

Since the original functions are given as real components, the sum $\vec{X}_1 + \vec{X}_2$ is given by $\text{Re}(\vec{X}) = A\cos(\omega t + \alpha)$. The sum of $\vec{X}_1$ and $\vec{X}_2$ can also be found using complex numbers:

$$\vec{X} = \vec{X}_1 + \vec{X}_2 = A_1 e^{i\omega t} + A_2 e^{i(\omega t + \theta)} = \left(A_1 + A_2 e^{i\theta}\right) e^{i\omega t}$$

$$= \left(A_1 + A_2\cos\theta + iA_2\sin\theta\right) e^{i\omega t}$$

$$= A e^{i\alpha} e^{i\omega t} = A e^{i(\omega t + \alpha)} \tag{1.43}$$

where $A$ and $\alpha$ are given by Eqs. (1.41) and (1.42).

---

**EXAMPLE 1.5**

Find the sum of the two harmonic motions $x_1(t) = 10\cos\omega t$ and $x_2(t) = 15\cos(\omega t + 2)$.

### Solution

*Method 1: By using trigonometric relations:* Since the circular frequency is the same for both $x_1(t)$ and $x_2(t)$, we express the sum as

$$x(t) = A\cos(\omega t + \alpha) = x_1(t) + x_2(t) \tag{E.1}$$

that is,

$$A(\cos\omega t \cos\alpha - \sin\omega t \sin\alpha) = 10\cos\omega t + 15\cos(\omega t + 2)$$
$$= 10\cos\omega t + 15(\cos\omega t \cos 2 - \sin\omega t \sin 2) \tag{E.2}$$

that is,

$$\cos\omega t(A\cos\alpha) - \sin\omega t(A\sin\alpha) = \cos\omega t(10 + 15\cos 2) - \sin\omega t(15\sin 2) \tag{E.3}$$

By equating the corresponding coefficients of $\cos\omega t$ and $\sin\omega t$ on both sides, we obtain

$$A\cos\alpha = 10 + 15\cos 2 \tag{E.4}$$

$$A\sin\alpha = 15\sin 2$$

$$A = \sqrt{(10 + 15\cos 2)^2 + (15\sin 2)^2}$$

$$= 14.1520$$

and

$$\alpha = \tan^{-1}\left(\frac{15\sin 2}{10 + 15\cos 2}\right) = 74.5796° \tag{E.5}$$

*Method 2: By using vectors:* For an arbitrary value of $\omega t$, the harmonic motions $x_1(t)$ and $x_2(t)$ can be denoted graphically as shown in Fig. 1.30. By adding them vectorially, the resultant vector $x(t)$ can be found to be

$$x(t) = 14.1520\cos(\omega t + 74.5796°) \tag{E.6}$$

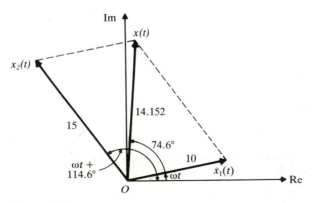

**Figure 1.30**

*Method 3: By using complex number representation:* The two harmonic motions can be denoted in terms of complex numbers:

$$x_1(t) = \text{Re}\left[A_1 e^{i\omega t}\right] \equiv \text{Re}\left[10 e^{i\omega t}\right]$$

$$x_2(t) = \text{Re}\left[A_2 e^{i(\omega t + 2)}\right] \equiv \text{Re}\left[15 e^{i(\omega t + 2)}\right] \qquad (E.7)$$

The sum of $x_1(t)$ and $x_2(t)$ can be expressed as

$$x(t) = \text{Re}\left[A e^{i(\omega t + \alpha)}\right] \qquad (E.8)$$

where $A$ and $\alpha$ can be determined using Eqs. (1.41) and (1.42) as $A = 14.1520$ and $\alpha = 74.5796°$.

### 1.10.3 Definitions

The following definitions are useful in dealing with harmonic motion.

**Cycle.** The movement of a vibrating body from its mean position to its extreme position in one direction, then to the mean, then to its extreme position in other direction, and back to mean is called a *cycle* of vibration. One revolution (i.e., angular displacement of $2\pi$ radians) of the pin $P$ in Fig. 1.25 or one revolution of the vector $\overrightarrow{OP}$ in Fig. 1.26 constitutes a cycle.

**Amplitude.** The maximum displacement of a vibrating body from its mean position is called the *amplitude* of vibration. In Figs. 1.25 and 1.26 the amplitude of vibration is equal to $A$.

**Period of Oscillation.** The time taken to complete one cycle of motion is known as the *period of oscillation* or *time period* and is denoted by $\tau$. It is equal to the time required for the vector $\overrightarrow{OP}$ in Fig. 1.26 to rotate through an angle of $2\pi$ and hence

$$\tau = \frac{2\pi}{\omega} \qquad (1.44)$$

**Frequency of Oscillation.** The number of cycles per unit time is called the *frequency of oscillation* or simply the *frequency* and is denoted by $f$. Thus

$$f = \frac{1}{\tau} = \frac{\omega}{2\pi} \qquad (1.45)$$

Since $2\pi$ is a constant, $\omega$ can also be used to denote frequency. $\omega$ is called the circular frequency to distinguish it from the linear frequency $f = \omega/2\pi$. $\omega$ denotes the angular velocity of the cyclic motion and is measured in radians per second.

**Phase Angle.** Consider two vibratory motions denoted by

$$x_1 = A_1 \sin \omega t \qquad (1.46)$$

$$x_2 = A_2 \sin(\omega t + \phi) \qquad (1.47)$$

The two harmonic motions given by Eqs. (1.46) and (1.47) are called *synchronous* because they have the same frequency or angular velocity $\omega$. Two synchronous oscillations need not have the same amplitude, and they need not attain their

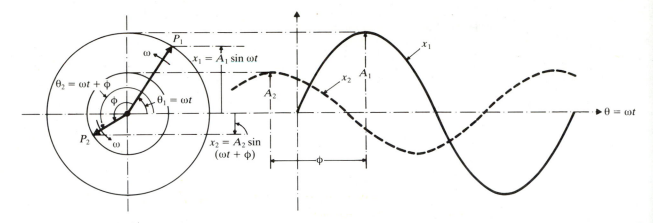

**Figure 1.31**   Phase difference between two vectors.

maximum values at the same time. The motions given by Eqs. (1.46) and (1.47) can be represented graphically as shown in Fig. 1.31. In this figure, the second vector $\overrightarrow{OP_2}$ leads the first one $\overrightarrow{OP_1}$ by an angle $\phi$, known as the *phase angle*. This means that the maximum of the second vector would occur $\phi$ radians earlier than that of the first vector. In place of maxima, any other corresponding points can be taken for finding the phase angle. In Eqs. (1.46) and (1.47) or in Fig. 1.31, the two vectors are said to have a *phase difference* of $\phi$.

**Natural Frequency.** If a system is disturbed and allowed to vibrate on its own, the frequency with which it oscillates without external forces or damping is known as its *natural frequency*. As will be seen later, a vibratory system having $n$ degrees of freedom will have, in general, $n$ distinct natural frequencies of vibration.

## 1.11   HARMONIC ANALYSIS

Although harmonic motion is simplest to handle, the motion of many vibratory systems is not harmonic. However, in many cases the vibrations are periodic—for example, the type shown in Fig. 1.32(a). Fortunately, any periodic function of time can be represented by Fourier series as an infinite sum of sine and cosine terms [1.28, 1.29]. If $x(t)$ is a periodic function with period $\tau$, its Fourier series representation is given by

$$x(t) = \frac{a_0}{2} + a_1\cos \omega t + a_2\cos 2\omega t + \cdots + b_1\sin \omega t + b_2\sin 2\omega t + \cdots$$

$$= \frac{a_0}{2} + \sum_{n=1}^{\infty} \left( a_n\cos n\omega t + b_n\sin n\omega t \right) \tag{1.48}$$

where $\omega = 2\pi/\tau$ is the fundamental frequency and $a_0, a_1, a_2, \ldots, b_1, b_2, \ldots$ are constant coefficients. To determine the coefficients $a_n$ and $b_n$, we multiply Eq. (1.48) by $\cos n\omega t$ and $\sin n\omega t$, respectively, and integrate over one period $\tau = 2\pi/\omega$, for example from 0 to $2\pi/\omega$. Then we notice that all terms except one on the right-hand side of the equation will be zero, and we obtain

$$a_0 = \frac{\omega}{\pi} \int_0^{2\pi/\omega} x(t)\,dt = \frac{2}{\tau} \int_0^{\tau} x(t)\cdot dt \tag{1.49}$$

$$a_n = \frac{\omega}{\pi} \int_0^{2\pi/\omega} x(t)\cos n\omega t\,dt = \frac{2}{\tau} \int_0^{\tau} x(t)\cos n\omega t\cdot dt \tag{1.50}$$

$$b_n = \frac{\omega}{\pi} \int_0^{2\pi/\omega} x(t)\sin n\omega t\,dt = \frac{2}{\tau} \int_0^{\tau} x(t)\sin n\omega t\cdot dt \tag{1.51}$$

The physical interpretation of Eq. (1.48) is that any periodic function can be represented as a sum of harmonic functions. Although the series in Eq. (1.48) is an infinite sum, we can approximate most periodic functions with the help of only a few harmonic functions. For example, the triangular wave of Fig. 1.32(a) can be represented closely by adding only three harmonic functions, as shown in Fig. 1.32(b).

Fourier series can also be represented by the sum of cosine terms only:

$$x(t) = c_0 + c_1\cos(\omega t - \phi_1) + c_2\cos(2\omega t - \phi_2) + \cdots \tag{1.52}$$

where

$$c_0 = a_0/2. \tag{1.53}$$

$$c_n = \left(a_n^2 + b_n^2\right)^{1/2} \tag{1.54}$$

and

$$\phi_n = \tan^{-1}\left(\frac{b_n}{a_n}\right) \tag{1.55}$$

The Fourier series can also be represented in terms of complex numbers by writing Eq. (1.48) as

$$x(t) = \frac{a_0}{2} + \sum_{n=1}^{\infty} \left\{ a_n\left(\frac{e^{in\omega t} + e^{-in\omega t}}{2}\right) + b_n\left(\frac{e^{in\omega t} - e^{-in\omega t}}{2i}\right) \right\} \tag{1.56}$$

**Figure 1.32** A periodic function.

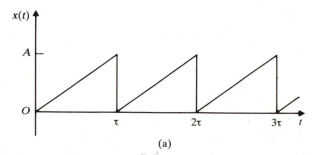

(a)

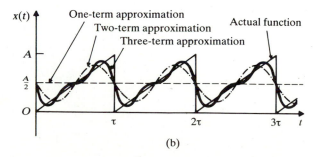

(b)

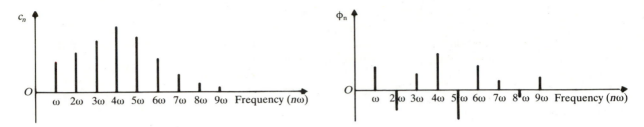

**Figure 1.33**   Frequency spectrum of a typical periodic function of time.

The harmonic functions $a_n \cos n\omega t$ or $b_n \sin n\omega t$ in Eq. (1.48) are called the *harmonics* of order $n$ of the periodic function $x(t)$. The harmonic of order $n$ has a period $\tau/n$. These harmonics can be plotted as vertical lines on a diagram of amplitude ($a_n$ and $b_n$ or $c_n$ and $\phi_n$) versus frequency ($n\omega$) called the *frequency spectrum* or *spectral diagram*. Figure 1.33 shows a typical frequency spectrum.

### 1.11.1   Numerical Computation of Coefficients

For very simple forms of the function $x(t)$, the integrals of Eqs. (1.49) to (1.51) can be evaluated easily. However, the integration becomes cumbersome if $x(t)$ does not have a simple form. In some practical applications, as in the case of experimental determination of the amplitude of vibration using a vibration transducer, the function $x(t)$ is not available in the form of a mathematical expression; only the values of $x(t)$ at a number of points $t_1, t_2, \ldots, t_N$ are available, as shown in Fig. 1.34. In these cases, the coefficients $a_n$ and $b_n$ of Eqs. (1.49) to (1.51) can be evaluated by using a numerical integration procedure like Simpson's rule [1.30].

**Figure 1.34**   Values of the periodic function $x(t)$ at discrete points $t_1, t_2, \ldots, t_N$.

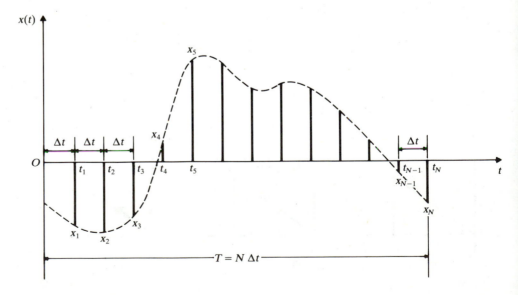

If $t_1, t_2, \ldots, t_N$ are assumed to be an even number of equidistant points over the period $\tau$ ($N$ = even) with the corresponding values of $x(t)$ given by $x_1 = x(t_1)$, $x_2 = x(t_2), \ldots, x_N = x(t_N)$, respectively, the coefficients $a_n$ and $b_n$ can be computed by applying Simpson's rule (by setting $\tau = N \cdot \Delta t$):

$$a_0 = \frac{2}{N} \sum_{i=1}^{N} x_i \tag{1.57}$$

$$a_n = \frac{2}{N} \sum_{i=1}^{N} x_i \cos \frac{2n\pi t_i}{\tau} \tag{1.58}$$

$$b_n = \frac{2}{N} \sum_{i=1}^{N} x_i \sin \frac{2n\pi t_i}{\tau} \tag{1.59}$$

---

### EXAMPLE 1.6

Find the Fourier series expansion of the function shown in Fig. 1.32(a).

**Solution.** The function $x(t)$ can be represented within the first cycle as

$$x(t) = A \frac{t}{\tau}, \quad 0 \le t \le \tau \tag{E.1}$$

where the period is given by $\tau = 2\pi/\omega$. To compute the Fourier coefficients $a_n$ and $b_n$, we use Eqs. (1.49) to (1.51):

$$a_0 = \frac{\omega}{\pi} \int_0^{2\pi/\omega} x(t)\, dt = \frac{\omega}{\pi} \int_0^{2\pi/\omega} A \frac{t}{\tau}\, dt = \frac{\omega}{\pi} \frac{A}{\tau} \left( \frac{t^2}{2} \right)_0^{2\pi/\omega} = A \tag{E.2}$$

$$a_n = \frac{\omega}{\pi} \int_0^{2\pi/\omega} x(t)\cos n\omega t \cdot dt = \frac{\omega}{\pi} \int_0^{2\pi/\omega} A \frac{t}{\tau} \cos n\omega t \cdot dt$$

$$= \frac{A\omega}{\pi\tau} \int_0^{2\pi/\omega} t \cos n\omega t \cdot dt = \frac{A}{2\pi^2} \left[ \frac{\cos n\omega t}{n^2} + \frac{\omega t \sin n\omega t}{n} \right]_0^{2\pi/\omega}$$

$$= 0, n = 1, 2, \ldots \tag{E.3}$$

$$b_n = \frac{\omega}{\pi} \int_0^{2\pi/\omega} x(t)\sin n\omega t \cdot dt = \frac{\omega}{\pi} \int_0^{2\pi/\omega} A \frac{t}{\tau} \sin n\omega t \cdot dt$$

$$= \frac{A\omega}{\pi\tau} \int_0^{2\pi/\omega} t \sin n\omega t \cdot dt = \frac{A}{2\pi^2} \left[ \frac{\sin n\omega t}{n^2} - \frac{\omega t \cos n\omega t}{n} \right]_0^{2\pi/\omega}$$

$$= -\frac{A}{n\pi}, \quad n = 1, 2, \ldots \tag{E.4}$$

Therefore the Fourier series expansion of $x(t)$ is

$$x(t) = \frac{A}{2} - \frac{A}{\pi}\sin \omega t - \frac{A}{2\pi}\sin 2\omega t - \cdots$$

$$= \frac{A}{\pi}\left[ \frac{\pi}{2} - \left\{ \sin \omega t + \frac{1}{2}\sin 2\omega t + \frac{1}{3}\sin 3\omega t + \cdots \right\} \right] \tag{E.5}$$

The first three terms of the series are shown plotted in Fig. 1.32(b). It can be seen that the approximation reaches the sawtooth shape even with a small number of terms.

---

### EXAMPLE 1.7

The pressure fluctuations of water in a pipe, measured at 0.01 second intervals, are given in Table 1.1. These fluctuations are repetitive in nature. Make a harmonic analysis of the pressure fluctuations and determine the first three harmonics of the Fourier series expansion.

***Solution.*** Since the given pressure fluctuations repeat every 0.12 sec, the period is $\tau = 0.12$ sec and the circular frequency of the first harmonic is $2\pi$ radians per 0.12 sec or $\omega = 2\pi/0.12 = 52.36$ rad/sec. As the number of observed values in each wave ($N$) is 12, we obtain from Eq. (1.57)

$$a_0 = \frac{2}{N} \sum_{i=1}^{N} p_i = \frac{1}{6} \sum_{i=1}^{12} p_i = 68166.7 \tag{E.1}$$

The coefficients $a_n$ and $b_n$ can be determined from Eqs. (1.58) and (1.59):

$$a_n = \frac{2}{N} \sum_{i=1}^{N} p_i \cos \frac{2n\pi t_i}{\tau} = \frac{1}{6} \sum_{i=1}^{12} p_i \cos \frac{2n\pi t_i}{0.12} \tag{E.2}$$

$$b_n = \frac{2}{N} \sum_{i=1}^{N} p_i \sin \frac{2n\pi t_i}{\tau} = \frac{1}{6} \sum_{i=1}^{12} p_i \sin \frac{2n\pi t_i}{0.12} \tag{E.3}$$

The computations involved in Eqs. (E.2) and (E.3) are shown in Table 1.2. From these calculations, the Fourier series expansion of the pressure fluctuations $p(t)$ can

### TABLE 1.1

| Time station, $i$ | Time (sec), $t_i$ | Pressure (kN/m$^2$), $p_i$ |
|---|---|---|
| 0 | 0 | 0 |
| 1 | 0.01 | 20 |
| 2 | 0.02 | 34 |
| 3 | 0.03 | 42 |
| 4 | 0.04 | 49 |
| 5 | 0.05 | 53 |
| 6 | 0.06 | 70 |
| 7 | 0.07 | 60 |
| 8 | 0.08 | 36 |
| 9 | 0.09 | 22 |
| 10 | 0.10 | 16 |
| 11 | 0.11 | 7 |
| 12 | 0.12 | 0 |

**TABLE 1.2**

| $i$ | $t_i$ | $p_i$ | $n=1$ | | $n=2$ | | $n=3$ | |
|---|---|---|---|---|---|---|---|---|
| | | | $p_i\cos\dfrac{2\pi t_i}{0.12}$ | $p_i\sin\dfrac{2\pi t_i}{0.12}$ | $p_i\cos\dfrac{4\pi t_i}{0.12}$ | $p_i\sin\dfrac{4\pi t_i}{0.12}$ | $p_i\cos\dfrac{6\pi t_i}{0.12}$ | $p_i\sin\dfrac{6\pi t_i}{0.12}$ |
| 1 | 0.01 | 20000 | 17320 | 10000 | 10000 | 17320 | 0 | 20000 |
| 2 | 0.02 | 34000 | 17000 | 29444 | −17000 | 29444 | −34000 | 0 |
| 3 | 0.03 | 42000 | 0 | 42000 | −42000 | 0 | 0 | −42000 |
| 4 | 0.04 | 49000 | −24500 | 42434 | −24500 | −42434 | 49000 | 0 |
| 5 | 0.05 | 53000 | −45898 | 26500 | 26500 | −45898 | 0 | 53000 |
| 6 | 0.06 | 70000 | −70000 | 0 | 70000 | 0 | −70000 | 0 |
| 7 | 0.07 | 60000 | −51960 | −30000 | 30000 | 51960 | 0 | −60000 |
| 8 | 0.08 | 36000 | −18000 | −31176 | −18000 | 31176 | 36000 | 0 |
| 9 | 0.09 | 22000 | 0 | −22000 | −22000 | 0 | 0 | 22000 |
| 10 | 0.10 | 16000 | 8000 | −13856 | −8000 | −13856 | −16000 | 0 |
| 11 | 0.11 | 7000 | 6062 | −3500 | 3500 | −6062 | 0 | −7000 |
| 12 | 0.12 | 0 | 0 | 0 | 0 | 0 | 0 | 0 |
| $\displaystyle\sum_{i=1}^{12}(\ )$ | | 409000 | −161976 | 49846 | 8500 | 21650 | −35000 | 14000 |
| $\dfrac{1}{6}\displaystyle\sum_{i=1}^{12}(\ )$ | | 68166.7 | −26996.0 | 8307.7 | 1416.7 | 3608.3 | −5833.3 | 2333.3 |

be obtained (see Eq. (1.48)):

$$p(t) = 34083.3 - 26996.0\cos 52.36t + 8307.7\sin 52.36t$$
$$+ 1416.7\cos 104.72t + 3608.3\sin 104.72t$$
$$- 5833.3\cos 157.08t + 2333.3\sin 157.08t$$
$$+ \cdots \; N/m^2 \tag{E.4}$$

## 1.12  VIBRATION LITERATURE

The literature on vibrations is large and diverse. Several textbooks are available [1.31], and dozens of technical periodicals regularly publish papers relating to vibrations. This is primarily because vibration affects so many disciplines, from science of materials to machinery analysis to spacecraft structures. Researchers in many fields must be attentive to vibration research.

The most widely circulated journals that publish papers relating to vibrations are *Journal of Vibration, Acoustics, Stress, and Reliability in Design; Journal of Applied Mechanics; Journal of Sound and Vibration; AIAA Journal; ASCE Journal of Engineering Mechanics; Earthquake Engineering and Structural Dynamics; Bulletin of the Japan Society of Mechanical Engineers*; *International Journal of Solids and Structures*; *International Journal for Numerical Methods in Engineering*; *Journal of the Acoustical Society of America*; *Sound and Vibration*; and *Vehicle System Dynamics*. Many of these journals are cited in the chapter references.

In addition, *Shock and Vibration Digest* and *Applied Mechanics Reviews* are monthly abstract journals containing brief discussions of nearly every published vibration paper. Formulas and solutions in vibration engineering can be readily found in references [1.32–1.34].

## 1.13  COMPUTER PROGRAM

A FORTRAN computer program, in the form of subroutine FORIER, is given for the harmonic analysis of a function $x(t)$. The arguments of this subroutine are as follows:

| | | |
|---|---|---|
| N | = | Number of equidistant points at which the values of $x(t)$ are known. Input data. |
| M | = | Number of Fourier coefficients to be computed. Input data. |
| TIME | = | Time period of the function $x(t)$. Input data. |
| X | = | Array of dimension N, containing the known values of $x(t)$. $X(I) = x(t_i)$. Input data. |
| T | = | Array of dimension N, containing the known values of $t$. $T(I) = t_i$. Input data. |
| AZERO | = | $a_0$ of Eq. (1.57). Output. |
| A | = | Array of dimension M, containing the computed values of $a_n$ of Eq. (1.58). Output. |
| B | = | Array of dimension M, containing the computed values of $b_n$ of Eq. (1.59). Output. |

To illustrate the use of subroutine FORIER, consider Example 1.7 with M = 5 instead of M = 3. Thus we have N = 12, M = 5, and TIME = 0.12. The main program that calls subroutine FORIER, subroutine FORIER itself, and the output of the program are given below.

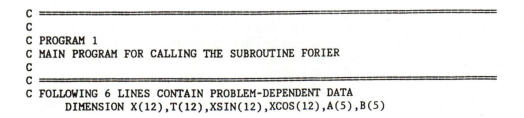

```
C ================================================================
C
C PROGRAM 1
C MAIN PROGRAM FOR CALLING THE SUBROUTINE FORIER
C
C ================================================================
C FOLLOWING 6 LINES CONTAIN PROBLEM-DEPENDENT DATA
      DIMENSION X(12),T(12),XSIN(12),XCOS(12),A(5),B(5)
```

```
      DATA N,M,TIME /12,5,0.12/
      DATA X /20000.0,34000.0,42000.0,49000.0,53000.0,70000.0,60000.0,
     2 36000.0,22000.0,16000.0,7000.0,0.0/
      DATA T /0.01,0.02,0.03,0.04,0.05,0.06,0.07,0.08,0.09,0.10,0.11,
     2 0.12/
C END OF PROBLEM-DEPENDENT DATA
      CALL FORIER (N,M,TIME,X,T,AZERO,A,B,XSIN,XCOS)
      PRINT 100
 100  FORMAT (//,46H FOURIER SERIES EXPANSION OF THE FUNCTION X(T),//)
      PRINT 200, N,M,TIME
 200  FORMAT (6H DATA:,//,37H NUMBER OF DATA POINTS IN ONE CYCLE =,I5,
     2 /,42H NUMBER OF FOURIER COEFFICIENTS REQUIRED =,I5,/,
     3 14H TIME PERIOD =,E15.8)
      PRINT 300, (T(I),I=1,N)
 300  FORMAT (/,33H TIME AT VARIOUS STATIONS, T(I) =,/,(4E15.8,1X))
      PRINT 400, (X(I),I=1,N)
 400  FORMAT (/,31H KNOWN VALUES OF X(I) AT T(I) =,/,(4E15.8,1X))
      PRINT 500
 500  FORMAT (//,29H RESULTS OF FOURIER ANALYSIS:,/)
      PRINT 600, AZERO
 600  FORMAT (8H AZERO =,2X,E15.8,//,31H VALUES OF I, A(I) AND B(I) ARE
     2 ,/)
      DO 700, I =1,M
 700  PRINT 800, I,A(I),B(I)
 800  FORMAT (I5,2X,E15.8,2X,E15.8)
      STOP
      END


C ==========================================================================
C
C SUBROUTINE FORIER
C
C ==========================================================================
      SUBROUTINE FORIER (N,M,TIME,X,T,AZERO,A,B,XSIN,XCOS)
      DIMENSION X(N),T(N),A(M),B(M),XSIN(N),XCOS(N)
      PI=3.1416
      SUMZ=0.0
      DO 100 I=1,N
 100  SUMZ=SUMZ+X(I)
      AZERO=2.0*SUMZ/REAL(N)
      DO 300 II=1,M
      SUMS=0.0
      SUMC=0.0
      DO 200 I=1,N
      THETA=2.0*PI*T(I)*REAL(II)/TIME
      XCOS(I)=X(I)*COS(THETA)
      XSIN(I)=X(I)*SIN(THETA)
      SUMS=SUMS+XSIN(I)
      SUMC=SUMC+XCOS(I)
 200  CONTINUE
      A(II)=2.0*SUMC/REAL(N)
      B(II)=2.0*SUMS/REAL(N)
 300  CONTINUE
      RETURN
      END
```

```
FOURIER SERIES EXPANSION OF THE FUNCTION X(T)


DATA:

NUMBER OF DATA POINTS IN ONE CYCLE =    12
NUMBER OF FOURIER COEFFICIENTS REQUIRED =     5
TIME PERIOD = 0.12000000E+00

TIME AT VARIOUS STATIONS, T(I) =
0.99999998E-02 0.20000000E-01 0.29999999E-01 0.39999999E-01
0.50000001E-01 0.59999999E-01 0.70000000E-01 0.79999998E-01
0.90000004E-01 0.10000000E+00 0.11000000E+00 0.12000000E+00

KNOWN VALUES OF X(I) AT T(I) =
0.20000000E+05 0.34000000E+05 0.42000000E+05 0.49000000E+05
0.53000000E+05 0.70000000E+05 0.60000000E+05 0.36000000E+05
0.22000000E+05 0.16000000E+05 0.70000000E+04 0.00000000E+00


RESULTS OF FOURIER ANALYSIS:

AZERO =    0.68166664E+05

VALUES OF I, A(I) AND B(I) ARE

    1   -0.26996299E+05    0.83075869E+04
    2    0.14166348E+04    0.36084932E+04
    3   -0.58332480E+04   -0.23334373E+04
    4   -0.58340521E+03    0.21650562E+04
    5   -0.21702822E+04   -0.64117188E+03
```

# REFERENCES

**1.1.** S. P. Timoshenko, *History of Strength of Materials*, McGraw-Hill, New York, 1953.

**1.2.** J. T. Cannon and S. Dostrovsky, *The Evolution of Dynamics: Vibration Theory from 1687 to 1742*, Springer Verlag, New York, 1981.

**1.3.** L. L. Bucciarelli and N. Dworsky, *Sophie Germain: An Essay in the History of the Theory of Elasticity*, D. Reidel Publishing Co., Dordrecht, Holland, 1980.

**1.4.** J. W. Strutt (Baron Rayleigh), *The Theory of Sound*, Dover, New York, 1945.

**1.5.** R. B. Lindsay, *Lord Rayleigh: The Man and His Work*, Pergamon Press, Oxford, 1970.

**1.6.** N. Minorsky, *Nonlinear Oscillations*, D. Van Nostrand, Princeton, N.J., 1962.

**1.7.** J. J. Stoker, *Nonlinear Vibrations*, Interscience Publishers, New York, 1950.

**1.8.** S. H. Crandall and W. D. Mark, *Random Vibration in Mechanical Systems*, Academic Press, New York, 1963.

**1.9.** J. D. Robson, *Random Vibration*, Edinburgh University Press, Edinburgh, 1964.

**1.10.** T. G. Butler and D. Michel, "NASTRAN—A summary of the functions and capabilities of the NASA structural analysis computer system," NASA SP-260, 1971.

**1.11.** S. S. Rao, *The Finite Element Method in Engineering*, Pergamon Press, Oxford, 1982.

**1.12.** D. Radaj et al., "Finite element analysis, an automobile engineer's tool," *International Conference on Vehicle Structural Mechanics: Finite Element Application to Design*, Society of Automotive Engineers, Detroit, 1974.

**1.13.** M. H. Richardson and K. A. Ramsey, "Integration of dynamic testing into the product design cycle," *Sound and Vibration*, Vol. 15, No. 11, November 1981, pp. 14–27.

**1.14.** M. J. Griffin and E. M. Whitham, "The discomfort produced by impulsive whole-body vibration," *Journal of the Acoustical Society of America*, Vol. 65, No. 5, 1980, pp. 1277–1284.

**1.15.** M. J. Griffin and C. H. Lewis, "A review of the effects of vibration on visual acuity and continuous manual control," *Journal of Sound and Vibration*, Vol. 56, No. 3, 1978, pp. 383–457.

**1.16.** J. E. Ruzicka, "Fundamental concepts of vibration control," *Sound and Vibration*, Vol. 5, No. 7, July 1971, pp. 16–23.

**1.17.** T. W. Black, "Vibratory finishing goes automatic" (Part 1: Types of machines; Part 2: Steps to automation), *Tool and Manufacturing Engineer*, July 1964, pp. 53–56; and August 1964, pp. 72–76.

**1.18.** E. O. Doebelin, *System Modeling and Response*, John Wiley, New York, 1980.

**1.19.** R. W. Fitzgerald, *Mechanics of Materials* (2nd Ed.), Addison-Wesley, Reading, Mass., 1982.

**1.20.** S. S. Rao, "Optimum design of bridge girders for electric overhead traveling cranes," *Journal of Engineering for Industry*, Vol. 100, No. 3, 1978, pp. 375–382.

**1.21.** F. Y. Chen, *Mechanics and Design of Cam Mechanisms*, Pergamon Press, New York, 1982.

**1.22.** S. H. Crandall, "The role of damping in vibration theory," *Journal of Sound and Vibration*, Vol. 11, No. 1, 1970, pp. 3–18.

**1.23.** C. W. Bert, "Material damping: An introductory review of mathematical models, measures, and experimental techniques," *Journal of Sound and Vibration*, Vol. 29, No. 2, 1973, pp. 129–153.

**1.24.** A. H. Burr, *Mechanical Analysis and Design*, Elsevier, New York, 1981.

**1.25.** J. M. Gasiorek and W. G. Carter, *Mechanics of Fluids for Mechanical Engineers*, Hart Publishing Co., New York, 1968.

**1.26.** A. Mironer, *Engineering Fluid Mechanics*, McGraw-Hill, New York, 1979.

**1.27.** P. Black, *Mechanics of Machines: A Course for Students*, Pergamon Press, Oxford, 1967.

**1.28.** E. Kreyszig, *Advanced Engineering Mathematics* (4th Ed.), John Wiley, New York, 1979.

**1.29.** M. H. Richardson, "Fundamentals of the discrete Fourier transform," *Sound and Vibration*, Vol. 12, No. 3, March 1978, pp. 40–46.

**1.30.** C. F. Gerald and P. O. Wheatley, *Applied Numerical Analysis* (3d Ed.), Addison-Wesley, Reading, Mass., 1984.

**1.31.** N. F. Rieger, "The literature of vibration engineering," *Shock and Vibration Digest*, Vol. 14, No. 1, January 1982, pp. 5–13.

**1.32.** R. D. Blevins, *Formulas for Natural Frequency and Mode Shape*, Van Nostrand Reinhold, New York, 1979.

**1.33.** W. D. Pilkey and P. Y. Chang, *Modern Formulas for Statics and Dynamics*, McGraw-Hill, New York, 1978.

**1.34.** C. M. Harris and C. E. Crede, *Shock and Vibration Handbook* (2nd Ed.), McGraw-Hill, New York, 1976.

## REVIEW QUESTIONS

**1.1.** What was Galileo's contribution to the development of vibration theory?

**1.2.** Give the names of two early investigators who derived the governing equation for the lateral vibration of prismatic bars.

**1.3.** Give two examples each of the bad and the good effects of vibration.

**1.4.** What are the three elementary parts of a vibrating system?

**1.5.** Define the degree of freedom of a vibrating system.

**1.6.** What is the difference between a discrete and a continuous system? Is it possible to solve any vibration problem as a discrete one?

**1.7.** What is the difference between free and forced vibration?

**1.8.** In vibration analysis, can we always disregard damping?

**1.9.** Can we identify a nonlinear vibration problem by looking at its governing differential equation?

**1.10.** What is the difference between deterministic and random vibration? Give two practical examples of each.

**1.11.** What methods are available for solving the governing equations of a vibration problem?

**1.12.** How do you connect several springs to increase the overall stiffness?

**1.13.** Define spring stiffness and damping constant.

**1.14.** What are the common types of damping?

**1.15.** What is the difference between harmonic motion and periodic motion?

**1.16.** State three different ways of expressing a periodic function in terms of its harmonics.

**1.17.** Define these terms: cycle, amplitude, phase angle, linear frequency, period, and natural frequency.

**1.18.** How are $\tau$, $\omega$, and $f$ related to each other?

**1.19.** How can we obtain the frequency, phase, and amplitude of a harmonic motion from the corresponding rotating vector?

**1.20.** How do you add two harmonic motions having different frequencies?

**1.21.** Suggest two methods for finding the time derivative of a harmonic motion.

**1.22.** What is harmonic analysis?

**1.23.** Give the names of two technical journals and two abstract journals for vibration research.

## PROBLEMS

The problem assignments are organized as follows:

| Problems | Section covered | Topic covered |
|---|---|---|
| 1.1–1.5, 1.7–1.9 | 1.7 | Spring elements |
| 1.6 | 1.7, 1.9 | Spring and damping elements |
| 1.10–1.14 | 1.8 | Mass elements |
| 1.15–1.17 | 1.9 | Damping elements |
| 1.18–1.31 | 1.10 | Harmonic motion |
| 1.32–1.42 | 1.11 | Harmonic analysis |
| 1.43–1.46 | 1.13 | Computer program |

**1.1.** Determine the equivalent spring constant of the system shown in Fig. 1.35.

**1.2.** In Fig. 1.36, find the equivalent spring constant of the system in the direction of $\theta$.

**1.3.** Find the equivalent torsional spring constant of the system shown in Fig. 1.37.

**1.4.** A machine of mass $m = 500$ kg is mounted on a simply supported steel beam of length $l = 2$ m, moment of inertia $I = 10^{-4}$ m$^4$, Young's modulus $E = 2.06 \times 10^{11}$ N/m$^2$, and negligible mass. To reduce the vertical deflection of the beam, a spring of stiffness $k$ is attached at the mid-span, as shown in Fig. 1.38. Determine the value of $k$ needed to reduce the deflection of the beam to one-third of its original value.

**1.5.** Four identical rigid bars—each of length $a$, cross sectional area $A$, and Young's modulus $E$—are connected to a spring of stiffness $k$ to form a structure for carrying a

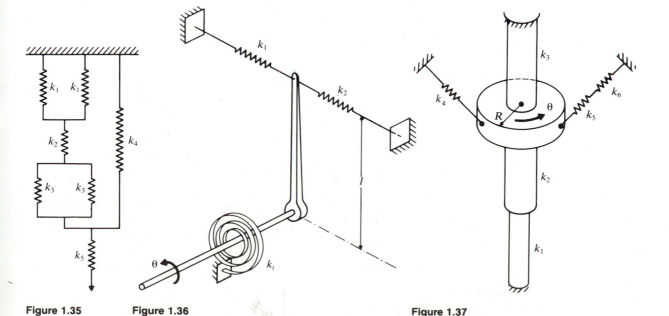

**Figure 1.35**          **Figure 1.36**                                      **Figure 1.37**

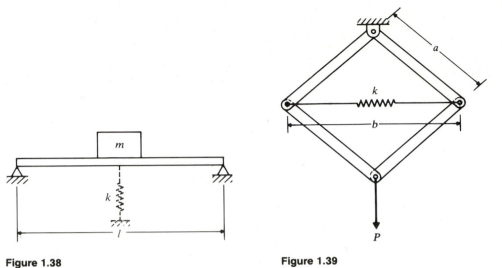

**Figure 1.38**

**Figure 1.39**

vertical load $P$, as shown in Fig. 1.39. Find the equivalent spring constant of the system, $k_{eq}$, disregarding the masses of the bars and the friction in the joints.

**1.6.** A simplified model of the Viking landing craft is shown in Fig. 1.40(a). Each of the three legs consists of a spring and a damper within a tube, so that the system will not bounce under a landing impact. The three legs are located symmetrically about the mid-vertical axis, and each leg makes an angle $\alpha$ with the vertical (Fig. 1.40(b)). The

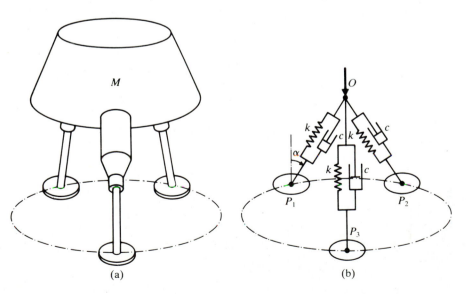

(a)                (b)

**Figure 1.40**

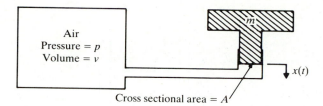

Cross sectional area = A

**Figure 1.41**

mass of the legs is negligible compared to the mass $M$ of the main body. If the length of each leg (spring) is $l$, find the equivalent spring stiffness and the equivalent damping constant of the landing craft in the vertical direction.

**1.7.** Figure 1.41 shows an air spring. This type of spring is generally used for obtaining very low natural frequencies while maintaining zero deflection under static loads. Find the spring constant of this air spring by assuming that the pressure $p$ and volume $v$ change adiabatically when the mass $m$ moves.

Hint: $pv^\gamma$ = constant for an adiabatic process, where $\gamma$ is the ratio of specific heats. For air, $\gamma = 1.4$.

**1.8.** Find the equivalent spring constant of the system shown in Fig. 1.42 in the direction of the load $P$.

**1.9.** Design a steel helical compression spring for a stiffness of 100 N/mm. The spring index $(D/d)$ is 8 and the number of active turns is 10.

**1.10.** In Fig. 1.43, find the equivalent mass of the rocker arm assembly, referred to the $x$ coordinate.

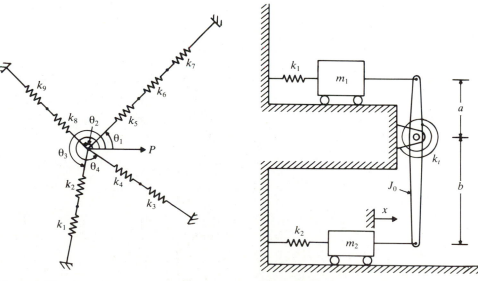

**Figure 1.42**                                   **Figure 1.43**

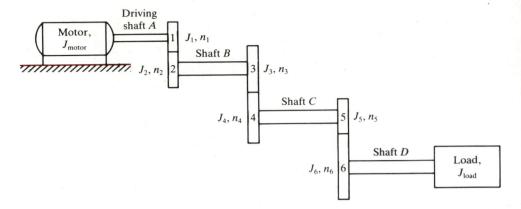

**Figure 1.44**

**1.11.** Find the equivalent mass moment of inertia of the gear train shown in Fig. 1.44, with reference to the driving shaft. In Fig. 1.44, $J_i$ and $n_i$ denote the mass moment of inertia and the number of teeth, respectively, of gear $i$, $i = 1, 2, \ldots, 6$.

**1.12.** In the pinion-and-rack system shown in Fig. 1.45, the weight of the rack is $W$ and the mass moment of inertia of each of the pinions is $J_0$. If the masses of the springs are negligible, find the equivalent mass of the system.

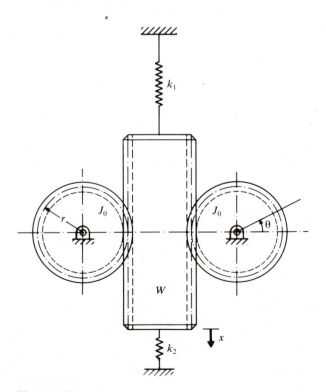

**Figure 1.45**

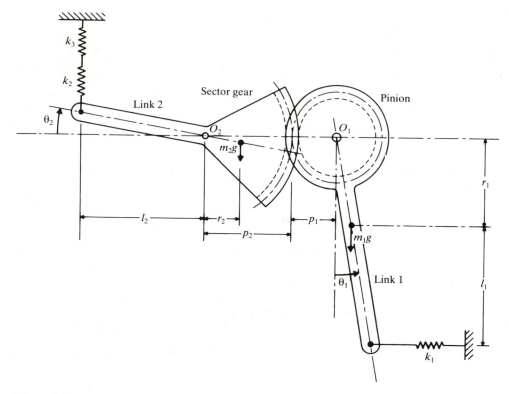

**Figure 1.46**

**1.13.** A pinion and a sector gear, located at the ends of links 1 and 2, are engaged together and rotate about $O_1$ and $O_2$, as shown in Fig. 1.46. If links 1 and 2 are connected to springs $k_1$, $k_2$, and $k_3$ as shown, find the equivalent torsional spring stiffness and equivalent mass moment of inertia of the system with reference to $\theta_1$. Assume the mass moment of inertia of link 1 (including the pinion) about $O_1$ as $J_1$ and that of link 2 (including the sector gear) about $O_2$ as $J_2$.

**1.14.** Two masses, having mass moments of inertia $J_1$ and $J_2$, are placed on rotating rigid shafts that are connected by gears, as shown in Fig. 1.47. If the number of teeth on

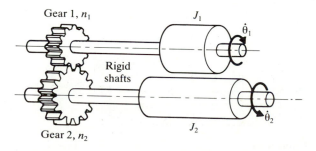

**Figure 1.47**  Rotational masses on geared shafts.

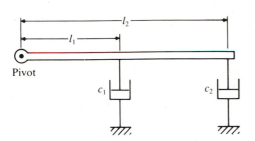

**Figure 1.48**  Dampers connected to a lever.

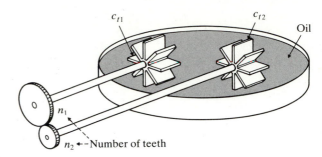

**Figure 1.49**  Dampers located on geared shafts.

gears 1 and 2 are $n_1$ and $n_2$, respectively, find the equivalent mass moment of inertia corresponding to $\theta_1$.

**1.15.** Find a single equivalent damping constant for the following cases:

   **i.** When two dampers are parallel.

   **ii.** When two dampers are in series.

   **iii.** When two dampers are connected to a lever (Fig. 1.48), and the equivalent damper is at the site $c_1$.

   **iv.** When two torsional dampers are located on geared shafts (Fig. 1.49), and the equivalent damper is at the location $c_{t1}$.

   Hint: The energy dissipated by a viscous damper in a cycle during harmonic motion is given by $\pi c \omega X^2$, where $c$ is the damping constant, $\omega$ is the frequency, and $X$ is the amplitude of oscillation.

**1.16.** A thin steel disc of radius $R$, thickness $h$, and mass polar moment of inertia $J_0$ is suspended by a steel wire so that it just touches a layer of oil in a shallow vessel, as shown in Fig. 1.50. This arrangement serves as a torsional damper when the steel disc makes angular oscillations in contact with the oil. If the thickness of the oil layer is $l$ and the viscosity of the oil is $\mu$, find the torsional damping coefficient of the system.

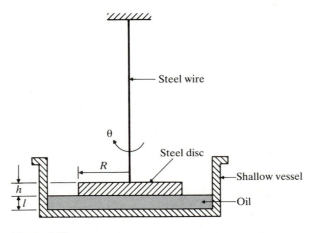

**Figure 1.50**

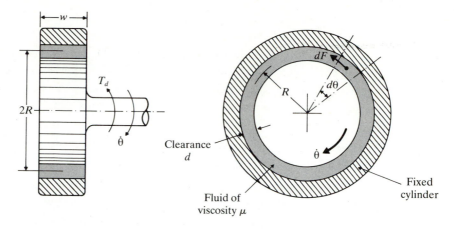

**Figure 1.51**   Torsional or rotary damper.

**1.17.** A viscous torsional damper can be made by using two concentric cylinders separated by a fluid of viscosity $\mu$, as shown in Fig. 1.51. If the outer cylinder is fixed and the inner one is made to rotate at an angular velocity $\dot{\theta}$, find the resulting torsional damping constant $c_t$.

Hint: The circumferential shear force produced over an arc length $d\theta$ is given by $dF = \mu Av/d$ where $A = wR\,d\theta$ is the surface area, $v = R\dot{\theta}$ is the peripheral velocity, and $d$ is the radial clearance.

**1.18.** Express the complex number $5 + 2i$ in the exponential form $Ae^{i\theta}$.

**1.19.** Add the two complex numbers $(1 + 2i)$ and $(3 - 4i)$ and express the result in the form $Ae^{i\theta}$.

**1.20.** Verify that the harmonic functions $x_1(t) = A_1\cos(\omega t + \phi)$ and $x_2(t) = A_2\sin(\omega t + \phi)$ satisfy the differential equation $d^2x/dt^2 + \omega^2 x = 0$.

**1.21.** A harmonic motion is described as $x(t) = A\cos(50t + \alpha)$ mm. The initial conditions are $x(0) = 3$ mm and $\dot{x}(0) = 1.0$ m/s.
  **i.** Find the constants $A$ and $\alpha$.
  **ii.** Express $x(t)$ in the form $x(t) = A_1\cos\omega t + A_2\sin\omega t$, and identify the constants $A_1$ and $A_2$.

**1.22.** Show that any linear combination of $\sin\omega t$ and $\cos\omega t$ such that $x(t) = A_1\cos\omega t + A_2\sin\omega t$ $(A_1, A_2 = \text{constants})$ represents a simple harmonic motion.

**1.23.** Find the sum of the two harmonic motions $x_1(t) = 5\cos(3t + 1)$ and $x_2(t) = 10\cos(3t + 2)$. Use:
  **i.** trigonometric relations
  **ii.** vector addition
  **iii.** complex number representation

**1.24.** If one of the components of the harmonic motion $x(t) = 10\sin(\omega t + 60°)$ is $x_1(t) = 5\sin(\omega t + 30°)$, find the other component.

**1.25.** Consider the two harmonic motions $x_1(t) = \frac{1}{2}\cos\frac{\pi}{2}t$ and $x_2(t) = \sin\pi t$. Is the sum $x_1(t) + x_2(t)$ a periodic motion? If so, what is its period?

**1.26.** Consider two harmonic motions of different frequencies: $x_1(t) = 2\cos 2t$ and $x_2(t) = \cos 3t$. Is the sum $x_1(t) + x_2(t)$ a harmonic motion? If so, what is its period?

**1.27.** Consider the two harmonic motions $x_1(t) = \frac{1}{2}\cos\frac{\pi}{2}t$ and $x_2(t) = \cos \pi t$. Is the difference $x(t) = x_1(t) - x_2(t)$ a harmonic motion? If so, what is its period?

**1.28.** Whenever two harmonic motions $x_1(t)$ and $x_2(t)$ having slightly different frequencies are combined, the amplitude of the resulting motion $x(t)$ varies between a maximum and a minimum value. Every time the amplitude of $x(t)$ reaches a maximum, there is said to be a *beat*. What are the maximum and minimum amplitudes of the combined motion $x(t) = x_1(t) + x_2(t)$ when $x_1(t) = 3\sin 30t$ and $x_2(t) = 3\sin 29t$? Also find the frequency of beats corresponding to $x(t)$.

**1.29.** A harmonic motion has an amplitude of 0.05 m and a frequency of 10 Hz. Find its period, maximum velocity, and maximum acceleration.

**1.30.** An accelerometer mounted on a building frame indicates that the frame is vibrating harmonically at 15 cps, with a maximum acceleration of 0.5 g. Determine the amplitude and the maximum velocity of the building frame.

**1.31.** In the vibration test of a machine, its maximum velocity and the maximum acceleration were measured as $\dot{x}_{max} = 0.01$ m/s and $\ddot{x}_{max} = 0.4$ g. Assuming that the machine vibrates harmonically at a single frequency, determine the frequency of vibration.

**1.32.** Prove that the complex Fourier series of Eq. (1.56) can be rewritten as

$$x(t) = \frac{1}{\tau} \sum_{n=-\infty}^{\infty} c_n e^{in\omega t}$$

where

$$c_n = \int_0^\tau x(t) e^{-in\omega t} \cdot dt$$

**1.33.** Prove that the sine Fourier components ($b_n$) are zero for even functions, that is, when $x(-t) = x(t)$. Also prove that the cosine Fourier components ($a_0$ and $a_n$) are zero for odd functions, that is, when $x(-t) = -x(t)$.

**1.34–1.39.** Find the Fourier series expansions of the periodic functions shown in Figs. 1.52 to 1.57. Also plot the corresponding frequency spectra.

**1.40.** The turning effort ($M_t$) of a six-cylinder four-stroke gasoline engine at various instants is given in the Table 1.3. Make a harmonic analysis of the turning effort. Find the amplitudes of the first three harmonics.

**1.41.** Conduct a harmonic analysis, including the first three harmonics, of the function given below:

| $t_i$ | 0.02 | 0.04 | 0.06 | 0.08 | 0.10 | 0.12 | 0.14 | 0.16 | 0.18 |
|-------|------|------|------|------|------|------|------|------|------|
| $x_i$ | 5 | 7 | 9 | 15 | 22 | 30 | 32 | 29 | 25 |

| $t_i$ | 0.20 | 0.22 | 0.24 | 0.26 | 0.28 | 0.30 | 0.32 |
|-------|------|------|------|------|------|------|------|
| $x_i$ | 18 | 18 | 22 | 24 | 21 | 7 | 4 |

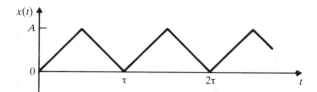

Figure 1.52

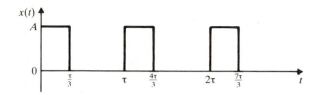

Figure 1.53

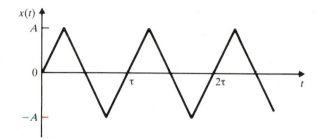

Figure 1.54

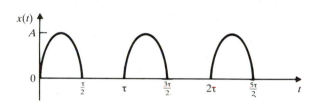

Figure 1.55

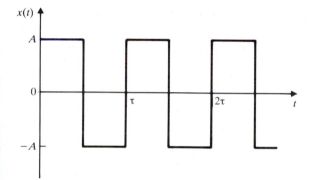

Figure 1.56

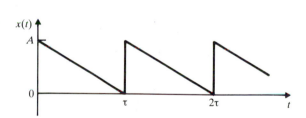

Figure 1.57

| TABLE 1.3 | | | | | |
|---|---|---|---|---|---|
| $t(s)$ | $M_t(N-m)$ | $t(s)$ | $M_t(N-m)$ | $t(s)$ | $M_t(N-m)$ |
| 0.00050 | 390 | 0.00450 | 950 | 0.00850 | 530 |
| 0.00100 | 410 | 0.00500 | 880 | 0.00900 | 500 |
| 0.00150 | 430 | 0.00550 | 820 | 0.00950 | 470 |
| 0.00200 | 460 | 0.00600 | 760 | 0.01000 | 450 |
| 0.00250 | 510 | 0.00650 | 700 | 0.01050 | 430 |
| 0.00300 | 590 | 0.00700 | 650 | 0.01100 | 410 |
| 0.00350 | 690 | 0.00750 | 600 | 0.01150 | 390 |
| 0.00400 | 810 | 0.00800 | 560 | 0.01200 | 380 |

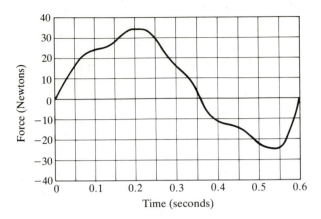

Figure 1.58

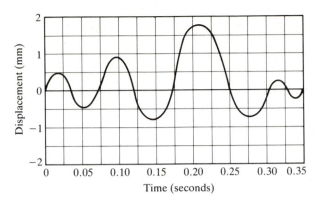

Figure 1.59

**1.42.** Make a harmonic analysis of the function shown in Fig. 1.58, including the first three harmonics.

**1.43.** Solve problem 1.40 using subroutine FORIER.

**1.44.** Solve problem 1.41 using subroutine FORIER.

**1.45.** Find the first six harmonics of the function shown in Fig. 1.58, using subroutine FORIER.

**1.46.** Use subroutine FORIER to conduct a harmonic analysis of the function shown in Fig. 1.59, including the first ten harmonics.

# CHAPTER 2

# Free Vibration of Single Degree of Freedom Systems

Sir Isaac Newton (1642–1727) was an English natural philosopher, a professor of mathematics at Cambridge University, and President of the Royal Society. His ''Principia Mathematica'' (1687), which deals with the laws and conditions of motion, is considered to be the greatest scientific work ever produced. The definitions of force, mass, and momentum, and his three laws of motion crop up continually in dynamics. Quite fittingly, the unit of force named ''Newton'' in SI units happens to be the approximate weight of an average apple, which inspired him to study the laws of gravity.

Courtesy of the Granger Collection

## 2.1  INTRODUCTION

The simplest possible vibratory system consists of a mass element connected to an elastic member, as shown in Fig. 2.1(a). For this single degree of freedom system, there is no external force applied to the mass; hence the motion resulting from an initial disturbance is a free vibration. Since there is no element that causes dissipation of energy during the motion of the mass, the amplitude of motion remains constant with time; it is an *undamped* system. In actual practice, except in a vacuum, the amplitude of free vibration diminishes gradually over time, due to the resistance offered by the surrounding medium (such as air). Such vibrations are said to be *damped*. The study of the free vibration of undamped and damped single degree of freedom systems is fundamental to the understanding of more advanced topics in vibrations.

Several mechanical and structural systems can be idealized as single degree of freedom systems. In most practical systems, the mass is distributed, but the behavior can be approximated by a single point mass. Similarly, the elastic component may appear complicated, but it can be idealized by a single spring. For the cam-and-follower system shown in Fig. 1.21, for example, the various masses were replaced by an equivalent mass ($m_{eq}$) in Example 1.3. The elements of the follower system (pushrod, rocker arm, valve, and valve spring) are all elastic but can be reduced to a single equivalent spring of stiffness $k_{eq}$. For a simple analysis, the cam-and-follower system can thus be idealized as a single degree of freedom spring-mass system, as shown in Fig. 2.2. Similarly, the building frame shown in Fig. 2.3(a) can be idealized as a spring-mass system, as shown in Fig. 2.3(b). In this case, since the spring constant $k$ is merely the ratio of force to deflection, it can be determined from the geometric and material properties of the columns. The mass of the idealized system is the same as that of the floor if we assume the mass of the columns to be negligible.

## 2.2  FREE VIBRATION OF AN UNDAMPED TRANSLATIONAL SYSTEM

### 2.2.1  Equation of Motion Using Newton's Second Law of Motion

**Spring-mass System in Horizontal Position.** Consider the undamped single degree of freedom system shown in Fig. 2.1(a). The mass is supported on frictionless rollers

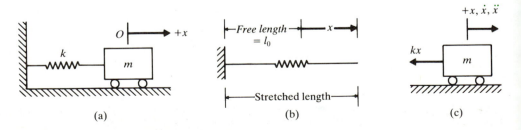

**Figure 2.1**  A spring-mass system in horizontal position.

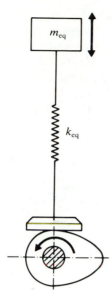

**Figure 2.2** Equivalent spring-mass system for the cam-and-follower system of Fig. 1.21.

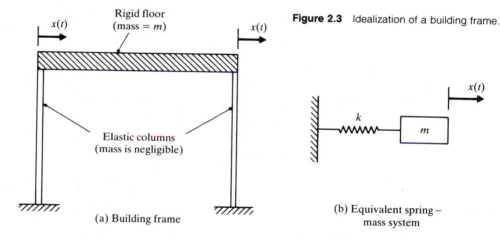

**Figure 2.3** Idealization of a building frame.

Rigid floor
(mass = $m$)

$x(t)$

$x(t)$

$x(t)$

$k$

$m$

Elastic columns
(mass is negligible)

(a) Building frame

(b) Equivalent spring –
mass system

and can have translatory motion in the horizontal direction. The unstretched length of the spring is $l_0$. Let the mass be displaced a distance $+x$ from its rest position. This results in a spring force $kx$, as shown in Fig. 2.1(c). Newton's second law states that

$$\text{mass} \times \text{acceleration} = \text{resultant force on the mass} \tag{2.1}$$

The application of Eq. (2.1) to the mass $m$ yields the equation of motion

$$m\ddot{x} = -kx$$

or

$$m\ddot{x} + kx = 0 \tag{2.2}$$

where $\ddot{x} = \dfrac{d^2x}{dt^2}$ is the acceleration of the mass.

**Spring-mass System in Vertical Position.** Consider the configuration of the spring-mass system shown in Fig. 2.4(a). The mass hangs at the lower end of a spring, which in turn is attached to a rigid support at its upper end. At rest the mass will hang in a position called the *static equilibrium position*, in which the upward spring force exactly balances the downward gravitational force on the mass. In this position the length of the spring is $l_0 + \delta_{st}$, where $\delta_{st}$ is the static deflection—the elongation due to the weight $W$ of the mass $m$. From Fig. 2.4(a), we find that, for static equilibrium,

$$W = mg = k\delta_{st} \tag{2.3}$$

where $g$ is the acceleration due to gravity. Let the mass be deflected a distance $+x$ from its static equilibrium position; then the spring force is $-k(x + \delta_{st})$, as shown in Fig. 2.4(c). The application of Newton's second law of motion to the mass $m$ gives

$$m\ddot{x} = -k(x + \delta_{st}) + W$$

and since $k\delta_{st} = W$, we obtain

$$m\ddot{x} + kx = 0 \tag{2.4}$$

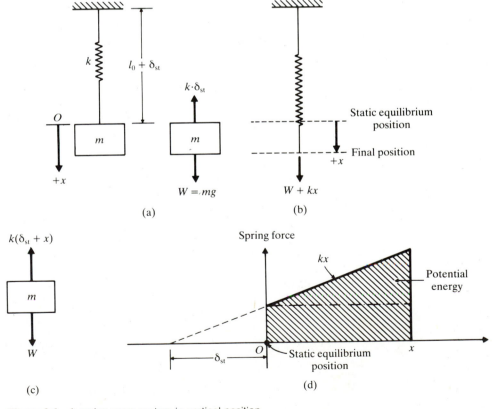

**Figure 2.4**    A spring-mass system in vertical position.

Notice that Eqs. (2.2) and (2.4) are identical. This indicates that when a mass moves in a vertical direction, we can ignore its weight, provided we measure $x$ from its static equilibrium position.

### 2.2.2 Equation of Motion Using the Principle of Conservation of Energy

Equation (2.2) can also be derived by using the conservation of energy principle. To apply this principle, first note that the system shown in Fig. 2.1(a) is conservative, since there is no energy dissipation due to damping. During vibration, the energy of the system is partly kinetic and partly potential. The kinetic energy $T$ is stored in the mass by virtue of its velocity, and the potential energy $U$ is stored in the spring by virtue of its elastic deformation. Due to the conservation of energy, we have

$$T + U = \text{constant}$$

or

$$\frac{d}{dt}(T + U) = 0 \tag{2.5}$$

The kinetic and potential energies are given by

$$T = \tfrac{1}{2}m\dot{x}^2 \tag{2.6}$$

and*

$$U = \tfrac{1}{2}kx^2 \tag{2.7}$$

Substitution of Eqs. (2.6) and (2.7) into Eq. (2.5) yields the desired equation

$$m\ddot{x} + kx = 0 \tag{2.2}$$

### 2.2.3 Solution

The solution of Eq. (2.2) can be found by assuming

$$x(t) = Ce^{st} \tag{2.8}$$

where $C$ and $s$ are constants to be determined. Substitution of Eq. (2.8) into Eq. (2.2) gives

$$C(ms^2 + k) = 0$$

Since $C$ cannot be zero, we have

$$ms^2 + k = 0 \tag{2.9}$$

---

*Equation (2.7) can also be derived by considering the weight of the mass (Fig. 2.4). Since the spring force is $mg$ at $x = 0$, the potential energy of the spring under the deformation $x$ will be $mgx + \tfrac{1}{2}kx^2$, as shown in Fig. 2.4(d). The potential energy of the system due to the change in elevation of the mass (note that $+x$ is downwards) is $-mgx$. Thus the net potential energy of the system about the static equilibrium position is given by

$U$ = potential energy of the spring + change in potential energy due to change in elevation of the mass $m$

$= mgx + \dfrac{1}{2}kx^2 - mgx = \dfrac{1}{2}kx^2$

and hence

$$s = \pm\left(-\frac{k}{m}\right)^{1/2} = \pm i\omega_n \tag{2.10}$$

where $i = (-1)^{1/2}$ and

$$\omega_n = \left(\frac{k}{m}\right)^{1/2} \tag{2.11}$$

Equation (2.9) is called the *auxiliary* or the *characteristic* equation corresponding to the differential equation (2.2). The two values of $s$ given by Eq. (2.10) are the roots of the characteristic equation, also known as the *eigenvalues* or the *characteristic values* of the problem. Since both values of $s$ satisfy Eq. (2.9), the general solution of Eq. (2.2) can be expressed as

$$x(t) = C_1 e^{i\omega_n t} + C_2 e^{-i\omega_n t} \tag{2.12}$$

where $C_1$ and $C_2$ are constants. By using the identities

$$e^{\pm i\alpha t} = \cos\alpha t \pm i\sin\alpha t$$

Eq. (2.12) can be rewritten as

$$x(t) = A_1\cos\omega_n t + A_2\sin\omega_n t \tag{2.13}$$

where $A_1$ and $A_2$ are new constants. The constants $C_1$ and $C_2$ or $A_1$ and $A_2$ can be determined from the initial conditions of the system. If the values of displacement $x(t)$ and velocity $\dot{x}(t) = \dfrac{dx}{dt}(t)$ are specified as $x_0$ and $\dot{x}_0$ at $t = 0$, we have, from Eq. (2.13),

$$x(t = 0) = A_1 = x_0$$
$$\dot{x}(t = 0) = \omega_n A_2 = \dot{x}_0 \tag{2.14}$$

Hence $A_1 = x_0$ and $A_2 = \dot{x}_0/\omega_n$. Thus the solution of Eq. (2.2) subject to the initial conditions of Eq. (2.14) is given by

$$x(t) = x_0\cos\omega_n t + \frac{\dot{x}_0}{\omega_n}\sin\omega_n t \tag{2.15}$$

### 2.2.4   Harmonic Motion

Equations (2.12), (2.13), and (2.15) are harmonic functions of time. The motion is symmetric about the equilibrium position of the mass $m$. The velocity is a maximum and the acceleration is zero each time the mass passes through this position. At the extreme displacements the velocity is zero and the acceleration is a maximum. Since this represents simple harmonic motion (see Sec. 1.10), the spring-mass system itself is called a *harmonic oscillator*. The quantity $\omega_n$, given by Eq. (2.11), represents the natural frequency of vibration of the system.

Equation (2.13) can be expressed in a different form by introducing the notation

$$A_1 = A\cos\phi$$
$$A_2 = A\sin\phi \tag{2.16}$$

where $A$ and $\phi$ are the new constants which can be expressed in terms of $A_1$ and $A_2$ as

$$A = \left( A_1^2 + A_2^2 \right) = \left[ x_0^2 + \left( \frac{\dot{x}_0}{\omega_n} \right)^2 \right]^{1/2} = \text{amplitude}$$

$$\phi = \tan^{-1}\left( \frac{A_2}{A_1} \right) = \tan^{-1}\left( \frac{\dot{x}_0}{x_0 \omega_n} \right) = \text{phase angle} \tag{2.17}$$

Introducing Eq. (2.16) into Eq. (2.13), the solution can be written as

$$x(t) = A\cos(\omega_n t - \phi) \tag{2.18}$$

By using the relations

$$A_1 = A_0 \sin\phi_0$$

$$A_2 = A_0 \cos\phi_0 \tag{2.19}$$

Eq. (2.13) can also be expressed as

$$x(t) = A_0 \sin(\omega_n t + \phi_0) \tag{2.20}$$

where

$$A_0 = A = \left[ x_0^2 + \left( \frac{\dot{x}_0}{\omega_n} \right)^2 \right]^{1/2} \tag{2.21}$$

and

$$\phi_0 = \tan^{-1}\left( \frac{x_0 \omega_n}{\dot{x}_0} \right) \tag{2.22}$$

The nature of harmonic oscillation can be represented graphically as in Fig. 2.5(a). If $\vec{A}$ denotes a vector of magnitude $A$ which makes an angle $\omega_n t - \phi$ with

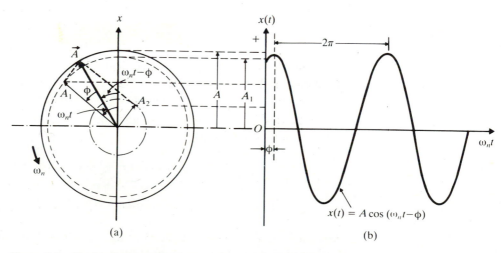

(a)                (b)

**Figure 2.5**  Graphical representation of the motion of a harmonic oscillator.

respect to the vertical $(x)$ axis, then the solution, Eq. (2.18), can be seen to be the projection of the vector $\vec{A}$ on the x-axis. The constants $A_1$ and $A_2$ of Eq. (2.13), given by Eq. (2.16), are merely the rectangular components of $\vec{A}$ along two orthogonal axes making angles $\phi$ and $-(\frac{\pi}{2} - \phi)$ with respect to the vector $\vec{A}$. Since the angle $\omega_n t - \phi$ is a linear function of time, it increases linearly with time; the entire diagram thus rotates anticlockwise at an angular velocity $\omega_n$. As the diagram (Fig. 2.5a) rotates, the projection of $\vec{A}$ onto the x-axis varies harmonically so that the motion repeats itself every time the vector $\vec{A}$ sweeps an angle of $2\pi$. The projection of $\vec{A}$, namely $x(t)$, is shown plotted in Fig. 2.5(b) as a function of time. The phase angle $\phi$ can also be interpreted as the angle between the origin and the first peak.

Note the following aspects of the spring-mass system:

1. If the spring-mass system is in a vertical position, as shown in Fig. 2.4(a), the circular natural frequency can be expressed as

$$\omega_n = \left( \frac{k}{m} \right)^{1/2} \tag{2.23}$$

The spring constant $k$ can be expressed in terms of the mass $m$ from Eq. (2.3) as

$$k = \frac{W}{\delta_{st}} = \frac{mg}{\delta_{st}} \tag{2.24}$$

Substitution of Eq. (2.24) into Eq. (2.11) yields

$$\omega_n = \left( \frac{g}{\delta_{st}} \right)^{1/2} \tag{2.25}$$

Hence the natural frequency in cycles per second and the natural period are given by

$$f_n = \frac{1}{2\pi} \left( \frac{g}{\delta_{st}} \right)^{1/2} = \frac{\omega_n}{2\pi} \tag{2.26}$$

$$T_n = \frac{1}{f_n} = 2\pi \left( \frac{\delta_{st}}{g} \right)^{1/2} = \frac{2\pi}{\omega_n} \tag{2.27}$$

Thus, when the mass vibrates in a vertical direction, we can compute the natural frequency and the period of vibration by simply measuring the static deflection $\delta_{st}$. It is not necessary that we know the spring stiffness $k$ and the mass $m$.

2. From Eq. (2.18), the velocity $\dot{x}(t)$ and the acceleration $\ddot{x}(t)$ of the mass $m$ at time $t$ can be obtained as

$$\dot{x}(t) = \frac{dx}{dt}(t) = \omega_n A \sin(\omega_n t - \phi) = \omega_n A \cos\left( \omega_n t - \phi + \frac{\pi}{2} \right)$$

$$\ddot{x}(t) = \frac{d^2 x}{dt^2}(t) = -\omega_n^2 A \cos(\omega_n t - \phi) = \omega_n^2 A \cos(\omega_n t - \phi + \pi) \tag{2.28}$$

Equation (2.28) shows that the velocity leads the displacement by $\frac{\pi}{2}$ and the acceleration leads the displacement by $\pi$.

**3.** If the initial displacement $(x_0)$ is zero, Eq. (2.18) becomes

$$x(t) = \frac{\dot{x}_0}{\omega_n}\cos\left(\omega_n t - \frac{\pi}{2}\right) = \frac{\dot{x}_0}{\omega_n}\sin\omega_n t \qquad (2.29)$$

On the other hand, if the initial velocity $(\dot{x}_0)$ is zero, the solution becomes

$$x(t) = x_0\cos\omega_n t \qquad (2.30)$$

---

### EXAMPLE 2.1

Find the natural frequency of vibration in the vertical direction of the overhead traveling crane shown in Fig. 1.16.

**Solution.** The equivalent spring constant of the crane (girder and cable) was derived in Example 1.1:

$$k_{eq} = \frac{2k_g k_c}{k_g + 2k_c} = \frac{48\pi EId^2}{96Il_2 + \pi d^2 l_1^3} \qquad (E.1)$$

The crane and the weight being lifted can now be modeled as a single degree of freedom system, as shown in Fig. 1.16(d). This leads to the natural frequency of the system:

$$\omega_n = \left(\frac{k_{eq}}{m}\right)^{1/2} = \left(\frac{k_{eq}g}{W}\right)^{1/2} = \left[\frac{48\pi gEId^2}{W(96Il_2 + \pi d^2 l_1^3)}\right]^{1/2} \qquad (E.2)$$

---

### EXAMPLE 2.2

Determine the natural frequency of the system shown in Fig. 2.6. Assume the pulleys to be frictionless and of negligible mass.

**Solution.** Since the pulleys are frictionless and massless, the tension in the rope is constant and is equal to the weight $W$ of the mass $m$. Thus the upward force acting on pulley 1 is $2W$, and the downward force acting on pulley 2 is $2W$. The center of pulley 1 moves up by a distance $2W/k_1$, and the center of pulley 2 moves down by $2W/k_2$. Thus the total movement of the mass $m$ is

$$2\left(\frac{2W}{k_1} + \frac{2W}{k_2}\right)$$

as the rope on either side of the pulley is free to move the mass downward. If $k_{eq}$ denotes the equivalent spring constant of the system,

$$\frac{\text{weight of the mass}}{\text{equivalent spring constant}} = \text{net displacement of the mass}$$

$$\frac{W}{k_{eq}} = 4W\left(\frac{1}{k_1} + \frac{1}{k_2}\right) = \frac{4W(k_1 + k_2)}{k_1 k_2}$$

$$k_{eq} = \frac{k_1 k_2}{4(k_1 + k_2)} \qquad (E.1)$$

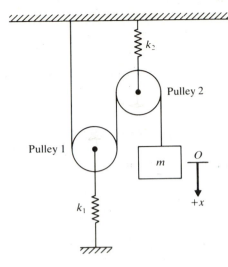

**Figure 2.6**

If the equation of motion of the mass is written as

$$m\ddot{x} + k_{eq}x = 0 \qquad \text{(E.2)}$$

the natural frequency is given by

$$\omega_n = \left(\frac{k_{eq}}{m}\right)^{1/2} = \left[\frac{k_1 k_2}{4m(k_1 + k_2)}\right]^{1/2} \text{rad/sec} \qquad \text{(E.3)}$$

or

$$f_n = \frac{\omega_n}{2\pi} = \frac{1}{4\pi}\left[\frac{k_1 k_2}{m(k_1 + k_2)}\right]^{1/2} \text{cycles/sec} \qquad \text{(E.4)}$$

## 2.3 FREE VIBRATION OF AN UNDAMPED TORSIONAL SYSTEM

If a rigid body oscillates about a specific reference axis, the resulting motion is called *torsional vibration*. In this case, the displacement of the body is measured in terms of an angular coordinate. In a torsional vibration problem, the restoring moment may be due to the torsion of an elastic member or to the unbalanced moment of a force or couple.

Figure 2.7 shows a disc, which has a polar mass moment of inertia $J_0$, mounted at one end of a solid circular shaft, the other end of which is fixed. Let the angular rotation of the disc about the axis of the shaft be $\theta$; $\theta$ also represents the angle of twist of the shaft. From the theory of torsion of circular shafts [2.1], we have the

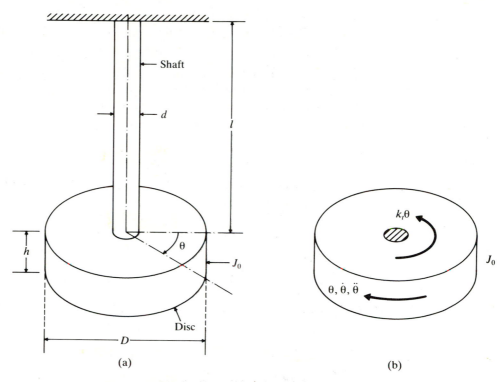

**Figure 2.7**   Torsional vibration of a disc.

relation

$$M_t = \frac{GJ\theta}{l} \tag{2.31}$$

where $M_t$ is the torque that produces the twist $\theta$, $G$ is the shear modulus, $l$ is the length of the shaft, $J$ is the polar moment of inertia of the cross section of the shaft given by

$$J = \frac{\pi d^4}{32} \tag{2.32}$$

and $d$ is the diameter of the shaft. If the disc is displaced by $\theta$ from its equilibrium position, the shaft provides a restoring torque of magnitude $M_t$. Thus the shaft acts as a torsional spring with a torsional spring constant

$$k_t = \frac{M_t}{\theta} = \frac{GJ}{l} = \frac{\pi G d^4}{32l} \qquad \frac{N \cdot m}{r \cdot a \cdot d} \tag{2.33}$$

### 2.3.1   Equation of Motion

The equation of the angular motion of the disc about its axis can be derived by using Newton's second law or the principle of conservation of energy. By considering the free body diagram of the disc (Fig. 2.7b), we can derive the equation of motion by applying Newton's second law of motion:

$$J_0\ddot{\theta} + k_t\theta = 0 \qquad (2.34)$$

which can be seen to be identical to Eq. (2.2) if the polar mass moment of inertia $J_0$, the angular displacement $\theta$, and the torsional spring constant $k_t$ are replaced by the mass $m$, the displacement $x$, and the linear spring constant $k$, respectively. Thus the natural circular frequency of the torsional system is

$$\omega_n = \left(\frac{k_t}{J_0}\right)^{1/2} \qquad (2.35)$$

and the period and frequency of vibration in cycles per second are

$$\tau_n = 2\pi\left(\frac{J_0}{k_t}\right)^{1/2} \qquad (2.36)$$

$$f_n = \frac{1}{2\pi}\left(\frac{k_t}{J_0}\right)^{1/2} \qquad (2.37)$$

Note the following aspects of this system:

1. If the cross section of the shaft supporting the disc is not circular, an appropriate torsional spring constant (as given on the inside cover) is to be used.

2. The polar mass moment of inertia of a disc is given by

$$J_0 = \frac{\rho h\pi D^4}{32} = \frac{WD^2}{8g}$$

   where $\rho$ is the mass density, $h$ is the thickness, $D$ is the diameter, and $W$ is the weight of the disc.

3. The torsional spring-inertia system shown in Fig. 2.7 is referred to as a *torsional pendulum*. One of the most important applications of a torsional pendulum is in a mechanical clock, where a ratchet and pawl convert the regular oscillation of a small torsional pendulum into the movements of the hands.

### 2.3.2   Solution

The general solution of Eq. (2.34) can be obtained, as in the case of Eq. (2.2):

$$\theta(t) = A_1\cos\omega_n t + A_2\sin\omega_n t \qquad (2.38)$$

where $\omega_n$ is given by Eq. (2.35), and $A_1$ and $A_2$ can be determined from the initial conditions. If

$$\theta(t=0) = \theta_0 \qquad \text{and} \qquad \dot{\theta}(t=0) = \frac{d\theta}{dt}(t=0) = \dot{\theta}_0 \qquad (2.39)$$

the constants $A_1$ and $A_2$ can be found:

$$A_1 = \theta_0$$

$$A_2 = \dot{\theta}_0/\omega_n \qquad (2.40)$$

Equation (2.38) can also be seen to represent a simple harmonic motion.

**EXAMPLE 2.3**

Any rigid body pivoted at a point other than its center of mass will oscillate about the pivot point under its own gravitational force. Such a system is known as a compound pendulum (Fig. 2.8). Find the natural frequency of such a system.

**Solution.** Let $O$ be the point of suspension and $G$ be the center of mass of the compound pendulum, as shown in Fig. 2.8. Let the rigid body oscillate in the $xy$ plane so that the coordinate $\theta$ can be used to describe its motion. Let $d$ denote the distance between $O$ and $G$, and $J_0$ the mass moment of inertia of the body about the $z$-axis (perpendicular to both $x$ and $y$). For a displacement $\theta$, the restoring torque (due to the weight of the body $W$) is $(Wd \sin \theta)$ and the equation of motion is

$$J_0 \ddot{\theta} + Wd \sin \theta = 0 \tag{E.1}$$

For small angles of oscillation, $\sin \theta \simeq \theta$. Hence Eq. (E.1) can be expressed as

$$J_0 \ddot{\theta} + Wd\theta = 0 \tag{E.2}$$

This gives the natural frequency of the compound pendulum:

$$\omega_n = \left(\frac{Wd}{J_0}\right)^{1/2} = \left(\frac{mgd}{J_0}\right)^{1/2} \tag{E.3}$$

Comparing Eq. (E.3) with the natural frequency of a simple pendulum, $\omega_n = (g/l)^{1/2}$ (see Problem 2.26), we can find the length of the equivalent simple pendulum:

$$l = \frac{J_0}{md} \tag{E.4}$$

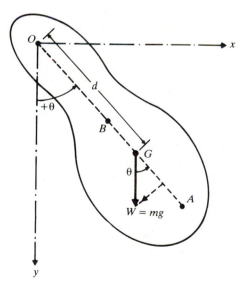

**Figure 2.8**

If $J_0$ is replaced by $mk_0^2$ where $k_0$ is the radius of gyration of the body about $O$, Eqs. (E.3) and (E.4) become

$$\omega_n = \left(\frac{gd}{k_0^2}\right)^{1/2} \tag{E.5}$$

$$l = \left(\frac{k_0^2}{d}\right) \tag{E.6}$$

If $k_G$ denotes the radius of gyration of the body about $G$, we have

$$k_0^2 = k_G^2 + d^2 \tag{E.7}$$

and Eq. (E.6) becomes

$$l = \left(\frac{k_G^2}{d} + d\right) \tag{E.8}$$

If the line $OG$ is extended to point $A$ such that

$$GA = \frac{k_G^2}{d} \tag{E.9}$$

Eq. (E.8) becomes

$$l = GA + d = OA \tag{E.10}$$

Hence, from Eq. (E.5), $\omega_n$ is given by

$$\omega_n = \left\{\frac{g}{\left(k_0^2/d\right)}\right\}^{1/2} = \left(\frac{g}{l}\right)^{1/2} = \left(\frac{g}{OA}\right)^{1/2} \tag{E.11}$$

This equation shows that, no matter whether the body is pivoted from $O$ or $A$, its natural frequency is the same. The point $A$ is called the *center of percussion*.

---

An automobile (shown in Fig. 2.9) can be considered as an example of a compound pendulum. If the front wheels strike a bump at point $O$, a reaction will be felt by the passengers unless the center of percussion of the vehicle is located at or near the rear axle (at point $A$). Similarly, if the rear wheels strike a bump at point $A$, some reaction will be felt unless the center of percussion is located at or near the front axle (at point $O$). Hence it is good design for an automobile to have its center of mass located at one axle and the center of percussion at the other axle.

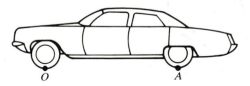

**Figure 2.9**

## 2.4 STABILITY CONDITIONS

Consider an inverted pendulum connected symmetrically by two independent springs, as shown in Fig. 2.10(a). Assume that the mass of the rod is negligible and that the springs are unstretched when the pendulum is vertical. When the pendulum is displaced by an angle $\theta$, the spring force in each spring is $ka \sin \theta$; the total spring force is $2ak \sin \theta$. The gravity force $W = mg$ acts vertically downwards. The moment about the point of rotation $O$ due to the angular acceleration $\ddot{\theta}$ is $J_0 \ddot{\theta} = ml^2 \ddot{\theta}$. Thus the equation of motion of the pendulum, for rotation about the point $O$, can be written as

$$ml^2 \ddot{\theta} + (2ka \sin \theta) a \cos \theta - W \sin \theta l = 0 \tag{2.41}$$

For small oscillations, Eq. (2.41) reduces to

$$ml^2 \ddot{\theta} + 2ka^2 \theta - Wl\theta = 0$$

$$\ddot{\theta} + \left( \frac{2ka^2 - Wl}{ml^2} \right) \theta = 0 \tag{2.42}$$

The solution of Eq. (2.42) depends on the sign of $(2ka^2 - Wl)/(ml^2)$, as discussed below.

**Case 1.** When $(2ka^2 - Wl)/(ml^2) > 0$, the solution of Eq. (2.42) can be expressed as

$$\theta(t) = A_1 \cos \omega_n t + A_2 \sin \omega_n t \tag{2.43}$$

where $A_1$ and $A_2$ are constants and

$$\omega_n = \left( \frac{2ka^2 - Wl}{ml^2} \right)^{1/2} \tag{2.44}$$

This represents the case of a stable oscillation.

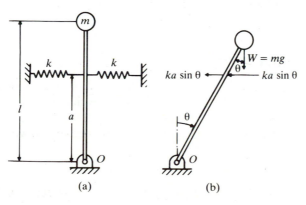

(a)          (b)

**Figure 2.10**

**Case 2.** When $(2ka^2 - Wl)/(ml^2) = 0$, Eq. (2.42) reduces to $\ddot{\theta} = 0$ and the solution can be obtained directly by integrating twice as

$$\theta(t) = C_1 t + C_2 \tag{2.45}$$

For the initial conditions $\theta(t = 0) = \theta_0$ and $\dot{\theta}(t = 0) = \dot{\theta}_0$, the solution becomes

$$\theta(t) = \dot{\theta}_0 t + \theta_0 \tag{2.46}$$

Equation (2.46) represents the case of equilibrium with a constant velocity, $\dot{\theta}_0$. Since $\theta$ is assumed to be small, Eq. (2.46) denotes the beginning of the constant-velocity motion. Equation (2.46) represents the static equilibrium state if $\dot{\theta}_0 = 0$. In this case, the pendulum remains in its original position, defined by $\theta = \theta_0$.

**Case 3.** When $(2ka^2 - Wl)/(ml^2) < 0$, we define

$$\alpha = \left( \frac{Wl - 2ka^2}{ml^2} \right)^{1/2}$$

and express the solution of Eq. (2.42) as

$$\theta(t) = B_1 e^{\alpha t} + B_2 e^{-\alpha t} \tag{2.47}$$

where $B_1$ and $B_2$ are constants. For the initial conditions $\theta(t = 0) = \theta_0$ and $\dot{\theta}(t = 0) = \dot{\theta}_0$, Eq. (2.47) becomes

$$\theta(t) = \frac{1}{2\alpha} \left[ \left( \alpha\theta_0 + \dot{\theta}_0 \right) e^{\alpha t} + \left( \alpha\theta_0 - \dot{\theta}_0 \right) e^{-\alpha t} \right] \tag{2.48}$$

For nonzero values of $\theta_0$ and $\dot{\theta}_0$, this represents a nonoscillatory motion in which $\theta$ increases exponentially with time. Since $\theta$ is assumed to be small, this solution represents only the beginning of the motion. This shows that the system is unstable. The physical reason for this is that the restoring moment due to the spring $(2ka^2\theta)$ is less than the nonrestoring moment due to gravity $(-Wl\theta)$. The restoring condition tends to bring the system to equilibrium, while the nonrestoring condition tends to move it away from the equilibrium configuration. In the vibration analysis of various systems, these factors regarding restoring and nonrestoring conditions and stability should be kept in mind.

## 2.5 ENERGY METHOD

For a single degree of freedom system, the equation of motion was derived using the energy method in Sec. 2.2.2. In this section, we shall use the energy method to find the natural frequencies of single degree of freedom systems. The principle of conservation of energy, in the context of an undamped vibrating system, can be restated as

$$T_1 + U_1 = T_2 + U_2 \tag{2.49}$$

where the subscripts 1 and 2 denote two different instants of time. Specifically, we use the subscript 1 to denote the time when the mass is passing through its static equilibrium position and choose $U_1 = 0$ as reference for the potential energy. If we let the subscript 2 indicate the time corresponding to the maximum displacement of the mass, we have $T_2 = 0$. Thus Eq. (2.49) becomes

$$T_1 + 0 = 0 + U_2 \tag{2.50}$$

If the system is undergoing harmonic motion, then $T_1$ and $U_2$ denote the maximum values of $T$ and $U$ respectively, and Eq. (2.50) becomes

$$T_{\text{max}} = U_{\text{max}} \tag{2.51}$$

The application of Eq. (2.51), which is also known as *Rayleigh's energy method*, gives the natural frequency of the system directly, as illustrated in the following examples.

### EXAMPLE 2.4

The exhaust from a single-cylinder four-stroke diesel engine is to be connected to a silencer, and the pressure therein is to be measured with a simple U-tube manometer (see Figure 2.11). Calculate the minimum length of the manometer tube so that the natural frequency of oscillation of the mercury column will be 3.5 times slower than the frequency of the pressure fluctuations in the silencer at an engine speed of 600 revolutions per minute. The frequency of pressure fluctuations in the silencer is equal to

$$\frac{\text{number of cylinders} \times \text{speed of the engine}}{2}$$

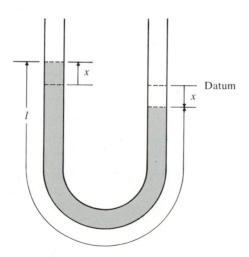

**Figure 2.11**

*Solution*

*1. Natural frequency of oscillation of the liquid column:* Let the datum in Fig. 2.11 be taken as the equilibrium position of the liquid. If the displacement of the liquid column from the equilibrium position is denoted by $x$, the change in potential energy is given by

$U$ = potential energy of raised liquid column + potential energy of depressed liquid column

= (weight of mercury raised × displacement of the c.g. of the segment) + (weight of mercury depressed × displacement of the c.g. of the segment)

$$= (Ax\gamma)\frac{x}{2} + (Ax\gamma)\frac{x}{2} = A\gamma x^2 \tag{E.1}$$

where $A$ is the cross sectional area of the mercury column and $\gamma$ is the specific weight of mercury. The change in kinetic energy is given by

$$T = \frac{1}{2}(\text{mass of mercury})(\text{velocity})^2$$

$$= \frac{1}{2}\frac{Al\gamma}{g}(\dot{x})^2 \tag{E.2}$$

where $l$ is the length of the mercury column. By assuming harmonic motion, we can write

$$x(t) = X\cos\omega_n t \tag{E.3}$$

where $X$ is the maximum displacement and $\omega_n$ is the natural frequency. By substituting Eq. (E.3) into Eqs. (E.1) and (E.2), we obtain

$$U = U_{max}\cos^2\omega_n t \tag{E.4}$$

$$T = T_{max}\sin^2\omega_n t \tag{E.5}$$

where

$$U_{max} = A\gamma X^2 \tag{E.6}$$

and

$$T_{max} = \frac{1}{2}\frac{A\gamma l\omega_n^2}{g}X^2 \tag{E.7}$$

By equating $U_{max}$ to $T_{max}$, we obtain the natural frequency:

$$\omega_n = \left(\frac{2g}{l}\right)^{1/2} \tag{E.8}$$

*2. Length of the mercury column:* The frequency of pressure fluctuations in the silencer

$$= \frac{1 \times 600}{2}$$

$$= 300 \text{ rev/min}$$

$$= \frac{300 \times 2\pi}{60} = 10\pi \text{ rad/sec} \tag{E.9}$$

Thus the frequency of oscillations of the liquid column in the manometer is $10\pi/3.5 = 9.0$ rad/sec. By using Eq. (E.8), we obtain

$$\left(\frac{2g}{l}\right)^{1/2} = 9.0 \tag{E.10}$$

or

$$l = \frac{2.0 \times 9.81}{(9.0)^2} = 0.243 \text{ m} \tag{E.11}$$

---

**EXAMPLE 2.5**

Determine the effect of the mass of the spring on the natural frequency of the spring-mass system shown in Fig. 2.12.

**Solution.** Let $l$ be the total length of the spring. If $x$ denotes the displacement of the lower end of the spring (or mass $m$), the displacement at distance $y$ from the support is given by $y(x/l)$. Similarly, if $\dot{x}$ denotes the velocity of the mass $m$, the velocity of a spring element located at distance $y$ from the support is given by $y(\dot{x}/l)$. The kinetic energy of the spring element of length $dy$ is

$$dT_s = \frac{1}{2}\left(\frac{m_s}{l}dy\right)\left(\frac{y\dot{x}}{l}\right)^2 \tag{E.1}$$

where $m_s$ is the mass of the spring. The total kinetic energy of the system can be expressed as

$$T = \text{kinetic energy of mass } (T_m) + \text{kinetic energy of spring } (T_s)$$

$$= \frac{1}{2}m\dot{x}^2 + \int_{y=0}^{l} \frac{1}{2}\left(\frac{m_s}{l}dy\right)\left(\frac{y^2\dot{x}^2}{l^2}\right)$$

$$= \frac{1}{2}m\dot{x}^2 + \frac{1}{2}\frac{m_s}{3}\dot{x}^2 \tag{E.2}$$

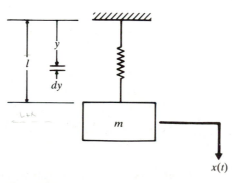

**Figure 2.12**

The total potential energy of the system is given by

$$U = \tfrac{1}{2}kx^2 \tag{E.3}$$

By assuming a harmonic motion

$$x(t) = X\cos\omega_n t \tag{E.4}$$

where $X$ is the maximum displacement of the mass and $\omega_n$ is the natural frequency, the maximum kinetic and potential energies can be expressed as

$$T_{\max} = \frac{1}{2}\left(m + \frac{m_s}{3}\right)X^2\omega_n^2 \tag{E.5}$$

$$U_{\max} = \frac{1}{2}kX^2 \tag{E.6}$$

By equating $T_{\max}$ and $U_{\max}$, we obtain the expression for the natural frequency:

$$\omega_n = \left(\frac{k}{m + \dfrac{m_s}{3}}\right)^{1/2} \quad \text{Mass-spring} \tag{E.7}$$

Thus the effect of the mass of the spring can be accounted by adding one third of its mass to the main mass [2.2].

## 2.6  FREE VIBRATION WITH VISCOUS DAMPING

### 2.6.1  Equation of Motion

As stated in Sec. 1.9, the viscous damping force $F$ is proportional to the velocity $\dot{x}$ or $v$ and can be expressed as

$$F = -c\dot{x} \tag{2.52}$$

where $c$ is the damping constant or coefficient of viscous damping and the negative sign indicates that the damping force is opposite to the direction of velocity. A single degree of freedom system with a viscous damper is shown in Fig. 2.13. If $x$ is measured from the equilibrium position of the mass $m$, the application of Newton's law yields the equation of motion: $ma$     $\Sigma F$

$$m\ddot{x} = -c\dot{x} - kx$$

or

$$m\ddot{x} + c\dot{x} + kx = 0 \tag{2.53}$$

### 2.6.2  Solution

To solve Eq. (2.53), we assume a solution in the form

$$x(t) = ce^{st} \tag{2.54}$$

constant

where $c$ and $s$ are undetermined constants. Inserting this function into Eq. (2.53)

$$x'(t) = sce^{st}$$
$$x''(t) = s^2ce^{st}$$

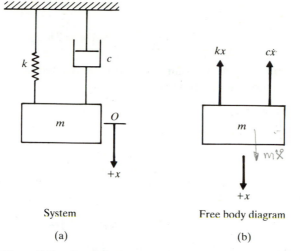

System                     Free body diagram

(a)                          (b)

**Figure 2.13**  Single degree of freedom system with viscous damper.

leads to the characteristic equation $\quad ms^2 ce^{st} + cs\,ce^{st} + k\,ce^{st} = 0$

$$ms^2 + cs + k = 0 \tag{2.55}$$

the roots of which are

$$s_{1,2} = \frac{-c \pm \sqrt{c^2 - 4mk}}{2m} = -\frac{c}{2m} \pm \sqrt{\left(\frac{c}{2m}\right)^2 - \frac{k}{m}} \tag{2.56}$$

These roots give two solutions to Eq. (2.53):

$$x_1(t) = e^{s_1 t} \qquad \text{and} \qquad x_2(t) = e^{s_2 t} \tag{2.57}$$

Thus the general solution of Eq. (2.53) is given by a linear combination of the two solutions $x_1(t)$ and $x_2(t)$:

$$x(t) = C_1 e^{s_1 t} + C_2 e^{s_2 t}$$

$$= C_1 e^{\left\{-\frac{c}{2m} + \sqrt{\left(\frac{c}{2m}\right)^2 - \frac{k}{m}}\right\} t} + C_2 e^{\left\{-\frac{c}{2m} - \sqrt{\left(\frac{c}{2m}\right)^2 - \frac{k}{m}}\right\} t} \tag{2.58}$$

where $C_1$ and $C_2$ are arbitrary constants to be determined from the initial conditions of the system.

**Critical Damping Constant and the Damping Ratio.** The critical damping $c_c$ is defined as the value of the damping constant $c$ for which the radical in Eq. (2.56) becomes zero:

$$\left(\frac{c_c}{2m}\right)^2 - \frac{k}{m} = 0$$

$$C_c \equiv \text{determinent} = 0$$

or

$$c_c = 2m\sqrt{\frac{k}{m}} = 2\sqrt{km} = 2m\omega_n \tag{2.59}$$

For any damped system, the <u>damping ratio $\zeta$</u> is defined as the ratio of the damping constant to the critical damping constant:

$$\zeta = c/c_c \tag{2.60}$$

Using Eqs. (2.60) and (2.59), we can write

$$\frac{c}{2m} = \frac{c}{c_c} \cdot \frac{c_c}{2m} = \zeta\omega_n \tag{2.61}$$

and hence

$$s_{1,2} = \left(-\zeta \pm \sqrt{\zeta^2 - 1}\right)\omega_n \tag{2.62}$$

Thus the solution, Eq. (2.58), can be written as

$$x(t) = C_1 e^{\left(-\zeta + \sqrt{\zeta^2-1}\right)\omega_n t} + C_2 e^{\left(-\zeta - \sqrt{\zeta^2-1}\right)\omega_n t} \tag{2.63}$$

The nature of the roots $s_1$ and $s_2$ and hence the behavior of the solution, Eq. (2.63), depends upon the magnitude of damping. It can be seen that the case $\zeta = 0$ leads to the undamped vibrations discussed in Sec. 2.2. Hence we assume that $\zeta \neq 0$ and consider the following three cases.

**Case 1.** Underdamped system ($\zeta < 1$ or $c < c_c$ or $c/2m < \sqrt{k/m}$). For this condition, ($\zeta^2 - 1$) is negative and the roots $s_1$ and $s_2$ can be expressed as

$$s_1 = \left(-\zeta + i\sqrt{1-\zeta^2}\right)\omega_n$$

$$s_2 = \left(-\zeta - i\sqrt{1-\zeta^2}\right)\omega_n$$

and the solution, Eq. (2.63), can be written in different forms:

$$\begin{aligned}
x(t) &= C_1 e^{\left(-\zeta + i\sqrt{1-\zeta^2}\right)\omega_n t} + C_2 e^{\left(-\zeta - i\sqrt{1-\zeta^2}\right)\omega_n t} \\
&= e^{-\zeta\omega_n t}\left\{C_1 e^{i\sqrt{1-\zeta^2}\,\omega_n t} + C_2 e^{-i\sqrt{1-\zeta^2}\,\omega_n t}\right\} \\
&= e^{-\zeta\omega_n t}\left\{(C_1 + C_2)\cos\sqrt{1-\zeta^2}\,\omega_n t + i(C_1 - C_2)\sin\sqrt{1-\zeta^2}\,\omega_n t\right\} \\
&= e^{-\zeta\omega_n t}\left\{C_1'\cos\sqrt{1-\zeta^2}\,\omega_n t + C_2'\sin\sqrt{1-\zeta^2}\,\omega_n t\right\} \\
&= X e^{-\zeta\omega_n t}\sin\left(\sqrt{1-\zeta^2}\,\omega_n t + \phi\right) \\
x_h(t) &= X_0 e^{-\zeta\omega_n t}\cos\left(\sqrt{1-\zeta^2}\,\omega_n t + \phi_0\right) \tag{2.64}
\end{aligned}$$

where $(C_1', C_2')$, $(X, \phi)$, and $(X_0, \phi_0)$ are arbitrary constants to be determined from the initial conditions.

For the initial conditions $x(t=0)=x_0$ and $\dot{x}(t=0)=\dot{x}_0$, $C_1'$ and $C_2'$ can be found:

$$C_1' = x_0 \qquad \text{and} \qquad C_2' = \frac{\dot{x}_0 + \zeta\omega_n x_0}{\sqrt{1-\zeta^2}\,\omega_n} \tag{2.65}$$

and hence the solution becomes

$$x(t) = e^{-\zeta\omega_n t}\left\{ x_0\cos\sqrt{1-\zeta^2}\,\omega_n t + \frac{\dot{x}_0 + \zeta\omega_n x_0}{\sqrt{1-\zeta^2}\,\omega_n}\sin\sqrt{1-\zeta^2}\,\omega_n t \right\} \tag{2.66}$$

The constants $(X,\phi)$ and $(X_0,\phi_0)$ can be expressed as

$$X = X_0 = \sqrt{(C_1')^2 + +(C_2')^2} \tag{2.67}$$

$$\phi = \tan^{-1}(C_1'/C_2') \tag{2.68}$$

$$\phi_0 = \tan^{-1}(C_2'/C_1') \tag{2.69}$$

The motion described by Eq. (2.66) is a harmonic motion of angular frequency $\sqrt{1-\zeta^2}\,\omega_n$, but because of the factor $e^{-\zeta\omega_n t}$, the amplitude decreases exponentially with time, as shown in Fig. 2.14. The quantity

$$\omega_d = \sqrt{1-\zeta^2}\,\omega_n \tag{2.70}$$

is called the *frequency of damped vibration*. It can be seen that the frequency of damped vibration $\omega_d$ is always less than the undamped natural frequency $\omega_n$. The decrease in the frequency of damped vibration with increasing amount of damping is shown in Fig. 2.15. The underdamped case is very important in the study of mechanical vibrations, as it is the only case which leads to an oscillatory motion [2.10].

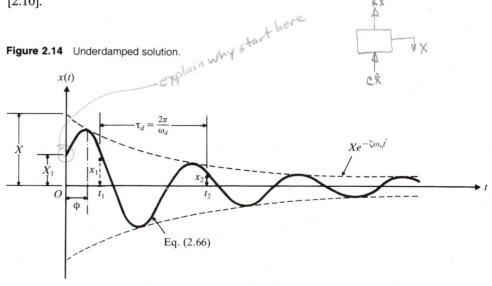

**Figure 2.14**   Underdamped solution.

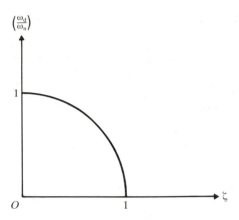

**Figure 2.15**   Variation of $\omega_d$ with damping.

**Case 2.** Critically damped system ($\zeta = 1$ or $c = c_c$ or $c/2m = \sqrt{k/m}$). In this case the two roots $s_1$ and $s_2$ in Eq. (2.62) are equal:

$$s_1 = s_2 = -\frac{c_c}{2m} = -\omega_n \tag{2.71}$$

Because of the repeated roots, the solution of Eq. (2.53) is given by [2.3]*

$$x(t) = (C_1 + C_2 t)e^{-\omega_n t} \tag{2.72}$$

The application of the initial conditions $x(t = 0) = x_0$ and $\dot{x}(t = 0) = \dot{x}_0$ for this case gives

$$C_1 = x_0$$

$$C_2 = \dot{x}_0 + \omega_n x_0 \tag{2.73}$$

and the solution becomes

$$x(t) = \left[ x_0 + (\dot{x}_0 + \omega_n x_0)t \right] e^{-\omega_n t} \tag{2.74}$$

It can be seen that the motion represented by Eq. (2.74) is *aperiodic* (i.e., non-periodic). Since $e^{-\omega_n t} \to 0$ as $t \to \infty$, the motion will eventually diminish to zero, as indicated in Fig. 2.16.

---

*Equation (2.72) can also be obtained by making $\zeta$ approach unity in the limit in Eq. (2.66). As $\zeta \to 1$, $\omega_d \to 0$; hence $\cos \omega_d t \to 1$ and $\sin \omega_d t \to \omega_d t$. Thus Eq. (2.66) yields

$$x(t) = e^{-\omega_n t}(C_1' + C_2' \omega_d t) = (C_1 + C_2 t)e^{-\omega_n t}$$

where $C_1 = C_1'$ and $C_2 = C_2' \omega_d$ are new constants.

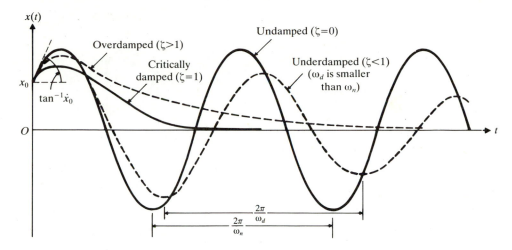

**Figure 2.16**  Comparison of motions with different types of damping.

**Case 3.** Overdamped system ($\zeta > 1$ or $c > c_c$ or $c/2m > \sqrt{k/m}$). As $\sqrt{\zeta^2 - 1} > 0$, Eq. (2.62) shows that the roots $s_1$ and $s_2$ are real and distinct and are given by

$$s_1 = \left(-\zeta + \sqrt{\zeta^2 - 1}\right)\omega_n < 0$$

$$s_2 = \left(-\zeta - \sqrt{\zeta^2 - 1}\right)\omega_n < 0$$

with $s_2 \ll s_1$. In this case, the solution, Eq. (2.63), can be expressed as

$$x(t) = C_1 e^{\left(-\zeta + \sqrt{\zeta^2 - 1}\right)\omega_n t} + C_2 e^{\left(-\zeta - \sqrt{\zeta^2 - 1}\right)\omega_n t} \tag{2.75}$$

For the initial conditions $x(t = 0) = x_0$ and $\dot{x}(t = 0) = \dot{x}_0$, the constants $C_1$ and $C_2$ can be obtained:

$$C_1 = \frac{x_0 \omega_n \left(\zeta + \sqrt{\zeta^2 - 1}\right) + \dot{x}_0}{2\omega_n \sqrt{\zeta^2 - 1}}$$

$$C_2 = \frac{-x_0 \omega_n \left(\zeta - \sqrt{\zeta^2 - 1}\right) - \dot{x}_0}{2\omega_n \sqrt{\zeta^2 - 1}} \tag{2.76}$$

Equation (2.75) shows that the motion is aperiodic regardless of the initial conditions imposed on the system. Since roots $s_1$ and $s_2$ are both negative, the motion diminishes exponentially with time, as shown in Fig. 2.16.

Note the following two aspects of these systems:

1. The nature of the roots $s_1$ and $s_2$ with varying values of damping $c$ or $\zeta$ can be shown in a complex plane. In Fig. 2.17, the horizontal and vertical axes are chosen as the real and imaginary axes. The semicircle represents the locus of the

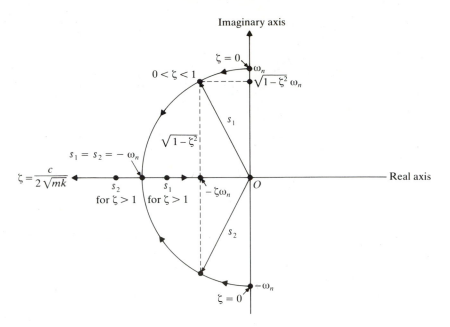

**Figure 2.17**   Locus of $s_1$ and $s_2$.

roots $s_1$ and $s_2$ for different values of $\zeta$ in the range $0 < \zeta < 1$. This figure permits us to see instantaneously the effect of the parameter $\zeta$ on the behavior of the system. We find that for $\zeta = 0$, we obtain the imaginary roots $s_1 = i\omega_n$ and $s_2 = -i\omega_n$, leading to the solution given in Eq. (2.12). For $0 < \zeta < 1$, the roots $s_1$ and $s_2$ are complex conjugate and are located symmetrically about the real axis. As the value of $\zeta$ approaches 1, both roots approach the point $-\omega_n$ on the real axis. If $\zeta > 1$, both roots lie on the real axis, one increasing and the other decreasing. In the limit when $\zeta \to \infty$, $s_1 \to 0$ and $s_2 \to -\infty$. The value $\zeta = 1$ can be seen to represent a transition stage, below which both roots are complex and above which both roots are real.

2. A critically damped system will have the smallest damping required for aperiodic motion; hence the mass returns to the position of rest in the shortest possible time without overshooting. The property of critical damping is used in many practical applications. For example, large guns have dashpots with critical damping value, so that they return to their original position after recoil in the minimum time without vibrating. If the damping provided were more than the critical value, some delay would be caused before the next firing.

### 2.6.3    Logarithmic Decrement
The logarithmic decrement represents the rate at which the amplitude of a free damped vibration decreases. It is defined as the natural logarithm of the ratio of any

two successive amplitudes. Let $t_1$ and $t_2$ denote the times corresponding to two consecutive amplitudes (displacements), measured one cycle apart for an under-damped system, as in Fig. 2.14. Using Eq. (2.64), we can form the ratio

$$\frac{x_1}{x_2} = \frac{X_0 e^{-\zeta\omega_n t_1}\cos(\omega_d t_1 + \phi_0)}{X_0 e^{-\zeta\omega_n t_2}\cos(\omega_d t_2 + \phi_0)} \tag{2.77}$$

But $t_2 = t_1 + \tau_d$ where $\tau_d = 2\pi/\omega_d$ is the period of damped vibration. Hence $\cos(\omega_d t_2 + \phi_0) = \cos(2\pi + \omega_d t_1 + \phi_0) = \cos(\omega_d t_1 + \phi_0)$, and Eq. (2.77) can be written as

$$\frac{x_1}{x_2} = \frac{e^{-\zeta\omega_n t_1}}{e^{-\zeta\omega_n(t_1 + \tau_d)}} = e^{\zeta\omega_n \tau_d} \tag{2.78}$$

The logarithmic decrement $\delta$ can be obtained from Eq. (2.78):

$$\delta = \ln\frac{x_1}{x_2} = \zeta\omega_n \tau_d = \zeta\omega_n \frac{2\pi}{\sqrt{1-\zeta^2}\,\omega_n} = \frac{2\pi\zeta}{\sqrt{1-\zeta^2}} = \frac{2\pi}{\omega_d}\cdot\frac{c}{2m} \tag{2.79}$$

For small damping, Eq. (2.79) can be approximated:

$$\delta \simeq 2\pi\zeta \qquad \text{if} \quad \zeta \ll 1 \tag{2.80}$$

Figure 2.18 shows the variation of the logarithmic decrement $\delta$ with $\zeta$ as given by Eqs. (2.79) and (2.80). It can be noticed that for values up to $\zeta = 0.3$, the two curves are difficult to distinguish.

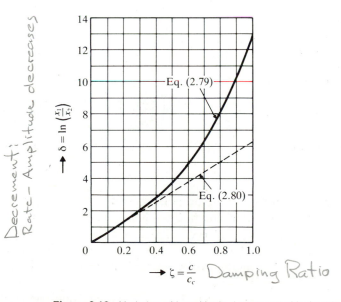

**Figure 2.18** Variation of logarithmic decrement with damping.

The logarithmic decrement is dimensionless and is actually another form of the dimensionless damping ratio $\zeta$. Once $\delta$ is known, $\zeta$ can be found by solving Eq. (2.79):

*To find "ζ"*
*from 2 consecutive*
*displacements $x_1, x_2$:*
*$\zeta(\delta)$*

$$\zeta = \frac{\delta}{\sqrt{(2\pi)^2 + \delta^2}} \qquad (2.81)$$

*$\delta = \ln \frac{x_1}{x_2}$*

*subst. $\delta$*

If we use Eq. (2.80) instead of Eq. (2.79), we have

*for small damping*

$$\zeta \approx \frac{\delta}{2\pi} \qquad (2.82)$$

If the damping in the given system is not known, we can determine it experimentally by measuring any two consecutive displacements $x_1$ and $x_2$. By taking the natural logarithm of the ratio of $x_1$ and $x_2$, we obtain $\delta$. By using Eq. (2.81), we can compute the damping ratio $\zeta$. In fact, the damping ratio $\zeta$ can also be found by measuring two displacements separated by any number of complete cycles. If $x_1$ and $x_{m+1}$ denote the amplitudes corresponding to times $t_1$ and $t_{m+1} = t_1 + m\tau_d$ where $m$ is an integer, we obtain

$$\frac{x_1}{x_{m+1}} = \frac{x_1}{x_2} \cdot \frac{x_2}{x_3} \cdot \frac{x_3}{x_4} \cdots \frac{x_m}{x_{m+1}} \qquad (2.83)$$

Since any two successive displacements separated by one cycle satisfy the equation

$$\frac{x_j}{x_{j+1}} = e^{\zeta \omega_n \tau_d} \qquad (2.84)$$

Eq. (2.83) becomes

$$\frac{x_1}{x_{m+1}} = \left(e^{\zeta \omega_n \tau_d}\right)^m = e^{m\zeta \omega_n \tau_d} \qquad (2.85)$$

Equations (2.85) and (2.79) yield

*for "m" successive cycles:*

$$\delta = \frac{1}{m} \ln\left(\frac{x_1}{x_{m+1}}\right) \qquad (2.86)$$

which can be substituted into Eq. (2.81) or Eq. (2.82) to obtain the viscous damping ratio $\zeta$.

### 2.6.4   Energy Dissipated in Viscous Damping

In a viscously damped system, the rate of change of energy with time $(dW/dt)$ is given by

*$(N) \times \left(\frac{m}{s}\right)$*     *$F = -c\dot{x}$*     *energy dissipates $f(t)$*

$$\frac{dW}{dt} = \text{force} \times \text{velocity} = Fv = -cv^2 = -c\left(\frac{dx}{dt}\right)^2 \qquad (2.87)$$

using Eq. (2.52). The negative sign in Eq. (2.87) denotes that energy dissipates with time. Assume a simple harmonic motion as $x(t) = X \sin \omega_d t$, where $X$ is the amplitude of motion and the energy dissipated in a complete cycle is given by*

$$\Delta W = \int_{t=0}^{(2\pi/\omega_d)} c\left(\frac{dx}{dt}\right)^2 dt = \int_0^{2\pi} cX^2\omega_d\cos^2\omega_d t \cdot d(\omega_d t)$$

$$= \pi c\omega_d X^2 \qquad (2.88)$$

This shows that the energy dissipated is proportional to the square of the amplitude of motion. It is to be noted that it is not a constant for given values of damping and amplitude, since $\Delta W$ is also a function of the frequency $\omega_d$.

Equation (2.88) is valid even when there is a spring of stiffness $k$ parallel to the viscous damper. To see this, consider the system shown in Fig. 2.19. The total force resisting motion can be expressed as

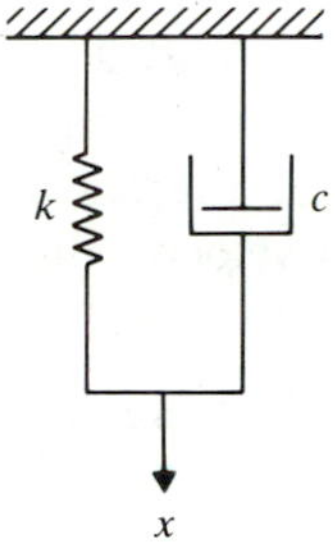

$$F = -kx - cv = -kx - c\dot{x} \qquad (2.89)$$

If we assume simple harmonic motion

$$x(t) = X\sin\omega_d t \qquad (2.90)$$

as before, Eq. (2.89) becomes

$$F = -kX\sin\omega_d t - c\omega_d X\cos\omega_d t \qquad (2.91)$$

The energy dissipated in a complete cycle will be

$$\Delta W = \int_{t=0}^{2\pi/\omega_d} Fv\,dt$$

$$= \int_0^{2\pi/\omega_d} kX^2\omega_d\sin\omega_d t\cdot\cos\omega_d t\cdot d(\omega_d t) + \int_0^{2\pi/\omega_d} c\omega_d X^2\cos^2\omega_d t\cdot d(\omega_d t)$$

$$= \pi c\omega_d X^2 \qquad (2.92)$$

which can be seen to be identical with Eq. (2.88). This result is to be expected, since the spring force will not do any net work over a complete cycle or any integral number of cycles.

We can also compute the fraction of the total energy of the vibrating system which is dissipated in each cycle of motion ($\Delta W/W$), as follows. The total energy of the system $W$ can be expressed either as the maximum potential energy ($\frac{1}{2}kX^2$) or as the maximum kinetic energy ($\frac{1}{2}mv_{max}^2 = \frac{1}{2}mX^2\omega_d^2$), the two being approximately equal for small values of damping. Thus

$$\frac{\Delta W}{W} = \frac{\pi c\omega_d X^2}{\frac{1}{2}m\omega_d^2 X^2} = 2\left(\frac{2\pi}{\omega_d}\right)\left(\frac{c}{2m}\right) = 2\delta \simeq 4\pi\zeta = \text{constant} \qquad (2.93)$$

using Eqs. (2.79) and (2.82). The quantity $\Delta W/W$ is called the *specific damping capacity* and is useful in comparing the damping capacity of engineering materials. Another quantity known as the *loss coefficient* is also used for comparing the

**Figure 2.19**

---

*In the case of a damped system, simple harmonic motion $x(t) = X\cos\omega_d t$ is possible only when the steady-state response is considered under a harmonic force of frequency $\omega_d$ (see Sec. 3.4). The loss of energy due to the damper is supplied by the excitation under steady state forced vibration [2.4].

damping capacity of engineering materials. The loss coefficient is defined as the ratio of the energy dissipated per radian and the total strain energy:

$$\text{loss coefficient} = \frac{(\Delta W/2\pi)}{W} = \frac{\Delta W}{2\pi W} \tag{2.94}$$

## EXAMPLE 2.6

An underdamped shock absorber is to be designed for a vehicle. The initial amplitude is to be reduced to one-fourth in the first half cycle (in Fig. 2.20, $x_{1.5} = x_1/4$). The mass of the vehicle is 500 kg and the damped period of vibration is to be 1 sec. Find the necessary stiffness and damping constants of the shock absorber. Also, if the clearance is 250 mm, find the minimum initial velocity that results in bottoming (i.e., striking a rubber base) of the system.

***Solution.*** Since $x_{1.5} = x_1/4$, $x_2 = x_{1.5}/4 = x_1/16$. Hence the logarithmic decrement becomes

$$\delta = \ln\left(\frac{x_1}{x_2}\right) = \ln(16) = 2.7726 = \frac{2\pi\zeta}{\sqrt{1-\zeta^2}} \tag{E.1}$$

The solution of Eq. (E.1) gives $\zeta = 0.4037$. The damped period of vibration is given to be 1 sec. Hence

$$1 = \tau_d = \frac{2\pi}{\omega_d} = \frac{2\pi}{\omega_n\sqrt{1-\zeta^2}}$$

$$\omega_n = \frac{2\pi}{\sqrt{1-(0.4037)^2}} = 6.8677 \text{ rad/sec}$$

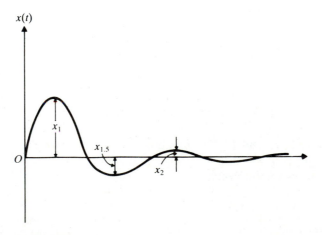

**Figure 2.20**

The critical damping constant can be obtained:

$$c_c = 2m\omega_n = 2(500)(6.8677) = 6867.7 \text{ N-s/m}$$

Thus the damping constant is given by

$$c = \zeta c_c = (0.4037)(6867.7) = 2772.4905 \text{ N-s/m}$$

and the stiffness by

$$k = m\omega_n^2 = (500)(6.8677)^2 = 23582.652 \text{ N/m}$$

The displacement of the mass will attain its maximum value at time $t_1$, given by

$$\sin \omega_d t_1 = \sqrt{1 - \zeta^2}$$

(See Problem 2.45.) This gives

$$\sin \omega_d t_1 = \sin 2\pi t_1 = \sqrt{1 - (0.4037)^2} = 0.9149$$

or

$$t_1 = \frac{\sin^{-1}(0.9149)}{2\pi} = 0.1839 \text{ sec}$$

The envelope passing through the maximum points (see Problem 2.45) is given by

$$x = \sqrt{1 - \zeta^2} \, X e^{-\zeta \omega_n t} \qquad \text{(E.2)}$$

Since $x = 250$ mm, Eq. (E.2) gives at $t_1$

$$0.25 = \sqrt{1 - (0.4037)^2} \, X e^{-(0.4037)(6.8677)(0.1839)}$$

or

$$X = 0.4550 \text{ m.}$$

The velocity of the mass can be obtained by differentiating the displacement

$$x(t) = X e^{-\zeta \omega_n t} \sin \omega_d t$$

as

$$\dot{x}(t) = X e^{-\zeta \omega_n t}(-\zeta \omega_n \sin \omega_d t + \omega_d \cos \omega_d t) \qquad \text{(E.3)}$$

When $t = 0$, Eq. (E.3) gives

$$\dot{x}(t = 0) = \dot{x}_0 = X\omega_d = X\omega_n\sqrt{1 - \zeta^2} = (0.4550)(6.8677)\left(\sqrt{1 - (0.4037)^2}\right)$$

$$= 2.8589 \text{ m/s}$$

## EXAMPLE 2.7

A sketch of a gun is shown in Fig. 2.21. When the gun is fired, high-pressure gases accelerate the projectile inside the barrel to a very high velocity. The reaction force pushes the gun barrel in the opposite direction of the projectile. Since it is desirable to bring the gun barrel to rest in the shortest time without oscillation, it is made to

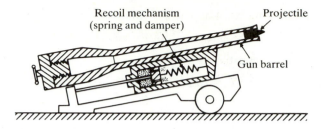

Recoil mechanism
(spring and damper)                    Projectile

Gun barrel

**Figure 2.21**

translate backwards against a critically damped spring-damper system called the *recoil mechanism*. In a particular case, the gun barrel and the recoil mechanism have a mass of 500 kg with a recoil spring of stiffness 10000 N/m. The gun recoils 0.4 m upon firing. Find (1) the critical damping coefficient of the damper, (2) the initial recoil velocity of the gun, and (3) the time taken by the gun to return to a position 0.1 m from its initial position.

*Solution.* 1. The undamped natural frequency of the system is

$$\omega_n = \sqrt{\frac{k}{m}} = \sqrt{\frac{10000}{500}} = 4.4721 \text{ rad/sec}$$

and the critical damping coefficient (Eq. (2.59)) of the damper is

$$c_c = 2m\omega_n = 2(500)(4.4721) = 4472.1 \text{ N-s/m}$$

2. The response of a critically damped system is given by Eq. (2.72):

$$x(t) = (C_1 + C_2 t)e^{-\omega_n t} \tag{E.1}$$

where $C_1 = x_0$ and $C_2 = \dot{x}_0 + \omega_n x_0$. The time $t_1$ at which $x(t)$ reaches a maximum value can be obtained by setting $\dot{x}(t) = 0$. The differentiation of Eq. (E.1) gives

$$\dot{x}(t) = C_2 e^{-\omega_n t} - \omega_n(C_1 + C_2 t)e^{-\omega_n t}$$

Hence $\dot{x}(t) = 0$ yields

$$t_1 = \left(\frac{1}{\omega_n} - \frac{C_1}{C_2}\right) \tag{E.2}$$

In this case, $x_0 = C_1 = 0$; hence Eq. (E.2) leads to $t_1 = 1/\omega_n$. Since the maximum value of $x(t)$ or the recoil distance is given to be $x_{max} = 0.4$ m, we have

$$x_{max} = x(t = t_1) = C_2 t_1 e^{-\omega_n t_1} = \frac{\dot{x}_0}{\omega_n} e^{-1} = \frac{\dot{x}_0}{e\omega_n}$$

or

$$\dot{x}_0 = x_{max}\omega_n e = (0.4)(4.4721)(2.7183) = 4.8626 \text{ m/s}$$

3. If $t_2$ denotes the time taken by the gun to return to a position 0.1 m from its initial position, we have

$$0.1 = C_2 t_2 e^{-\omega_n t_2} = 4.8626 t_2 e^{-4.4721 t_2} \qquad (\text{E.3})$$

The solution of Eq. (E.3) gives $t_2 = 0.8258$ sec.

## 2.7 FREE VIBRATION WITH COULOMB DAMPING

In many mechanical systems, *Coulomb* or *dry-friction* dampers are used because of their mechanical simplicity and convenience [2.5]. Also in vibrating structures, whenever the components slide relative to each other, dry-friction damping appears internally. As stated in Sec. 1.9, Coulomb damping arises when bodies slide on dry surfaces. Coulomb's law of dry friction states that when two bodies are in contact, the force required to produce sliding is proportional to the normal force acting in the plane of contact. Thus the friction force $F$ is given by

$$F = \mu N \qquad (2.95)$$

where $N$ is the normal force and $\mu$ is the coefficient of friction. The friction force acts in a direction opposite to the direction of velocity. Coulomb damping is sometimes called *constant damping*, since the damping force is independent of the displacement and velocity; it depends only on the normal force $N$ between the sliding surfaces.

### 2.7.1 Equation of Motion

Consider a single degree of freedom system with dry friction as shown in Fig. 2.22(a). Since the friction force varies with the direction of velocity, we need to consider two cases, as indicated in Figs. 2.22(b) and (c). For the half cycle during which the motion is from left to right (Fig. 2.22(b)), the equation of motion can be obtained using Newton's second law:

$$m\ddot{x} = -kx - \mu N \qquad \text{or} \qquad m\ddot{x} + kx = -\mu N \qquad (2.96)$$

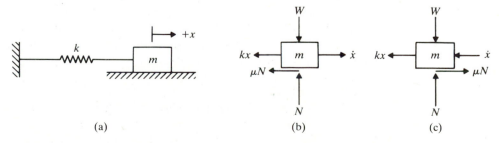

**Figure 2.22** Spring-mass system with Coulomb damping.

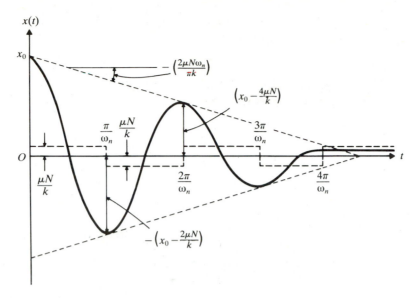

**Figure 2.23**   Motion of the mass with Coulomb damping.

This is a second order nonhomogeneous differential equation. The solution can be verified by substituting Eq. (2.97) into Eq. (2.96).

$$x(t) = A_1\cos \omega_n t + A_2\sin \omega_n t - \frac{\mu N}{k} \qquad (2.97)$$

where $\omega_n = \sqrt{k/m}$, and $A_1$ and $A_2$ are constants whose values depend on the initial conditions of this half cycle. Similarly, the equation of motion in the other half cycle, during which the mass moves from right to left, can be obtained from Fig. 2.22(c):

$$- kx + \mu N = m\ddot{x} \qquad \text{or} \qquad m\ddot{x} + kx = \mu N \qquad (2.98)$$

The solution of Eq. (2.98) is given by

$$x(t) = A_3\cos \omega_n t + A_4\sin \omega_n t + \frac{\mu N}{k} \qquad (2.99)$$

where $A_3$ and $A_4$ are constants to be found from the initial conditions of this half cycle. The term $\mu N/k$ appearing in Eqs. (2.97) and (2.99) is a constant representing the virtual displacement of the spring under the force $\mu N$, if it were applied as a static force. Equations (2.97) and (2.99) indicate that in each half cycle the motion is harmonic, with the equilibrium position changing from $\mu N/k$ to $-(\mu N/k)$ every half cycle as shown in Fig. 2.23.

### 2.7.2   Solution

To see the motion characteristics of the system more clearly, let us assume the initial conditions to be

$$x(t = 0) = x_0$$
$$\dot{x}(t = 0) = 0 \qquad (2.100)$$

That is, the system starts with zero velocity and displacement $x_0$ at $t = 0$. Since $x = x_0$ at $t = 0$, the motion starts from right to left. Using Eqs. (2.99) and (2.100), we can evaluate the constants $A_3$ and $A_4$:

$$A_3 = x_0 - \frac{\mu N}{k}, \qquad A_4 = 0$$

Thus Eq. (2.99) becomes

$$x(t) = \left(x_0 - \frac{\mu N}{k}\right)\cos \omega_n t + \frac{\mu N}{k} \qquad (2.101)$$

This solution is valid for half the cycle only, i.e., for $0 \leq t \leq \pi/\omega_n$. When $t = \pi/\omega_n$, the mass will be at its extreme left position and its displacement from equilibrium position can be found from Eq. (2.101):

$$x\left(t = \frac{\pi}{\omega_n}\right) = \left(x_0 - \frac{\mu N}{k}\right)\cos \pi + \frac{\mu N}{k} = -\left(x_0 - \frac{2\mu N}{k}\right) \qquad (2.102)$$

Since the motion started with a displacement of $x = x_0$ and in a half cycle the value of $x$ became $-(x_0 - (2\mu N/k))$, the reduction in magnitude of $x$ in time $\pi/\omega_n$ is $2\mu N/k$.

In the second half cycle, the mass moves from left to right, so Eq. (2.97) is to be used. The initial conditions for this half cycle are

$$x(t = 0) = \text{value of } x \text{ at } t = \frac{\pi}{\omega_n} \text{ in Eq. (2.101)} = -\left(x_0 - \frac{2\mu N}{k}\right)$$

and

$$\dot{x}(t = 0) = \text{value of } \dot{x} \text{ at } t = \frac{\pi}{\omega_n} \text{ in Eq. (2.101)}$$

$$= \left\{\text{value of } -\omega_n\left(x_0 - \frac{\mu N}{k}\right)\sin \omega_n t \text{ at } t = \frac{\pi}{\omega_n}\right\} = 0$$

Thus the constants in Eq. (2.97) become

$$A_1 = -x_0 + \frac{3\mu N}{k}, \qquad A_2 = 0$$

so that Eq. (2.97) can be written as

$$x(t) = \left(-x_0 + \frac{3\mu N}{k}\right)\cos \omega_n t - \frac{\mu N}{k} \qquad (2.103)$$

This equation is valid only for the second half cycle, that is, for $\pi/\omega_n \leq t \leq 2\pi/\omega_n$. At the end of this half cycle the value of $x(t)$ is

$$x\left(t = \frac{\pi}{\omega_n}\right) \text{ in Eq. (2.103)} = x_0 - \frac{4\mu N}{k}$$

and

$$\dot{x}\left(t = \frac{\pi}{\omega_n}\right) \text{ in Eq. (2.103)} = 0$$

These become the initial conditions for the third half cycle, and the procedure can be continued until the motion stops. Motion stops at the end of that half cycle in which the amplitude $x$ becomes smaller than $\mu N/k$. At such a point, the spring force $(kx)$ will be less than the friction force $(\mu N)$, so the motion ceases. This occurs at the end of the fourth half cycle in Fig. 2.23. It can be seen that in each successive cycle, the amplitude of motion is reduced by the amount $4\mu N/k$, so the amplitudes at the end of any two consecutive cycles are related:

$$X_m = X_{m-1} - \frac{4\mu N}{k} \tag{2.104}$$

As the amplitude is reduced by an amount $4\mu N/k$ in one cycle (i.e., in time $2\pi/\omega_n$), the slope of the enveloping straight lines (shown dotted) in Fig. 2.23 is

$$-\left(\frac{4\mu N}{k}\right)\bigg/\left(\frac{2\pi}{\omega_n}\right) = -\left(\frac{2\mu N\omega_n}{\pi k}\right)$$

Further, the rest position is usually displaced from equilibrium $(x = 0)$ position and represents a permanent displacement in which the friction force is locked in. Slight tapping or shaking of the mass will usually make it come to its equilibrium position.

Note the following characteristics of Coulomb damping:

1. The natural frequency of the system remains unaltered in Coulomb damping, in contrast to the other two types of damping.

2. With the other two types of damping, the motion theoretically continues forever, perhaps with an infinitesimally small amplitude. In the case of Coulomb damping, however, the system comes to rest after some time.

---

**EXAMPLE 2.8**

A mass of 1 kg is attached to a spring having a stiffness of 4000 N/m. The mass slides on a horizontal surface, the coefficient of friction between the mass and the surface being 0.1. Determine the frequency of vibration of the system and the amplitude after one cycle if the initial amplitude is 5 mm. Also determine the final rest position.

*Solution.* The frequency of vibration of the mass will be the same as the undamped frequency

$$\omega_n = \sqrt{\frac{k}{m}} = \left(\frac{4000}{1}\right)^{1/2} = 63.2456 \text{ rad/sec}$$

The reduction in the amplitude after each cycle is given by

$$\frac{4\mu N}{k} = \frac{4\mu(mg)}{k} = \frac{4(0.1)(1 \times 9.807)}{4000} = 0.00098 \text{ m}$$

Thus the amplitude after one cycle will be equal to $0.00500 - 0.00098 = 0.00402$ m.

The final rest position can be found from the equation $F = kx$:

$$x = \frac{F}{k} = \frac{\mu N}{k} = \frac{\mu(mg)}{k} = \frac{0.1(1 \times 9.807)}{4000} = 0.00025 \text{ m}$$

where $x$ is the distance from the mean position of the vibrating mass.

## 2.8 FREE VIBRATION WITH HYSTERETIC DAMPING

As stated earlier, the damping caused by the friction between internal planes that slip or slide as deformation takes place is called hysteretic (or solid or structural) damping. This type of damping causes a hysteresis loop to be formed in the stress-strain or force-displacement curve, as shown in Fig. 2.24. All elastic materials exhibit this behavior; rubberlike material shows a large loop, and metals such as steel exhibit a very small loop. The energy loss in one loading and unloading cycle is equal to the area within the hysteresis loop of the force-displacement curve.*

It has been found experimentally that the energy loss per cycle due to internal friction is independent of the frequency but is approximately proportional to the square of the amplitude. It is also considered to be proportional to the stiffness $k$ of the member. The energy loss per cycle $\Delta W$ can be expressed as

$$\Delta W = \pi k \beta X^2 \tag{2.105}$$

where $\beta$ is a dimensionless hysteresis damping constant of the material [2.9]. Thus $\beta$ is related to the property of the material, while $k\beta$ is related to the size and shape of the material as well as the material properties. The factor $\pi$ is included in Eq. (2.105) for the sake of similarity to the expression for the energy dissipated in viscous damping under harmonic motion (see Eq. (2.88)).

In many situations, $\Delta W$ is small, and the motion can be considered to be very nearly harmonic in form. In these cases, the decrease in amplitude per cycle can be determined from a consideration of the energy. For example, in Fig. 2.25, the energy loss for a quarter cycle is assumed to be $\frac{1}{4}(\pi k \beta X_j^2)$, where $X_j$ is the amplitude of the part in $j$th cycle. Thus the energy at points $A$ and $B$ are related:

$$\frac{kX_j^2}{2} - \frac{\pi k \beta X_j^2}{4} - \frac{\pi k \beta X_{j+0.5}^2}{4} = \frac{kX_{j+0.5}^2}{2}$$

$$\frac{X_j}{X_{j+0.5}} = \sqrt{\frac{2 + \pi\beta}{2 - \pi\beta}} \tag{2.106}$$

---

*In the case of the stress-strain curve, the area enclosed by the hysteresis loop represents the energy loss per cycle per unit volume of the material [2.6–2.8].

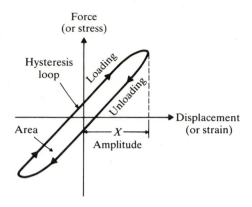

**Figure 2.24**   Hysteresis loop for elastic materials.

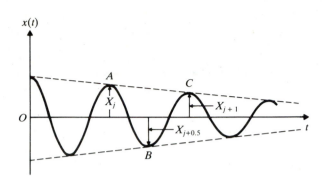

**Figure 2.25**

Similarly, for the next half cycle from $B$ to $C$, we have

$$\frac{X_{j+0.5}}{X_{j+1}} = \sqrt{\frac{2 + \pi\beta}{2 - \pi\beta}} \tag{2.107}$$

Multiplication of Eqs. (2.106) and (2.107) gives

$$\frac{X_j}{X_{j+1}} = \frac{2 + \pi\beta}{2 - \pi\beta} = \frac{2 - \pi\beta + 2\pi\beta}{2 - \pi\beta} \simeq 1 + \pi\beta = \text{constant} \tag{2.108}$$

where $\beta$ is very small.

Since the ratio of successive amplitudes is constant, the decay process will be exponential. The logarithmic decrement is defined as

$$\delta = \ln\left(\frac{X_j}{X_{j+1}}\right) \simeq \ln(1 + \pi\beta) \simeq \pi\beta \tag{2.109}$$

Equation (2.109) can also be used for the experimental determination of the hysteretic damping constant $\beta$. The procedure can also be applied to a built-up structure. By measuring the successive amplitudes, the value of $\beta$ can be computed by using Eq. (2.109) or (2.108). The value of $\beta$ so computed will include the effect of friction between adjacent parts as well as that within the material of the structure. Since the motion is assumed to be approximately harmonic, the corresponding frequency is defined by

$$\omega = \sqrt{\frac{k}{m}}$$

and the motion itself may be replaced by an equivalent viscously damped motion that would exhibit the same characteristics. The equivalent viscous damping ratio $\zeta_{eq}$ can be found by equating the approximate relations for the logarithmic decrement $\delta$

for the two cases:

$$\delta \simeq 2\pi\zeta_{eq} \simeq \pi\beta$$

$$\zeta_{eq} = \frac{\beta}{2} \qquad (2.110)$$

Thus the equivalent damping constant $c_{eq}$ is given by

$$c_{eq} = c_c \cdot \zeta_{eq} = 2\sqrt{mk} \cdot \frac{\beta}{2} = \beta\sqrt{mk} = \frac{\beta k}{\omega} \qquad (2.111)$$

Note that the method of finding an equivalent viscous damping coefficient for a structurally damped system is valid only for harmonic excitation. The above analysis assumes that the system responds approximately harmonically at the frequency $\omega$.

---

## EXAMPLE 2.9

The experimental measurements on a structure gave the force-deflection data shown in Fig. 2.26. From this data, estimate the hysteretic damping constant $\beta$ and the logarithmic decrement $\delta$.

**Solution.** The energy dissipated in each full load cycle is given by the area enclosed by the hysteresis curve. This area can be found by counting the squares enclosed by

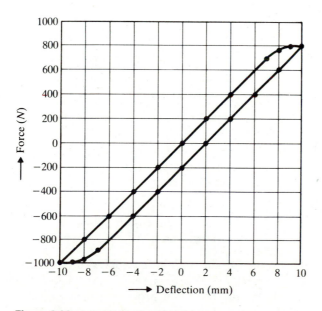

**Figure 2.26** Load-deflection curve.

the hysteresis loop. In the present case, this area is approximately equal to 18 half-squares, that is, 9 squares or $9 \times 200 \times 2/1000 = 3.6$ N-m. From Eq. (2.105), we have

$$\Delta W = \pi k \beta X^2 = 3.6 \text{ N-m} \qquad \text{(E.1)}$$

Since the maximum deflection $X$ is 0.01 m and the slope of the force-deflection curve is $k \simeq 100$ N/mm $= 10^5$ M/m, the hysteretic damping constant $\beta$ is given by

$$\beta = \frac{\Delta W}{\pi k X^2} = \frac{3.6}{\pi (10^5)(0.01^2)} = 0.11459 \qquad \text{(E.2)}$$

The logarithmic decrement can be found

$$\delta = \pi \beta = \pi (0.11459) = 0.36 \qquad \text{(E.3)}$$

## 2.9 COMPUTER PROGRAM

A FORTRAN computer program, in the form of subroutine FREVIB, is given for the free vibration analysis of a viscously damped single degree of freedom system. The system may be underdamped, critically damped, or overdamped. The arguments of this subroutine are as follows:

| | | |
|---|---|---|
| M | = | Mass. Input data. |
| K | = | Spring stiffness. Input data. |
| C | = | Damping constant. Input data. |
| X0 | = | Value of displacement of mass at time 0. Input data. |
| XD0 | = | Value of velocity at time 0. Input data. |
| N | = | Number of time steps at which the value of $x(t)$ is to be printed. Input data. |
| DELT | = | Time interval between consecutive time steps ($\Delta t$). Input data. |
| X, XD, XDD | = | Arrays of dimension $N$ each, which contain computed values of displacement, velocity, and acceleration. $X(I) = x(t_i)$, $XD(I) = \dot{x}(t_i)$, |
| XDD(I) | = | $\ddot{x}(t_i)$. Output. |
| T | = | Array of dimension $N$ which contains the values of time. $T(I) = t_i$. Output. |

To illustrate the use of the subroutine FREVIB, we consider an example with $m = 450$ kg, $k = 26519.2$ N/m, $c = 1000$ N-s/m, $x_0 = 0.539567$ m, $\dot{x}_0 = 1.0$ m/s, $\Delta t = 0.25$ s, and $N = 10$. The main program which calls FREVIB, the subroutine FREVIB, and the output of the program are given below.

```
C =========================================================
C
C PROGRAM 2
C MAIN PROGRAM FOR CALLING FREVIB
C
C =========================================================
      REAL M,K
C     THE FOLLOWING 3 LINES CONTAIN PROBLEM-DEPENDENT DATA
      DIMENSION X(10),XD(10),XDD(10),T(10)
      DATA M,K,C,X0,XD0,N,DELT/
     2 450.0,26519.2,1000.0,0.539567,1.0,10,0.25/
C     END OF PROBLEM-DEPENDENT DATA
      CALL FREVIB (M,K,C,X0,XD0,N,DELT,X,XD,XDD,T,II)
      PRINT 100
 100  FORMAT (/,24H FREE VIBRATION ANALYSIS,/,
     2 37H OF A SINGLE DEGREE OF FREEDOM SYSTEM,//,5H DATA)
      PRINT 200, M,K,C,X0,XD0,N,DELT
 200  FORMAT (/,7H M   =,E15.8,/,7H K   =,E15.8,/,7H C   =,E15.8,/,
     2 7H X0  =,E15.8,/,7H XD0 =,E15.8,/,7H N   =,I5,/,7H DELT =,
     3 E15.8)
      IF (II .EQ. 1) PRINT 500
      IF (II .EQ. 2) PRINT 600
      IF (II .EQ. 3) PRINT 700
      IF (II .EQ. 4) PRINT 800
      PRINT 900
      DO 300 I=1,N
      PRINT 400, I,T(I),X(I),XD(I),XDD(I)
 400  FORMAT (I5,4E15.6)
 300  CONTINUE
 500  FORMAT (//,19H SYSTEM IS UNDAMPED)
 600  FORMAT (//,23H SYSTEM IS UNDER DAMPED)
 700  FORMAT (//,28H SYSTEM IS CRITICALLY DAMPED)
 800  FORMAT (//,22H SYSTEM IS OVER DAMPED)
 900  FORMAT (//,9H RESULTS:,//,3X,2H I,3X,8H TIME(I),7X,5H X(I),10X,
     2 6H XD(I),9X,7H XDD(I),/)
      STOP
      END
C =========================================================
C
C SUBROUTINE FREVIB
C
C =========================================================
      SUBROUTINE FREVIB (M,K,C,X0,XD0,N,DELT,X,XD,XDD,T,II)
      DIMENSION X(N),XD(N),XDD(N),T(N)
      REAL M,K
      OMN=SQRT(K/M)
C     UNDAMPED SYSTEM
      IF (ABS(C) .GT. 1.0E-06) GO TO 100
      II=1
      OMN=SQRT(K/M)
      A=SQRT(X0**2+(XD0/OMN)**2)
      PHI=ATAN(XD0/(X0*OMN))
      DO 10 I=1,N
      IF (I .GT. 1) GO TO 20
```

```
           T(I)=DELT
           GO TO 30
    20     T(I)=T(I-1)+DELT
    30     TT=T(I)
           X(I)=A*COS(OMN*TT-PHI)
           XD(I)=A*OMN*COS(OMN*TT-PHI+1.5708)
           XDD(I)=-(C*XD(I)+K*X(I))/M
    10     CONTINUE
           GO TO 500
   100     CCRIT=2.0*SQRT(K*M)
           XAI=C/CCRIT
           IF (XAI - 1.0) 200,300,400
C          UNDERDAMPED SYSTEM
   200     II=2
           OMD=SQRT(1.0-(XAI**2))*OMN
           CP1=X0
           CP2=(XD0+XAI*OMN*X0)/OMD
           A=SQRT(CP1**2+CP2**2)
           PHI=ATAN(CP1/CP2)
           DO 110 I=1,N
           IF (I .GT. 1) GO TO 120
           T(I)=DELT
           GO TO 130
   120     T(I)=T(I-1)+DELT
   130     TT=T(I)
           X(I)=A*EXP(-XAI*OMN*TT)*SIN(OMD*TT+PHI)
           XD(I)=A*EXP(-XAI*OMN*TT)*(OMD*COS(OMD*TT+PHI)-XAI*OMN*SIN(OMD*
          2 TT+PHI))
           XDD(I)=-(C*XD(I)+K*X(I))/M
   110     CONTINUE
           GO TO 500
C          CRITICALLY DAMPED SYSTEM
   300     II=3
           DO 210 I=1,N
           IF (I .GT. 1) GO TO 220
           T(I)=DELT
           GO TO 230
   220     T(I)=T(I-1)+DELT
   230     TT=T(I)
           X(I)=(X0+(XD0+OMN*X0)*TT)*EXP(-OMN*TT)
           XD(I)=-(X0+(XD0+OMN*X0)*TT)*OMN*EXP(-OMN*TT)+(XD0+OMN*X0)*
          2 EXP(-OMN*TT)
           XDD(I)=-(C*XD(I)+K*X(I))/M
   210     CONTINUE
           GO TO 500
C          OVERDAMPED SYSTEM
   400     II=4
           X1=SQRT(XAI**2-1.0)
           C1=(X0*OMN*(XAI+X1)+XD0)/(2.0*OMN*X1)
           C2=(-X0*OMN*(XAI-X1)-XD0)/(2.0*OMN*X1)
           DO 310 I=1,N
           IF (I .GT. 1) GO TO 320
           T(I)=DELT
           GO TO 330
```

```
320   T(I)=T(I-1)+DELT
330   TT=T(I)
      X(I)=C1*EXP((-XAI+X1)*OMN*TT)+C2*EXP((-XAI-X1)*OMN*TT)
      XD(I)=C1*(-XAI+X1)*OMN*EXP((-XAI+X1)*OMN*TT)
    2 +C2*(-XAI-X1)*OMN*EXP((-XAI-X1)*OMN*TT)
      XDD(I)=-(C*XD(I)+K*X(I))/M
310   CONTINUE
500   RETURN
      END
```

FREE VIBRATION ANALYSIS
OF A SINGLE DEGREE OF FREEDOM SYSTEM

DATA

```
M    = 0.45000000E+03
K    = 0.26519199E+05
C    = 0.10000000E+04
X0   = 0.53956699E+00
XD0  = 0.10000000E+01
N    =    10
DELT = 0.25000000E+00
```

SYSTEM IS UNDER DAMPED

RESULTS:

| I | TIME(I) | X(I) | XD(I) | XDD(I) |
|---|---------|------|-------|--------|
| 1 | 0.250000E+00 | 0.192649E-01 | -0.335069E+01 | 0.631066E+01 |
| 2 | 0.500000E+00 | -0.318985E+00 | 0.106230E+01 | 0.164376E+02 |
| 3 | 0.750000E+00 | 0.144699E+00 | 0.140377E+01 | -0.116468E+02 |
| 4 | 0.100000E+01 | 0.112366E+00 | -0.129493E+01 | -0.374428E+01 |
| 5 | 0.125000E+01 | -0.137887E+00 | -0.173140E+00 | 0.851065E+01 |
| 6 | 0.150000E+01 | 0.285635E-02 | 0.827508E+00 | -0.200724E+01 |
| 7 | 0.175000E+01 | 0.777184E-01 | -0.304712E+00 | -0.390293E+01 |
| 8 | 0.200000E+01 | -0.395868E-01 | -0.326002E+00 | 0.305736E+01 |
| 9 | 0.225000E+01 | -0.252620E-01 | 0.334008E+00 | 0.746487E+00 |
| 10 | 0.250000E+01 | 0.350478E-01 | 0.239573E-01 | -0.211866E+01 |

## REFERENCES

**2.1.** R. W. Fitzgerald, *Mechanics of Materials* (2nd ed.), Addison-Wesley, Reading, Mass., 1982.

**2.2.** W. Zambrano, "A brief note on the determination of the natural frequencies of a spring-mass system," *International Journal of Mechanical Engineering Education*, Vol. 9, October 1981, pp. 331–334; Vol. 10, July 1982, p. 216.

**2.3.** E. Kreyszig, *Advanced Engineering Mathematics* (4th Ed.), John Wiley, New York, 1979.

**2.4.** S. H. Crandall, "The role of damping in vibration theory," *Journal of Sound and Vibration*, Vol. 11, 1970, pp. 3–18.

**2.5.** D. Sinclair, "Frictional vibrations," *Journal of Applied Mechanics*, Vol. 22, 1955, pp. 207–214.

**2.6.** E. E. Unger, "The status of engineering knowledge concerning the damping of built-up structures," *Journal of Sound and Vibration*, Vol. 26, 1973, pp. 141–154.

**2.7.** W. Pinsker, "Structural damping," *Journal of the Aeronautical Sciences*, Vol. 16, 1949, p. 699.

**2.8.** R. H. Scanlan and A. Mendelson, "Structural damping," *AIAA Journal*, Vol. 1, 1963, pp. 938–939.

**2.9.** T. H. Pian, "Structural damping," (Chapter 5) in *Random Vibration*, S. H. Crandall (Ed.), Technology Press and John Wiley, New York, 1958.

**2.10.** T. K. Caughey and M. E. J. O'Kelly, "Effect of damping on the natural frequencies of linear dynamic systems," *Journal of the Acoustical Society of America*, Vol. 33, 1961, pp. 1458–1461.

## REVIEW QUESTIONS

**2.1.** Suggest a method for determining the damping constant of a highly damped vibrating system that uses viscous damping.

**2.2.** Can you apply the results of Sec. 2.2 to systems where the restoring force is not proportional to the displacement, i.e., where $k$ is not a constant?

**2.3.** State the quantities corresponding to $m$, $c$, $k$, and $x$ for a torsional system.

**2.4.** What effect does a decrease in mass have on the frequency of a system?

**2.5.** What effect does a decrease in the stiffness of the system have on the natural period?

**2.6.** Why does the amplitude of free vibration gradually diminish in practical systems?

**2.7.** Why is it important to find the natural frequency of a vibrating system?

**2.8.** How many arbitrary constants must a general solution to a second order differential equation have? How are these constants determined?

**2.9.** Can the energy method be used to find the differential equation of motion of all single degree of freedom systems?

**2.10.** What assumptions are made in finding the natural frequency of a single degree of freedom system using the energy method?

**2.11.** Is the frequency of a damped free vibration smaller or greater than the natural frequency of the system?

**2.12.** What is the use of logarithmic decrement?

**2.13.** Is hysteresis damping a function of the maximum stress?

**2.14.** What is critical damping and what is its importance?

**2.15.** What happens to the energy dissipated by damping?

**2.16.** What is equivalent viscous damping? Is the equivalent viscous damping factor a constant?

**2.17.** What is the reason for studying the vibration of a single degree of freedom system?

**2.18.** How can you find the natural frequency of a system by measuring its static deflection?

**2.19.** Give two practical applications of a torsional pendulum.

**2.20.** Define these terms: damping ratio, logarithmic decrement, loss coefficient, and specific damping capacity.

**2.21.** In what ways is the response of a system with Coulomb damping different from that of systems with other types of damping?

## PROBLEMS

The problem assignments are organized as follows:

| Problem | Section covered | Topic covered |
|---------|-----------------|---------------|
| 2.1–2.25 | √2.2 | Undamped systems |
| 2.26–2.37 | 2.3 | Undamped torsional systems |
| 2.38–2.44 | 2.5 | Energy method |
| 2.45–2.59 | 2.6 | Systems with viscous damping |
| 2.60–2.64 | 2.7 | Systems with Coulomb damping |
| 2.65–2.68 | 2.8 | Systems with hysteretic damping |
| 2.69–2.72 | 2.9 | Computer program |

**2.1.** A machine is placed on a rubber mounting to isolate it from its foundation. If the dead weight of the machine compresses the mounting 5 mm, determine the natural frequency of the system.

**2.2.** A spring-mass system has a natural period of 0.21 sec. What will be the new period if the spring constant is increased by 50%?

**2.3.** A spring-mass system has a natural frequency of 10 Hz. When the spring constant is reduced by 800 N/m, the frequency is altered by 45%. Determine the mass and spring constant of the original system.

**2.4.** A helical spring, when fixed at one end and loaded at the other, requires a force of 100 N to produce an elongation of 10 mm. The ends of the spring are now rigidly fixed, one end vertically above the other, and a mass of 10 kg is attached at the middle point of its length. Determine the time taken to complete one vibration cycle when the mass is set vibrating in the vertical direction.

**2.5.** The maximum velocity attained by the mass of a simple harmonic oscillator is 10 cm/sec, and the period of oscillation is 2 sec. If the mass is released with an initial displacement of 2 cm, find (a) the amplitude, (b) the initial velocity, (c) the maximum acceleration, and (d) the phase angle.

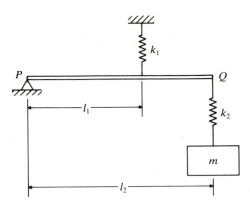

**Figure 2.27**

**2.6.** Find the natural frequency of vibration of the system shown in Fig. 2.27. Assume that the bar $PQ$ is rigid and weightless.

**2.7.** A car having a mass of 2000 kg deflects its springs 0.02 m under static conditions. Determine the natural frequency of the car in the vertical direction by assuming damping to be negligible.

**2.8.** Find the natural frequency of vibration of a spring-mass system arranged on an inclined plane, as shown in Fig. 2.28.

**2.9.** Find the natural frequency of the system shown in Fig. 2.29 with and without the spring at the mid-span of the elastic beam.

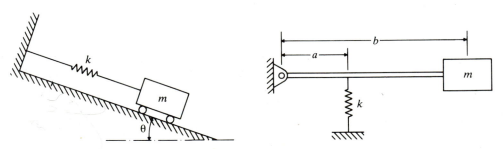

**Figure 2.28**                      **Figure 2.29**

**2.10.** The natural frequency of a spring-mass system is found to be 2 Hz. When an additional mass of 1 kg is added to the original mass $m$, the natural frequency is reduced to 1 Hz. Find the spring constant $k$ and the mass $m$.

**2.11.** A mass $M$ is attached to two identical springs of stiffness $k$. It hangs at rest as shown in Fig. 2.30. Another mass $m$ is then released from a height $l$ and falls on mass $M$, adhering to it without rebounding. Find the equation of motion and the natural frequency of vibration of the system.

**2.12.** Find the natural frequency of the system shown in Fig. 2.31. The friction and the mass of the pulleys are negligible.

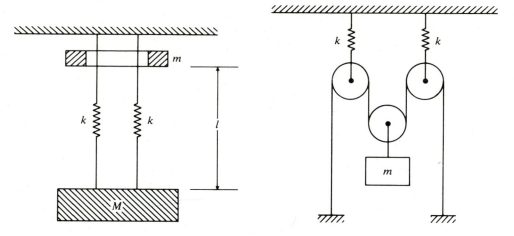

**Figure 2.30**                                 **Figure 2.31**

**2.13.** Derive the expression for the natural frequency of the system shown in Fig. 2.32. Assume that beam 3 is simply supported at its ends.

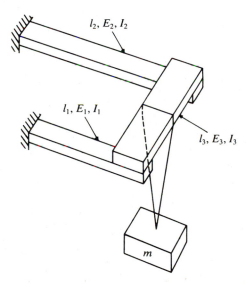

**Figure 2.32**

**2.14.** The cage of an elevator having a mass of 1000 kg is being lowered at a uniform velocity of 2 m/sec. It is supported by a steel cable of diameter 0.01 m. The upper end of the supporting cable is suddenly stopped when its length is 20 m. Find the period and amplitude of the ensuing vibration.

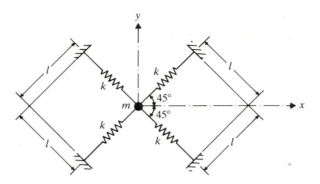

**Figure 2.33**

**2.15.** Figure 2.33 shows a small mass $m$ restrained by four linearly elastic springs, each of which has a stiffness constant $k$, an unstretched length $l$, and an angle of orientation of 45° with respect to the $x$-axis. Determine the equation of motion for small displacements of the mass in the $x$ direction.

**2.16.** A mass $m$ is supported by two sets of springs oriented at 45° and 120° with respect to the $X$ axis, as shown in Fig. 2.34. A third pair of springs, with a stiffness of $k_3$ each, is to be designed so as to make the system have a constant natural frequency while vibrating in any direction $x$. Determine the necessary spring stiffness $k_3$ and the orientation of the springs with respect to the $X$ axis.

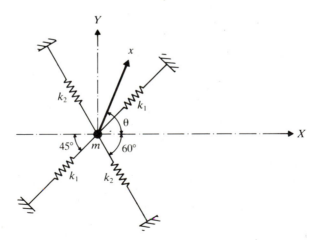

**Figure 2.34**

**2.17.** A mass $m$ is attached to a cord which is under a tension $T$, as shown in Fig. 2.35. Assuming that the tension $T$ remains unchanged when the mass is displaced normal to

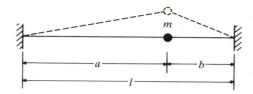

**Figure 2.35**

the cord, (a) write the differential equation of motion for small transverse vibrations and (b) find the natural frequency of vibration.

**2.18.** The schematic diagram of a centrifugal governor is shown in Fig. 2.36. The length of each rod is $l$, the mass of each ball is $m$ and the free length of the spring is $h$. If the shaft speed is $\omega$, determine the equilibrium position and the frequency for small oscillations about this position.

**2.19.** A square platform $PQRS$ and a car which it is supporting have a combined mass of $M$. The platform is suspended by four elastic wires from a fixed point $O$, as indicated in Fig. 2.37. The vertical distance between the point of suspension $O$ and the horizontal equilibrium position of the platform is $h$. If the side of the platform is $a$ and the stiffness of each wire is $k$, determine the period of vertical vibration of the platform.

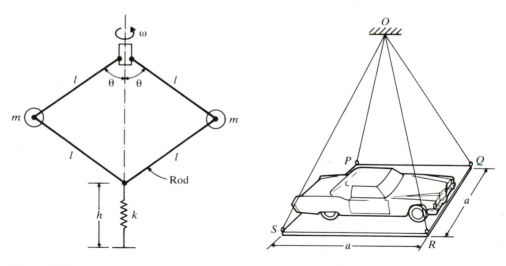

**Figure 2.36**

**Figure 2.37**

**2.20.** Determine the natural frequency of vibrations of an engine foundation located on an elastic ground, as shown in Fig. 2.38. The combined weight of the engine and the foundation is $W = 15000$ N, the area of the base of the foundation is 2 m², and the specific stiffness of the ground (stiffness of the ground per unit area) is 0.16 N/m³.

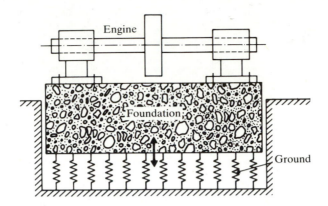

**Figure 2.38**

**2.21.** The boom $AB$ of the crane shown in Fig. 2.39 is equivalent to a slender uniform steel rod of length 10 m and area of cross-section 2500 mm². A 1000-kg mass is suspended while the truck is stationary. The cable $CDEBF$ is made of steel and has a cross sectional area of 100 mm². Assuming the mass as well as the friction of the pulley to be negligible, determine the natural frequency of vibration of the mass in the vertical direction.

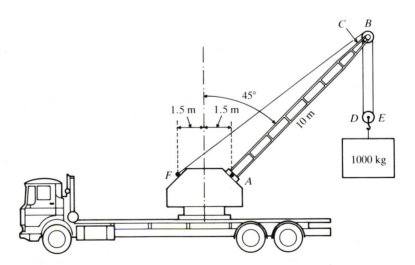

**Figure 2.39**

**2.22.** Figure 2.40 shows the Galileo orbiter sent by NASA for exploring the planet Jupiter. The magnetometer has a length of 9 m and a mass of 7 kg. If the natural frequency of bending vibration of the magnetometer is 2 Hz, estimate its bending stiffness. Consider the magnetometer as a cantilever beam.

**Figure 2.40**

**2.23.** A single-story building frame is composed of two identical columns and a rigid floor, as shown in Fig. 2.41. The girder, including the roof, has a weight of $W$; each of the columns has negligible weight with a bending rigidity $EI$. The columns are fixed against rotation at the lower ends. Determine the natural frequency of horizontal vibration of the girder by assuming the connection between the girder and the upper ends of the columns to be (a) pivoted as shown in Fig. 2.41(a), and (b) fixed against rotation as shown in Fig. 2.41(b).

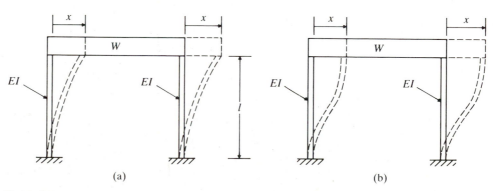

**Figure 2.41**

**2.24.** A pulley 250 mm in diameter drives a second pulley 1000 mm in diameter by means of a belt (see Fig. 2.42). The moment of inertia of the driven pulley is 0.2 kg-m². The belt connecting these pulleys is represented by two springs, each of stiffness $k$. For what value of $k$ will the natural frequency be 6 Hz?

**2.25.** A flywheel is mounted on a vertical shaft, as shown in Fig. 2.43. The shaft has a diameter $d$ and length $l$ and is fixed at both ends. The flywheel has a weight of $W$ and a radius of gyration of $r$. Find the natural frequency of the longitudinal, the transverse, and the torsional vibration of the system.

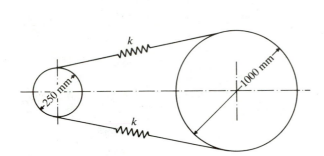

**Figure 2.42**

**Figure 2.43**

**2.26.** Derive an expression for the natural frequency of the simple pendulum shown in Fig. 1.5. Determine the period of oscillation of a simple pendulum having a mass $m = 5$ kg and a length $l = 0.5$ m.

**2.27.** A mass $m$ is attached at the end of a bar of negligible mass and is made to vibrate in three different configurations, as indicated in Figs. 2.44(a) to (c). Find the configuration corresponding to the highest natural frequency.

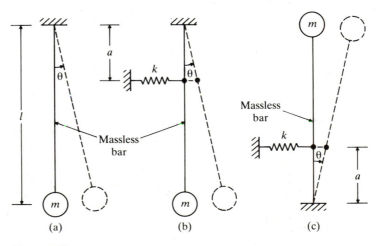

**Figure 2.44**

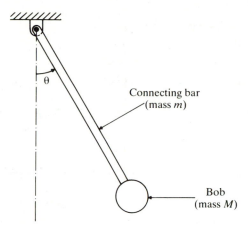

**Figure 2.45**

**2.28.** Find the natural frequency of the pendulum shown in Fig. 2.45 when the mass of the connecting bar is not negligible compared to the mass of the pendulum bob.

**2.29.** The rigid body shown in Fig. 2.8 is found to have a natural frequency of 75 cycles/min while making angular oscillations about an axis passing through point $O$ and perpendicular to the paper. When the rotation axis is moved from $O$ to some point $B$ along the line $OG$, the frequency of oscillation is found to be 50 cycles/min. Determine the distance $OB$ if the distance $OG$ is 0.1 m.

**2.30.** A prismatic bar $PQ$ of weight $W$ is suspended by two identical wires and is made to perform small rotational oscillations in the horizontal plane about the central axis $AB$ (see Fig. 2.46). Find the angular frequency of these oscillations.

**2.31.** A steel shaft of 0.05 m diameter and 2 m length is fixed at one end and carries at the other end a steel disc of 1 m diameter and 0.1 m thickness, as shown in Fig. 2.7. Find the natural frequency of torsional vibration of the system.

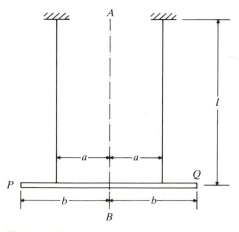

**Figure 2.46**

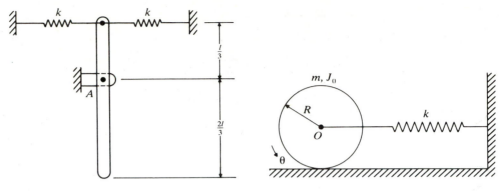

**Figure 2.47**

**Figure 2.48**

**2.32.** A uniform slender rod of mass $m$ and length $l$ is hinged at point $A$ and is attached to two springs at the top end, as shown in Fig. 2.47. Find the natural frequency of the system if $k = 2000$ N/m, $m = 10$ kg, and $l = 5$ m.

**2.33.** A cylinder of mass $m$ and mass moment of inertia $J_0$ is free to roll without slipping but is restrained by a spring of stiffness $k$, as shown in Fig. 2.48. Find its natural frequency of vibration.

**2.34.** A semicircular cylinder of radius $R$ oscillates on a flat surface without slipping, as shown in Fig. 2.49. By assuming the angular displacements to be small, find an expression for its natural frequency of oscillation. The radius of gyration of the cylinder about the centroidal axis is $r$. The mass center is located at a distance $d = 4R/3\pi$ from the axis of the cylinder at $O$.

**2.35.** Two identical masses $m$ are pin-connected at the ends of a rigid link of length $l$. One end of the link is free to move in a horizontal slot and the other end is free to move in a vertical slot, as shown in Fig. 2.50. By assuming the slots to be frictionless, find the natural frequency of vibration of the system for small amplitudes.

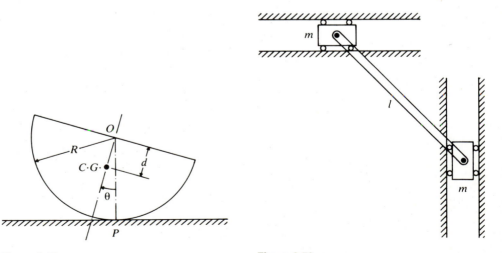

**Figure 2.49**

**Figure 2.50**

**2.36.** If the pendulum of Problem 2.26 is placed in a rocket moving vertically with an acceleration of 5 m/s², what will be its period of oscillation?

**2.37.** Find the equation of motion of the uniform rigid bar *OA* of length *l* and mass *m* shown in Fig. 2.51. Also find its natural frequency.

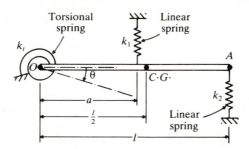

**Figure 2.51**

**2.38.** Solve problem 2.6 using Rayleigh's method.

**2.39.** Solve problem 2.12 using Rayleigh's method.

**2.40.** Find the natural frequency of the system shown in Fig. 2.33.

**2.41.** Solve problem 2.20 using Rayleigh's method.

**2.42.** Solve problem 2.32 using Rayleigh's method.

**2.43.** Solve problem 2.37 using Rayleigh's method.

**2.44.** A solid wooden cylinder of radius *r* is partially immersed in a water tub, as shown in Fig. 2.52. The cylinder is depressed slightly and released. Find the natural frequency of oscillation of the cylinder if it stays upright all the time. What will be the frequency if the water is replaced by an oil of specific gravity 1.2? Use Rayleigh's method.

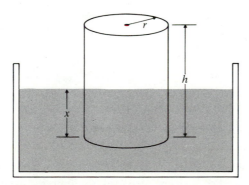

**Figure 2.52**

**2.45.** Assuming that the phase angle is zero, show that the response $x(t)$ of an underdamped single degree of freedom system reaches a maximum value when

$$\sin \omega_d t = \sqrt{1 - \zeta^2}$$

and a minimum value when

$$\sin \omega_d t = -\sqrt{1 - \zeta^2}$$

Also show that the equations of the curves passing through the maximum and minimum values of $x(t)$ are given, respectively, by

$$x = \sqrt{1 - \zeta^2} \, X e^{-\zeta \omega_n t}$$

and

$$x = -\sqrt{1 - \zeta^2} \, X e^{-\zeta \omega_n t}$$

**2.46.** Derive an expression for the time at which the response of a critically damped system will attain its maximum value. Also find the expression for the maximum response.

**2.47.** In Fig. 2.53(a), an automobile of mass 1500 kg is supported on four shock absorbers (dampers) and four springs. The static deflection of the car on the springs due to its own weight is 50 mm. Find the damping constant of each of the four shock absorbers necessary in order to achieve critical damping. Assume that the car moves only in the vertical direction and can be modeled as a single degree of freedom system, as shown in Fig. 2.53(b).

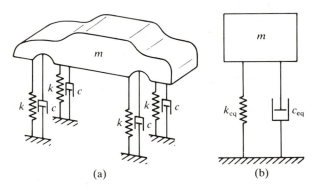

(a)                    (b)

**Figure 2.53**

**2.48.** A piston having a mass of 4 kg is supported on a helical spring. It vibrates at a frequency of 5 Hz in vacuum and 4.5 Hz when immersed in an oil-filled cylinder. Find the damping constant.

**2.49.** The ratio of successive amplitudes is found to be $12:1$ in a viscous damping apparatus. If the amount of damping is doubled, what will be the ratio of successive amplitudes?

**2.50.** A shock absorber is being designed to limit its overshoot to 15% of its initial displacement when released. Determine the necessary damping ratio $\zeta_0$. If $\zeta$ is made equal to $3\zeta_0/4$, what will be the overshoot?

**2.51.** For a spring-mass-damper system, $m = 50$ kg and $k = 5000$ N/m. Find the following: (a) critical damping constant $c_c$, (b) damped natural frequency when $c = c_c/2$, and (c) logarithmic decrement.

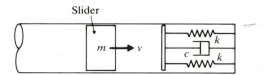

**Figure 2.54**

**2.52.** A slider of mass $m = 1$ kg travels in a cylinder with a velocity $v = 80$ m/s and engages a spring-damper system, as shown in Fig. 2.54. If the stiffness of the spring is $k = 40$ N/mm and the damping constant is $c = 0.2$ N-s/mm, find the maximum displacement of the slider after engaging the springs and damper. How much time does it take to reach the maximum displacement?

**2.53.** A torsional pendulum has a natural frequency of 200 cycles/min when vibrating in vacuum. The mass moment of inertia of the disc is 0.2 kg-m². It is then immersed in oil and its natural frequency is found to be 180 cycles/min. Determine the damping constant. If the disc, when placed in oil, is given an initial displacement of 2°, find its displacement at the end of the first cycle.

**2.54.** A body vibrating with viscous damping makes 5 complete oscillations per second, and in 50 cycles its amplitude diminishes to 10%. Determine the logarithmic decrement and the damping ratio. In what proportion will the period of vibration be decreased if damping is removed?

**2.55.** The door of a house has a height of 2 m, width of 1 m, thickness of 50 mm, and mass of 50 kg. The door opens against a torsion spring and a viscous damper, as shown in Fig. 2.55. If the spring constant of the torsion spring is 15 N-m/rad, find the damping constant necessary to provide critical damping in the return swing of the door. If the door is initially opened 75° and released, how long will it take for the door to come to within 5° of closing?

**2.56.** Figure 2.56 shows a simplified model of a mechanical tennis ball throwing machine. The ball is thrown by first giving an initial displacement to the block $M$ and the ball $m$ and then suddenly releasing them. The ball remains in contact with the block until $x$ becomes zero (i.e., during the first quarter cycle), at which time the block engages the damper shown in the figure. (a) If $M = 1$ kg, $m = 0.25$ kg, and $k = 50$ N/mm, find the initial displacement of the block and ball ($x_0$) necessary in order to make the velocity of the ball equal to 25 m/sec at the end of the first quarter cycle of motion. (b) If the damping constant of the damper is 30 N-s/m, find the response of the block during and after the first quarter cycle of motion.

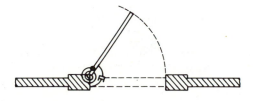

**Figure 2.55**

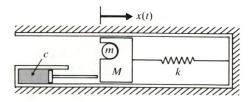

**Figure 2.56**

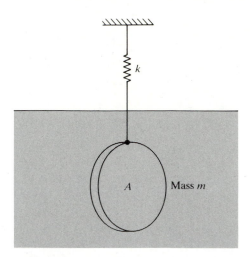

**Figure 2.57**          **Figure 2.58**

**2.57.** A thin plate of mass $m$ and area $A$ is attached to the end of a spring of stiffness $k$. It is allowed to oscillate in the vertical direction in a viscous fluid, as shown in Fig. 2.57. If $\tau_1$ denotes the natural period of undamped vibration (that is, with the system vibrating in vacuum), and $\mu$ is the viscosity of the fluid, show that the damped period of vibration of the system $\tau_2$ (that is, when the plate is immersed in the fluid) can be expressed as

$$\tau_2 = \tau_1 \left/ \sqrt{1 - \left(\frac{\mu A \tau_1}{2 \pi m}\right)^2}\right.$$

*Hint:* The damping force on the plate is given by $F_d = 2 A \mu \dot{x}$ where $2A$ is the total surface area of the plate and $\dot{x}$ is its velocity.

**2.58.** A large gun with a supporting base weighs 10000 N and has a recoil system with a spring stiffness of 40000 N/m and a critically damped dashpot. The recoil distance is 0.5 m. Determine (a) the initial recoil velocity, (b) the time to return to a position 0.05 m from the initial position, and (c) the displacement at $t = 0.25$ sec.

**2.59.** The pointer of an ammeter is made of steel and has a length of $l = 50$ mm, a width of $w = 4$ mm, and a thickness of $h = 1$ mm. The restoring spring has a spring constant of $k_t = 1$ N-mm/rad. A dashpot is connected at a radius of $a = 10$ mm for providing critical damping, as shown in Fig. 2.58. During a measurement, the ammeter showed 10 amperes. Find the time required for the pointer to return to 1 ampere after the current is disconnected.

**2.60.** A mass, connected to a spring, is constrained to move on a horizontal rough surface. Figure 2.59 shows a record of the motion. Determine the coefficient of friction between the mass and the surface.

**2.61.** A single degree of freedom system consists of a mass of 20 kg and a spring of stiffness 4000 N/m. The amplitudes of successive cycles are found to be 40, 36, 32, 28,... mm. Determine the nature and magnitude of the damping force and the frequency of the damped vibration.

**2.62.** A mass of 20 kg slides back and forth on a dry surface due to the action of a spring having a stiffness of 10 N/mm. After four complete cycles, the amplitude has been

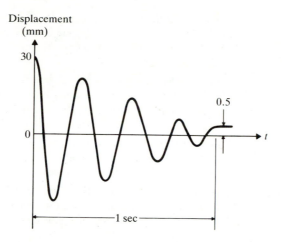

Figure 2.59

found to be 100 mm. What is the average coefficient of friction between the two surfaces if the original amplitude was 150 mm? How much time has elapsed during the four cycles?

**2.63.** A 10-kg mass is connected to a spring of stiffness 3000 N/m and is released after giving an initial displacement of 100 mm. Assuming that the mass moves on a horizontal surface as shown in Fig. 2.22(a), determine the position at which the mass comes to rest. Assume the coefficient of friction between the mass and the surface to be 0.12.

**2.64.** A connecting rod of mass $m$ is suspended on a cylinder, which fits loosely in the wrist pin bearing and is made to oscillate as shown in Fig. 2.60. The coefficient of friction

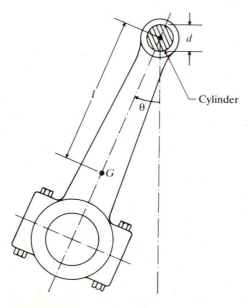

Figure 2.60

between the cylinder and the bearing is $\mu$. (a) Derive an expression for the decrease in the angle of swing for each cycle caused by the friction. (b) Find the solution for $\theta(t)$ if the connecting rod is released from an angle $\theta_0$. (c) Determine the number of cycles after which the motion ceases.

**2.65.** A structure is observed to behave as a single degree of freedom system of mass 1 kg and stiffness 2 N/m. The ratio of successive amplitudes is found to be 1.1. Determine the value of the hysteretic damping constant $\beta$, the equivalent viscous damping constant $c_{eq}$, and the energy loss per cycle for an amplitude of 10 mm.

**2.66.** A helical spring of stiffness 200 N/m supports a mass of 2 kg. The mass is displaced initially by 30 mm and released. If the amplitude is found to be 20 mm after 100 cycles of motion, estimate the hysteretic damping constant $\beta$.

**2.67.** A tall building frame deflects 1 mm under a load of 1000 N, and the deflection pattern is found to be similar to the first mode shape, for which the resonant frequency is 15 Hz and $\zeta = 0.1$. Find the hysteretic damping constant $\beta$ and the equivalent damping constant.

**2.68.** Through experimental measurements, the following force-deflection data is obtained for a built-up structure. Determine the hysteretic damping constant $\beta$ of the structure.

| Force (N) | Deflection (mm) |
|---|---|
| 0 | 0 |
| 100 | 0.2 |
| 200 | 0.5 |
| 400 | 1.5 |
| 600 | 2.4 |
| 800 | 4.1 |
| 1000 | 6.0 |
| 1200 | 8.0 |
| 1100 | 7.8 |
| 1000 | 7.6 |
| 800 | 7.0 |
| 600 | 5.5 |
| 400 | 4.0 |
| 200 | 2.4 |
| 100 | 1.5 |
| 0 | 0 |

**2.69–2.72.** Find the free vibration response of a viscously damped single degree of freedom system with $m = 4$ kg, $k = 2500$ N/m, $x_0 = 100$ mm, $\dot{x}_0 = -10$ m/s, $\Delta t = 0.01$ s, and $N = 50$ using the subroutine FREVIB for the following conditions:

(a) $c = 0$
(b) $c = 100$ N-s/m
(c) $c = 200$ N-s/m
(d) $c = 400$ N-s/m

# CHAPTER 3

# Harmonically Excited Vibration

Charles Augustin de Coulomb (1736–1806) was a French military engineer and physicist. His early work on statics and mechanics was presented in his great memoir "The Theory of Simple Machines" in 1779, which describes the effect of resistance and the so-called "Coulomb's law of proportionality" between friction and normal pressure. In 1784, he obtained the correct solution to the problem of the small oscillations of a body subjected to torsion. He is well known for his laws of force for electrostatic and magnetic charges. His name is remembered through the unit of electric charge.

Courtesy of Brown Brothers

## 3.1 INTRODUCTION

A dynamic system is often subjected to some type of external force or excitation, called the *forcing* or *exciting function*. This excitation is usually time-dependent. It may be harmonic, nonharmonic but periodic, nonperiodic, or random in nature. The response of a system to a harmonic excitation is called *harmonic response*. The nonperiodic excitation may have a long or short duration. The response of a dynamic system to suddenly applied nonperiodic excitations is called *transient response*.

In this chapter, we shall consider the dynamic response of a single degree of freedom system under harmonic excitations. The most general form of a harmonic excitation has the form $F(t) = F_0\cos(\omega t + \phi)$ or $F(t) = F_0\sin(\omega t + \phi)$, where $F_0$ is a constant denoting the amplitude of the excitation, $\omega$ is the frequency of the harmonic variation, and $\phi$ is the phase angle. The value of $\phi$ depends on the value of $F(t)$ at $t = 0$ and is usually taken to be zero.

## 3.2 EQUATION OF MOTION

If a force $F(t)$ acts on a viscously damped spring-mass system as shown in Fig. 3.1, the equation of motion can be obtained using Newton's second law:

$$m\ddot{x} + c\dot{x} + kx = F(t) \tag{3.1}$$

Since this equation is nonhomogeneous, its general solution $x(t)$ is given by the sum of the homogeneous solution, $x_h(t)$, and the particular solution, $x_p(t)$. The homogeneous solution, which is the solution of the homogeneous equation

$$m\ddot{x} + c\dot{x} + kx = 0 \tag{3.2}$$

represents the free vibration of the system and was discussed in Chap. 2. As seen in

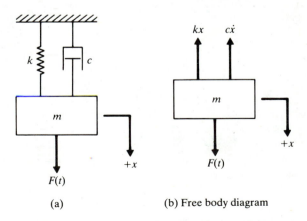

(a)

(b) Free body diagram

**Figure 3.1**  A spring-mass-damper system.

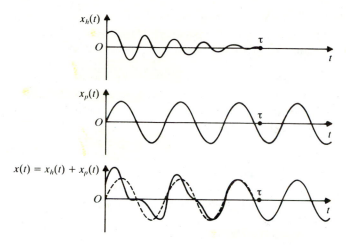

**Figure 3.2** Homogeneous, particular, and general solutions of Eq. (3.1) for an underdamped case.

Sec. 2.6.2, this free vibration dies out with time under each of the three possible conditions of damping (underdamping, critical damping, and overdamping) and under all possible initial conditions. Thus the general solution of Eq. (3.1) eventually reduces to the particular solution $x_p(t)$, which represents the steady-state vibration. The steady-state motion is present as long as the forcing function is present. The variations of homogeneous, particular, and general solutions with time for a typical case are shown in Fig. 3.2. It can be seen that $x_h(t)$ dies out and $x(t)$ becomes $x_p(t)$ after some time ($\tau$ in Fig. 3.2). The part of the motion which dies out due to damping (the free vibration part) is called *transient*. The rate at which the transient motion decays depends on the values of the system parameters $k$, $c$, and $m$. In this chapter, except in Sec. 3.3, we ignore the transient motion and derive only the particular solution of Eq. (3.1), which represents the steady-state response, under harmonic forcing functions.

## 3.3 RESPONSE OF AN UNDAMPED SYSTEM UNDER HARMONIC FORCE

Before studying the response of a damped system, we consider an undamped system subjected to a harmonic force, for the sake of simplicity. If a force $F(t) = F_0 \cos \omega t$ acts on the mass $m$ of an undamped system, the equation of motion, Eq. (3.1), reduces to

$$m\ddot{x} + kx = F_0 \cos \omega t \tag{3.3}$$

The homogeneous solution of this equation is given by

$$x_h(t) = C_1 \cos \omega_n t + C_2 \sin \omega_n t \tag{3.4}$$

where $\omega_n = (k/m)^{1/2}$ is the natural frequency of the system. Because the exciting force $F(t)$ is harmonic, the particular solution $x_p(t)$ is also harmonic and has the same frequency $\omega$. Thus we assume a solution in the form

$$x_p(t) = X\cos\omega t \tag{3.5}$$

where $X$ is a constant that denotes the maximum amplitude of $x_p(t)$. By substituting Eq. (3.5) into Eq. (3.3) and solving for $X$, we obtain

$$\text{max. ampl.} \qquad X = \frac{F_0}{k - m\omega^2} \tag{3.6}$$

Thus the total solution of Eq. (3.3) is

$$x(t) = C_1\cos\omega_n t + C_2\sin\omega_n t + \frac{F_0}{k - m\omega^2}\cos\omega t \tag{3.7}$$

Using the initial conditions $x(t=0) = x_0$ and $\dot{x}(t=0) = \dot{x}_0$, we find that

$$C_1 = x_0 - \frac{F_0}{k - m\omega^2}, \qquad C_2 = \frac{\dot{x}_0}{\omega_n} \tag{3.8}$$

and hence

$$x(t) = \left(x_0 - \frac{F_0}{k - m\omega^2}\right)\cos\omega_n t + \left(\frac{\dot{x}_0}{\omega_n}\right)\sin\omega_n t + \left(\frac{F_0}{k - m\omega^2}\right)\cos\omega t \tag{3.9}$$

The maximum amplitude $X$ in Eq. (3.6) can also be expressed as

$$\frac{X}{\delta_{st}} = \frac{1}{1 - \left(\dfrac{\omega}{\omega_n}\right)^2} \tag{3.10}$$

where $\delta_{st} = F_0/k$ denotes the static deflection of the mass under a force $F_0$. The quantity $X/\delta_{st}$ represents the ratio of the dynamic to the static amplitude of motion and is called the *magnification factor*, *amplification factor*, or *amplitude ratio*. Depending on the value of $\omega/\omega_n$, three different cases can be identified for the steady-state response of the system.

**Case 1.** When $0 < \omega/\omega_n < 1$, the denominator in Eq. (3.10) is positive and the response is given by Eq. (3.5) without change. The harmonic response of the system $x_p(t)$ is said to be in phase with the external force as shown in Fig. 3.3.

**Case 2.** When $\omega/\omega_n > 1$, the denominator in Eq. (3.10) is negative, and the steady-state solution can be expressed as

$$x_p(t) = -X\cos\omega t \tag{3.11}$$

where the amplitude of motion is positive and is defined by

$$X = \frac{\delta_{st}}{\left(\dfrac{\omega}{\omega_n}\right)^2 - 1}$$

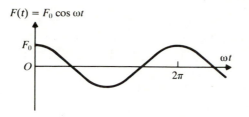

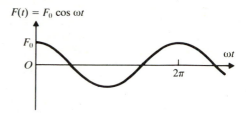

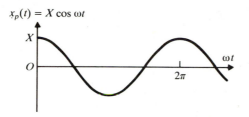

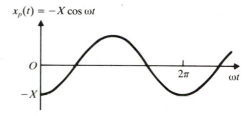

**Figure 3.3**                          **Figure 3.4**

Thus $x_p(t)$ and $F(t)$ can be seen to be of opposite sign; the response is said to be 180° out of phase with the external force. Further, as $\omega/\omega_n \to \infty$, $X \to 0$. Thus the response of the system to a harmonic force of very high frequency is close to zero. The response is shown in Fig. 3.4.

**Case 3.** When $\omega/\omega_n = 1$, the amplitude $X$ becomes infinite. This condition, for which the forcing frequency $\omega$ is equal to the natural frequency of the system $\omega_n$, is called *resonance.* To find the response for this condition, we rewrite Eq. (3.9) as

$$x(t) = x_0 \cos \omega_n t + \frac{\dot{x}_0}{\omega_n} \sin \omega_n t + \delta_{st} \left[ \frac{\cos \omega t - \cos \omega_n t}{1 - \left(\dfrac{\omega}{\omega_n}\right)^2} \right] \qquad (3.12)$$

Since the last term of this equation takes an indefinite form for $\omega = \omega_n$, we apply L'Hospital's rule [3.1] to evaluate the limit of this term:

$$\lim_{\omega \to \omega_n} \left[ \frac{\cos \omega t - \cos \omega_n t}{1 - \left(\dfrac{\omega}{\omega_n}\right)^2} \right] = \lim_{\omega \to \omega_n} \left[ \frac{\dfrac{d}{d\omega}(\cos \omega t - \cos \omega_n t)}{\dfrac{d}{d\omega}\left(1 - \dfrac{\omega^2}{\omega_n^2}\right)} \right]$$

$$= \lim_{\omega \to \omega_n} \left[ \frac{t \sin \omega t}{2 \dfrac{\omega}{\omega_n^2}} \right] = \frac{\omega_n t}{2} \sin \omega_n t \qquad (3.13)$$

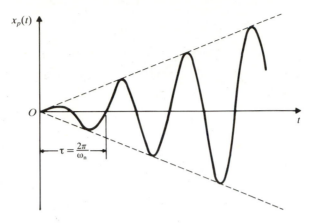

**Figure 3.5**

Thus the response of the system at resonance becomes

for $\dfrac{\omega}{\omega_n} = 1$

$$x(t) = x_0\cos \omega_n t + \frac{\dot{x}_0}{\omega_n}\sin \omega_n t + \frac{\delta_{st}\omega_n t}{2}\sin \omega_n t \qquad (3.14)$$

It can be seen from Eq. (3.14) that at resonance, $x(t)$ increases indefinitely. The last term of Eq. (3.14) is shown in Fig. 3.5, from which the amplitude of the response can be seen to increase linearly with time.

### 3.3.1   Total Response
The total response of the system, Eq. (3.7) or Eq. (3.9), can also be expressed as

$$x(t) = A\cos(\omega_n t - \phi) + \frac{\delta_{st}}{1 - \left(\dfrac{\omega}{\omega_n}\right)^2}\cos \omega t; \qquad \text{for } \frac{\omega}{\omega_n} < 1 \qquad (3.15)$$

$$x(t) = A\cos(\omega_n t - \phi) - \frac{\delta_{st}}{1 - \left(\dfrac{\omega}{\omega_n}\right)^2}\cos \omega t; \qquad \text{for } \frac{\omega}{\omega_n} > 1 \qquad (3.16)$$

where $A$ and $\phi$ can be determined as in the case of Eq. (2.18). Thus the complete motion can be expressed as the sum of two cosine curves of different frequency. In Eq. (3.15), the forcing frequency $\omega$ is smaller than the natural frequency, and the total response is shown in Fig. 3.6(a). In Eq. (3.16), the forcing frequency is greater than the natural frequency, and the total response appears as shown in Fig. 3.6(b).

### 3.3.2   Beating Phenomenon
If the forcing frequency is close to, but not exactly equal to, the natural frequency of the system, a phenomenon known as *beating* may occur. In this kind of vibration, the amplitude builds up and then diminishes in a regular pattern. In order to explain

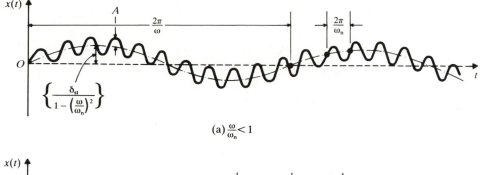

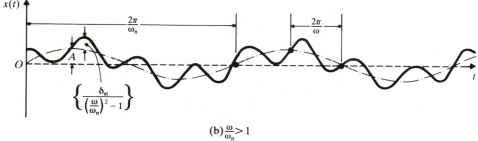

**Figure 3.6**

the beating phenomenon, consider the solution given by Eq. (3.9). If the initial conditions are taken as $x_0 = \dot{x}_0 = 0$, Eq. (3.9) reduces to

$$x(t) = \frac{(F_0/m)}{\omega_n^2 - \omega^2}(\cos \omega t - \cos \omega_n t)$$

$$= \frac{(F_0/m)}{\omega_n^2 - \omega^2}\left[2\sin\frac{\omega + \omega_n}{2}t \cdot \sin\frac{\omega_n - \omega}{2}t\right] \qquad (3.17)$$

Let the forcing frequency $\omega$ be slightly less than the natural frequency:

$$\omega_n - \omega = 2\varepsilon \qquad (3.18)$$

where $\varepsilon$ is a very small positive quantity. Then

$$\omega + \omega_n \simeq 2\omega \qquad (3.19)$$

Multiplying Eqs. (3.18) and (3.19), we get

$$\omega_n^2 - \omega^2 = 4\varepsilon\omega \qquad (3.20)$$

Use of Eqs. (3.18) to (3.20) in Eq. (3.17) gives

$$x(t) = \left(\frac{F_0/m}{2\varepsilon\omega}\sin \varepsilon t\right)\sin \omega t \qquad (3.21)$$

Since $\varepsilon$ is small, the function $\sin \varepsilon t$ varies slowly; its period, equal to $2\pi/\varepsilon$, is large. Thus Eq. (3.21) may be seen as representing vibration with period $2\pi/\omega$ and of

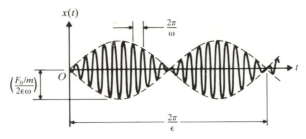

**Figure 3.7**

variable amplitude equal to

$$\left(\frac{F_0/m}{2\varepsilon\omega}\right)\sin \varepsilon t$$

It can also be observed that the $\sin \omega t$ curve will go through several cycles, while $\sin \varepsilon t$ waves go through a single cycle, as shown in Fig. 3.7. Notice the manner in which the amplitude builds up and dies down continuously. The time between the points of zero amplitude or the points of maximum amplitude is called the *period of beating* ($\tau_b$) and is given by

$$\tau_b = \frac{2\pi}{2\varepsilon} = \frac{2\pi}{\omega_n - \omega} \tag{3.22}$$

---

### EXAMPLE 3.1

An electric motor is driving two gears at both ends. The inertia of the gears is very large compared to that of the motor's rotor. If a torque $M_t = M_{t_0}\cos \omega t$ acts on the rotor due to some high electrical harmonic, find the amplitude of the resulting torsional vibration. Assume for the purpose of torsional vibration analysis that the ends of the shaft are clamped, as in Fig. 3.8(a). Numerical data: $M_{t_0} = 100$ N-m, $\omega = 200$ Hz, mass moment of inertia of motor's rotor ($J_0$) $= 0.04$ kg-m$^2$, diameter of the shaft $d = 0.025$ m, length of the shaft on either side of the motor $l = 0.5$ m, shear modulus of the shaft $= 0.8 \times 10^{11}$ N/m$^2$.

***Solution.*** Due to the flexibility of the shaft, the motor's rotor can be considered to be connected to torsional springs on either side. The torsional spring constant of the shaft lying on either side of the motor is given by

$$k_t = \frac{G\pi d^4}{32l} = 6135.9 \text{ N-m/rad} \tag{E.1}$$

Since both the springs try to resist the angular deformation of the rotor, as shown in Fig. 3.8(b), the total spring constant will be $2k_t$. The equation of motion of the rotor

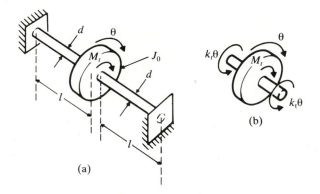

**Figure 3.8**

can be written as

$$J_0\ddot{\theta} + 2k_t\theta = M_t = M_{to}\cos\omega t \tag{E.2}$$

The steady-state solution of Eq. (E.2) can be obtained by substituting $M_{to}$, $2k_t$, and $J_0$ for $F_0$, $k$, and $m$, respectively, in Eq. (3.5):

$$\theta_s(t) = \Theta\cos\omega t \tag{E.3}$$

where $\Theta$ is the amplitude of vibration given by

$$\Theta = \frac{M_{to}}{2k_t - J_0\omega^2} = \frac{100}{2(6135.9) - 0.04(200\times2\pi)} = 0.008182 \text{ rad} \tag{E.4}$$

## 3.4 RESPONSE OF A DAMPED SYSTEM UNDER HARMONIC FORCE    9-25

If the forcing function is given by $F(t) = F_0\cos\omega t$, the equation of motion becomes

$$m\ddot{x} + c\dot{x} + kx = F_0\cos\omega t \tag{3.23}$$

The particular solution of Eq. (3.23) is also expected to be harmonic; we assume it in the form*

Steady State Response    $$x_p(t) = X\cos(\omega t - \phi) \tag{3.24}$$

where $X$ and $\phi$ are constants to be determined. $X$ and $\phi$ denote the amplitude and phase angle of the response, respectively. By substituting Eq. (3.24) into Eq. (3.23), we arrive at

$$X[(k - m\omega^2)\cos(\omega t - \phi) - c\omega\sin(\omega t - \phi)] = F_0\cos\omega t \tag{3.25}$$

---

*Alternatively, we can assume $x_p(t)$ to be of the form $x_p(t) = C_1\cos\omega t + C_2\sin\omega t$, which also involves two constants $C_1$ and $C_2$. But the final result will be same in both the cases.

Using the trigonometric relations

$$\cos(\omega t - \phi) = \cos \omega t \cos \phi + \sin \omega t \sin \phi$$
$$\sin(\omega t - \phi) = \sin \omega t \cos \phi - \cos \omega t \sin \phi \tag{3.26}$$

in Eq. (3.25) and equating the coefficients of $\cos \omega t$ and $\sin \omega t$ on both sides of the resulting equation, we obtain

$$X\big[(k - m\omega^2)\cos\phi + c\omega \sin\phi\big] = F_0$$
$$X\big[(k - m\omega^2)\sin\phi - c\omega \cos\phi\big] = 0 \tag{3.27}$$

Solution of Eqs. (3.27) gives

$$\checkmark \quad X = \frac{F_0}{\big[(k - m\omega^2)^2 + c^2\omega^2\big]^{1/2}} \tag{3.28}$$

and

$$\checkmark \quad \phi = \tan^{-1}\left(\frac{c\omega}{k - m\omega^2}\right) \tag{3.29}$$

By inserting the expressions of $X$ and $\phi$ from Eqs. (3.28) and (3.29) into Eq. (3.24) we obtain the particular solution of Eq. (3.23). Figure 3.9 shows typical plots of the forcing function and (steady-state) response. Dividing both the numerator and denominator of Eq. (3.28) by $k$ and making the following substitutions

$$\omega_n = \sqrt{\frac{k}{m}} = \text{undamped natural frequency,}$$

$$\zeta = \frac{c}{c_c} = \frac{c}{2m\omega_n}; \quad \frac{c}{m} = 2\zeta\omega_n,$$

$$\delta_{st} = \frac{F_0}{k} = \text{static deflection under the force } F_0, \text{ and}$$

$$r = \frac{\omega}{\omega_n} = \text{frequency ratio}$$

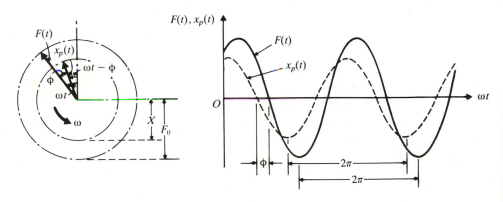

**Figure 3.9**   Graphical representation of forcing function and response.

we obtain

Amplitude
Ratio

$$\frac{X}{\delta_{st}} = \frac{1}{\left\{\left[1 - \left(\frac{\omega}{\omega_n}\right)^2\right]^2 + \left[2\zeta\frac{\omega}{\omega_n}\right]^2\right\}^{1/2}} = \frac{1}{\sqrt{(1-r^2)^2 + (2\zeta r)^2}} \tag{3.30}$$

and

$$\phi = \tan^{-1}\left\{\frac{2\zeta\frac{\omega}{\omega_n}}{1 - \left(\frac{\omega}{\omega_n}\right)^2}\right\} = \tan^{-1}\left(\frac{2\zeta r}{1-r^2}\right) \tag{3.31}$$

As stated in Sec. 3.3, the quantity $X/\delta_{st}$ is called the *magnification factor*, *amplification factor*, or *amplitude ratio*. The variation of $X/\delta_{st}$ and $\phi$ with the frequency ratio $r$ and the damping ratio $\zeta$ are shown in Fig. 3.10. The following observations can be made from Eqs. (3.30) and (3.31) and from Fig. 3.10:

1. For an undamped system ($\zeta = 0$), Eq. (3.31) shows that the phase angle $\phi = 0$ and Eq. (3.30) reduces to Eq. (3.10).

2. The damping reduces the amplitude ratio for all values of the forcing frequency.

3. The reduction of the amplitude ratio in the presence of damping is very significant at or near resonance.

**Figure 3.10**   Variation of $X$ and $\phi$ with frequency ratio $r$.

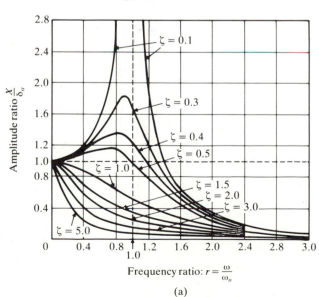

(a)

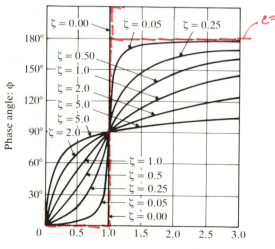

(b)

4. With damping, the maximum amplitude ratio (see Problem 3.9) occurs when

$$r = \sqrt{1 - 2\zeta^2} \qquad \text{or} \qquad \omega = \omega_n\sqrt{1 - 2\zeta^2} \qquad (3.32)$$

which is lower than the undamped natural frequency $\omega_n$ and the damped natural frequency $\omega_d = \omega_n\sqrt{1 - \zeta^2}$.

5. The maximum value of $X$ (when $r = \sqrt{1 - 2\zeta^2}$) is given by

$$\left(\frac{X}{\delta_{st}}\right)_{max} = \frac{1}{2\zeta\sqrt{1 - \zeta^2}} \qquad (3.33)$$

and the value of $X$ at $\omega = \omega_n$ by

$$\left(\frac{X}{\delta_{st}}\right)_{\omega = \omega_n} = \frac{1}{2\zeta} \qquad (3.34)$$

Equation (3.33) can be used for the experimental determination of the measure of damping present in the system. In a vibration test, if the maximum amplitude of the response, $(X)_{max}$, is measured, the damping ratio of the system can be found using Eq. (3.33). Conversely, if the amount of damping is known, one can make an estimate of the maximum amplitude of vibration.

6. For $\zeta > 1/\sqrt{2}$, the graph of $X$ has no peaks and for $\zeta = 0$, there is a discontinuity at $r = 1$.

7. The phase angle depends on the system parameters $m$, $c$, and $k$ and the forcing frequency $\omega$, but not on the amplitude $F_0$ of the forcing function.

8. The phase angle $\phi$ by which the response $x(t)$ or $X$ lags the forcing function $F(t)$ or $F_0$ will be very small for small values of $r$. For very large values of $r$, the phase angle approaches 180° asymptotically. Thus the amplitude of vibration will be in phase with the exciting force for $r \ll 1$ and out of phase for $r \gg 1$. The phase angle at resonance will be 90° for all values of damping ($\zeta$).

9. Below resonance ($\omega < \omega_n$), the phase angle increases with increase in damping. Above resonance ($\omega > \omega_n$), the phase angle decreases with increase in damping.

### 3.4.1   Total Response

The complete solution is given by $x(t) = x_h(t) + x_p(t)$ where $x_h(t)$ is given by Eq. (2.64). Thus

*Total Response :* ★   $$x(t) = \underbrace{X_0 e^{-\zeta\omega_n t}\cos(\omega_d t + \phi_0)}_{X_h(t)} + \underbrace{X\cos(\omega t - \phi)}_{X_p(t)} \qquad (3.35)$$

where

$$\omega_d = \sqrt{1 - \zeta^2} \cdot \omega_n \qquad (3.36)$$

$$r = \frac{\omega}{\omega_n} \qquad (3.37)$$

$X$ and $\phi$ are given by Eqs. (3.30) and (3.31), respectively, and $X_0$ and $\phi_0$ can be determined from the initial conditions.

### 3.4.2 Quality Factor and Bandwidth

For small values of damping ($\zeta < 0.05$), we can take

$$\left(\frac{X}{\delta_{st}}\right)_{max} \simeq \left(\frac{X}{\delta_{st}}\right)_{\omega = \omega_n} = \frac{1}{2\zeta} = Q \tag{3.38}$$

The value of the amplitude ratio at resonance is also called $Q$ *factor* or *quality factor* of the system, in analogy with some electrical-engineering applications, such as the tuning circuit of a radio, where the interest lies in an amplitude at resonance that is as large as possible [3.2]. The points $R_1$ and $R_2$, where the amplification factor falls to $Q/\sqrt{2}$, are called *half power points* because the power absorbed ($\Delta W$) by the damper (or by the resistor in an electrical circuit), responding harmonically at a given frequency, is proportional to the square of the amplitude (see Eq. (2.88)):

$$\Delta W = \pi c \omega X^2 \tag{3.39}$$

The difference between the frequencies associated with the half power points $R_1$ and $R_2$ is called the *bandwidth* of the system (see Fig. 3.11). For small values of $\zeta$, the bandwidth $\Delta \omega$ is given by

$$\Delta \omega = \omega|_{R_2} - \omega|_{R_1} \equiv \omega_2 - \omega_1 \simeq 2\zeta \omega_n \tag{3.40}$$

Combining Eqs. (3.38) and (3.40), we obtain

$$Q \simeq \frac{1}{2\zeta} \simeq \frac{\omega}{\omega_2 - \omega_1} \tag{3.41}$$

It can be seen that the quality factor $Q$ can be used for estimating the equivalent viscous damping in a mechanical system.

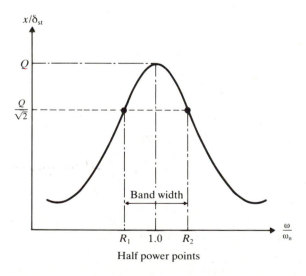

**Figure 3.11** Harmonic response curve showing half power points and bandwidth.

## 3.5 RESPONSE OF A DAMPED SYSTEM
## UNDER $F(t) = F_0 e^{i\omega t}$

Let the harmonic forcing function be represented in complex form as $F(t) = F_0 e^{i\omega t}$ so that the equation of motion becomes

$$m\ddot{x} + c\dot{x} + kx = F_0 e^{i\omega t} \tag{3.42}$$

Since the actual excitation is given only by the real part of $F(t)$, the response will also be given only by the real part of $x(t)$ where $x(t)$ is a complex quantity satisfying the differential equation (3.42). $F_0$ in Eq. (3.42) is, in general, a complex number. By assuming the particular solution $x_p(t)$

$$x_p(t) = Xe^{i\omega t} \tag{3.43}$$

we obtain, by substituting Eq. (3.43) into Eq. (3.42),*

$$X = \frac{F_0}{(k - m\omega^2) + ic\omega} \tag{3.44}$$

Multiplying the numerator and denominator on the right side of Eq. (3.44) by $[(k - m\omega^2) - ic\omega]$ and separating the real and imaginary parts, we obtain

$$X = F_0 \left[ \frac{k - m\omega^2}{(k - m\omega^2)^2 + c^2\omega^2} - i\frac{c\omega}{(k - m\omega^2)^2 + c^2\omega^2} \right] \tag{3.45}$$

Using the relation, $x + iy = Ae^{i\phi}$ where $A = \sqrt{x^2 + y^2}$ and $\tan\phi = y/x$, Eq. (3.45) can be expressed as

$$X = \frac{F_0}{\left[ (k - m\omega^2)^2 + c^2\omega^2 \right]^{1/2}} e^{-i\phi} \tag{3.46}$$

where

$$\phi = \tan^{-1}\left( \frac{c\omega}{k - m\omega^2} \right) \tag{3.47}$$

Thus the steady-state solution, Eq. (3.43), becomes

$$x_p(t) = \frac{F_0}{\left[ (k - m\omega^2)^2 + (c\omega)^2 \right]^{1/2}} e^{i(\omega t - \phi)} \tag{3.48}$$

**Frequency Response.** Equation (3.44) can be rewritten in the form

$$\frac{kX}{F_0} = \frac{1}{1 - r^2 + i2\zeta r} \equiv H(i\omega) \tag{3.49}$$

where $H(i\omega)$ is known as the *complex frequency response* of the system. The absolute

---

*Equation (3.44) can be written as $Z(i\omega)X = F_0$ where $Z(i\omega) = -m\omega^2 + i\omega c + k$ is called the *mechanical impedance* of the system [3.8].

value of $H(i\omega)$ given by

$$|H(i\omega)| = \left| \frac{kX}{F_0} \right| = \frac{1}{\left[ (1 - r^2)^2 + (2\zeta r)^2 \right]^{1/2}} \tag{3.50}$$

denotes the magnification factor defined in Eq. (3.30). Recalling that $e^{i\phi} = \cos\phi + i\sin\phi$, we can show that Eqs. (3.49) and (3.50) are related:

$$H(i\omega) = |H(i\omega)| e^{-i\phi} \tag{3.51}$$

where $\phi$ is given by Eq. (3.47), which can also be expressed as

$$\phi = \tan^{-1}\left( \frac{2\zeta r}{1 - r^2} \right) \tag{3.52}$$

Thus Eq. (3.48) can be expressed as

$$x_p(t) = \frac{F_0}{k} |H(i\omega)| e^{i(\omega t - \phi)} \tag{3.53}$$

If $F(t) = F_0 \cos\omega t$, the corresponding steady-state solution is given by the real part of Eq. (3.48):

$$x_p(t) = \frac{F_0}{\left[ (k - m\omega^2)^2 + (c\omega)^2 \right]^{1/2}} \cos(\omega t - \phi)$$

$$= \mathrm{Re}\left[ \frac{F_0}{k} H(i\omega) e^{i\omega t} \right] = \mathrm{Re}\left[ \frac{F_0}{k} |H(i\omega)| e^{i(\omega t - \phi)} \right] \tag{3.54}$$

which can be seen to be the same as Eq. (3.24). Similarly, if $F(t) = F_0 \sin\omega t$, the corresponding steady-state solution is given by the imaginary part of Eq. (3.48):

$$x_p(t) = \frac{F_0}{\left[ (k - m\omega^2)^2 + (c\omega)^2 \right]^{1/2}} \sin(\omega t - \phi)$$

$$= \mathrm{Im}\left[ \frac{F_0}{k} |H(i\omega)| e^{i(\omega t - \phi)} \right] \tag{3.55}$$

**Complex Vector Representation of Harmonic Motion.** The harmonic excitation and the response of the damped system to that excitation can be represented graphically in the complex plane, and interesting interpretation can be given to the resulting diagram. We first differentiate Eq. (3.53) with respect to time and obtain

$$\text{velocity} = \dot{x}_p(t) = i\omega \frac{F_0}{k} |H(i\omega)| e^{i(\omega t - \phi)} = i\omega x_p(t)$$

$$\text{acceleration} = \ddot{x}_p(t) = (i\omega)^2 \frac{F_0}{k} |H(i\omega)| e^{i(\omega t - \phi)} = -\omega^2 x_p(t) \tag{3.56}$$

Because $i$ can be expressed as

$$i = \cos\frac{\pi}{2} + i\sin\frac{\pi}{2} = e^{i\pi/2} \tag{3.57}$$

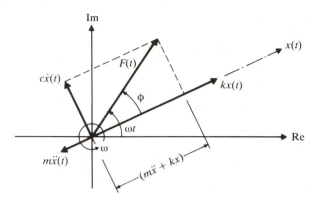

**Figure 3.12**   Representation of Eq. (3.42) in a complex plane.

we can conclude that the velocity leads the displacement by the phase angle $\pi/2$ and that it is multiplied by $\omega$. Similarly, $-1$ can be written as

$$-1 = \cos \pi + i \sin \pi = e^{\pi/2} \tag{3.58}$$

Hence the acceleration leads the displacement by the phase angle $\pi$, and it is multiplied by $\omega^2$.

Thus the various terms of the equation of motion (3.42) can be represented in the complex plane, as shown in Fig. 3.12. The interpretation of this figure is that the sum of the complex vectors $m\ddot{x}(t)$, $c\dot{x}(t)$, and $kx(t)$ balances $F(t)$, which is precisely what is required to satisfy Eq. (3.42). It is to be noted that the entire diagram rotates with angular velocity $\omega$ in the complex plane. If only the real part of the response is to be considered, then the entire diagram must be projected onto the real axis. Similarly, if only the imaginary part of the response is to be considered, then the diagram must be projected onto the imaginary axis. In Fig. 3.12, notice that the force $F(t) = F_0 e^{i\omega t}$ is represented as a vector located at an angle $\omega t$ to the real axis. This implies that $F_0$ is real. If $F_0$ is also complex, then the force vector $F(t)$ will be located at an angle of $(\omega t + \psi)$, where $\psi$ is some phase angle introduced by $F_0$. In such a case, all the other vectors, namely, $m\ddot{x}$, $c\dot{x}$, and $kx$ will be shifted by the same angle $\psi$. This is equivalent to multiplying both sides of Eq. (3.42) by $e^{i\psi}$.

## 3.6 RESPONSE OF A DAMPED SYSTEM UNDER THE HARMONIC MOTION OF THE BASE

Sometimes the base or support of a spring-mass-damper system undergoes harmonic motion, as shown in Fig. 3.13(a). Let $y(t)$ denote the displacement of the base and $x(t)$ the displacement of the mass from its static equilibrium position at time $t$. Then the net elongation of the spring is $x - y$ and the relative velocity between the two ends of the damper is $\dot{x} - \dot{y}$. From the free-body diagram shown in Fig. 3.13(b), we

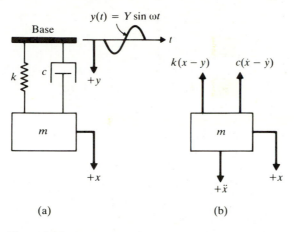

**Figure 3.13**   Base excitation.

obtain the equation of motion:

$$m\ddot{x} + c(\dot{x} - \dot{y}) + k(x - y) = 0 \tag{3.59}$$

If $y(t) = Y \sin \omega t$, Eq. (3.59) becomes

$$m\ddot{x} + c\dot{x} + kx = A \sin \omega t + B \cos \omega t \tag{3.60}$$

where $A = kY$ and $B = c\omega Y$. This shows that giving excitation to the base is equivalent to applying harmonic force of magnitude $(kY \sin \omega t + c\omega Y \cos \omega t)$ to the mass. By using the solutions given in Eqs. (3.54) and (3.55), the steady-state response of the mass can be expressed as

$$x_p(t) = \frac{kY \sin(\omega t - \phi_1)}{\left[(k - m\omega^2)^2 + (c\omega)^2\right]^{1/2}} + \frac{\omega c Y \cos(\omega t - \phi_1)}{\left[(k - m\omega^2)^2 + (c\omega)^2\right]^{1/2}} \tag{3.61}$$

The phase angle $\phi_1$ will be the same for both the terms because it depends on the values of $m$, $c$, $k$, and $\omega$, but not on the amplitude of the excitation. Equation (3.61) can be rewritten as

$$x_p(t) = X \cos(\omega t - \phi_1 - \phi_2)$$

$$\equiv Y \left[\frac{k^2 + (c\omega)^2}{(k - m\omega^2)^2 + (c\omega)^2}\right]^{1/2} \cos(\omega t - \phi_1 - \phi_2) \tag{3.62}$$

where the amplitude of the response $x_p(t)$ is given by

$$\frac{X}{Y} = \left[\frac{k^2 + (c\omega)^2}{(k - m\omega^2)^2 + (c\omega)^2}\right]^{1/2} = \left[\frac{1 + (2\zeta r)^2}{(1 - r^2)^2 + (2\zeta r)^2}\right]^{1/2} \tag{3.63}$$

and $\phi_1$ and $\phi_2$ by

$$\phi_1 = \tan^{-1}\left(\frac{c\omega}{k - m\omega^2}\right) = \tan^{-1}\left(\frac{2\zeta r}{1 - r^2}\right)$$

$$\phi_2 = \tan^{-1}\left(\frac{k}{c\omega}\right) = \tan^{-1}\left(\frac{1}{2\zeta r}\right) \tag{3.64}$$

The ratio $X/Y$ is called the *transmissibility.*

Note that if the harmonic excitation of the base is expressed in complex form as $y(t) = \text{Re}(Ye^{i\omega t})$, the response of the system can be expressed as

$$x_p(t) = \text{Re}\left\{\left(\frac{1 + i2\zeta r}{1 - r^2 + i2\zeta r}\right)Ye^{i\omega t}\right\} \tag{3.65}$$

and the transmissibility as

$$\frac{X}{Y} = \left[1 + (2\zeta r)^2\right]^{1/2} \cdot |H(i\omega)| \tag{3.66}$$

where $|H(i\omega)|$ is given by Eq. (3.50).

### 3.6.1   Force Transmitted

In Fig. 3.13(b), the force carried by the support $F$ must be due to the spring and dashpot which are connected to it. It can be determined as follows:

$$F = k(x - y) + c(\dot{x} - \dot{y}) = -m\ddot{x} \tag{3.67}$$

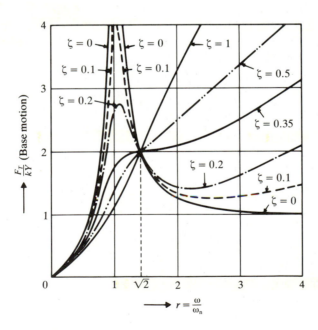

**Figure 3.14**

From Eq. (3.62), Eq. (3.67) can be written as

$$F = m\omega^2 X \cos(\omega t - \phi_1 - \phi_2) = F_T \cos(\omega t - \phi_1 - \phi_2) \qquad (3.68)$$

where $F_T$ is the amplitude or maximum value of the transmitted force given by

$$\frac{F_T}{kY} = r^2 \left[ \frac{1 + (2\zeta r)^2}{(1 - r^2)^2 + (2\zeta r)^2} \right]^{1/2} \qquad (3.69)$$

It can be noticed that the transmitted force is in phase with the motion of the mass $x(t)$. The variation of the force transmitted to the base with the frequency ratio $r$ is shown in Fig. 3.14 for different values of $\zeta$.

### 3.6.2 Relative Motion

If $z = x - y$ denotes the motion of the mass relative to the base, the equation of motion, Eq. (3.59), can be rewritten as

$$m\ddot{z} + c\dot{z} + kz = -m\ddot{y} = m\omega^2 Y \sin \omega t \qquad (3.70)$$

The steady-state solution of Eq. (3.70) is given by

$$z(t) = \frac{m\omega^2 Y \sin(\omega t - \phi_1)}{\left[ (k - m\omega^2)^2 + (c\omega)^2 \right]^{1/2}} = Z \sin(\omega t - \phi_1) \qquad (3.71)$$

where $Z$, the amplitude of $z(t)$, can be expressed as

$$Z = \frac{m\omega^2 Y}{\sqrt{(k - m\omega^2)^2 + (c\omega)^2}} = Y \frac{r^2}{\sqrt{(1 - r^2)^2 + (2\zeta r)^2}} \qquad (3.72)$$

and $\phi_1$ by Eq. (3.64). The ratio $Z/Y$ is shown graphically in Fig. 3.15.

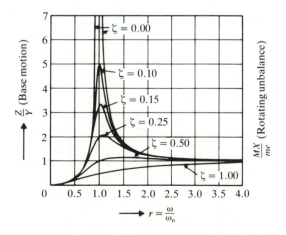

**Figure 3.15** Variation of $(Z/Y)$ or $(MX/me)$ with frequency ratio $r = (\omega/\omega_n)$.

---

**EXAMPLE 3.2**

Figure 3.16 shows a simple model of a motor vehicle traveling over a rough road. It is assumed that 1) the vehicle vibrates only in the vertical direction, 2) the spring constant of the tires is infinite so that the roughness is transmitted directly to the suspension system of the vehicle, and 3) the tires do not leave the road surface. The vehicle has a mass of 1200 kg when fully loaded and 300 kg when empty. The spring constant for the total suspension system is 400 kN/m and the damping ratio, when the vehicle is fully loaded, is $\zeta = 0.5$. If the vehicle speed is 100 km/hr, determine the displacement amplitude of the vehicle when fully loaded and empty. Assume that the road surface is sinusoidal with an amplitude of $Y = 0.05$ m and a period of 6 m.

***Solution.*** The excitation frequency $f$ can be found by dividing the vehicle speed by the length of one cycle of road roughness:

$$f = \left( \frac{100 \times 1000}{3600} \right) \cdot \frac{1}{6} = 4.6296 \text{ Hz}$$

The damping ratio is given by

$$\zeta = \frac{c}{2\sqrt{km}}$$

Since $k$ and $c$ are constant, $\zeta$ varies inversely with $\sqrt{m}$ : if the mass is decreased by a factor of 4, the damping ratio will be increased by a factor of 2. Thus $\zeta$ will become 1.0 when the vehicle is empty. The displacement amplitude of the vehicle can be found from Eq. (3.63). The results are shown in the following table.

| Quantity | When vehicle is fully loaded | When vehicle is empty |
|---|---|---|
| Natural frequency, $\omega_n = \sqrt{\dfrac{k}{m}}$ | $\omega_n = \left( \dfrac{400 \times 10^3}{1200} \right)^{1/2}$ $= 18.2574 \text{ rad/sec}$ $= 2.9057 \text{ Hz}$ | $\omega_n = \left( \dfrac{400 \times 10^3}{300} \right)^{1/2}$ $= 36.5148 \text{ rad/sec}$ $= 5.8114 \text{ Hz}$ |
| Frequency ratio, $r$ | $r = 4.6296/2.9057$ $= 1.5933$ | $r = 0.7966$ |
| Damping ratio, $\zeta$ | $\zeta = 0.5$ | $\zeta = 1.0$ |
| Amplitude ratio, $\dfrac{X}{Y} =$ Eq. (3.63) | $\dfrac{X}{Y} = 0.8493$ | $\dfrac{X}{Y} = 1.1039$ |
| Amplitude of the vehicle, $X$ | $X = 0.0425$ m | $X = 0.0552$ m |

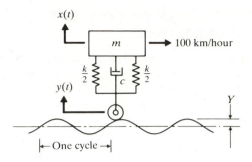

**Figure 3.16**   Vehicle moving over a rough road.

---

### EXAMPLE 3.3

A vibrating system with a moving base has a mass of 300 kg, a spring of stiffness 40,000 N/m and a damper with an unknown damping constant. It is observed that the mass vibrates with an amplitude of 10 mm when the support oscillates with an amplitude of 2.5 mm at the natural frequency of the supported system. Find (1) the damping constant of the system, (2) the dynamic force amplitude on the base, and (3) the amplitude of the displacement of the mass relative to the base.

*Solution*

**(1)** At resonance ($\omega = \omega_n$ or $r = 1$), Eq. (3.63) gives

$$\frac{X}{Y} = \frac{0.010}{0.0025} = 4 = \left[\frac{1 + (2\zeta)^2}{(2\zeta)^2}\right]^{1/2} \tag{E.1}$$

The solution of Eq. (E.1) gives $\zeta = 0.1291$. The damping constant is given by

$$c = \zeta \cdot c_c = \zeta 2\sqrt{km} = 0.1291 \times 2 \times \sqrt{40000 \times 300} = 894.431 \text{ N-s/m} \tag{E.2}$$

**(2)** The dynamic force amplitude carried by the support at $r = 1$ can be found from Eq. (3.69):

$$F_T = Yk\left[\frac{1 + 4\zeta^2}{4\zeta^2}\right]^{1/2} = kX = 40000 \times 0.01 = 400 \text{ N} \tag{E.3}$$

**(3)** The amplitude of the relative displacement of the mass at $r = 1$ can be obtained from Eq. (3.72):

$$Z = \frac{Y}{2\zeta} = \frac{0.0025}{2 \times 0.1291} = 0.00968 \text{ m} \tag{E.4}$$

It can be noticed that $X = 0.01$ m, $Y = 0.0025$ m, and $Z = 0.00968$ m; therefore, $Z \neq X - Y$. This is due to the phase difference of $x$, $y$, and $z$.

## 3.7 RESPONSE OF A DAMPED SYSTEM UNDER ROTATING UNBALANCE

Unbalance in rotating machinery is one of the main causes of vibration. A simplified model of such a machine is shown in Fig. 3.17. The total mass of the machine is $M$, and there are two eccentric masses $m/2$ rotating in opposite directions with a constant angular velocity $\omega$. The centrifugal force $(me\omega^2)/2$ due to each mass will cause excitation of the mass $M$. We consider two equal masses $m/2$ rotating in opposite directions in order to have the horizontal components of excitation of the two masses cancel each other. However, the vertical components of excitation add together and act along the axis of symmetry $A-A$ in Fig. 3.17. If the angular position of the masses is measured from a horizontal position, the total vertical component of the excitation is always given by $F(t) = me\omega^2\sin\omega t$. The equation of motion can be derived by the usual procedure:

$$M\ddot{x} + c\dot{x} + kx = me\omega^2\sin\omega t \tag{3.73}$$

The solution of this equation will be identical to Eq. (3.55) if we replace $m$ and $F_0$ by $M$ and $me\omega^2$ respectively. This solution can also be expressed as

$$x_p(t) = X\sin(\omega t - \phi) = \mathrm{Im}\left[\frac{me}{M}\left(\frac{\omega}{\omega_n}\right)^2 |H(i\omega)|e^{i(\omega t - \phi)}\right] \tag{3.74}$$

where $\omega_n = \sqrt{k/M}$ and $X$ and $\phi$ denote the amplitude and the phase angle of vibration given by

$$X = \frac{me\omega^2}{\left[(k - M\omega^2)^2 + (c\omega)^2\right]^{1/2}} = \frac{me}{M}\left(\frac{\omega}{\omega_n}\right)^2 |H(i\omega)| \tag{3.75}$$

and

$$\phi = \tan^{-1}\left(\frac{c\omega}{k - M\omega^2}\right) \tag{3.76}$$

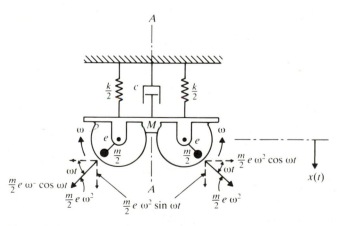

**Figure 3.17**   Rotating unbalanced masses.

By defining $\zeta = c/c_c$ and $c_c = 2M\omega_n$, Eqs. (3.75) and (3.76) can be rewritten as

$$\frac{MX}{me} = \frac{r^2}{\left[(1-r^2)^2 + (2\zeta r)^2\right]^{1/2}} = r^2 |H(i\omega)| \qquad (3.77)$$

and

$$\phi = \tan^{-1}\left(\frac{2\zeta r}{1-r^2}\right) \qquad (3.78)$$

The variation of $MX/me$ with $r$ for different values of $\zeta$ is shown in Fig. 3.15. On the other hand, the graph of $\phi$ versus $r$ remains as in Fig. 3.10(b). The following observations can be made from Eq. (3.77) and Fig. 3.15:

1. All the curves begin at zero amplitude. The amplitude near resonance ($\omega = \omega_n$) is markedly affected by damping. Thus if the machine is to be run near resonance, damping should be introduced purposefully to avoid dangerous amplitudes.

2. At very high speeds ($\omega$ large), $MX/me$ is almost unity, and the effect of damping is negligible.

3. The maximum of $MX/me$ occurs when

$$\frac{d}{dr}\left(\frac{MX}{me}\right) = 0 \qquad \text{Slope is } 0 \qquad (3.79)$$

The solution of Eq. (3.79) gives

$$r = \frac{1}{\sqrt{1-2\zeta^2}} > 1$$

Accordingly, the peaks occur to the right of the resonance value of $r = 1$.

---

### EXAMPLE 3.4

A variable-speed electric motor has a radial clearance of 1 mm between the stator and the rotor. The rotor has a mass of 25 kg and has an unbalance ($me$) of 5 kg-mm. The rotor is mounted on a steel shaft midway between the two bearings. The distance between the bearings is 2 m. The operating speed of the machine varies from 600 to 6000 rpm. Determine the diameter of the shaft so that the rotor is always clear of the stator at any operating speed within the specified range. Assume that the damping is negligible.

*Solution.* The maximum amplitude of the shaft (rotor) due to rotating unbalance can be obtained from Eq. (3.75) by setting $c = 0$ as

$$X = \frac{me\omega^2}{(k - M\omega^2)} = \frac{me\omega^2}{k(1-r^2)} \qquad (E.1)$$

where $me = 5$ kg-mm, $M = 25$ kg, and the limiting value of $X = 1$ mm. The value of

$\omega$ ranges from

$$600 \text{ rpm} = 600 \times \frac{2\pi}{60} = 20\pi \text{ rad/sec}$$

to

$$6000 \text{ rpm} = 6000 \times \frac{2\pi}{60} = 200\pi \text{ rad/sec}$$

while the natural frequency of the system is given by

$$\omega_n = \sqrt{\frac{k}{M}} = \sqrt{\frac{k}{25}} = 0.2\sqrt{k} \text{ rad/sec} \qquad \text{(E.2)}$$

if $k$ is in N/m. For $\omega = 20\pi$ rad/sec, Eq. (E.1) gives

$$0.001 = \frac{(5.0 \times 10^{-3}) \times (20\pi)^2}{k\left[1 - \dfrac{(20\pi)^2}{0.04\,k}\right]} = \frac{2\pi^2}{k - 10000\pi^2}$$

$$k = 12000\pi^2 \text{ N/m} \qquad \text{(E.3)}$$

For $\omega = 200\pi$ rad/sec, Eq. (E.1) gives

$$0.001 = \frac{(5.0 \times 10^{-3}) \times (200\pi)^2}{k\left[1 - \dfrac{(200\pi)^2}{0.04\,k}\right]} = \frac{200\pi^2}{k - 1000000\pi^2}$$

$$k = 1200000\pi^2 \text{ N/m} \qquad \text{(E.4)}$$

We know that the amplitude of vibration of the rotating shaft can be minimized by making $\omega \gg \omega_n$. This means that $\omega_n$ must be made small compared to $\omega$—that is, $k$ must be made small. This can be achieved by selecting the value of $k$ as $12000\pi^2$ N/m. Since the stiffness of a simply supported beam supporting a load at the center is given by

$$k = \frac{48EI}{l^3} = \frac{48E}{l^3}\left(\frac{\pi d^4}{64}\right) \qquad \text{(E.5)}$$

the diameter of the beam (shaft) can be found:

$$d^4 = \frac{64kl^3}{48\pi E} = \frac{64(12000\pi^2)(2^3)}{48\pi(2.06 \times 10^{11})} = 195.2062 \times 10^{-8} \text{ m}^4$$

$$d = 3.7379 \times 10^{-2} \text{ m} = 37.379 \text{ mm} \qquad \text{(E.6)}$$

## 3.8 FORCED VIBRATION WITH COULOMB DAMPING

For a single degree of freedom system with Coulomb or dry friction damping, subjected to a harmonic force $F(t) = F_0 \sin \omega t$ as in Fig. 3.18, the equation of motion

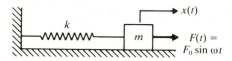

**Figure 3.18**

is given by

$$m\ddot{x} + kx \pm \mu N = F(t) = F_0 \sin \omega t \tag{3.80}$$

where the sign of the friction force ($\mu N$) changes with the direction of motion. The exact solution of Eq. (3.80) is quite involved. However, we can expect that if the dry friction damping force is large, the motion of the mass will be discontinuous. If the dry friction damping force is small compared to the forcing function $F(t)$, we can obtain an approximate solution by finding an equivalent viscous damping ratio. To find an equivalent viscous damping ratio, we equate the energy dissipated due to dry friction to the energy dissipated by an equivalent viscous damper during a full cycle of motion. If the amplitude of motion is denoted as $X$, the energy dissipated by the friction force $\mu N$ in a quarter cycle is $\mu N X$. Hence in a full cycle, the energy dissipated by dry friction damping is given by

$$\Delta W = 4\mu N X \tag{3.81}$$

If the equivalent viscous damping constant is denoted as $c_{eq}$, the energy dissipated during a full cycle (see Eq. (2.92)) will be

$$\Delta W = \pi c_{eq} \omega X^2 \tag{3.82}$$

By equating Eqs. (3.81) and (3.82), we obtain

$$c_{eq} = \frac{4\mu N}{\pi \omega X} \tag{3.83}$$

Thus the steady-state response is given by

$$x_p(t) = X \sin(\omega t - \phi) \tag{3.84}$$

where the amplitude $X$ can be found from Eq. (3.55):

$$X = \frac{F_0}{\left[(k - m\omega^2)^2 + (c_{eq}\omega)^2\right]^{1/2}} = \frac{(F_0/k)}{\left[\left(1 - \frac{\omega^2}{\omega_n^2}\right)^2 + \left(2\zeta_{eq}\frac{\omega}{\omega_n}\right)^2\right]^{1/2}} \tag{3.85}$$

with

$$\zeta_{eq} = \frac{c_{eq}}{c_c} = \frac{c_{eq}}{2m\omega_n} = \frac{4\mu N}{2m\omega_n \pi \omega X} = \frac{2\mu N}{\pi m \omega \omega_n X} \tag{3.86}$$

By substituting Eq. (3.86) into Eq. (3.85), we obtain

$$X = \frac{(F_0/k)}{\left[\left(1 - \frac{\omega^2}{\omega_n^2}\right)^2 + \left(\frac{4\mu N}{\pi k X}\right)^2\right]^{1/2}} \tag{3.87}$$

Since the amplitude $X$ occurs on both sides of Eq. (3.87), we can square both sides and solve for $X$ to obtain

$$X = \frac{F_0}{k}\left[\frac{1 - \left(\frac{4\mu N}{\pi F_0}\right)^2}{\left(1 - \frac{\omega^2}{\omega_n^2}\right)^2}\right]^{1/2} \tag{3.88}$$

This equation provides an approximation for $X$ as long as $(F_0/\mu N) > 4/\pi$. If $(F_0/\mu N) < 4/\pi$, the numerator of the fraction under the radical becomes negative and $X$ becomes imaginary, so the approximate solution cannot be used. In such a case the exact analysis is required [3.3]. The phase angle $\phi$ appearing in Eq. (3.84) can be found using Eq. (3.47):

$$\phi = \tan^{-1}\left(\frac{c\omega}{k - m\omega^2}\right) = \tan^{-1}\left(\frac{2\zeta\frac{\omega}{\omega_n}}{1 - \frac{\omega^2}{\omega_n^2}}\right)$$

$$= \tan^{-1}\left\{\frac{\frac{4\mu N}{\pi k X}}{\left(1 - \frac{\omega^2}{\omega_n^2}\right)}\right\} = \tan^{-1}\left\{\frac{\frac{4\mu N}{\pi F_0}}{1 - \frac{\omega^2}{\omega_n^2}} \cdot \frac{\left[\left(1 - \frac{\omega^2}{\omega_n^2}\right)^2\right]^{1/2}}{1 - \left(\frac{4\mu N}{\pi F_0}\right)^2}\right\} \tag{3.89}$$

Although $\tan\phi$ is a constant for a given value of $F_0/\mu N$, it will take a positive value if $\omega/\omega_n < 1$ and a negative value if $\omega/\omega_n > 1$. Thus we have

$$\phi = \tan^{-1}\left[\frac{\pm\frac{4\mu N}{\pi F_0}}{\left\{1 - \left(\frac{4\mu N}{\pi F_0}\right)^2\right\}^{1/2}}\right] \tag{3.90}$$

This equation shows that the phase angle is discontinuous at $\omega/\omega_n = 1$ (resonance) in the case of dry friction damping.

Equation (3.88) shows that friction serves to limit the amplitude of forced vibration for $\omega/\omega_n \neq 1$. However, at resonance ($\omega/\omega_n = 1$), the amplitude becomes

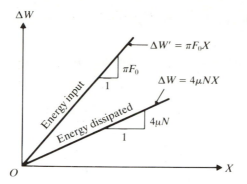

**Figure 3.19**

infinite. This can be explained as follows. The energy directed into the system over one cycle when it is excited harmonically at resonance is

$$\Delta W' = \int_{\text{cycle}} F \cdot dx = \int_0^\tau F \frac{dx}{dt} \, dt$$

$$= \int_0^{\tau = 2\pi/\omega} F_0 \sin \omega t \cdot \left[ \omega X \cos(\omega t - \phi) \right] dt \tag{3.91}$$

Since $\phi = 90°$ at resonance, Eq. (3.91) becomes

$$\Delta W' = F_0 X \omega \int_0^{2\pi/\omega} \sin^2 \omega t \, dt = \pi F_0 X \tag{3.92}$$

The energy dissipated from the system is given by Eq. (3.81). Since $\pi F_0 X > 4\mu N X$ for $X$ to be real-valued, $\Delta W' > \Delta W$ at resonance (see Fig. 3.19). Thus more energy is directed into the system per cycle than is dissipated per cycle. This extra energy is used to build up the amplitude of vibration. For the nonresonant condition $(\omega/\omega_n \neq 1)$, the energy input can be found from Eq. (3.91):

$$\Delta W' = \omega F_0 X \int_0^{2\pi/\omega} \sin \omega t \cos(\omega t - \phi) \, dt = \pi F_0 X \sin \phi \tag{3.93}$$

Due to the presence of $\sin \phi$ in Eq. (3.93), the input energy curve in Fig. 3.19 is brought into coincidence with the dissipated energy curve, so the amplitude is limited. Thus the phase of the motion limits the amplitude of the motion.

The periodic response of a spring-mass system with Coulomb damping subjected to base excitation is given in Refs. [3.10, 3.11].

---

### EXAMPLE 3.5

A spring-mass system, having a mass of 10 kg and a spring of stiffness of 2000 N/m, vibrates on a horizontal surface. The coefficient of friction is 0.12. When subjected to a harmonic force of frequency 2 Hz, the mass develops a sustained oscillation with an amplitude of 40 mm. Find the amplitude of the harmonic force applied to the mass.

***Solution.*** The vertical force (weight) of the mass is $N = mg = 10 \times 9.81 = 98.1$ N. The natural frequency is

$$\omega_n = \sqrt{\frac{k}{m}} = \sqrt{\frac{2000}{10}} = 14.1421 \text{ rad/sec}$$

and the frequency ratio is

$$\frac{\omega}{\omega_n} = \frac{2 \times 2\pi}{14.1421} = 0.8885$$

The amplitude of vibration $X$ is given by Eq. (3.88):

$$X = \frac{F_0}{k} \left[ \frac{1 - \left( \frac{4\mu N}{\pi F_0} \right)^2}{\left\{ 1 - \left( \frac{\omega}{\omega_n} \right)^2 \right\}^2} \right]^{1/2}$$

$$0.04 = \frac{F_0}{2000} \left[ \frac{1 - \left\{ \frac{4(0.12)(98.1)}{\pi F_0} \right\}^2}{(1 - 0.8885^2)^2} \right]^{1/2}$$

The solution of this equation gives $F_0 = 39.6547$ N.

## 3.9 FORCED VIBRATION WITH HYSTERESIS DAMPING

Consider a single degree of freedom system with hysteresis damping and subjected to a harmonic force $F(t) = F_0 \sin \omega t$, as indicated in Fig. 3.20. The equation of motion of the mass can be derived, using Eq. (2.111), as

$$m\ddot{x} + \frac{\beta k}{\omega} \dot{x} + kx = F_0 \sin \omega t \tag{3.94}$$

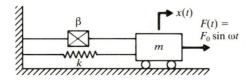

**Figure 3.20**

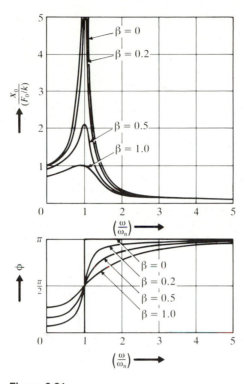

**Figure 3.21**

where $(\beta k/\omega)\dot{x}$ denotes the damping force.* Since the solution of Eq. (3.94) is quite complex for a general forcing function $F(t)$, a harmonic force was assumed to simplify the analysis. The steady-state solution of Eq. (3.94) can be assumed:

$$x_p(t) = X\sin(\omega t - \phi) \qquad (3.95)$$

By substituting Eq. (3.95) into Eq. (3.94), we obtain

$$X = \frac{F_0}{k\left[\left(1 - \dfrac{\omega^2}{\omega_n^2}\right)^2 + \beta^2\right]^{1/2}} \qquad (3.96)$$

and

$$\phi = \tan^{-1}\left[\frac{\beta}{\left(1 - \dfrac{\omega^2}{\omega_n^2}\right)}\right] \qquad (3.97)$$

Equations (3.96) and (3.97) are shown plotted in Fig. 3.21 for several values of $\beta$. A

---

*In contrast to viscous damping, this damping force is also a function of the forcing frequency $\omega$ (see Sec. 2.8).

comparison of Fig. 3.21 with Fig. 3.10 for viscous damping reveals the following:

1. The amplitude ratio

$$\frac{X}{(F_0/k)}$$

attains its maximum value of $F_0/k\beta$ at the resonant frequency ($\omega = \omega_n$) in the case of hysteresis damping, while it occurs at a frequency below resonance ($\omega < \omega_n$) in the case of viscous damping.

2. The phase angle $\phi$ has a value of $\tan^{-1}(\beta)$ at $\omega = 0$ in the case of hysteresis damping, while it has a value of zero at $\omega = 0$ in the case of viscous damping. This indicates that the response can never be in phase with the forcing function in the case of hysteresis damping.

Note that if the harmonic excitation is assumed to be $F(t) = F_0 e^{i\omega t}$ in Fig. 3.20, the equation of motion becomes

$$m\ddot{x} + \frac{\beta k}{\omega}\dot{x} + kx = F_0 e^{i\omega t} \tag{3.98}$$

In this case, the response $x(t)$ is also a harmonic function involving the factor $e^{i\omega t}$. Hence $\dot{x}(t)$ is given by $i\omega x(t)$, and Eq. (3.98) becomes

$$m\ddot{x} + k(1 + i\beta)x = F_0 e^{i\omega t} \tag{3.99}$$

where the quantity $k(1 + i\beta)$ is called the *complex stiffness* or *complex damping* [3.7]. The steady-state solution of Eq. (3.99) is given by the real part of

$$x(t) = \frac{F_0 e^{i\omega t}}{k\left[1 - \left(\dfrac{\omega}{\omega_n}\right)^2 + i\beta\right]} \tag{3.100}$$

## 3.10 FORCED MOTION WITH OTHER TYPES OF DAMPING

Viscous damping is the simplest form of damping to use in practice, since it leads to linear equations of motion. In the cases of Coulomb and hysteretic damping, we defined equivalent viscous damping coefficients to simplify the analysis. Even for a more complex form of damping, we define an equivalent viscous damping coefficient, as illustrated in the following examples. The practical use of equivalent damping is discussed in Ref. [3.12].

### EXAMPLE 3.6

*Quadratic* or *velocity squared damping* is present whenever a body moves in a fluid such as air or water with moderate speeds on the order of 5 to 20 m/s. Determine the equivalent viscous damping coefficient and the amplitude of the steady-state vibration of a single degree of freedom system having quadratic damping.

*Solution.* The damping force is assumed to be

$$F_d = \pm a(\dot{x})^2 \tag{E.1}$$

where $a$ is a constant, $\dot{x}$ is the relative velocity across the damper, and the negative sign must be used in Eq. (E.1) when $\dot{x}$ is positive and vice versa. The energy dissipated per cycle during harmonic motion $x(t) = X\sin\omega t$ is given by

$$\Delta W = 2\int_{-X}^{X} a(\dot{x})^2 \, dx = 2X^3 \int_{-\pi/2}^{\pi/2} a\omega^2 \cos^3\omega t \, d(\omega t) = \frac{8}{3}\omega^2 a X^3 \tag{E.2}$$

By equating this energy to the energy dissipated in an equivalent viscous damper (see Eq. (2.92)):

$$\Delta W = \pi c_{eq}\omega X^2 \tag{E.3}$$

we obtain the equivalent viscous damping coefficient ($c_{eq}$):

$$c_{eq} = \frac{8}{3\pi}a\omega X \tag{E.4}$$

It can be noted that $c_{eq}$ is not a constant but varies with $\omega$ and $X$. The amplitude of the steady-state response can be found from Eq. (3.30):

$$\frac{X}{\delta_{st}} = \frac{1}{\sqrt{(1-r^2)^2 + (2\zeta_{eq}r)^2}} \tag{E.5}$$

where $r = \omega/\omega_n$ and

$$\zeta_{eq} = \frac{c_{eq}}{c_c} = \frac{c_{eq}}{2m\omega_n} \tag{E.6}$$

Using Eqs. (E.4) and (E.6), Eq. (E.5) can be solved to obtain

$$X = \frac{3\pi m}{8ar^2}\left[ -\frac{(1-r^2)^2}{2} + \sqrt{\frac{(1-r^2)^4}{4} + \left(\frac{8ar^2\delta_{st}}{3\pi m}\right)^2} \right]^{1/2} \tag{E.7}$$

---

### EXAMPLE 3.7

A single degree of freedom system vibrates under a harmonic force $F_0\sin\omega t$. It is acted upon by several different forms of damping. Derive an equation for the equivalent viscous damping constant.

*Solution.* Let $\Delta W_1, \Delta W_2, \Delta W_3, \ldots$ denote the energies dissipated per cycle by the various forms of dampers. By equating the total energy dissipated to that of the equivalent viscous damper, we obtain

$$\pi c_{eq}\omega X^2 = \Delta W_1 + \Delta W_2 + \Delta W_3 + \cdots$$

Thus

$$c_{eq} = \frac{\displaystyle\sum_i \Delta W_i}{\pi \omega X^2}$$

## 3.11 SELF-EXCITED VIBRATION AND INSTABILITY

The force acting on a vibrating system is usually external to the system and independent of the motion. In this case, even when the system is clamped and the motion is stopped, the force continues to act, since it is independent of the motion. On the other hand, there are systems for which the exciting force is a function of the displacement, velocity, or acceleration of the system. Such systems are called *self-excited* vibrating systems since the motion itself produces the exciting force. In this case, if the system is clamped, the exciting force disappears. The instability of rotating shafts, electric oscillations of electronic equipment, the flutter of airplane wings, flow-induced vibrations in power plants, automobile wheel shimmy and aerodynamically induced motion of bridges are typical examples of self-excited vibrations.

A system is *dynamically stable* if the motion or displacement converges or remains steady with time. On the other hand, if the amplitude of displacement increases continuously (diverges) with time, it is said to be *dynamically unstable*. If the self-excitation increases the energy, the motion diverges and the system becomes unstable. To see the circumstances which lead to instability, we consider the equation of motion of a single degree of freedom system under a specific exciting force $F(t) = F_0 \dot{x}$:

$$m\ddot{x} + c\dot{x} + kx = F_0 \dot{x} \tag{3.101}$$

Equation (3.101) can be rewritten as

$$\ddot{x} + \left(\frac{c - F_0}{m}\right)\dot{x} + \frac{k}{m}x = 0 \tag{3.102}$$

If a solution of the form $x(t) = Ce^{st}$, where $C$ is a constant, is assumed, Eq. (3.102) gives the auxiliary equation

$$s^2 + \left(\frac{c - F_0}{m}\right)s + \frac{k}{m} = 0 \tag{3.103}$$

from which we can find

$$s_{1,2} = -\left(\frac{c - F_0}{2m}\right) \pm \sqrt{\left(\frac{c - F_0}{2m}\right)^2 - \frac{k}{m}} \tag{3.104}$$

First consider the case of positive damping ($c > F_0$) in Eq. (3.102). In this case, Eq. (3.102) is similar to the equation of motion of a damped free system (discussed in

Sec. 2.6.1). It can be verified that the system will be stable for all possible values of

$$\left[\left(\frac{c-F_0}{2m}\right)^2 - \frac{k}{m}\right]$$

Next, consider the condition of negative damping $(c < F_0)$ in Eq. (3.102). We can have three types of solution, depending on the sign of the quantity under the radical in Eq. (3.104), as stated below:

**Case 1:**

$$\left(\frac{c-F_0}{2m}\right)^2 > \frac{k}{m}$$

In this case both $s_1$ and $s_2$ are real and positive, and the total solution is given by

$$x(t) = C_1 e^{s_1 t} + C_2 e^{s_2 t} \qquad (3.105)$$

This denotes a diverging nonoscillatory motion, so the system will be unstable.

**Case 2:**

$$\left(\frac{c-F_0}{2m}\right)^2 = \frac{k}{m}$$

Here $s_1$ and $s_2$ are identical and are real and positive. The total solution (see Eq. (2.72)) is given by

$$x(t) = (C_1 + C_2 t) e^{s_1 t} \qquad (3.106)$$

This represents a diverging nonoscillatory solution, so the system will be unstable.

**Case 3:**

$$\left(\frac{c-F_0}{2m}\right)^2 < \frac{k}{m}$$

For this case, $s_1$ and $s_2$ are complex conjugate and can be expressed as

$$s_{1,2} = \left(\frac{F_0 - c}{2m}\right) \pm i \sqrt{\frac{k}{m} - \left(\frac{c - F_0}{2m}\right)^2} \qquad (3.107)$$

The solution (see Eq. (2.64)) can be expressed as:

$$x(t) = X e^{\left(\frac{F_0 - c}{2m}\right)t} \sin\left\{\sqrt{\frac{k}{m} - \left(\frac{c - F_0}{2m}\right)^2}\, t + \phi\right\} \qquad (3.108)$$

Since the exponent is positive, Eq. (3.108) denotes a diverging oscillatory solution and hence the system is unstable. Thus the condition for dynamic stability of the system can be stated as

$$F_0 \le c \qquad (3.109)$$

**General Approach.** In the above analysis, the dynamic instability was caused by the negative value of $c$. As will be seen in Example 3.8, dynamic instability can also occur due to negative values of $k$. For a multidegree of freedom system, we may not be able to study the instability on a purely physical basis. In such cases, we must study the problem mathematically. We shall now consider the general mathematical approach for a single degree of freedom system. The auxiliary equation of a spring-mass-damper system can be written as

$$s^2 + a_1 s + a_2 = 0 \tag{3.110}$$

The roots of this equation are

$$s_{1,2} = \frac{-a_1 \pm \sqrt{a_1^2 - 4a_2}}{2} \tag{3.111}$$

Since the solution was assumed to be $x(t) = Ce^{st}$ in deriving Eq. (3.110), the motion will be diverging and aperiodic if the roots $s_1$ and $s_2$ are real and positive. We can observe that if both $a_1$ and $a_2$ are positive, $s_1$ and $s_2$ will not have positive real values, so dynamic instability cannot occur. The other possibility is that $s_1$ and $s_2$ may be complex conjugates. In this case the roots can be expressed as

$$s_1 = p + iq$$
$$s_2 = p - iq \tag{3.112}$$

where $p$ and $q$ are real numbers. By rewriting the auxiliary equation as

$$(s - s_1)(s - s_2) = s^2 - (s_1 + s_2)s + s_1 s_2 = 0 \tag{3.113}$$

we obtain

$$a_1 = -(s_1 + s_2) = -2p$$
$$a_2 = s_1 s_2 = p^2 + q^2 \tag{3.114}$$

Note from Eq. (3.114) that $p$ must be negative for the vibration to be convergent. This means that $a_1$ must be positive. Thus, considering all the possibilities, we can state that the system will be dynamically stable if $a_1$ and $a_2$ are positive.

---

**EXAMPLE 3.8**

A pipe of length $l$, cross-sectional area $A$, flexural rigidity $EI$, and mass $m$ is simply supported at both ends, as shown in Fig. 3.22. If a fluid of mass density $\rho$ flows through the pipe, determine the velocity of the fluid $v$ at which the system will become unstable.

**Figure 3.23**

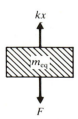

**Figure 3.22**

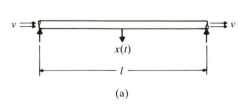

(a)

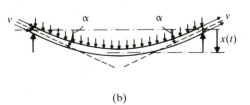

(b)

***Solution.*** We idealize the pipe as a single degree of freedom system. When the equivalent mass of the pipe $m_{eq} = m/2$, assumed to be located at the center, moves by a distance $x$, the spring force due to the elasticity of the pipe is $kx$, where the spring constant $k$ of the pipe can be found from the deflection formula of a simply supported beam under a uniformly distributed load [3.19]:

$$k = \frac{384EI}{5l^3} \tag{E.1}$$

The force induced due to a fluid velocity $v$ is $\rho A v^2$, and the component of this force due to a change in the direction of the fluid flow is

$$F = 2\rho A v^2 \sin\alpha \simeq 2\rho A v^2 \alpha \tag{E.2}$$

for small angles $\alpha$. The slope of the simply supported pipe under a uniformly distributed load $w$ can be expressed as

$$\alpha = \frac{wl^3}{24EI} = \frac{16}{5l}\left(\frac{5wl^4}{384EI}\right) = \frac{16x}{5l} \tag{E.3}$$

The equation of motion of the equivalent mass (Fig. 3.23) can be written as

$$m_{eq}\ddot{x} + kx = F \tag{E.4}$$

When Eqs. (E.1), (E.2), and (E.3) are used, Eq. (E.4) takes the form

$$\frac{m}{2}\ddot{x} + \left(\frac{384EI}{5l^3} - 2\rho A v^2 \frac{16}{5l}\right)x = 0 \tag{E.5}$$

This shows that the system is unstable when the coefficient of $x$ in Eq. (E.5) becomes negative, that is, when

$$v > \sqrt{\frac{12EI}{\rho A l^2}} \tag{E.6}$$

---

## 3.12  VIBRATION-MEASURING INSTRUMENTS

The purpose of a vibration-measuring instrument is to indicate an output or response which reproduces an input such as the displacement amplitude, velocity, or acceleration of vibration as closely as possible [3.14–3.18]. Figure 3.24 shows the

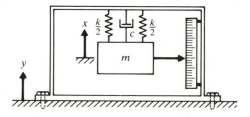

**Figure 3.24**  A typical vibration-measuring instrument.

essential elements of a basic vibration-measuring instrument. It consists of a mass $m$ supported by means of a system of springs of total spring constant $k$ inside a cage, which is to be fastened to the vibrating body. A dashpot of damping constant $c$ is connected in parallel with the springs. With this arrangement, the upper ends of the springs and the dashpot will have the same motion as the cage (which is to be measured, $y$) and their vibration excites the suspended mass into motion. Then the displacement of the mass relative to the cage, $z = x - y$, where $x$ denotes the vertical displacement of the suspended mass, can be measured if we attach a pointer to the mass and a scale to the cage, as shown in Fig. 3.24.

The equation of motion of the mass $m$ can be written as

$$m\ddot{x} + c(\dot{x} - \dot{y}) + k(x - y) = 0 \tag{3.115}$$

By defining the relative displacement $z$ as

$$z = x - y \tag{3.116}$$

Eq. (3.115) can be written as

$$m\ddot{z} + c\dot{z} + kz = -m\ddot{y} \tag{3.117}$$

If we assume the motion $y$ to be harmonic,

$$y(t) = Y \sin \omega t \tag{3.118}$$

Eq. (3.117) becomes

$$m\ddot{z} + c\dot{z} + kz = m\omega^2 Y \sin \omega t \tag{3.119}$$

This equation is identical to Eq. (3.70); hence the steady-state solution is given by

$$z(t) = Z \sin(\omega t - \phi) \tag{3.120}$$

where $Z$ and $\phi$ are given by (see Eqs. (3.72) and (3.64)):

$$Z = \frac{Y\omega^2}{\left[(k - m\omega^2)^2 + c^2\omega^2\right]^{1/2}} = \frac{r^2 Y}{\left[(1 - r^2)^2 + (2\zeta r)^2\right]^{1/2}} \tag{3.121}$$

$$\phi = \tan^{-1}\left(\frac{c\omega}{k - m\omega^2}\right) = \tan^{-1}\left(\frac{2\zeta r}{1 - r^2}\right) \tag{3.122}$$

$$r = \frac{\omega}{\omega_n} \tag{3.123}$$

and

$$\zeta = \frac{c}{2m\omega_n} \tag{3.124}$$

The variations of $Z$ and $\phi$ with respect to $r$ are shown in Figs. 3.25 and 3.26. As will be seen later, the type of instrument is determined by the useful range of the frequencies, indicated in Fig. 3.25.

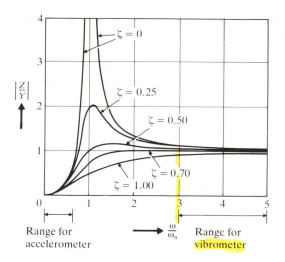

Range for accelerometer

Range for vibrometer

**Figure 3.25** Response of a vibration-measuring instrument.

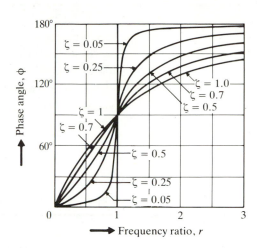

**Figure 3.26**

### 3.12.1 Vibrometer

A vibrometer or a seismometer is an instrument that measures the displacement of a vibrating body. We notice from Eq. (3.121) that if

$$\frac{r^2}{\left[(1-r^2)^2+(2\zeta r)^2\right]^{1/2}} \simeq 1 \tag{3.125}$$

then

$$z(t) = Y\sin(\omega t - \phi) \tag{3.126}$$

A comparison of Eq. (3.126) with $y(t) = Y\sin\omega t$ shows that $z(t)$ gives directly the motion $y(t)$ except for the phase lag $\phi$. This means that the recorded displacement $z(t)$ lags behind the displacement being measured $y(t)$ by time $t' = \phi/\omega$. If the base displacement $y(t)$ consists of a single harmonic component, the time lag is unimportant, and we get a proper record of $y(t)$.

In order to satisfy the requirement of Eq. (3.125), the natural frequency of the vibrometer must be low compared to that of the vibration to be measured (i.e., $r$ must be large). Since the natural frequency $\omega_n = \sqrt{k/m}$ must be low, the mass must be large and the springs need to have a low stiffness. This results in a bulky instrument, which is not desirable in many applications. Damping need not be present to satisfy Eq. (3.125), but its presence will improve the range of application of the instrument. It can be seen from Fig. 3.25 that for $\zeta = 0.7$, $Z \simeq Y$ if $r$ is greater than 3. Thus the instrument shown in Fig. 3.24 can be used as a vibrometer by making the value of $r$ very large. If the actual value of $r = \omega/\omega_n$ is not very large, $Z$ will not be equal to $Y$. The true value of $Y$ can be computed by using Eq. (3.121), as indicated in the following example.

**EXAMPLE 3.9**

A vibrometer having a natural frequency of 4 rad/sec and $\zeta = 0.2$ is attached to a machine part that performs a harmonic motion. If the difference between the maximum and the minimum recorded values is 8 mm, find the amplitude of motion of the vibrating part when its frequency is 40 rad/sec.

*Solution.* The amplitude of the recorded motion $Z$ is 4 mm. For $\zeta = 0.2$, $\omega = 40.0$ rad/sec, and $\omega_n = 4$ rad/sec, $r = 10.0$, and Eq. (3.121) gives

$$ Z = \frac{Y(10)^2}{\left[(1-10^2)^2 + \{2(0.2)(10)\}^2\right]^{1/2}} = 1.0093Y $$

Thus the amplitude of vibration of the machine part is $Y = Z/1.0093 = 3.9631$ mm.

### 3.12.2    Accelerometer

An accelerometer is an instrument that measures the acceleration of a vibrating body. Accelerometers are widely used for vibration measurements [3.6] and also to record earthquakes. From the accelerometer record, the velocity and displacements are obtained by integration. Equation (3.120) gives

$$ -z(t)\omega_n^2 = \frac{1}{\left[(1-r^2)^2 + (2\zeta r)^2\right]^{1/2}}\left\{-Y\omega^2 \sin(\omega t - \phi)\right\} \tag{3.127} $$

If

$$ \frac{1}{\left[(1-r^2)^2 + (2\zeta r)^2\right]^{1/2}} \simeq 1 \tag{3.128} $$

Eq. (3.127) becomes

$$ -z(t)\omega_n^2 \simeq -Y\omega^2 \sin(\omega t - \phi) \tag{3.129} $$

By comparing Eq. (3.129) with $\ddot{y}(t) = -Y\omega^2 \sin \omega t$, we find that the term $z(t)\omega_n^2$ gives the acceleration of the base $\ddot{y}$, except for the phase lag $\phi$. The instrument can be calibrated so that the record gives directly the value of $\ddot{y} = -z(t)\omega_n^2$. The time by which the record lags the acceleration is given by $t' = \phi/\omega$. If $\ddot{y}$ consists of a single harmonic component, the time lag will not be of importance.

The value of the expression on the left-hand side of Eq. (3.128) is shown plotted in Fig. 3.27. It can be seen that the requirement of Eq. (3.128) is satisfied by $\zeta = 0.7$ for $0 \le r \le 0.6$. Since $r$ is small, the natural frequency of the instrument is high compared to the frequency of vibration to be measured. From the relation $\omega_n = \sqrt{k/m}$, we find that the mass is small and the spring is short (for a large value of $k$), so the instrument will be small in size. Due to their small size and high sensitivity, accelerometers are preferred in vibration measurements. If Eq. (3.128) is not satisfied, the factor

$$ \frac{1}{\left[(1-r^2)^2 + (2\zeta r)^2\right]^{1/2}} $$

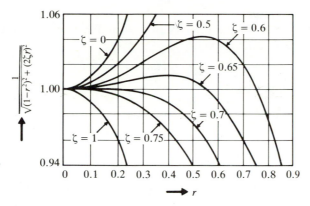

**Figure 3.27**

can be used to find the correct value of the acceleration measured as illustrated in the following example.

## EXAMPLE 3.10

Find the spring stiffness and the damping coefficient of an accelerometer that can measure vibrations in the range of 0 to 40 Hz with a maximum error of 0.5%. Assume the suspended mass $m$ to be 0.05 kg.

***Solution.*** From Eq. (3.121), the accuracy in measuring acceleration can be obtained:

$$\frac{1}{\left[(1-r^2)^2+(2\zeta r)^2\right]^{1/2}} = 0.995$$

or

$$r^4 - 2r^2(1-2\zeta^2) - 0.0100755 = 0 \tag{E.1}$$

By arbitrarily selecting a value of $\zeta = 0.7$, Eq. (E.1) becomes

$$r^4 - 0.04r^2 - 0.0100755 = 0$$

which gives the positive value of $r^2$ as 0.122350. Hence $r = 0.349785$. The maximum value of the forcing frequency is $\omega = 2\pi f = 2\pi(40) = 251.328$ rad/sec, so

$$\omega_n = \frac{\omega}{r} = \frac{251.328}{0.349785} = 718.520350 \text{ rad/sec}$$

The spring stiffness is given by

$$k = m\omega_n^2 = (0.05)(718.520350)^2 = 25813.57400 \text{ N/m}$$

and the damping constant by

$$c = 2m\omega_n\zeta = 2(0.05)(718.520350)(0.7) = 50.29642 \text{ N-s/m}.$$

### 3.12.3    Velometer

A velometer measures the velocity of a vibrating body. Equation (3.118) gives the velocity of the vibrating body:

$$\dot{y}(t) = \omega Y \cos \omega t \qquad (3.130)$$

and Eq. (3.120) gives

$$\dot{z}(t) = \frac{r^2 \omega Y}{\left[(1 - r^2)^2 + (2\zeta r)^2\right]^{1/2}} \cos(\omega t - \phi) \qquad (3.131)$$

If

$$\frac{r^2}{\left[(1 - r^2)^2 + (2\zeta r)^2\right]^{1/2}} \simeq 1 \qquad (3.132)$$

then

$$\dot{z}(t) \simeq \omega Y \cos(\omega t - \phi) \qquad (3.133)$$

A comparison of Eqs. (3.130) and (3.133) shows that, except for the phase difference $\phi$, $\dot{z}(t)$ gives directly $\dot{y}(t)$, provided that Eq. (3.132) holds true. In order to satisfy Eq. (3.132), $r$ must be very large. In case Eq. (3.132) is not satisfied, then the velocity of the vibrating body can be computed using Eq. (3.121).

---

### EXAMPLE 3.11

A velometer has a damping ratio of $\zeta = 0.6$ and a natural frequency of 20 Hz. Find the lowest frequency of vibration of a vibrating body for which the velocity may be measured with an error of 1%.

***Solution.***  Since the maximum permissible error is given as 1%, we have

$$\frac{r^2}{\left[(1 - r^2)^2 + (2\zeta r)^2\right]^{1/2}} = 1.01$$

or

$$0.019704 r^4 - 2r^2(1 - 2\zeta^2) + 1 = 0 \qquad (E.1)$$

With $\zeta = 0.6$, Eq. (E.1) becomes

$$0.019704 r^4 - 0.56 r^2 + 1 = 0$$

which has the roots

$$r^2 = 1.914713, \, 26.505916 \qquad \text{or} \qquad r = 1.383732, \, 5.148390$$

Since the large value of $r$ gives the minimum frequency for a specified error, we obtain

$$f = r f_n = 5.148390 \times 20 = 102.9678 \text{ Hz}$$

---

### 3.12.4    Phase Distortion

As shown by Eq. (3.120), all vibration-measuring instruments exhibit phase lag. Thus the response or record of the instrument lags behind the motion it measures.

The time lag is given by the phase angle divided by the frequency $\omega$. The time lag is unimportant if the motion consists of a single harmonic wave, but if the motion has a complex wave form, the record will be different from the motion being measured. This is due to the fact that a complex wave contains several harmonic components of different frequencies, and each component of the output wave of the instrument lags by a different amount depending on the value of its frequency. The distortion in the wave form of the recorded signal is called the *phase distortion* or *phase-shift error*.

Let the complex wave being measured be given by the series

$$y(t) = a_1\cos \omega t + a_2\cos 2\omega t + \cdots + b_1\sin \omega t + b_2\sin 2\omega t + \cdots \quad (3.134)$$

If the displacement is measured using a vibrometer, its response to each component of the series is given by an equation similar to Eq. (3.120) so that the output of the vibrometer becomes

$$z(t) = a_1\cos(\omega t - \phi_1) + a_2\cos(2\omega t - \phi_2) + \cdots$$
$$+ b_1\sin(\omega t - \phi_1) + b_2\sin(2\omega t - \phi_2) + \cdots \quad (3.135)$$

where

$$\tan\phi_j = \frac{2\zeta\left(j\dfrac{\omega}{\omega_n}\right)}{1-\left(j\dfrac{\omega}{\omega_n}\right)^2} ; \quad j = 1, 2, \ldots \quad (3.136)$$

Since $\omega/\omega_n$ is large for this instrument, we can find from Fig. 3.26 that $\phi_j \simeq \pi$, $j = 1, 2, \ldots$ and Eq. (3.135) becomes

$$z(t) \simeq -(a_1\cos \omega t + a_2\cos 2\omega t + \cdots + b_1\sin \omega t + b_2\sin 2\omega t + \cdots)$$
$$\simeq -y(t) \quad (3.137)$$

Thus the output record will be simply opposite to the motion being measured. This is unimportant and can easily be corrected. It can be noted from Fig. 3.26 that damping is undesirable in the instrument so far as phase distortion is concerned.

By using a similar reasoning, we can show, in the case of a velometer, that

$$\dot{z}(t) \simeq -\dot{y}(t) \quad (3.138)$$

for the case of a complex wave. For an accelerometer, the phase distortion is of a different nature; the phase lag depends on the frequency of the component being measured. The acceleration curve to be measured can be expressed, using Eq. (3.134), as

$$\ddot{y}(t) = -a_1\omega^2\cos \omega t - a_2(2\omega)^2\cos 2\omega t - \cdots$$
$$- b_1\omega^2\sin \omega t - b_2(2\omega)^2\sin 2\omega t - \cdots \quad (3.139)$$

The response of the instrument to each component can be found as in Eq. (3.135), and so

$$-\omega^2\ddot{z}(t) = -a_1\omega^2\cos(\omega t - \phi_1) - a_2(2\omega)^2\cos(2\omega t - \phi_2) - \cdots$$
$$- b_1\omega^2\sin(\omega t - \phi_1) - a_2(2\omega)^2\sin(2\omega t - \phi_2) - \cdots \quad (3.140)$$

where the phase lags $\phi_j$ are different for different components of the series in Eq.

(3.140). Since the phase lag $\phi$ varies almost linearly from $0°$ at $r = 0$ to $90°$ at $r = 1$ for $\zeta = 0.7$ (see Fig. 3.26), we can express $\phi$ as

$$\phi \simeq \alpha r = \alpha \frac{\omega}{\omega_n} = \beta \omega \tag{3.141}$$

where $\alpha$ and $\beta$ are constants. The time lag is given by

$$t' = \frac{\phi}{\omega} = \frac{\beta \omega}{\omega} = \beta \tag{3.142}$$

This shows that the time lag of the accelerometer is independent of the frequency for any component, provided that the frequency lies in the range $0 \leq r \leq 1$. Since each component of the output is offset by the same amount of time $\beta$ with respect to the acceleration component being measured, we have, from Eq. (3.140),

$$
\begin{aligned}
-\omega^2 \ddot{z}(t) = {} & -a_1 \omega^2 \cos(\omega t - \omega \beta) - a_2 (2\omega)^2 \cos(2\omega t - 2\omega \beta) \\
& - \cdots - b_1 \omega^2 \sin(\omega t - \omega \beta) - b_2 (2\omega)^2 \sin(2\omega t - 2\omega \beta) \\
& - \cdots \\
= {} & -a_1 \omega^2 \cos \omega \tau - a_2 (2\omega)^2 \cos 2\omega \tau - \cdots \\
& - b_1 \omega^2 \sin \omega \tau - b_2 (2\omega)^2 \sin 2\omega \tau - \cdots
\end{aligned}
\tag{3.143}
$$

where $\tau = t - \beta$. Although Eq. (3.143) assumes that the highest frequency involved, $n\omega$, is less than $\omega_n$, no significant phase distortion occurs in practice, even when some of the higher order frequencies are larger than $\omega_n$. The reason is that, generally, only the first few components are important to approximate even a complex wave form; the amplitudes of the higher modes are small and contribute very little to the form of the curve. Thus the output record of the instrument is a true representation of the acceleration being measured.

### 3.12.5   Common Types of Instruments

The output of the vibration-measuring device shown in Fig. 3.24 is the relative mechanical motion of the mass. For high-speed operations and convenience, the motion is often converted into an electrical signal. Such a conversion device is called a *transducer*. Some of the schemes used for this conversion are shown in Fig. 3.28. Figure 3.28(a) shows an arrangement for converting the mechanical output to an electrical signal. Here a permanent magnet is fixed to the base of the instrument, and coil is wound around the mass. As the mass (and hence the coil) vibrates, magnetic flux lines are cut and electric voltage is generated in the coil circuit. The voltage generated is proportional to the velocity of the mass relative to the base, and so the instrument gives the relative velocity as output.

Figure 3.28(b) shows the use of electrical resistance strain gauges. The strain gauge experiences the same strain as induced in the element onto which it is cemented. The strain induced by the vibrating mass causes the length and cross section of the wire to change, so the electrical resistance of the gauge changes. By proper calibration of the circuit, the strain induced by vibration and thus the displacement amplitude can be obtained directly.

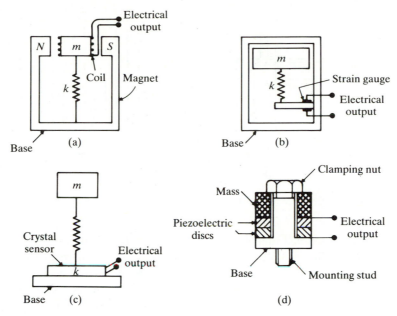

**Figure 3.28**

Figure 3.28(c) shows an accelerometer based on the principle of piezoelectric effect [3.5]. The principle involved here is that when certain crystals such as quartz are strained, they become polarized so that a difference in electric potential can be measured. The measured voltage is proportional to the pressure or force acting on the element and, with proper calibration, gives the vibration record directly. The crystal serves both as a spring and a sensor, as shown in Fig. 3.28(d).

In all the accelerometers, the mass may be enclosed and filled with a viscous fluid to provide the desired amount of damping.

### 3.12.6 Frequency-Measuring Instruments

Most frequency-measuring instruments are of the mechanical type and are based on the principle of resonance. Two kinds of instruments are discussed in the following paragraphs: the Fullarton tachometer and the Frahm tachometer.

**Single-reed Instrument or Fullarton Tachometer.** This instrument consists of a variable-length cantilever strip with a mass attached at one of its ends. The other end of the strip is clamped, and its free length can be changed by means of a screw mechanism (see Fig. 3.29(a)). Since each length of the strip corresponds to a different natural frequency, the reed is marked along its length in terms of its natural frequency. In practice, the clamped end of the strip is pressed against the vibrating body, and the screw mechanism is manipulated to alter its free length until the free end shows the largest amplitude of vibration. At that instant, the excitation frequency is equal to the natural frequency of the centilever; it can be read directly from the strip.

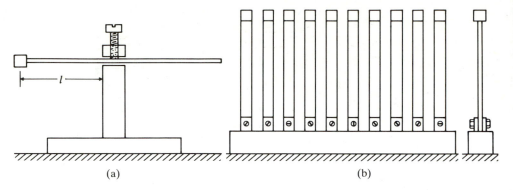

**Figure 3.29**   Frequency-measuring instruments.

**Multireed-Instrument or Frahm Tachometer.** This instrument consists of a number of cantilevered reeds carrying small masses at their free ends (see Fig. 3.29(b)). Each reed has a different natural frequency and is marked accordingly. Using a number of reeds makes it possible to cover a wide frequency range. When the instrument is mounted on a vibrating body, the reed whose natural frequency is nearest the unknown frequency of the body vibrates with the largest amplitude. The frequency of the vibrating body can be found from the known frequency of the vibrating reed.

## 3.13  COMPUTER PROGRAM

A FORTRAN computer program, in the form of subroutine HARESP, is given for finding the steady-state response of a viscously damped single degree of freedom system under the harmonic force $F_0 \cos \omega t$ or $F_0 \sin \omega t$. The arguments of the subroutine are as follows:

| | | |
|---|---|---|
| XM | = | Mass. Input data. |
| XC | = | Damping constant. Input data. |
| XK | = | Spring stiffness. Input data. |
| F0 | = | Amplitude of the force. Input data. |
| OM | = | Forcing frequency. Input data. |
| IC | = | Integer for identifying the nature of the force. IC = 1 for cosine variation and IC = 0 for sine variation. Input data. |
| N | = | Number of time steps in a cycle at which the response is to be printed. Input data. |
| X, XD, XDD | = | Arrays of dimension $N$ each, which contain the computed values of displacement, velocity, and acceleration. X(I) = $x(t_i)$, XD(I) = $\dot{x}(t_i)$, XDD(I) = $\ddot{x}(t_i)$. Output. |
| XAMP | = | Amplitude of the response (Eq. (3.46)). Output. |
| XPHI | = | Phase angle of the response (Eq. (3.47)). Output. |

To illustrate the use of subroutine HARESP, an example is considered with $m = 5$ kg, $c = 20$ N-s/m, $k = 500$ N/s, $F_0 = 250$ N, $\omega = 40$ rad/s, $N = 20$, and $F(t) = F_0 \sin \omega t$. The main program which calls HARESP, subroutine HARESP, and the output of the program are given below.

```
C ====================================================================
C
C PROGRAM 3
C MAIN PROGRAM WHICH CALLS HARESP
C
C ====================================================================
C FOLLOWING 2 LINES CONTAIN PROBLEM-DEPENDENT DATA
      DIMENSION X(20),XD(20),XDD(20)
      DATA XM,XC,XK,F0,OM,N,IC/5.0,20.0,500.0,250.0,40.0,20,0/
C END OF PROBLEM-DEPENDENT DATA
      CALL HARESP (XM,XC,XK,F0,OM,IC,N,X,XD,XDD,XAMP,XPHI)
      PRINT 100
  100 FORMAT (//,40H STEADY STATE RESPONSE OF AN UNDERDAMPED,/,
     2 53H SINGLE DEGREE OF FREEDOM SYSTEM UNDER HARMONIC FORCE)
      PRINT 200, XM,XC,XK,F0,OM,IC,N
  200 FORMAT (//,12H GIVEN DATA:,/,5H XM =,E15.8,/,5H XC =,E15.8,/,
     2 5H XK =,E15.8,/,5H F0 =,E15.8,/,5H OM =,E15.8,/,5H IC =,I2,/,
     3 5H N  =,I2)
      PRINT 300
  300 FORMAT (//,10H RESPONSE:,//,5H    I ,3X,5H X(I),12X,6H XD(I),
     2 11X,7H XDD(I),/)
      DO 400 I=1,N
  400 PRINT 500,I,X(I),XD(I),XDD(I)
  500 FORMAT (I4,2X,E15.8,2X,E15.8,2X,E15.8)
      STOP
      END
C ====================================================================
C
C SUBROUTINE HARESP
C
C ====================================================================
      SUBROUTINE HARESP (XM,XC,XK,F0,OM,IC,N,X,XD,XDD,XAMP,XPHI)
      DIMENSION X(N),XD(N),XDD(N)
      OMN=SQRT(XK/XM)
      XAI=XC/(2.0*XM*OMN)
      DST=F0/XK
      R=OM/OMN
      XAMP=DST/SQRT((1.0-R**2)**2+(2.0*XAI*R)**2)
      XPHI=ATAN(2.0*XAI*R/(1.0-R**2))
      DELT=2.0*3.1416/(OM*REAL(N))
      IF (IC .EQ. 0) GO TO 20
      TIME=0.0
      DO 10 I=1,N
      TIME=TIME+DELT
      X(I)=XAMP*COS(OM*TIME-XPHI)
      XD(I)=-XAMP*OM*SIN(OM*TIME-XPHI)
   10 XDD(I)=-XAMP*(OM**2)*COS(OM*TIME-XPHI)
      RETURN
   20 TIME=0.0
      DO 30 I=1,N
```

```
        TIME=TIME+DELT
        X(I)=XAMP*SIN(OM*TIME-XPHI)
        XD(I)=XAMP*OM*COS(OM*TIME-XPHI)
   30   XDD(I)=-XAMP*(OM**2)*SIN(OM*TIME-XPHI)

RETURN
END
```

```
STEADY STATE RESPONSE OF AN UNDERDAMPED
SINGLE DEGREE OF FREEDOM SYSTEM UNDER HARMONIC FORCE


GIVEN DATA:
XM = 0.50000000E+01
XC = 0.20000000E+02
XK = 0.50000000E+03
F0 = 0.25000000E+03
OM = 0.40000000E+02
IC = 0
N  =20


RESPONSE:

    I      X(I)            XD(I)           XDD(I)

    1    0.13528203E-01   0.12103548E+01  -0.21645124E+02
    2    0.22216609E-01   0.98389733E+00  -0.35546574E+02
    3    0.28730286E-01   0.66112888E+00  -0.45968460E+02
    4    0.32431632E-01   0.27364409E+00  -0.51890614E+02
    5    0.32958329E-01  -0.14062698E+00  -0.52733330E+02
    6    0.30258821E-01  -0.54113245E+00  -0.48414116E+02
    7    0.24597352E-01  -0.88866800E+00  -0.39355762E+02
    8    0.16528117E-01  -0.11492138E+01  -0.26444986E+02
    9    0.68409811E-02  -0.12972662E+01  -0.10945570E+02
   10   -0.35157942E-02  -0.13183327E+01   0.56252708E+01
   11   -0.13528424E-01  -0.12103508E+01   0.21645479E+02
   12   -0.22216786E-01  -0.98389095E+00   0.35546856E+02
   13   -0.28730409E-01  -0.66112041E+00   0.45968655E+02
   14   -0.32431684E-01  -0.27363408E+00   0.51890697E+02
   15   -0.32958303E-01   0.14063700E+00   0.52733284E+02
   16   -0.30258721E-01   0.54114151E+00   0.48413952E+02
   17   -0.24597190E-01   0.88867509E+00   0.39355507E+02
   18   -0.16527895E-01   0.11492189E+01   0.26444633E+02
   19   -0.68407385E-02   0.12972684E+01   0.10945182E+02
   20    0.35160405E-02   0.13183316E+01  -0.56256652E+01
```

## REFERENCES

**3.1.** G. B. Thomas and R. L. Finney, *Calculus and Analytic Geometry* (6th Ed.), Addison-Wesley, Reading, Mass., 1984.

**3.2.** J. W. Nilsson, *Electric Circuits*, Addison-Wesley, Reading, Mass., 1983.

**3.3.** J. P. Den Hartog, "Forced vibrations with combined Coulomb and viscous friction," *Journal of Applied Mechanics* (Transactions of ASME), Vol. 53, 1931, pp. APM 107–115.

**3.4.** R. D. Belvins, *Flow-Induced Vibration*, Van Nostrand Reinhold, New York, 1977.

**3.5.** M. C. Kokmeci, "Dynamic applications of piezoelectric crystals, Part II: Theoretical studies," *Shock and Vibration Digest*, Vol. 15, April 1983, pp. 15–26; and "Part III: Experimental studies," Vol. 15, May 1983, pp. 11–22.

**3.6.** D. E. Weiss, "Design and application of accelerometers," *Proceedings of the Society for Experimental Stress Analysis*, Vol. IV, No. 2, 1947, p. 89.

**3.7.** N. O. Myklestad, "The concept of complex damping," *Journal of Applied Mechanics*, Vol. 19, 1952, pp. 284–286.

**3.8.** R. Plunkett (Ed.), *Mechanical Impedance Methods for Mechanical Vibrations*, American Society of Mechanical Engineers, New York, 1958.

**3.9.** M. W. Wambsganss, "Understanding flow-induced vibrations, Part I: Basic concepts; fluid forcing functions," *Sound and Vibration*, Vol. 10, November 1976, pp. 18–23.

**3.10.** B. Westermo and F. Udwadia, "Periodic response of a sliding oscillator system to harmonic excitation," *Earthquake Engineering and Structural Dynamics*, Vol. 11, No. 1, 1983, pp. 135–146.

**3.11.** M. S. Hundal, "Response of a base excited system with Coulomb viscous friction," *Journal of Sound and Vibration*, Vol. 64, 1979, pp. 371–378.

**3.12.** J. P. Bandstra, "Comparison of equivalent viscous damping and nonlinear damping in discrete and continuous vibrating systems," *Journal of Vibration, Acoustics, Stress, and Reliability in Design*, Vol. 105, 1983, pp. 382–392.

**3.13.** M. P. Paidoussis and N. T. Issid, "Dynamic stability of pipes carrying fluid," *Journal of Sound and Vibration*, Vol. 33, 1974, pp. 267–294.

**3.14.** G. M. Hieber et al., "Understanding and measuring the shock response spectrum, Part 1," *Sound and Vibration*, Vol. 8, March 1974, pp. 42–49.

**3.15.** M. P. Blake and W. S. Mitchell (Eds.), *Vibration and Acoustic Measurement Handbook*, Haydon Book Co., Rockelle Park, N.J., 1972.

**3.16.** R. Plunkett, "Shock and vibration instrumentation," *Shock and Vibration Digest*, Vol. 14, September 1982, pp. 3–5.

**3.17.** G. F. Lang, "Understanding vibration measurements," *Sound and Vibration*, Vol. 10, March 1976, pp. 26–37.

**3.18.** A. J. Pretlove, "Some current methods in vibration measurement," in B. O. Skipp (Ed.), *Vibration in Civil Engineering*, pp. 95–110, Butterworth, London, 1966.

**3.19.** R. W. Fitzgerald, *Mechanics of Materials* (2nd Ed.), Addison-Wesley, Reading, Mass., 1982.

## REVIEW QUESTIONS

**3.1.** How are the amplitude, frequency, and phase of a steady-state vibration related to those of the applied harmonic force?

**3.2.** Explain why a constant force on the vibrating mass has no effect on the steady-state vibration.

**3.3.** Define the term *magnification factor*. How is the magnification factor related to the frequency ratio?

**3.4.** What will be the frequency of the applied force with respect to the natural frequency of the system if the magnification factor is less than unity?

**3.5.** What are the amplitude and the phase angle of the response of a viscously damped system in the neighborhood of resonance?

**3.6.** Is the phase angle corresponding to the peak amplitude of a viscously damped system ever larger than 90°?

**3.7.** Why is damping considered only in the neighborhood of resonance in most cases?

**3.8.** Show the various terms in the forced equation of motion of a viscously damped system in a vector diagram.

**3.9.** What happens to the response of an undamped system at resonance?

**3.10.** Define these terms: beating, quality factor, transmissibility, complex stiffness, quadratic damping.

**3.11.** Give a physical explanation of why the magnification factor is nearly equal to 1 for small values of $r$ and is small for large values of $r$.

**3.12.** Will the force transmitted to the base of a spring-mounted machine decrease with the addition of damping?

**3.13.** How does the force transmitted to the base change as the speed of the machine increases?

**3.14.** If a vehicle vibrates badly while moving on a uniformly bumpy road, will a change in the speed improve the condition?

**3.15.** Is it possible to find the maximum amplitude of a damped forced vibration for any value of $r$ by equating the energy dissipated by damping to the work done by the external force?

**3.16.** What assumptions are made about the motion of a forced vibration with nonviscous damping in finding the amplitude?

**3.17.** Is it possible to find the approximate value of the amplitude of a damped forced vibration without considering damping at all? If so, under what circumstances?

**3.18.** Is dry friction effective in limiting the resonant amplitude?

**3.19.** How do you find the response of a viscously damped system under rotating unbalance?

**3.20.** What is the frequency of the response of a viscously damped system when the external force is $F_0 \sin \omega t$? Is this response harmonic?

**3.21.** What is the difference between the peak amplitude and the resonant amplitude?

**3.22.** Why is viscous damping used in most cases rather than other types of damping?

**3.23.** What is self-excited vibration?

**3.24.** Describe the basic principle used in a vibration-measuring instrument.

**3.25.** Can we use a vibration-measuring instrument to measure the amplitude of a vibration whose frequency is lower than that of the instrument?

**3.26.** Describe the working principle of the following instruments: vibrometer, velometer, accelerometer, tachometer.

**3.27.** What is phase-shift error?

## PROBLEMS

The problem assignments are organized as follows:

| Problems | Section covered | Topic covered |
|---|---|---|
| 3.1–3.6 | 3.3 | Undamped systems |
| 3.7–3.16 | 3.4, 3.5 | Damped systems |
| 3.17–3.24 | 3.6 | Base excitation |
| 3.25–3.32 | 3.7 | Rotating unbalance |
| 3.33–3.37 | 3.8 | Response under Coulomb damping |
| 3.38–3.39 | 3.9 | Response under hysteretic damping |
| 3.40–3.46 | 3.10 | Response under other types of damping |
| 3.47–3.61 | 3.12 | Vibration measuring instruments |
| 3.62–3.65 | 3.13 | Computer program |

**3.1.** An undamped system consists of a mass weighing 50 N and a spring of stiffness 4000 N/m. It is acted on by a harmonic force of amplitude 60 N and frequency 6 Hz. Find (i) the displacement of the spring due to the weight of the mass, (ii) the static displacement of the spring due to the maximum applied force, and (iii) the amplitude of forced motion of the mass.

**3.2.** A spring-mass system is harmonically forced near resonance, resulting in a beating phenomenon. The forcing frequency is 39.8 Hz and the natural frequency is 40.0 Hz. Find the beat period of the motion.

**3.3.** An undamped system consists of a body weighing 100 N supported by an elastic member having a modulus of 2000 N/m. The body is driven at resonance by a harmonic force $F(t) = 25 \cos \omega t$ Newtons. Determine the amplitude of the forced motion at the end of (i) $\frac{1}{4}$ cycle, (ii) $2\frac{1}{2}$ cycles, and (iii) $5\frac{3}{4}$ cycles.

**3.4.** A mass $m$ is suspended from a spring having a modulus of 4000 N/m and is driven by a harmonic force having an amplitude of 50 N and a frequency of 4 Hz. The amplitude of the forced motion of the mass is observed to be 20 mm. Find the value of $m$.

**3.5.** An undamped single degree of freedom system consists of a mass of 10 kg and a spring of stiffness 4000 N/m. A harmonic force of amplitude 250 N and frequency $\omega$ acts on the mass and causes it to vibrate with a forced amplitude of 100 mm. Find the value of $\omega$.

**3.6.** In Fig. 3.1(a), a periodic force $F(t) = F_0 \cos \omega t$ is applied directly to the spring at its mid-length. Find the steady-state response of the mass $m$. Assume $c = 0$.

**3.7.** A spring-mass-damper system is subjected to a harmonic force. At resonance the amplitude is measured as 15 mm, and at a frequency 0.85 times the resonant frequency the amplitude is measured as 12 mm. Find the damping ratio of the system.

**3.8.** The mass of a vibrating system weighs 20 N and is made to vibrate in a viscous medium. Determine the damping ratio and the damping coefficient when a harmonic exciting force of 30 N results in a resonant amplitude of 15 mm with a period of 0.2 s.

**3.9.** Show that, for a spring-mass-damper system, the peak amplitude occurs at the frequency ratio $r = \sqrt{1 - 2\zeta^2}$. Also show that the peak amplitude is equal to

$$\left( \frac{X}{\delta_{st}} \right)_{max} = \frac{1}{2\zeta\sqrt{1 - \zeta^2}}$$

**3.10.** For the system shown in Fig. 3.30, $x$ and $y$ represent, respectively, the absolute displacements of the mass $m$ and the end $Q$ of the spring. End $Q$ is moved according to the relation $y(t) = Y\cos \omega t$. (i) Derive the equation of motion of the mass $m$. (ii) Find the steady-state motion of the mass $m$. (iii) Determine the force transmitted to the support at $P$.

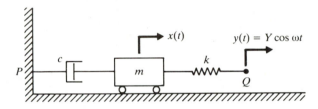

**Figure 3.30**

**3.11.** For the system shown in Fig. 3.31, $x$ denotes the absolute displacement of the mass and $y$ represents the absolute displacement of the end $Q$ of the dashpot. End $Q$ is moved according to the relation $y(t) = Y\cos \omega t$. (i) Write the equation of motion of the mass $m$. (ii) Determine the steady-state motion of the mass $m$. (iii) Find the force transmitted to the support at $P$.

**3.12.** In the system shown in Fig. 3.32, the force $F(t)$ applied to the mass moves it according to the relation $x(t) = X\cos \omega t$. (i) Write the differential equation for the dynamic condition of mass $m$. (ii) Find the differential equation and its solution for the motion of point $Q$. (iii) Determine the force transmitted to the support at $P$.

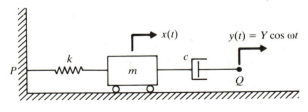

**Figure 3.31**

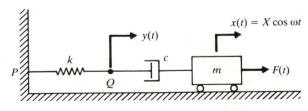

**Figure 3.32**

**3.13.** Show that, for small amounts of damping, the damping ratio $\zeta$ can be expressed as

$$\zeta = \frac{\omega_2 - \omega_1}{\omega_2 + \omega_1}$$

where $\omega_2$ and $\omega_1$ denote the frequencies on either side of resonance where the amplitude is $(1/\sqrt{2})$ of its maximum value.

**3.14.** For a vibrating system, $m = 10$ kg, $k = 2500$ N/m, and $c = 45$ N-s/m. A harmonic force of amplitude 180 N and frequency 3.5 Hz acts on the mass. If the initial displacement and velocity of the mass are 15 mm and 5 m/s, find the complete solution representing the motion of the mass.

**3.15.** Find the ratio of the maximum amplitude to the resonant amplitude for the steady-state motion of a viscously damped system under a harmonic force.

**3.16.** A torsional system is composed of a disc with a mass moment of inertia $J_0 = 6$ kg-m², a shaft having a torsional spring constant $k_t = 14000$ N-m/rad, and a torsional damper of damping constant $c_t = 210$ N-m-s/rad. A harmonic torque of amplitude 450 N-m produces a steady angular oscillation of amplitude 2°. (i) Find the frequency of the applied torque. (ii) Determine the maximum torque transmitted to the support.

**3.17.** Find the steady-state motion of the mass $m$ if the upper end of the spring is subjected to harmonic acceleration, as shown in Fig. 3.33.

**3.18.** Find the displacement of the mass $m$ shown in Fig. 3.33 when the ground (base) acceleration is given by $\ddot{x}_g = 100 \sin \omega t$ mm/s². Assume $m = 2$ kg, $k = 100$ N/m, $\omega = 25$ rad/sec, and $x_g(t = 0) = \dot{x}_g(t = 0) = x(t = 0) = \dot{x}(t = 0) = 0$.

**3.19.** A wheel rolls with a constant horizontal velocity of $v = 50$ m/s on a wavy surface, as shown in Fig. 3.34. Find the amplitude of the forced vertical vibration of the mass $m$ attached to the axle of the wheel by a spring. The static deflection of the spring due to the weight of mass $m$ is $\delta_{st} = 100$ mm and the wavy surface is defined by $y = 20 \sin \pi z / 10$ mm.

**3.20.** The amplitude of free vibration of the cantilever beam shown in Fig. 3.35 is found to decrease from 30 mm to 10 mm in 10 cycles. Find the maximum amplitude of vibration

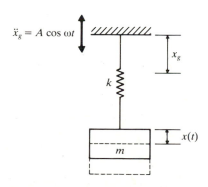

**Figure 3.33**  A spring-mass system subjected to base acceleration.

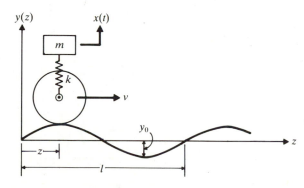

**Figure 3.34**

$y(t) = 2 \sin \omega t$

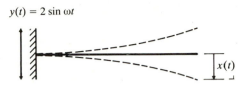

$x(t)$

**Figure 3.35**

of the beam at resonance if the base is subjected to a harmonic vibration of amplitude 2 mm.

**3.21.** If the upper end of the spring in Fig. 3.33 is subjected to a vertical harmonic motion with frequency $\omega = 200$ rad/sec and amplitude $X_g = 15$ mm, find the amplitude of the mass $m$. Assume the static deflection of the mass as $\delta_{st} = 40$ mm.

**3.22.** A car has a vertical natural frequency of 2 Hz. It is driven along a road whose elevation varies approximately sinusoidally. The distance from peak to trough is 0.2 m and the distance along the road between the peaks is 35 m. Assuming the car as a single degree of freedom system and damping ratio of the shock absorbers as 0.15, determine the amplitude of vibration of the car at a speed of 60 km/hour.

**3.23.** Figure 3.34 denotes a simplified diagram of a spring-supported vehicle traveling over an uneven road. Derive the equation for the amplitude of the mass $m$ as a function of the speed $v$, and then find the most unfavorable speed.

**3.24.** Derive Eq. (3.69).

**3.25.** Figure 3.36 shows a cam-and-follower mechanism. The cam is an eccentric circular disc of radius $r = 60$ mm and eccentricity $e = 20$ mm, which rotates at an angular velocity of $\omega = 250$ rad/sec. If the mass of the follower is $m = 2$ kg, determine the spring constant

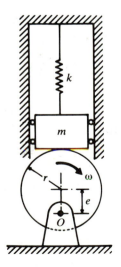

**Figure 3.36**

$k$ required to maintain contact between the cam and the follower at all times. Disregard the effect of gravity.

**3.26.** An electric motor is mounted on a steel table. The deflection of the table under the weight of the motor is observed to be 4 mm. The mass of the motor added to the effective mass of the table is 50 kg. The rotating parts of the motor have a mass of 10 kg and have an eccentricity of 12 mm. During free vibration, it is observed that a displacement of 40 mm is reduced to 2 mm in 1 sec. Find the amplitude of motion if the operating speed of the motor is 1750 rpm. Assume the damping to be viscous.

**3.27.** A counter-rotating eccentric weight exciter is used to produce forced oscillation of a single degree of freedom system, as indicated in Fig. 3.17. By varying the speed of rotation, a resonant amplitude of 12 mm was observed. When the speed of rotation is increased considerably beyond the resonant frequency, the amplitude is found to approach a fixed value of 1 mm. Determine the damping ratio of the system.

**3.28.** Figure 3.37 shows a spring-mass system in which one fifth of the mass is in the form of a rotor that rotates at an angular velocity $\omega$ with an eccentricity of 4 mm. The coefficient of dry friction between the mass and the ground is 0.12. If the total mass is 20 kg and the spring stiffness is 2000 N/m, determine the amplitude of motion at (i) $\omega = 5$ rad/sec, and (ii) $\omega = 150$ rad/sec.

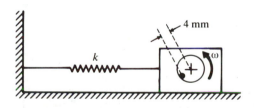

**Figure 3.37**

**3.29.** The sketch of a helicopter is shown in Fig. 3.38. The tail section of the helicopter can be idealized as a cantilever beam of length $l$ and bending rigidity $EI$. The combined mass of the gearbox and tail rotor blades is 50 kg and is represented as a concentrated mass at the end of the cantilever, as shown in Fig. 3.38(b). The natural frequency of the tail section is found to be 8 Hz. If one of the tail rotor blades suddenly falls off in flight, the remaining blades can be considered as an unbalanced rotating mass located at a distance of 0.15 m from the axis of rotation. If the blades rotate at 1200 rpm, determine the forced response of the tail section of the helicopter when one blade falls off. Assume the mass of one blade as 4 kg and the equivalent viscous damping ratio of the tail section as $\zeta = 0.15$. Note that when one blade falls off, the mass at the end of the beam (and thus its natural frequency) changes.

**3.30.** The decay of free vibration of a viscously damped vibrating system is shown in Fig. 3.39. Determine (i) the amplitude ratio of vibration, $(MX/me)$, at resonance, and (ii) the amplitude ratio at the operating speed of 1750 rpm if the system has a rotating unbalance.

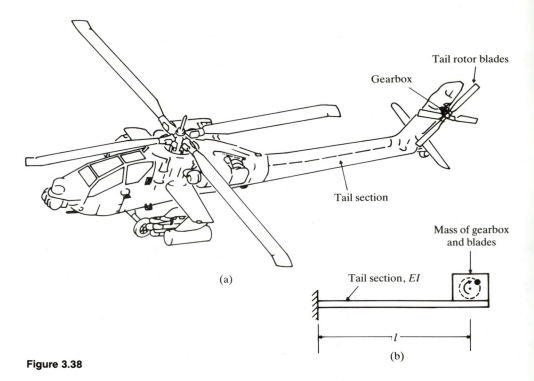

(a)

(b)

**Figure 3.38**

**3.31.** A machine of mass 380 kg is supported on springs with a static deflection of 45 mm. The damping of the system is negligible. If the machine has a rotating unbalance of 0.15 kg-m, find (i) the dynamic amplitude at 1750 rpm and (ii) the force transmitted to the ground at this speed.

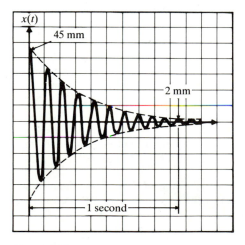

**Figure 3.39**

**3.32.** A simply supported beam carries an electric motor of mass 50 kg and speed 1200 rpm at its mid-span, as shown in Fig. 3.40. Due to unbalance in the machine, the rotor sets up a rotating force of magnitude $F_0 = 5000$ N. Find the amplitude of steady-state vibrations by disregarding the mass of the beam. Assume $l = 5$ m, $E = 2.07 \times 10^{11}$ Pa, and $I = 10^{-4}$ m$^4$.

Find X

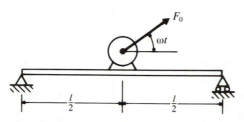

**Figure 3.40**

**3.33.** Derive Eq. (3.93).

**3.34.** A spring-mass system is subjected to Coulomb damping. When a harmonic force of amplitude 120 N and frequency 2.5173268 Hz is applied, the system is found to oscillate with a steady amplitude of 75 mm. Determine the average coefficient of dry friction if $m = 2$ kg and $k = 2100$ N/m.

**3.35.** If the spring in Problem 2.63 has a stiffness of 15 kN/m, determine the amplitude of steady-state motion when a harmonic force of amplitude 15 N and frequency 25 Hz is applied to the mass.

**3.36.** The connecting rod of Problem 2.64 is subjected to a harmonically varying torque of amplitude 0.25 N-m and frequency 5 Hz. Find the approximate amplitude of the steady-state motion. Assume $\mu = 0.08$, $d = 50$ mm, $m = 4$ kg, $J_0 = 0.2$ kg-m$^2$, and $\omega_n = 6$ Hz.

**3.37.** Derive the equation of motion of the mass $m$ shown in Fig. 3.41 when the pressure in the cylinder fluctuates sinusoidally. The spring of stiffness $k_1$ is initially under a tension of $T_0$, and the coefficient of friction between the mass and the surface is $\mu$.

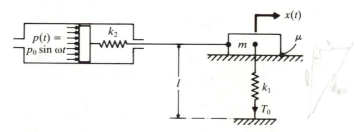

**Figure 3.41**

**3.38.** A framed structure shows hysteresis damping characteristics. A load of 5000 N on the structure causes a static displacement of 0.05 m, and a supported machine that is subjected to a harmonic force of amplitude 350 N produces a resonant amplitude of 0.11 m. Find (i) the hysteresis damping constant $\beta$, (ii) the energy dissipated per cycle at resonance, (iii) the steady-state amplitude at one-half the resonant frequency, and (iv) the steady-state amplitude at twice the resonant frequency.

**3.39.** The energy dissipated in hysteresis damping per cycle under harmonic excitation can be expressed in the general form

$$\Delta W = \pi b k X^{\gamma} \tag{E.1}$$

where $\gamma$ is an exponent ($\gamma = 2$ was considered in Secs. 2.8 and 3.9), and $b$ is a coefficient of dimension (meter)$^{2-\gamma}$. A spring-mass system having $k = 60$ kN/m vibrates under hysteresis damping. When excited harmonically at resonance, the steady-state amplitude is found to be 40 mm for an energy input of 3.8 N-m. When the resonant energy input is increased to 9.5 N-m, the amplitude is found to be 60 mm. Determine the values of $b$ and $\gamma$ in Eq. (E.1).

**3.40.** Derive Eq. (E.7) of Example 3.6.

**3.41.** Show that the loss factor is proportional to the frequency and is independent of the amplitude for viscous damping.

**3.42.** The energy loss per cycle is usually a function of the amplitude and the frequency. State the condition under which the logarithmic decrement is independent of the amplitude.

**3.43.** Show that the energy dissipated per cycle for viscous damping can be expressed as

$$\Delta W = \frac{\pi F_0^2}{k} \cdot \frac{2\zeta r}{\left(1 - r^2\right)^2 + \left(2\zeta r\right)^2}$$

**3.44.** A force $F(t) = 5 \sin \pi t$ Newtons acts on a mass whose displacement is given by $x(t) = 3 \sin(\pi t - \pi/3)$ meters. Find the work done (i) during the first 1 second, and (ii) during the first 4 seconds.

**3.45.** Find the equivalent viscous damping coefficient of a damper that offers a damping force of $F_d = c(\dot{x})^n$, where $c$ and $n$ are constants and $\dot{x}$ is the relative velocity across the damper. Also, find the amplitude of vibration.

**3.46.** Show that for a system with both viscous and Coulomb damping the approximate value of the steady-state amplitude is given by

$$X^2 \left[ k^2 (1 - r^2)^2 + c^2 \omega^2 \right] + X \frac{8 \mu N c \omega}{\pi} + \left( \frac{16 \mu^2 N^2}{\pi^2} - F_0^2 \right) = 0$$

**3.47.** Determine the maximum percent error that occurs in measurement in the range $3 \leq r < \infty$ for a vibrometer having a damping ratio of $\zeta = 0$. Also find the value of $r$ at which this occurs.

**3.48.** Solve Problem 3.47 for a damping ratio of $\zeta = 0.5$.

**3.49.** Solve Problem 3.47 for a damping ratio of $\zeta = 0.7$.

3.50. Determine the maximum percent error that occurs in measurement in the range $0 \leq r \leq 0.6$ for an accelerometer having a damping ratio of $\zeta = 0$. Also find the value of $r$ at which this occurs.

3.51. Solve Problem 3.50 for $\zeta = 0.5$.

3.52. Solve Problem 3.50 for $\zeta = 0.75$.

3.53. The vibration of a structure is given by

$$x(t) = 20 \sin 4\pi t + 10 \sin 8\pi t + 5 \sin 12\pi t$$

where $x$ is in mm and $t$ is in sec. Determine the record of this vibration that would be indicated by a vibrometer having a natural frequency of 0.5 Hz and a damping ratio of $\zeta = 0.3$. Discuss the accuracy of this record.

3.54. A machine is vibrating according to the relation

$$x(t) = 20 \sin 6.5\pi t + 5 \sin 19.5\pi t$$

where $x$ is in mm and $t$ is in sec. An accelerometer having a damping ratio of $\zeta = 0.6$ and a natural frequency $\omega_n = 45$ rad/sec is attached to the machine. Determine the output of the instrument assuming that it is calibrated to read the acceleration directly in mm/s$^2$.

3.55. Find the phase shift and time shift of each harmonic component of a complex wave. Discuss the phase distortion of the acceleration record.

3.56. An accelerometer, having an undamped frequency of 80 Hz and a damping constant of 8.0 N-s/m, is attached to a vibrating structure. When the structure vibrates with an acceleration amplitude of 7.5 m/s$^2$ and a frequency of 50 Hz, the instrument records the acceleration amplitude as 8.0 m/s$^2$. Determine the suspended mass and the spring constant of the accelerometer.

3.57. The static deflection of the mass of a vibrometer is 20 mm. The instrument records a relative amplitude of 0.02 mm when attached to a machine vibrating at a frequency of 100 Hz. Determine (i) the amplitude, (ii) the maximum velocity, and (iii) the maximum acceleration of the machine.

3.58. A vibration pickup has a natural frequency of 5 Hz and a damping ratio of $\zeta = 0.5$. Find the lowest frequency that can be measured with a 1% error.

3.59. Determine the suspended mass and the spring constant required for an accelerometer if the maximum error is to be limited to 3% for measurements in the frequency range of 0 to 75 Hz. Assume the damping constant as 50 N-s/m.

3.60. An accelerometer is to be designed to have an optimum range of frequency ratio for a maximum accelerometer error of 3%. Determine (i) the required damping ratio and (ii) the optimum range of the frequency ratio.

3.61. A vibrometer is used to measure the vibration of an engine whose operating speed range is from 500 to 2000 rpm. The vibration consists of two harmonics. The amplitude distortion must be less than 3%. Find the natural frequency of the vibrometer if (i) the damping is negligible and (ii) the damping ratio is $\zeta = 0.6$.

3.62. Use subroutine HARESP to find the steady-state response of a torsional system with $J_0 = 6$ kg-m$^2$, $c_t = 210$ N-m-s/rad, $k_t = 14000$ N-m/rad, and $F(t) = 450 \sin 10t$ N-m.

**3.63.** Write a subroutine called TOTALR for finding the complete solution (homogeneous part plus particular integral) of a single degree of freedom system. Use this program to find the solution of Problem 3.14.

**3.64.** Find the steady-state solution of a single degree of freedom system with $m = 10$ kg, $c = 45$ N-s/m, $k = 2500$ N/m, $F(t) = 180\cos 20t$ N, $x_0 = 0$, and $\dot{x}_0 = 10$ m/s, using subroutine HARESP.

**3.65.** Write a computer program for finding the total response of a spring-mass-viscous damper system subjected to base excitation. Use this program to find the solution of a problem with $m = 2$ kg, $c = 10$ N-s/m, $k = 100$ N/m, $y(t) = 0.1\sin 25t$ m, $x_0 = 10$ mm, and $\dot{x}_0 = 5$ m/s.

# CHAPTER 4

# Vibration under General Forcing Conditions

Jean Baptiste Joseph Fourier (1768–1830) was a French mathematician and a professor at the Ecole Polytechnique in Paris. His works on heat flow, published in 1822, and on trigonometric series are well known. The expansion of a periodic function in terms of harmonic functions has been named after him as the ''Fourier series.''

Courtesy The Bettmann Archive, Inc.

## 4.1  INTRODUCTION

This chapter deals with the vibration of a viscously damped single degree of freedom system under general forcing conditions. If the excitation is periodic but not harmonic, it can be replaced by a sum of harmonic functions using the harmonic analysis procedure discussed in Sec. 1.11. By the principle of superposition, the response of the system can then be determined by superposing the responses due to the individual harmonic forcing functions. On the other hand, if the system is excited by a suddenly applied nonperiodic excitation, the response is transient, since steady-state oscillations are not usually produced. The transient response of a system can be found using what is known as the *convolution integral*.

## 4.2  RESPONSE UNDER A GENERAL PERIODIC FORCE

When the external force $F(t)$ is periodic with period $\tau = 2\pi/\omega$, it can be expanded in a Fourier series (see Sec. 1.11):

$$F(t) = \frac{a_0}{2} + \sum_{j=1}^{\infty} a_j \cos j\omega t + \sum_{j=1}^{\infty} b_j \sin j\omega t \tag{4.1}$$

where

$$a_j = \frac{2}{\tau} \int_0^{\tau} F(t) \cos j\omega t \cdot dt, \qquad j = 0, 1, 2, \ldots \tag{4.2}$$

and

$$b_j = \frac{2}{\tau} \int_0^{\tau} F(t) \sin j\omega t \cdot dt, \qquad j = 1, 2, \ldots \tag{4.3}$$

The equation of motion of the system can be expressed as

$$m\ddot{x} + c\dot{x} + kx = F(t) = \frac{a_0}{2} + \sum_{j=1}^{\infty} a_j \cos j\omega t + \sum_{j=1}^{\infty} b_j \sin j\omega t \tag{4.4}$$

The right-hand side of this equation is a constant plus a sum of harmonic functions. Using the principle of superposition, the steady-state solution of Eq. (4.4) is the sum of the steady-state solutions of the following equations:

$$m\ddot{x} + c\dot{x} + kx = \frac{a_0}{2} \tag{4.5}$$

$$m\ddot{x} + c\dot{x} + kx = a_j \cos j\omega t \tag{4.6}$$

$$m\ddot{x} + c\dot{x} + kx = b_j \sin j\omega t \tag{4.7}$$

Noting that the solution of Eq. (4.5) is given by

$$x_p(t) = \frac{a_0}{2k} \qquad (4.8)$$

and using the results of Sec. 3.4, we can express the solutions of Eqs. (4.6) and (4.7), respectively, as

$$x_p(t) = \frac{(a_j/k)}{\sqrt{(1 - j^2 r^2)^2 + (2\zeta jr)^2}} \cos(j\omega t - \phi_j) \qquad (4.9)$$

$$x_p(t) = \frac{(b_j/k)}{\sqrt{(1 - j^2 r^2)^2 + (2\zeta jr)^2}} \sin(j\omega t - \phi_j) \qquad (4.10)$$

where

$$\phi_j = \tan^{-1}\left(\frac{2\zeta jr}{1 - j^2 r^2}\right) \qquad (4.11)$$

and

$$r = \frac{\omega}{\omega_n} \qquad (4.12)$$

Thus the complete steady-state solution of Eq. (4.4) is given by

$$x_p(t) = \frac{a_0}{2k} + \sum_{j=1}^{\infty} \frac{(a_j/k)}{\sqrt{(1 - j^2 r^2)^2 + (2\zeta jr)^2}} \cos(j\omega t - \phi_j)$$

$$+ \sum_{j=1}^{\infty} \frac{(b_j/k)}{\sqrt{(1 - j^2 r^2)^2 + (2\zeta jr)^2}} \sin(j\omega t - \phi_j) \qquad (4.13)$$

It can be seen from the solution, Eq. (4.13), that the amplitude and phase shift corresponding to the $j$th term depend on $j$. If $j\omega = \omega_n$, for any $j$, the amplitude of the corresponding harmonic will be comparatively large. This will be particularly true for small values of $j$ and $\zeta$. Further, as $j$ becomes larger, the amplitude becomes smaller, so that such terms tend to vanish. Thus the first few terms are usually sufficient to obtain the response fairly accurately.

The solution given by Eq. (4.13) denotes the steady-state response of the system. The transient part of the solution arising from the initial conditions can also be included to find the complete solution. However, the evaluation of the arbitrary constants becomes awkward, since it involves setting the value of the complete solution and its time derivative equal to the specified values of the initial displacement and velocity. This results in rather complicated coefficients for the transient part of the total solution.

**EXAMPLE 4.1**

A piston of mass 0.25 kg is placed in a smooth cylinder of internal diameter 50 mm, as shown in Fig. 4.1(a). One end of the cylinder contains a spring of stiffness 2500 N/m and a viscous damper of damping constant 10 N-s/m and is open to the atmosphere. Find the steady-state response of the piston when the pressure on the right-hand side of the cylinder varies periodically, as shown in Fig. 4.1(b).

***Solution.*** The piston can be considered as a mass connected to a spring and a damper on one side and subjected to a forcing function $F(t)$ on the other side. The forcing function can be expressed as

$$F(t) = Ap(t) \tag{E.1}$$

where $A$ is the cross sectional area of the piston, given by

$$A = \frac{\pi(50)^2}{4} = 625\pi \text{ mm}^2 = 0.000625\pi \text{ m}^2 \tag{E.2}$$

and $p(t)$ is the pressure acting on the piston at any instant $t$. Since $p(t)$ is periodic with period $\tau = 2$ seconds and $A$ is a constant, $F(t)$ is also a periodic function of period $\tau = 2$ seconds. The frequency of the forcing function is $\omega = (2\pi/\tau) = \pi$ rad/sec. $F(t)$ can be expressed in a Fourier series as:

$$F(t) = \frac{a_0}{2} + a_1 \cos \omega t + a_2 \cos 2\omega t + \cdots$$
$$+ b_1 \sin \omega t + b_2 \sin 2\omega t + \cdots \tag{E.3}$$

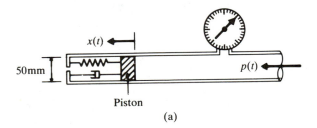

(a)

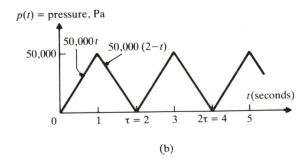

(b)

**Figure 4.1**

where $a_j$ and $b_j$ are given by Eqs. (4.2) and (4.3). Since the function $F(t)$ is given by

$$F(t) = \begin{cases} 50000At & \text{for } 0 \le t \le \dfrac{\tau}{2} \\ 50000A(2-t) & \text{for } \dfrac{\tau}{2} \le t \le \tau \end{cases} \tag{E.4}$$

the Fourier coefficients $a_j$ and $b_j$ can be computed with the help of Eqs. (4.2) and (4.3):

$$a_0 = \frac{2}{2}\left[\int_0^1 50000At\,dt + \int_1^2 50000A(2-t)\,dt\right] = 50000\,A \tag{E.5}$$

$$a_1 = \frac{2}{2}\left[\int_0^1 50000At\cos\pi t\,dt + \int_1^2 50000A(2-t)\cos\pi t\,dt\right]$$
$$= -\frac{2\times10^5 A}{\pi^2} \tag{E.6}$$

$$b_1 = \frac{2}{2}\left[\int_0^1 50000At\sin\pi t\,dt + \int_1^2 50000A(2-t)\sin\pi t\,dt\right] = 0 \tag{E.7}$$

$$a_2 = \frac{2}{2}\left[\int_0^1 50000At\cos2\pi t\,dt + \int_1^2 50000A(2-t)\cos2\pi t\,dt\right] = 0 \tag{E.8}$$

$$b_2 = \frac{2}{2}\left[\int_0^1 50000At\sin2\pi t\,dt + \int_1^2 50000A(2-t)\sin2\pi t\,dt\right] = 0 \tag{E.9}$$

$$a_3 = \frac{2}{2}\left[\int_0^1 50000At\cos3\pi t\,dt + \int_1^2 50000A(2-t)\cos3\pi t\,dt\right]$$
$$= -\frac{2\times10^5 A}{9\pi^2} \tag{E.10}$$

$$b_3 = \frac{2}{2}\left[\int_0^1 50000At\sin3\pi t\,dt + \int_1^2 50000A(2-t)\sin3\pi t\,dt\right] = 0 \tag{E.11}$$

Likewise, we can obtain $a_4 = a_6 = \cdots = b_4 = b_5 = b_6 = \cdots = 0$. By considering only the first three harmonics, the forcing function can be approximated:

$$F(t) \simeq 25000A - \frac{2\times10^5 A}{\pi^2}\cos\omega t - \frac{2\times10^5 A}{9\pi^2}\cos3\omega t \tag{E.12}$$

The steady-state response of the piston to the forcing function of Eq. (E.12) can be expressed as

$$x_p(t) = \frac{25000A}{k} - \frac{(2\times10^5 A/(k\pi^2))}{\sqrt{(1-r^2)^2+(2\zeta r)^2}}\cos(\omega t - \phi_1)$$
$$- \frac{(2\times10^5 A/(9k\pi^2))}{\sqrt{(1-9r^2)^2+(6\zeta r)^2}}\cos(3\omega t - \phi_3) \tag{E.13}$$

The natural frequency of the piston is given by

$$\omega_n = \sqrt{\frac{k}{m}} = \sqrt{\frac{2500}{0.25}} = 100 \text{ rad/sec} \qquad (E.14)$$

and the forcing frequency $\omega$ by

$$\omega = \frac{2\pi}{\tau} = \frac{2\pi}{2} = \pi \text{ rad/sec} \qquad (E.15)$$

Thus the frequency ratio can be obtained:

$$r = \frac{\omega}{\omega_n} = \frac{\pi}{100} = 0.031416 \qquad (E.16)$$

and the damping ratio:

$$\zeta = \frac{c}{c_c} = \frac{c}{2m\omega_n} = \frac{10.0}{2(0.25)(100)} = 0.2 \qquad (E.17)$$

The phase angles $\phi_1$ and $\phi_3$ can be computed as follows:

$$\phi_1 = \tan^{-1}\left(\frac{2\zeta r}{1-r^2}\right) = \tan^{-1}\left(\frac{2\times0.2\times0.031416}{1-0.031416^2}\right) = 0.0125664 \text{ rad} \qquad (E.18)$$

and

$$\phi_3 = \tan^{-1}\left(\frac{6\zeta r}{1-9r^2}\right) = \tan^{-1}\left(\frac{6\times0.2\times0.031416}{1-9(0.031416)^2}\right) = 0.0380483 \text{ rad} \qquad (E.19)$$

In view of Eqs. (E.2) and (E.14) to (E.19), the solution can be written as

$$x_p(t) = 0.019635 - 0.015930\cos(\pi t - 0.0125664)$$
$$- 0.0017828\cos(3\pi t - 0.0380483) \text{ m} \qquad (E.20)$$

## 4.3  RESPONSE UNDER A PERIODIC FORCE OF IRREGULAR FORM

In some cases, the forcing function may be found experimentally and may be quite irregular. In such cases, the irregular functions can be represented graphically but not as explicit functions $F(t)$. Sometimes, only the values of $F(t)$ at a number of points $t_1, t_2, \ldots, t_N$ are available. An example of such an irregular periodic forcing function is shown in Fig. 4.2. In this case it is possible to find the Fourier coefficients by using a numerical integration procedure, as described in Sec. 1.11. If $F_1, F_2, \ldots, F_N$ denote the values of $F(t)$ at $t_1, t_2, \ldots, t_N$, respectively, where $N$ denotes an even number of equidistant points in one time period $\tau$ ($\tau = N\Delta t$), Simpson's formula

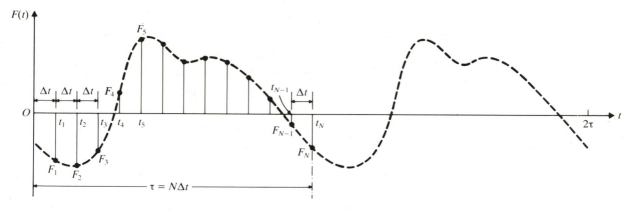

**Figure 4.2**

[4.1] gives

$$a_0 = \frac{2}{N} \sum_{i=1}^{N} F_i \tag{4.14}$$

$$a_j = \frac{2}{N} \sum_{i=1}^{N} F_i \cos \frac{2j\pi t_i}{\tau}, \qquad j = 1, 2, \ldots \tag{4.15}$$

$$b_j = \frac{2}{N} \sum_{i=1}^{N} F_i \sin \frac{2j\pi t_i}{\tau}, \qquad j = 1, 2, \ldots \tag{4.16}$$

Once the Fourier coefficients $a_0$, $a_j$, and $b_j$ are known, the steady-state response of the system can be found using Eq. (4.13) with

$$r = \left( \frac{2\pi}{\tau \omega_n} \right)$$

## EXAMPLE 4.2

Find the steady-state response of the piston in Example 4.1 if the pressure fluctuations in the tube are found to be periodic. The values of pressure measured at 0.01 second intervals in one cycle are given below.

| time, $t_i$ (seconds) | 0 | 0.01 | 0.02 | 0.03 | 0.04 | 0.05 | 0.06 | 0.07 | 0.08 | 0.09 | 0.10 | 0.11 | 0.12 |
|---|---|---|---|---|---|---|---|---|---|---|---|---|---|
| $p_i = p(t_i)$ (kN/m²) | 0 | 20 | 34 | 42 | 49 | 53 | 70 | 60 | 36 | 22 | 16 | 7 | 0 |

***Solution.*** The Fourier analysis of the pressure fluctuations (see Example 1.7) gives the result

$$p(t) = 34083.3 - 26996.0 \cos 52.36t + 8307.7 \sin 52.36t$$
$$+ 1416.7 \cos 104.72t + 3608.3 \sin 104.72t$$
$$- 5833.3 \cos 157.08t + 2333.3 \sin 157.08t + \cdots \, \text{N}/\text{m}^2 \qquad \text{(E.1)}$$

Other quantities needed for the computation are

$$\omega = \frac{2\pi}{\tau} = \frac{2\pi}{0.12} = 52.36 \text{ rad}/\text{sec}$$

$$\omega_n = 100 \text{ rad}/\text{sec}$$

$$r = \frac{\omega}{\omega_n} = 0.5236$$

$$\zeta = 0.2$$

$$A = 0.000625\pi \text{ m}^2$$

$$\phi_1 = \tan^{-1}\left(\frac{2\zeta r}{1 - r^2}\right) = \tan^{-1}\left(\frac{2 \times 0.2 \times 0.5236}{1 - 0.5236^2}\right) = 16.1°$$

$$\phi_2 = \tan^{-1}\left(\frac{4\zeta r}{1 - 4r^2}\right) = \tan^{-1}\left(\frac{4 \times 0.2 \times 0.5236}{1 - 4 \times 0.5236^2}\right) = -77.01°$$

$$\phi_3 = \tan^{-1}\left(\frac{6\zeta r}{1 - 9r^2}\right) = \tan^{-1}\left(\frac{6 \times 0.2 \times 0.5236}{1 - 9 \times 0.5236^2}\right) = -23.18°$$

The steady-state response of the piston can be expressed, using Eq. (4.13), as

$$x_p(t) = \frac{34083.3A}{k} - \frac{(26996.0A/k)}{\sqrt{(1 - r^2)^2 + (2\zeta r)^2}} \cos(52.36t - \phi_1)$$

$$+ \frac{(8309.7A/k)}{\sqrt{(1 - r^2)^2 + (2\zeta r)^2}} \sin(52.36t - \phi_1)$$

$$+ \frac{(1416.7A/k)}{\sqrt{(1 - 4r^2)^2 + (4\zeta r)^2}} \cos(104.72t - \phi_2)$$

$$+ \frac{(3608.3A/k)}{\sqrt{(1 - 4r^2)^2 + (4\zeta r)^2}} \sin(104.72t - \phi_2)$$

$$- \frac{(5833.3A/k)}{\sqrt{(1 - 9r^2)^2 + (6\zeta r)^2}} \cos(157.08t - \phi_3)$$

$$+ \frac{(2333.3A/k)}{\sqrt{(1 - 9r^2)^2 + (6\zeta r)^2}} \sin(157.08t - \phi_3)$$

## 4.4 RESPONSE UNDER NONPERIODIC FORCE

We have seen that periodic forces of any general wave form can be represented by Fourier series as a superposition of harmonic components of various frequencies. The response of a linear system is then found by superposing the harmonic response to each of the exciting forces. When the exciting force $F(t)$ is nonperiodic, such as that due to the blast from an explosion, a different method of calculating the response is required. Various methods can be used to find the response of the system to an arbitrary excitation. Some of these methods are as follows:

1. by representing the excitation by a Fourier integral;
2. by using the method of convolution integral;
3. by using the method of Laplace transformation;
4. by first approximating $F(t)$ by a suitable interpolation model and then using a numerical procedure;
5. by numerically integrating the equations of motion.

We shall discuss Methods 2, 3, and 4 in the following sections and Method 5 in Chap. 10.

## 4.5 CONVOLUTION INTEGRAL

A nonperiodic exciting force usually has a magnitude that varies with time; it acts for a specified period of time and then stops. The simplest form of such a force is the impulsive force. An impulsive force is one that has a large magnitude $F$ and acts for a very short period of time $\Delta t$. From dynamics we know that impulse can be measured by finding the change in momentum of the system caused by it [4.2]. If $\dot{x}_1$ and $\dot{x}_2$ denote the velocities of the mass $m$ before and after the application of the impulse, we have

$$\text{Impulse} = F\Delta t = m\dot{x}_2 - m\dot{x}_1 \qquad (4.17)$$

*momentum change*

By designating the magnitude of the impulse $F\Delta t$ by $\underset{\sim}{F}$, we can write, in general,

$$\text{Magnitude} \quad \underset{\sim}{F} = \int_t^{t+\Delta t} F\,dt \qquad (4.18)$$

A unit impulse $(\underset{\sim}{f})$ is defined as

$$\underset{\sim}{f} = \lim_{\Delta t \to 0} \int_t^{t+\Delta t} F\,dt = F\cdot dt = 1 \qquad (4.19)$$

It can be seen that in order for $F\cdot dt$ to have a finite value, $F$ tends to infinity (since $dt$ tends to zero). Although the unit impulse function has no physical meaning, it is a convenient tool in our present analysis.

### 4.5.1 Response to an Impulse

We first consider the response of a single degree of freedom system to an impulse excitation; this case is important in studying the response under more general

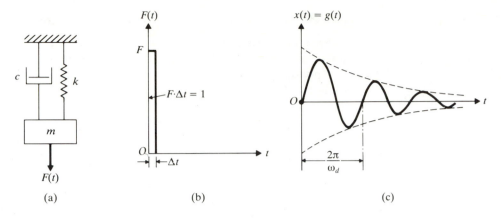

**Figure 4.3**

excitations. Consider a viscously damped spring-mass system subjected to a unit impulse at $t = 0$, as shown in Figs. 4.3(a) and (b). For an underdamped system, the solution of the equation of motion

$$m\ddot{x} + c\dot{x} + kx = 0 \tag{4.20}$$

is given by Eq. (2.66) as follows:

for underdamped & r<1

$$x(t) = e^{-\zeta\omega_n t}\left\{ x_0 \cos\omega_d t + \frac{\dot{x}_0 + \zeta\omega_n x_0}{\omega_d}\sin\omega_d t \right\} \tag{4.21}$$

where

$$\zeta = \frac{c}{2m\omega_n} \tag{4.22}$$

$$\omega_d = \omega_n\sqrt{1 - \zeta^2} = \sqrt{\frac{k}{m} - \left(\frac{c}{2m}\right)^2} \tag{4.23}$$

$$\omega_n = \sqrt{\frac{k}{m}} \tag{4.24}$$

If the mass is at rest before the unit impulse is applied ($x = \dot{x} = 0$ for $t < 0$ or at $t = 0^-$), we obtain, from the impulse-momentum relation,

$$\text{Impulse} = f = 1 = m\dot{x}(t = 0) - m\dot{x}(t = 0^-) = m\dot{x}_0 \tag{4.25}$$

Thus the initial conditions are given by

$$x(t = 0) = x_0 = 0$$

$$\dot{x}(t = 0) = \dot{x}_0 = \frac{1}{m} \tag{4.26}$$

In view of Eq. (4.26), Eq. (4.21) reduces to

Impulse Response
Function $g(t)$

$$x(t) = g(t) = \frac{e^{-\zeta\omega_n t}}{m\omega_d}\sin\omega_d t \tag{4.27}$$

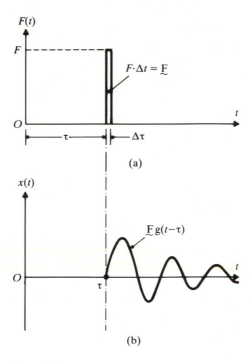

**Figure 4.4**

Equation (4.27) gives the response of a single degree of freedom system to a unit impulse, which is also known as the *impulse response function*, denoted by $g(t)$. The function $g(t)$, Eq. (4.27), is shown in Fig. 4.3(c).

If the magnitude of the impulse is $\underline{F}$ instead of unity, the initial velocity $\dot{x}_0$ is $\underline{F}/m$ and the response of the system becomes

$$x(t) = \frac{\underline{F}e^{-\zeta\omega_n t}}{m\omega_d}\sin\omega_d t = \underline{F}g(t) \tag{4.28}$$

If the impulse $\underline{F}$ is applied at an arbitrary time $t = \tau$, as shown in Fig. 4.4(a), it will change the velocity at $t = \tau$ by an amount $\underline{F}/m$. Assuming that $x = 0$ until the impulse is applied, the displacement $x$ at any subsequent time $t$, caused by a change in the velocity at time $\tau$, is given by Eq. (4.28) with $t$ replaced by the time elapsed after the application of the impulse, that is, $t - \tau$. Thus we obtain

$$x(t) = \underline{F}g(t - \tau) \tag{4.29}$$

This is shown in Fig. 4.4(b).

### 4.5.2 Response to General Forcing Condition

Now we consider the response of the system under an arbitrary external force $F(t)$, shown in Fig. 4.5. This force may be assumed to be made up of a series of impulses

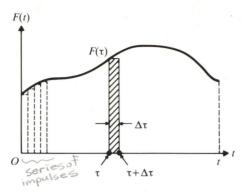

**Figure 4.5**   An arbitrary (nonperiodic) forcing function.

of varying magnitude. Assuming that at time $\tau$, the force $F(\tau)$ acts on the system for a short period of time $\Delta\tau$, the impulse acting at $t = \tau$ is given by $F(\tau)\Delta\tau$. At any time $t$, the elapsed time since the impulse is $t - \tau$, so the response of the system at $t$ due to this impulse alone is given by Eq. (4.29) with $F = F(\tau)\Delta\tau$:

$$\Delta x(t) = F(\tau)\,\Delta\tau\,g(t - \tau) \qquad (4.30)$$

The total response at time $t$ can be found by summing all the responses due to the elementary impulses acting at all times $\tau$:

$$x(t) \simeq \sum F(\tau)g(t - \tau)\,\Delta\tau \qquad (4.31)$$

Letting $\Delta\tau \to 0$ and replacing the summation by integration, we obtain

$$x(t) = \int_0^t F(\tau)g(t - \tau)\,d\tau \qquad (4.32)$$

By substituting Eq. (4.27) into Eq. (4.32), we obtain

$$x(t) = \frac{1}{m\omega_d}\int_0^t F(\tau)e^{-\zeta\omega_n(t-\tau)}\sin\omega_d(t - \tau)\,d\tau \qquad (4.33)$$

which represents the response of an underdamped single degree of freedom system to the arbitrary excitation $F(t)$. Note that Eq. (4.33) does not consider the effect of initial conditions of the system. The integral in Eq. (4.32) or Eq. (4.33) is called the *convolution* or *Duhamel integral*. In many cases the function $F(t)$ has a form that permits an explicit integration of Eq. (4.33). In case such integration is not possible, it can be evaluated numerically without much difficulty, as illustrated in Sec. 4.8 and Chap. 10. An elementary discussion of the Duhamel integral in vibration analysis is given in Ref. [4.6].

### 4.5.3   Response to Base Excitation

If a spring-mass-damper system is subjected to an arbitrary base excitation described by its displacement, velocity, or acceleration, the equation of motion can be expressed in terms of the relative displacement of the mass $z = x - y$ as follows (see Sec. 3.6.2)

$$m\ddot{z} + c\dot{z} + kz = -m\ddot{y} \tag{4.34}$$

This equation is similar to the equation

$$m\ddot{x} + c\dot{x} + kx = F \tag{4.35}$$

with the variable $z$ replacing $x$ and the term $-m\ddot{y}$ replacing the forcing function $F$. Hence all of the results derived for the force-excited system are applicable to the base-excited system also for $z$ when the term $F$ is replaced by $-m\ddot{y}$. For an underdamped system subjected to base excitation, the relative displacement can be found from Eq. (4.33):

$$z(t) = -\frac{1}{\omega_d} \int_0^t \ddot{y}(\tau) e^{-\zeta \omega_n (t-\tau)} \sin \omega_d (t - \tau) \, d\tau \tag{4.36}$$

## EXAMPLE 4.3

Determine the response of a single degree of freedom system subjected to the step force shown in Fig. 4.6(a).

**Solution.** By noting that $F(t) = F_0$, we can write Eq. (4.33) as

$$x(t) = \frac{F_0}{m\omega_d} \int_0^t e^{-\zeta \omega_n (t-\tau)} \sin \omega_d (t - \tau) \, d\tau$$

$$= \frac{F_0}{m\omega_d} \left[ e^{-\zeta \omega_n (t-\tau)} \left\{ \frac{\zeta \omega_n \sin \omega_d (t - \tau) + \omega_d \cos \omega_d (t - \tau)}{(\zeta \omega_n)^2 + (\omega_d)^2} \right\} \right]_{\tau=0}^{t}$$

$$= \frac{F_0}{k\sqrt{1 - \zeta^2}} \left[ \sqrt{1 - \zeta^2} - e^{-\zeta \omega_n t} \cos(\omega_d t - \phi) \right] \tag{E.1}$$

where

$$\phi = \tan^{-1} \left( \frac{\zeta}{\sqrt{1 - \zeta^2}} \right) \tag{E.2}$$

This response is shown in Fig. 4.6(b). If the system is undamped ($\zeta = 0$ and $\omega_d = \omega_n$), Eq. (E.1) reduces to

$$x(t) = \frac{F_0}{k} [1 - \cos \omega_n t] \tag{E.3}$$

Equation (E.3) is shown graphically in Fig. 4.6(c). It can be seen that if the load is instantaneously applied to an undamped system, a maximum displacement of twice the static displacement will be attained, that is, $x_{max} = 2F_0/k$.

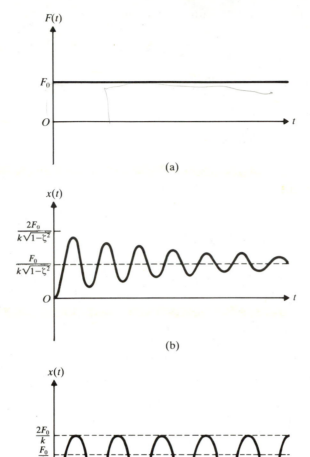

**Figure 4.6**

---

**EXAMPLE 4.4**

Find the response of a viscously damped system to the step function shown in Fig. 4.7.

***Solution.*** Since the forcing function starts at $t = t_0$ instead of at $t = 0$, the response can be obtained from Eq. (E.1) of Example 4.3 by replacing $t$ by $t - t_0$. This gives

$$x(t) = \frac{F_0}{k\sqrt{1 - \zeta^2}} \left[ \sqrt{1 - \zeta^2} - e^{-\zeta\omega_n(t - t_0)}\cos\{\omega_d(t - t_0) - \phi\} \right] \qquad \text{(E.1)}$$

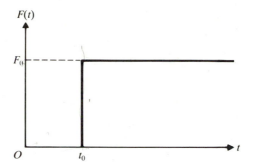

$F(t)$

$F_0$

$O$    $t_0$    $t$

**Figure 4.7**

If the system is undamped, Eq. (E.1) reduces to

$$x(t) = \frac{F_0}{k}\left[1 - \cos\omega_n(t - t_0)\right] \tag{E.2}$$

## EXAMPLE 4.5

Find the response of a damped single degree of freedom system to a rectangular pulse, shown in Fig. 4.8(a).

*Solution.* This forcing function can be considered as the sum of a step function of magnitude $+F_0$ beginning at $t = 0$ and a second step function of magnitude $-F_0$ starting at time $t = t_0$. Thus the response of the system can be obtained by subtracting Eq. (E.1) of Example 4.4 from Eq. (E.1) of Example 4.3. This gives

$$x(t) = \frac{F_0 e^{-\zeta\omega_n t}}{k\sqrt{1-\zeta^2}}\left[-\cos(\omega_d t - \phi) + e^{\zeta\omega_n t_0}\cos\{\omega_d(t - t_0) - \phi\}\right] \tag{E.1}$$

with

$$\phi = \tan^{-1}\left(\frac{\zeta}{\sqrt{1-\zeta^2}}\right) \tag{E.2}$$

To see the vibration response graphically, we consider the system as undamped, so that Eq. (E.1) reduces to

$$x(t) = \frac{F_0}{k}\left[\cos\omega_n(t - t_0) - \cos\omega_n t\right] \tag{E.3}$$

This response is shown in Fig. 4.8(b) for two cases: (1) $t_0 > \tau_n/2$, and (2) $t_0 < \tau_n/2$ where $t_0$ is the duration of the rectangular pulse and $\tau_n$ is the natural time period of the system. It can be seen that the peak occurs during the forced vibration era (that

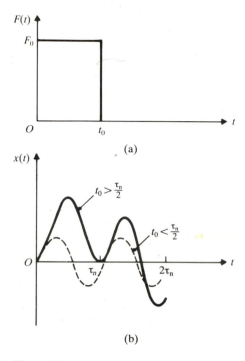

**Figure 4.8**

is, prior to $t_0$) for $t_0 > \tau_n/2$ while the peak occurs in the residual vibration era (that is, after $t_0$) if $t_0 < \tau_n/2$.

---

### EXAMPLE 4.6

Determine the response of a spring-mass-damper system to the linearly increasing force, known as the *ramp function*, shown in Fig. 4.9(a). The rate of increase of the force $F$ per unit time is $\delta F$.

***Solution.*** The forcing function in this case is given by $F(\tau) = \delta F \tau$. By substituting this into Eq. (4.33), we obtain
$$= (\text{Force Rate})(\text{time})$$

$$x(t) = \frac{\delta F}{m\omega_d} \int_0^t \tau e^{-\zeta\omega_n(t-\tau)} \sin\omega_d(t-\tau)\, d\tau$$

$$= \frac{\delta F}{m\omega_d} \int_{0^-}^t (t-\tau) e^{-\zeta\omega_n(t-\tau)} \sin\omega_d(t-\tau)(-d\tau)$$

$$- \frac{\delta F \cdot t}{m\omega_d} \int_0^t e^{-\zeta\omega_n(t-\tau)} \sin\omega_d(t-\tau)(-d\tau)$$

These integrals can be evaluated and the response expressed as follows. (See Problem

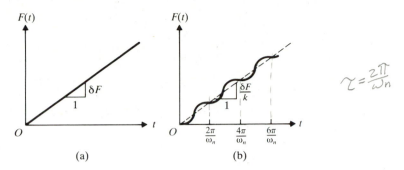

**Figure 4.9**

4.26.)

$$x(t) = \frac{\delta F}{k}\left[t - \frac{2\zeta}{\omega_n} + e^{-\zeta\omega_n t}\left(\frac{2\zeta}{\omega_n}\cos\omega_d t - \left\{\frac{\omega_d^2 - \zeta^2\omega_n^2}{\omega_n^2\omega_d}\right\}\sin\omega_d t\right)\right] \qquad (E.1)$$

For an undamped system, Eq. (E.1) reduces to

$$x(t) = \frac{\delta F}{\omega_n k}[\omega_n t - \sin\omega_n t] \qquad (E.2)$$

Figure 4.9(b) shows the response given by Eq. (E.2).

## EXAMPLE 4.7

A building frame is modeled as an undamped single degree of freedom system (Fig. 4.10(a)). Find the response of the frame if it is subjected to a blast loading represented by the triangular pulse shown in Fig. 4.10(b).

***Solution.*** The forcing function is given by

$$F(\tau) = F_0\left(1 - \frac{\tau}{t_0}\right) \qquad \text{for } 0 \le \tau \le t_0 \qquad (E.1)$$

$$F(\tau) = 0 \qquad \text{for } t_0 < \tau \qquad (E.2)$$

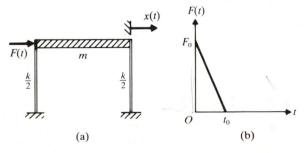

(a)                                              (b)

**Figure 4.10**

Equation (4.33) gives, for an undamped system,

$$x(t) = \frac{1}{m\omega_n} \int_0^t F(\tau) \sin \omega_n (t - \tau) \, d\tau \qquad \text{(E.3)}$$

*Response during $0 \le t \le t_0$:* Using Eq. (E.1) for $F(\tau)$ in Eq. (E.3) gives

$$x(t) = \frac{F_0}{m\omega_n^2} \int_0^t \left(1 - \frac{\tau}{t_0}\right) [\sin \omega_n t \cos \omega_n \tau - \cos \omega_n t \sin \omega_n \tau] \, d(\omega_n \tau)$$

$$= \frac{F_0}{k} \sin \omega_n t \int_0^t \left(1 - \frac{\tau}{t_0}\right) \cos \omega_n \tau \cdot d(\omega_n \tau)$$

$$- \frac{F_0}{k} \cos \omega_n t \int_0^t \left(1 - \frac{\tau}{t_0}\right) \sin \omega_n \tau \cdot d(\omega_n \tau) \qquad \text{(E.4)}$$

By noting that integration by parts gives

$$\int \tau \cos \omega_n \tau \cdot d(\omega_n \tau) = \tau \sin \omega_n \tau + \frac{1}{\omega_n} \cos \omega_n \tau \qquad \text{(E.5)}$$

and

$$\int \tau \sin \omega_n \tau \cdot d(\omega_n \tau) = -\tau \cos \omega_n \tau + \frac{1}{\omega_n} \sin \omega_n \tau \qquad \text{(E.6)}$$

Eq. (E.4) can be written as

$$x(t) = \frac{F_0}{k} \left\{ \sin \omega_n t \left[ \sin \omega_n t - \frac{t}{t_0} \sin \omega_n t - \frac{1}{\omega_n t_0} \cos \omega_n t + \frac{1}{\omega_n t_0} \right] \right.$$

$$\left. - \cos \omega_n t \left[ -\cos \omega_n t + 1 + \frac{t}{t_0} \cos \omega_n t - \frac{1}{\omega_n t_0} \sin \omega_n t \right] \right\} \qquad \text{(E.7)}$$

Simplifying this expression, we obtain

$$x(t) = \frac{F_0}{k} \left[ 1 - \frac{t}{t_0} - \cos \omega_n t + \frac{1}{\omega_n t_0} \sin \omega_n t \right] \qquad \text{(E.8)}$$

*Response during $t > t_0$:* Here also we use Eq. (E.1) for $F(\tau)$, but the upper limit of integration in Eq. (E.3) will be $t_0$, since $F(\tau) = 0$ for $\tau > t_0$. Thus the response can be found from Eq. (E.7) by setting $t = t_0$ within the square brackets. This results in

$$x(t) = \frac{F_0}{k\omega_n t_0} [(1 - \cos \omega_n t_0) \sin \omega_n t - (\omega_n t_0 - \sin \omega_n t_0) \cos \omega_n t] \qquad \text{(E.9)}$$

## 4.6   RESPONSE SPECTRUM

A shock is a sudden application of a force or other form of disturbance that results in the transient response of a system. The severity of the shock can be measured in terms of the maximum value of the response of the system. In general, the maximum response depends not only on the nature of the loading but also on the dynamic characteristics of the system. In engineering practice, all types of shock excitations are categorized by taking a single degree of freedom undamped oscillator (spring-mass system) as a standard system and comparing the responses obtained with various shock excitations. The concept of the response spectrum has proved very useful in design [4.2]. A review of recent literature on shock and seismic response spectra in engineering design is given in Ref. [4.7].

The response spectrum is a diagram that shows the maximum peak response of a single degree of freedom system as a function of the natural frequency or natural period of the system.* Different types of shock excitations result in different response spectra. According to the definition, the response spectrum is found from a single point on the time-response curve, which by itself is an incomplete piece of information and thus does not uniquely define the input shock. It is possible for similar response spectra to result from two different shock excitations. Despite this limitation, the concept of the response spectrum is used extensively in engineering design.

The response of an undamped system to an arbitrary excitation $F(t)$ can be found from Eq. (4.33). The peak response of the system can be expressed as

$$x(t)\Big|_{\max} = \frac{1}{m\omega_n} \int_0^t F(\tau)\sin\omega_n(t-\tau)\,d\tau\Big|_{\max} \tag{4.37}$$

In general, some characteristic time $t_0$, such as the duration of the pulse, can be associated with the shock excitation $F(t)$. The maximum value of the response is usually plotted as a nondimensional quantity as a function of $\omega_n$ or $t_0/\tau_n$, where $\omega_n$ and $\tau_n$ denote the natural frequency and natural period of the system. The following example illustrates the construction of a response spectrum.

---

### EXAMPLE 4.8

Find the undamped response spectrum for a forcing function that increases linearly from 0 to $F_0$ in time $t_0$ and remains constant thereafter (see Fig. 4.11(a)).

*Solution.* The forcing function $F(t)$ can be considered as the sum of two ramp functions $F_0 t/t_0$, the second of which is negative and delayed by the time $t_0$, as shown in Fig. 4.11(b). The response of an undamped spring-mass system for the first ramp function can be obtained from Eq. (E.2) of Example 4.6, with $\delta F = F_0/t_0$:

$$x_1(t) = \frac{F_0}{k}\left(\frac{t}{t_0} - \frac{\sin\omega_n t}{\omega_n t_0}\right) \qquad \text{for } t \geq 0 \tag{E.1}$$

The response of the system to the second ramp function starting at $t_0$ can be

---

*The response spectrum can also be viewed as a plot of the maximum response for all possible single degree of freedom systems for a specified forcing function.

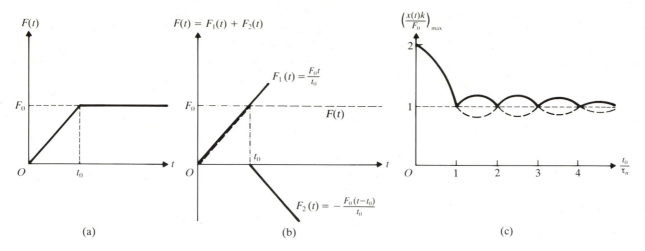

**Figure 4.11**

obtained from Eq. (E.1) by replacing $t$ by $t - t_0$ and $F_0$ by $-F_0$:

$$x_2(t) = -\frac{F_0}{k}\left\{\frac{t - t_0}{t_0} - \frac{\sin \omega_n(t - t_0)}{\omega_n t_0}\right\} \qquad \text{for } t \geq t_0 \qquad \text{(E.2)}$$

By adding Eqs. (E.1) and (E.2), the total response of the system can be found:

$$x(t) = x_1(t) + x_2(t) = \frac{F_0}{k}\left\{1 + \frac{\sin \omega_n(t - t_0) - \sin \omega_n t}{\omega_n t_0}\right\} \qquad \text{for } t \geq t_0 \qquad \text{(E.3)}$$

For maximum response, we differentiate $x(t)$ given by Eq. (E.3) with respect to $t$ and equate it to zero. By inspecting Eqs. (E.1) and (E.3), we find that the maximum response always occurs after time $t_0$. Hence, only Eq. (E.3) is of interest. By setting the derivative of Eq. (E.3) equal to zero, we obtain

$$\tan \omega_n t = \frac{1 - \cos \omega_n t_0}{\sin \omega_n t_0} \qquad \text{(E.4)}$$

If $t = t_m$ denotes the time at which the maximum response occurs, Eq. (E.4) gives

$$\tan \omega_n t_m = \frac{1 - \cos \omega_n t_0}{\sin \omega_n t_0} \qquad \text{(E.5)}$$

As $\omega_n t_m$ must be greater than $\pi$, Eq. (E.5) leads to

$$\sin \omega_n t_m = -\left\{\frac{(1 - \cos \omega_n t_0)^{1/2}}{\sqrt{2}}\right\} \qquad \text{(E.6)}$$

$$\cos \omega_n t_m = -\left\{\frac{\sin \omega_n t_0}{\sqrt{2}(1 - \cos \omega_n t_0)^{1/2}}\right\} \qquad \text{(E.7)}$$

By using Eqs. (E.6) and (E.7) in Eq. (E.3), the maximum response can be expressed as

$$\left(\frac{xk}{F_0}\right)_{max} = 1 + \frac{\sqrt{2}}{\omega_n t_0}(1 - \cos \omega_n t_0)^{1/2} \tag{E.8}$$

Equation (E.8) is shown plotted against $t_0/\tau_n = t_0\omega_n/2\pi$, where $\tau_n$ is the natural period of the system, in Fig. 4.11(c).

### 4.6.1 Response Spectrum for Base Excitation

In the design of machinery and structures subjected to a ground shock, such as that caused by an earthquake, the response spectrum corresponding to the base excitation is useful. If the base of a damped single degree of freedom system is subjected to an acceleration $\ddot{y}(t)$, the equation of motion, in terms of the relative displacement $z = x - y$, is given by Eq. (4.34) and the response $z(t)$ by Eq. (4.36). In the case of a ground shock, the velocity response spectrum is generally used. The displacement and acceleration spectra are then expressed in terms of the velocity spectrum. For a harmonic oscillator (an undamped system under free vibration), we notice that

$$\ddot{x}|_{max} = -\omega_n^2 x|_{max} \quad \text{and} \quad \dot{x}|_{max} = \omega_n x|_{max}$$

Thus the acceleration and displacement spectra $S_a$ and $S_d$ can be obtained in terms of the velocity spectrum $(S_v)$:

$$S_d = \frac{S_v}{\omega_n}, \qquad S_a = \omega_n S_v \tag{4.38}$$

To consider damping in the system, if we assume that the maximum relative displacement occurs after the shock pulse has passed, the subsequent motion must be harmonic. In such a case, we can use Eq. (4.38). The fictitious velocity associated with this apparent harmonic motion is called the *pseudo velocity* and its response spectrum, $S_v$, is called the *pseudo spectrum*. The velocity spectra of damped systems are used extensively in earthquake analysis.

To find the relative velocity spectrum, we differentiate Eq. (4.36) and obtain*

$$\dot{z}(t) = -\frac{1}{\omega_d}\int_0^t \ddot{y}(\tau)e^{-\zeta\omega_n(t-\tau)}\left[-\zeta\omega_n\sin\omega_d(t-\tau) + \omega_d\cos\omega_d(t-\tau)\right]d\tau \tag{4.39}$$

Equation (4.39) can be rewritten as

$$\dot{z}(t) = \frac{e^{-\zeta\omega_n t}}{\sqrt{1-\zeta^2}}\sqrt{P^2 + Q^2}\sin(\omega_d t - \phi) \tag{4.40}$$

---

*The following relation is used in deriving Eq. (4.39) from Eq. (4.36):

$$\frac{d}{dt}\int_0^t f(t,\tau)\,d\tau = \int_0^t \frac{\partial f}{\partial t}(t,\tau)\cdot d\tau + f(t,\tau)\Big|_{\tau=t}$$

where

$$P = \int_0^t \ddot{y}(\tau) e^{\zeta \omega_n \tau} \cos \omega_d \tau \cdot d\tau \qquad (4.41)$$

$$Q = \int_0^t \ddot{y}(\tau) e^{\zeta \omega_n \tau} \sin \omega_d \tau \cdot d\tau \qquad (4.42)$$

and

$$\phi = \tan^{-1} \left\{ \frac{-\left( P\sqrt{1 - \zeta^2} + Q\zeta \right)}{\left( P\zeta - Q\sqrt{1 - \zeta^2} \right)} \right\} \qquad (4.43)$$

The velocity response spectrum, $S_v$, can be obtained from Eq. (4.40):

$$S_v = |\dot{z}(t)|_{max} = \left| \frac{e^{-\zeta \omega_n t}}{\sqrt{1 - \zeta^2}} \sqrt{P^2 + Q^2} \right|_{max} \qquad (4.44)$$

Thus the pseudo response spectra are given by

$$S_d = |z|_{max} = \frac{S_v}{\omega_n}; \quad S_v = |\dot{z}|_{max}; \quad S_a = |\ddot{z}|_{max} = \omega_n S_v \qquad (4.45)$$

## EXAMPLE 4.9

An undamped system is subjected to a base acceleration, as shown Fig. 4.12. Find the response spectrum of $z$.

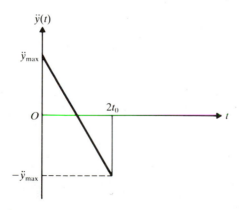

**Figure 4.12**

*Solution.* The base acceleration can be expressed as

$$\ddot{y}(t) = \ddot{y}_{max}\left(1 - \frac{t}{t_0}\right) \quad \text{for } 0 \le t \le 2t_0 \tag{E.1}$$

$$\ddot{y}(t) = 0 \quad \text{for } 2t_0 < t \tag{E.2}$$

*Response during $0 \le t \le 2t_0$:* By substituting Eq. (E.1) into Eq. (4.36), the response can be expressed, for an undamped system, as

$$z(t) = -\frac{1}{\omega_n}\ddot{y}_{max}\left[\int_0^t \left(1 - \frac{\tau}{t_0}\right)(\sin \omega_n t \cos \omega_n \tau - \cos \omega_n t \sin \omega_n \tau) \, d\tau\right] \tag{E.3}$$

This equation is the same as Eq. (E.4) of Example 4.7 except that $(-\ddot{y}_{max})$ appears in place of $F_0/m$. Hence $z(t)$ can be written, using Eq. (E.8) of Example 4.7, as

$$z(t) = -\frac{\ddot{y}_{max}}{\omega_n^2}\left[1 - \frac{t}{t_0} - \cos \omega_n t + \frac{1}{\omega_n t_0}\sin \omega_n t\right] \tag{E.4}$$

To find the maximum response, $z_{max}$, we have

$$\dot{z}(t) = -\frac{\ddot{y}_{max}}{t_0 \omega_n^2}\left[-1 + \omega_n t_0 \sin \omega_n t + \cos \omega_n t\right] = 0 \tag{E.5}$$

This equation gives the time, $t_m$, at which $z_{max}$ occurs:

$$t_m = \frac{2}{\omega_n}\tan^{-1}(\omega_n t_0) \tag{E.6}$$

By substituting Eq. (E.6) into Eq. (E.4), the maximum response can be found:

$$z_{max} = -\frac{\ddot{y}_{max}}{\omega_n^2}\left[1 - \frac{t_m}{t_0} - \cos \omega_n t_m + \frac{1}{\omega_n t_0}\sin \omega_n t_m\right] \tag{E.7}$$

*Response during $2t_0 < t$:* Since there is no excitation during this time, we can use the solution of the free vibration problem (Eq. (2.15)):

$$z(t) = z_0\cos \omega_n t + \left(\frac{\dot{z}_0}{\omega_n}\right)\sin \omega_n t \tag{E.8}$$

provided that we take the initial displacement and initial velocity as

$$z_0 = z(t = 2t_0) \quad \text{and} \quad \dot{z}_0 = \dot{z}(t = 2t_0) \tag{E.9}$$

using Eq. (E.7). The maximum of $z(t)$ given by Eq. (E.8) is simply

$$z_{max} = \left[z_0^2 + \left(\frac{\dot{z}_0}{\omega_n}\right)^2\right]^{1/2} \tag{E.10}$$

where $z_0$ and $\dot{z}_0$ are computed as indicated in Eq. (E.9).

## 4.7 LAPLACE TRANSFORMATION

The Laplace transform method can be used to find the response of a system under any type of excitation, including the harmonic and periodic types. This method can be used for the efficient solution of linear differential equations, particularly those with constant coefficients [4.3]. It permits the conversion of differential equations into algebraic ones, which are easier to manipulate. The major advantages of the method are that it can treat discontinuous functions without any particular difficulty and that it automatically takes into account the initial conditions.

The Laplace transform of a function $x(t)$, denoted symbolically as $\bar{x}(s) = \mathscr{L}x(t)$, is defined as

$$\bar{x}(s) = \mathscr{L}x(t) = \int_0^\infty e^{-st}x(t)\,dt \tag{4.46}$$

where $s$ is, in general, a complex quantity and is called the *subsidiary variable*. The function $e^{-st}$ is called the *kernel* of the transformation. Since the integration is with respect to $t$, the transformation gives a function of $s$. In order to solve a vibration problem using the Laplace transform method, the following steps are necessary:

1. Write the equation of motion of the system.

2. Transform each term of the equation, using known initial conditions.

3. Solve for the transformed response of the system.

4. Obtain the desired solution (response) by using inverse Laplace transformation.

In order to solve the forced vibration equation

$$m\ddot{x} + c\dot{x} + kx = F(t) \tag{4.47}$$

by the Laplace transform method, it is necessary to find the transforms of the derivatives

$$\dot{x}(t) = \frac{dx}{dt}(t) \qquad \text{and} \qquad \ddot{x}(t) = \frac{d^2x}{dt^2}(t)$$

These can be found as follows:

$$\mathscr{L}\frac{dx}{dt}(t) = \int_0^\infty e^{-st}\frac{dx}{dt}(t)\,dt \tag{4.48}$$

This can be integrated by parts to obtain

$$\mathscr{L}\frac{dx}{dt}(t) = e^{-st}x(t)\Big|_0^\infty + s\int_0^\infty e^{-st}x(t)\,dt = s\bar{x}(s) - x(0) \tag{4.49}$$

where $x(0) = x_0$ is the initial displacement of the mass $m$. Similarly, the Laplace transform of the second derivative of $x(t)$ can be obtained:

$$\mathscr{L}\frac{d^2x}{dt^2}(t) = \int_0^\infty e^{-st}\frac{d^2x}{dt^2}(t)\,dt = s^2\bar{x}(s) - sx(0) - \dot{x}(0) \tag{4.50}$$

where $\dot{x}(0) = \dot{x}_0$ is the initial velocity of the mass $m$. Since the Laplace transform of the force $F(t)$ is given by

$$\bar{F}(s) = \mathscr{L}F(t) = \int_0^\infty e^{-st}F(t)\, dt \tag{4.51}$$

we can transform both sides of Eq. (4.47) and obtain, using Eqs. (4.46) and (4.48) to (4.51),

$$m\mathscr{L}\ddot{x}(t) + c\mathscr{L}\dot{x}(t) + k\mathscr{L}x(t) = \mathscr{L}F(t)$$

or

$$(ms^2 + cs + k)\bar{x}(s) = \bar{F}(s) + m\dot{x}(0) + (ms + c)x(0) \tag{4.52}$$

where the right-hand side of Eq. (4.52) can be regarded as a generalized transformed excitation.

For the present, we take $\dot{x}(0)$ and $x(0)$ as zero, which is equivalent to ignoring the homogeneous solution of the differential equation (4.47). Then the ratio of the transformed excitation to the transformed response $\bar{Z}(s)$ can be expressed as

$$\bar{Z}(s) = \frac{\bar{F}(s)}{\bar{x}(s)} = ms^2 + cs + k \tag{4.53}$$

The function $\bar{Z}(s)$ is known as the *generalized impedance* of the system. The reciprocal of the function $\bar{Z}(s)$ is called the *admittance* or *transfer function* of the system and is denoted as $\bar{Y}(s)$:

$$\bar{Y}(s) = \frac{1}{\bar{Z}(s)} = \frac{\bar{x}(s)}{\bar{F}(s)} = \frac{1}{ms^2 + cs + k} = \frac{1}{m(s^2 + 2\zeta\omega_n s + \omega_n^2)} \tag{4.54}$$

It can be seen that by letting $s = i\omega$ in $\bar{Y}(s)$ and multiplying by $k$, we obtain the complex frequency response $H(i\omega)$ defined in Eq. (3.49). Equation (4.54) can also be expressed as

$$\bar{x}(s) = \bar{Y}(s)\bar{F}(s) \tag{4.55}$$

which indicates that the transfer function can be regarded as an algebraic operator that operates on the transformed force to yield the transformed response.

To find the desired response $x(t)$ from $\bar{x}(s)$, we have to take the inverse Laplace transform of $\bar{x}(s)$, which can be defined symbolically as

$$x(t) = \mathscr{L}^{-1}\bar{x}(s) = \mathscr{L}^{-1}\bar{Y}(s)\bar{F}(s) \tag{4.56}$$

In general, the operator $\mathscr{L}^{-1}$ involves a line integral in the complex domain. Fortunately, we need not evaluate these integrals separately for each problem; such integrations have been carried out for various common forms of the function $F(t)$ and tabulated [4.4]. One such table is given in Appendix B. In order to find the solution using Eq. (4.56), we usually look for ways of decomposing $\bar{x}(s)$ into a combination of simple functions whose inverse transformations are available in Laplace transform tables. We can decompose $\bar{x}(s)$ conveniently by the method of partial fractions.

In the above discussion, we ignored the homogeneous solution by assuming $x(0)$ and $\dot{x}(0)$ as zero. We now consider the general solution by taking the initial conditions as $x(0) = x_0$ and $\dot{x}(0) = \dot{x}_0$. From Eq. (4.52), the transformed response $\bar{x}(s)$ can be obtained:

$$\bar{x}(s) = \frac{\bar{F}(s)}{m\left(s^2 + 2\zeta\omega_n s + \omega_n^2\right)} + \frac{s + 2\zeta\omega_n}{s^2 + 2\zeta\omega_n s + \omega_n^2}x_0 + \frac{1}{s^2 + 2\zeta\omega_n s + \omega_n^2}\dot{x}_0 \qquad (4.57)$$

We can obtain the inverse transform of $\bar{x}(s)$ by considering each term on the right side of Eq. (4.57) separately. We also make use of the following relation [4.4]:

$$\mathscr{L}^{-1}\bar{f}_1(s)\bar{f}_2(s) = \int_0^t f_1(\tau)f_2(t - \tau)\,d\tau \qquad (4.58)$$

By considering the first term on the right side of Eq. (4.57) as $\bar{f}_1(s)\bar{f}_2(s)$, where

$$\bar{f}_1(s) = \bar{F}(s) \quad \text{and} \quad \bar{f}_2(s) = \frac{1}{m\left(s^2 + 2\zeta\omega_n s + \omega_n^2\right)}$$

and by noting that $f_1(t) = \mathscr{L}^{-1}\bar{f}_1(s) = F(t)$, we obtain*

$$\mathscr{L}^{-1}\bar{f}_1(s)\bar{f}_2(s) = \frac{1}{m\omega_d}\int_0^t F(\tau)e^{-\zeta\omega_n(t-\tau)}\sin\omega_d(t - \tau)\,d\tau \qquad (4.59)$$

Considering the second term on the right side of Eq. (4.57), we find the inverse transform of the coefficient of $x_0$ from the table in Appendix B:

$$\mathscr{L}^{-1}\left(\frac{s + 2\zeta\omega_n}{s^2 + 2\zeta\omega_n s + \omega_n^2}\right) = \frac{1}{\sqrt{1 - \zeta^2}}e^{-\zeta\omega_n t}\sin(\omega_d t + \phi_1) \qquad (4.60)$$

where

$$\phi_1 = \cos^{-1}(\zeta) \qquad (4.61)$$

Finally, the inverse transform of the coefficient of $\dot{x}_0$ in the third term on the right side of Eq. (4.57) can be obtained from the table in Appendix B:

$$\mathscr{L}^{-1}\left[\frac{1}{\left(s^2 + 2\zeta\omega_n s + \omega_n^2\right)}\right] = \frac{1}{\omega_d}e^{-\zeta\omega_n t}\sin\omega_d t \qquad (4.62)$$

Using Eqs. (4.57), (4.59), (4.60), and (4.62), the general solution of Eq. (4.47) can be expressed as

$$x(t) = \frac{x_0}{\left(1 - \zeta^2\right)^{1/2}}e^{-\zeta\omega_n t}\sin(\omega_d t + \phi_1) + \frac{\dot{x}_0}{\omega_d}e^{-\zeta\omega_n t}\sin\omega_d t$$

$$+ \frac{1}{m\omega_d}\int_0^t F(\tau)e^{-\zeta\omega_n(t-\tau)}\sin\omega_d(t - \tau)\,d\tau \qquad (4.63)$$

---

*The inverse transform of $\bar{f}_2(s)$ is obtained from the Laplace transform table in Appendix B.

**EXAMPLE 4.10**

Find the response of a spring-mass-damper system when $\zeta < 1$ and the system is subjected to the force

$$F(t) = \begin{cases} F_0 & \text{for } 0 \le t \le t_0 \\ 0 & \text{for } t > t_0 \end{cases}$$

**Solution.** By taking the Laplace transform of the governing differential equation, Eq. (4.47), we obtain Eq. (4.57), using Appendix B, with

$$\overline{F}(s) = \mathscr{L}F(t) = \frac{F_0(1 - e^{-t_0 s})}{s} \tag{E.1}$$

Thus Eq. (4.57) can be written as

$$\overline{x}(s) = \frac{F_0(1 - e^{-t_0 s})}{ms(s^2 + 2\zeta\omega_n s + \omega_n^2)} + \frac{s + 2\zeta\omega_n}{s^2 + 2\zeta\omega_n s + \omega_n^2} x_0$$

$$+ \frac{1}{s^2 + 2\zeta\omega_n + \omega_n^2}\dot{x}_0$$

$$= \frac{F_0}{m\omega_n^2} \frac{1}{s\left(\dfrac{s^2}{\omega_n^2} + \dfrac{2\zeta s}{\omega_n} + 1\right)} - \frac{F_0}{m\omega_n^2} \frac{e^{-t_0 s}}{s\left(\dfrac{s^2}{\omega_n^2} + \dfrac{2\zeta s}{\omega_n} + 1\right)}$$

$$+ \frac{x_0}{\omega_n^2} \frac{s}{\left(\dfrac{s^2}{\omega_n^2} + \dfrac{2\zeta s}{\omega_n} + 1\right)} + \left(\frac{2\zeta x_0}{\omega_n} + \frac{\dot{x}_0}{\omega_n^2}\right) \frac{1}{\left(\dfrac{s^2}{\omega_n^2} + \dfrac{2\zeta s}{\omega_n} + 1\right)} \tag{E.2}$$

The inverse transform of Eq. (E.2) can be expressed by using the results in Appendix B as

$$x(t) = \frac{F_0}{m\omega_n^2}\left[1 - \frac{e^{-\zeta\omega_n t}}{\sqrt{1 - \zeta^2}} \sin\left\{\omega_n\sqrt{1 - \zeta^2}\, t + \phi_1\right\}\right]$$

$$- \frac{F_0}{m\omega_n^2}\left[1 - \frac{e^{-\zeta\omega_n(t - t_0)}}{\sqrt{1 - \zeta^2}} \sin\left\{\omega_n\sqrt{1 - \zeta^2}\,(t - t_0) + \phi_1\right\}\right]$$

$$- \frac{x_0}{\omega_n^2}\left[\frac{\omega_n^2 e^{-\zeta\omega_n t}}{\sqrt{1 - \zeta^2}} \sin\left\{\omega_n\sqrt{1 - \zeta^2}\, t - \phi_1\right\}\right]$$

$$+ \left(\frac{2\zeta x_0}{\omega_n} + \frac{\dot{x}_0}{\omega_n^2}\right)\left[\frac{\omega_n}{\sqrt{1 - \zeta^2}} e^{-\zeta\omega_n t}\sin\left(\omega_n\sqrt{1 - \zeta^2}\, t\right)\right] \tag{E.3}$$

where $\phi_1$ is given by Eq. (4.61). Thus the response of the spring-mass-damper system can be expressed as

$$x(t) = \frac{F_0}{m\omega_n^2\sqrt{1-\zeta^2}}\left[-e^{-\zeta\omega_n t}\sin\left(\omega_n\sqrt{1-\zeta^2}\,t + \phi_1\right)\right.$$

$$\left. + e^{-\zeta\omega_n(t-t_0)}\sin\left\{\omega_n\sqrt{1-\zeta^2}\,(t-t_0) + \phi_1\right\}\right]$$

$$- \frac{x_0}{\sqrt{1-\zeta^2}}e^{-\zeta\omega_n t}\sin\left(\omega_n\sqrt{1-\zeta^2}\,t - \phi_1\right)$$

$$+ \frac{(2\zeta\omega_n x_0 + \dot{x}_0)}{\omega_n\sqrt{1-\zeta^2}}e^{-\zeta\omega_n t}\sin\left(\omega_n\sqrt{1-\zeta^2}\,t\right) \tag{E.4}$$

Although the first part of Eq. (E.4) is expected to be the same as Eq. (E.1) of Example 4.5, it is difficult to see the equivalence in the present form of Eq. (E.4). However, for the undamped system, Eq. (E.4) reduces to

$$x(t) = \frac{F_0}{m\omega_n^2}\left[-\sin\left(\omega_n t + \frac{\pi}{2}\right) + \sin\left\{\omega_n(t-t_0) + \frac{\pi}{2}\right\}\right]$$

$$- x_0\sin\left(\omega_n t - \frac{\pi}{2}\right) + \frac{\dot{x}_0}{\omega_n}\sin\omega_n t$$

$$= \frac{F_0}{k}\left[\cos\omega_n(t-t_0) - \cos\omega_n t\right] + x_0\cos\omega_n t + \frac{\dot{x}_0}{\omega_n}\sin\omega_n t \tag{E.5}$$

The first or steady-state part of Eq. (E.5) can be seen to be identical to Eq. (E.3) of Example 4.5.

## 4.8 RESPONSE TO IRREGULAR FORCING CONDITIONS USING NUMERICAL METHODS

In the previous sections, it was assumed that the forcing functions $F(t)$ are available as functions of time in an explicit manner. In many practical problems, however, the forcing functions $F(t)$ are not available in the form of analytical expressions. When a forcing function is determined experimentally, $F(t)$ may be known as an irregular curve. Sometimes only the values of $F(t) = F_i$ at a series of points $t = t_i$ may be available, in the form of a diagram or a table. In such cases it may be feasible to replace the data with polynomials or some such formulas by using curve-fitting techniques and then to use (in the case of a single degree of freedom system) the Duhamel integral, Eq. (4.33), to find the response. However, a more generalized method of finding the response involves using a numerical approach, assuming a suitable type of variation of the function $F(t)$ during any time step. We shall present this numerical approach in this section, using several types of interpolation functions

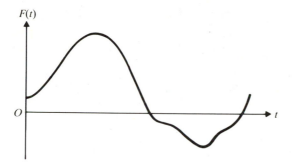

**Figure 4.13**

for $F(t)$. The direct numerical integration of the equations of motion is discussed in Chap. 10.

**Method 1.** Let the function $F(t)$ vary with time in an arbitrary manner, as indicated in Fig. 4.13. This forcing function can be approximated by a series of step functions of various magnitudes and starting at different instants, as shown in Fig. 4.14. The first step function has a magnitude $\Delta F_1$ and starts at time $t = t_1 = 0$, the second one has a magnitude $\Delta F_2$ and starts at time $t = t_2$, etc. The response of the system in any time interval $t_{j-1} \leq t \leq t_j$ due to the step functions $\Delta F_i$ ($i = 1, 2, \ldots, j-1$) can be found, using the results of Example 4.3:

$$x(t) = \frac{1}{k} \sum_{i=1}^{j-1} \Delta F_i \left[ 1 - e^{-\zeta \omega_n (t - t_i)} \left\{ \cos \omega_d (t - t_i) + \frac{\zeta \omega_n}{\omega_d} \sin \omega_d (t - t_i) \right\} \right] \qquad (4.64)$$

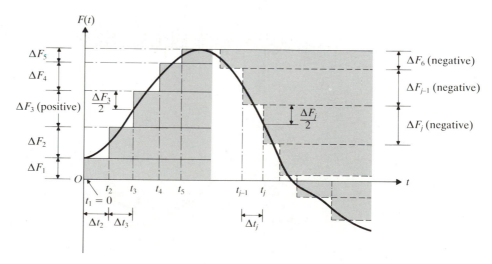

**Figure 4.14**

Thus the response of the system at $t = t_j$ becomes

$$x_j = \frac{1}{k} \sum_{i=1}^{j-1} \Delta F_i \left[ 1 - e^{-\zeta \omega_n (t_j - t_i)} \left\{ \cos \omega_d (t_j - t_i) + \frac{\zeta \omega_n}{\omega_d} \sin \omega_d (t_j - t_i) \right\} \right] \qquad (4.65)$$

Notice that the step function $\Delta F_i$ of step $i$ is positive if the slope of the $F$-versus-$t$ curve is positive, and it is negative if the slope of the $F$-versus-$t$ curve is negative, as indicated in Fig. 4.14. To find the response $x(t)$ accurately, the steps taken should be small. Also, the errors involved in the area under the $F(t)$ curve should be made approximately self-compensating; that is, the hatched areas lying above the $F(t)$ curve should be approximately equal to the unhatched areas lying below the $F(t)$ curve in Fig. 4.14. This can be achieved if we assume, after the first step, that the steps start at instants when the ordinates of the curve are at the mid-heights of the steps, as shown in Fig. 4.14.

**Method 2.** Instead of approximating the $F(t)$ curve by a succession of step functions, we can approximate it by a series of rectangular impulses $F_i$, as shown in Fig. 4.15. These impulses $F_i$ are positive or negative, depending on whether the curve $F(t)$ lies above or below the time $(t)$ axis. As in the previous case, the magnitudes of $F_i$ should be selected as the ordinates of the $F(t)$ curve at the midpoints of the time intervals $\Delta t_i$, as shown in Fig. 4.15. The response of the system in any time interval $t_{j-1} \leq t \leq t_j$ can be found by adding the response due to $F_j$ (applied in the interval $\Delta t_j$) to the response existing at $t = t_{j-1}$ (initial condition). This gives

$$x(t) = \frac{F_j}{k} \left[ 1 - e^{-\zeta \omega_n (t - t_{j-1})} \left\{ \cos \omega_d (t - t_{j-1}) + \frac{\zeta \omega_n}{\omega_d} \sin \omega_d (t - t_{j-1}) \right\} \right]$$

$$+ e^{-\zeta \omega_n (t - t_{j-1})} \left\{ x_{j-1} \cos \omega_d (t - t_{j-1}) + \frac{\dot{x}_{j-1} + \zeta \omega_n x_{j-1}}{\omega_d} \sin \omega_d (t - t_{j-1}) \right\}$$

$$\qquad (4.66)$$

By substituting $t = t_j$ in Eq. (4.66) the response of the system at the end of the

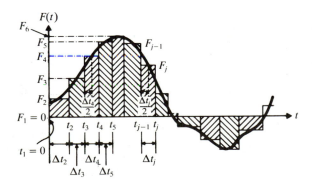

**Figure 4.15**

interval $\Delta t_j$ can be obtained:

$$x_j = \frac{F_j}{k}\left[1 - e^{-\zeta\omega_n \Delta t_j}\left\{\cos\omega_d \cdot \Delta t_j + \frac{\zeta\omega_n}{\omega_d}\sin\omega_d \cdot \Delta t_j\right\}\right]$$

$$+ e^{-\zeta\omega_n \Delta t_j}\left\{x_{j-1}\cos\omega_d \cdot \Delta t_j + \frac{\dot{x}_{j-1} + \zeta\omega_n x_{j-1}}{\omega_d}\sin\omega_d \cdot \Delta t_j\right\} \qquad (4.67)$$

By differentiating Eq. (4.66) with respect to $t$ and substituting $t = t_j$, we obtain the velocity $\dot{x}_j$ at the end of the interval $\Delta t_j$:

$$\dot{x}_j = \frac{F_j\omega_d}{k}e^{-\zeta\omega_n \Delta t_j}\left(1 + \frac{\zeta^2\omega_n^2}{\omega_d^2}\right)\sin\omega_d \cdot \Delta t_j + \omega_d e^{-\zeta\omega_n \cdot \Delta t_j}$$

$$\times\left\{-x_{j-1}\sin\omega_d \cdot \Delta t_j + \frac{\dot{x}_{j-1} + \zeta\omega_n x_{j-1}}{\omega_d}\cos\omega_d \cdot \Delta t_j\right.$$

$$\left. - \frac{\zeta\omega_n}{\omega_d}\left[x_{j-1}\cos\omega_d \cdot \Delta t_j + \frac{\dot{x}_{j-1} + \zeta\omega_n x_{j-1}}{\omega_d}\sin\omega_d \cdot \Delta t_j\right]\right\} \qquad (4.68)$$

Equations (4.67) and (4.68) represent recurrence relations for computing the response at the end of $j$th time step. They also provide the initial conditions of $x_j$ and $\dot{x}_j$ at the beginning of step $j + 1$. These equations may be repetitively applied to find the time histories of displacement and velocity of the system.

**Method 3.** In the piecewise-constant types of interpolations discussed in Methods 1 and 2, it is not always possible to balance the errors in area above and below the $F(t)$ curve. Therefore, higher-order interpolations are sometimes used. If we use a piecewise linear type of interpolation, the approximation of the $F(t)$ curve will appear as shown in Fig. 4.16. In this case the response of the system in the time

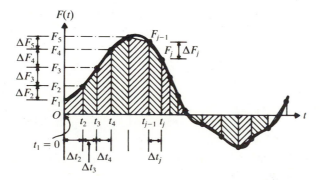

**Figure 4.16**

interval $t_{j-1} \le t \le t_j$ is given by

$$
x(t) = \frac{\Delta F_j}{k \cdot \Delta t_j} \left[ t - t_{j-1} - \frac{2\zeta}{\omega_n} + e^{-\zeta \omega_n (t - t_{j-1})} \right.
$$

$$
\times \left\{ \frac{2\zeta}{\omega_n} \cos \omega_d (t - t_{j-1}) \right.
$$

$$
\left. \left. - \frac{\omega_d^2 - \zeta^2 \omega_n^2}{\omega_n^2 \omega_d} \sin \omega_d (t - t_{j-1}) \right\} \right]
$$

$$
+ \frac{F_{j-1}}{k} \left[ 1 - e^{-\zeta \omega_n (t - t_{j-1})} \left\{ \cos \omega_d (t - t_{j-1}) + \frac{\zeta \omega_n}{\omega_d} \sin \omega_d (t - t_{j-1}) \right\} \right]
$$

$$
+ e^{-\zeta \omega_n (t - t_{j-1})} \left[ x_{j-1} \cos \omega_d (t - t_{j-1}) + \frac{\dot{x}_{j-1} + \zeta \omega_n x_{j-1}}{\omega_d} \sin \omega_d (t - t_{j-1}) \right]
$$

$$
\tag{4.69}
$$

where $\Delta F_j = F_j - F_{j-1}$. Notice that the first part of Eq. (4.69) corresponds to the solution of the ramp function (Example 4.6). By setting $t = t_j$ in Eq. (4.69), we obtain the response at the end of the interval $\Delta t_j$:

$$
x_j = \frac{\Delta F_j}{k \Delta t_j} \left[ \Delta t_j - \frac{2\zeta}{\omega_n} + e^{-\zeta \omega_n \Delta t_j} \left( \frac{2\zeta}{\omega_n} \cos \omega_d \cdot \Delta t_j - \frac{\omega_d^2 - \zeta^2 \omega_n^2}{\omega_n^2 \omega_d} \sin \omega_d \cdot \Delta t_j \right) \right]
$$

$$
+ \frac{F_{j-1}}{k} \left[ 1 - e^{-\zeta \omega_n \cdot \Delta t_j} \left( \cos \omega_d \cdot \Delta t_j + \frac{\zeta \omega_n}{\omega_d} \sin \omega_d \cdot \Delta t_j \right) \right]
$$

$$
+ e^{-\zeta \omega_n \cdot \Delta t_j} \left[ x_{j-1} \cos \omega_d \cdot \Delta t_j + \frac{\dot{x}_{j-1} + \zeta \omega_n x_{j-1}}{\omega_d} \sin \omega_d \cdot \Delta t_j \right] \tag{4.70}
$$

By differentiating Eq. (4.69) with respect to $t$ and substituting $t = t_j$, we obtain the velocity at the end of the interval:

$$
\dot{x}_j = \frac{\Delta F_j}{k \Delta t_j} \left[ 1 - e^{-\zeta \omega_n \cdot \Delta t_j} \left( \cos \omega_d \cdot \Delta t_j + \frac{\zeta \omega_n}{\omega_d} \sin \omega_d \cdot \Delta t_j \right) \right]
$$

$$
+ \frac{F_{j-1}}{k} e^{-\zeta \omega_n \cdot \Delta t_j} \frac{\omega_n^2}{\omega_d} \cdot \sin \omega_d \cdot \Delta t_j + e^{-\zeta \omega_n \cdot \Delta t_j}
$$

$$
\left[ \dot{x}_{j-1} \cos \omega_d \cdot \Delta t_j - \frac{\zeta \omega_n}{\omega_d} \left( \dot{x}_{j-1} + \frac{\omega_n}{\zeta} x_{j-1} \right) \sin \omega_d \cdot \Delta t_j \right] \tag{4.71}
$$

Equations (4.70) and (4.71) are the recurrence relations analogous to Eqs. (4.67) and (4.68) for rectangular impulses.

**EXAMPLE 4.11**

Find the response of a spring-mass-damper system subjected to the forcing function

$$F(t) = F_0\left(1 - \sin\frac{\pi t}{2t_0}\right) \tag{E.1}$$

in the interval $0 \le t \le t_0$, using a numerical procedure. Assume $F_0 = 1$, $k = 1$, $m = 1$, $\zeta = 0.1$, and $t_0 = \tau_n/2$, where $\tau_n$ denotes the natural period of vibration given by

$$\tau_n = \frac{2\pi}{\omega_n} = \frac{2\pi}{(k/m)^{1/2}} = 2\pi \tag{E.2}$$

The values of $x$ and $\dot{x}$ at $t = 0$ are zero.

***Solution.*** Figure 4.17 shows the forcing function of Eq. (E.1) and Figs. 4.18 to 4.21 show four different ways of idealizing the forcing function using vertical strips. In all these idealizations, the time interval 0 to $t_0$ is divided into ten equal steps $\Delta t_i$ with

$$\Delta t_i = \frac{t_0}{10} = \frac{\pi}{10}; \qquad i = 2, 3, \dots, 11 \tag{E.3}$$

In Figs. 4.18 to 4.20, rectangular impulses are used; in Fig. 4.21, piecewise linear (trapezoidal) impulses are used. In the case of Figs. 4.18, 4.19, and 4.20, the magnitudes of the impulses are computed by the ordinates of the $F(t)$ curve at the beginning of the step, at the end of the step, and at the middle of the step, respectively. The numerical results are given in Table 4.1. As can be expected from the idealizations, the results obtained by idealizations 1 and 2 (Figs. 4.18 and 4.19)

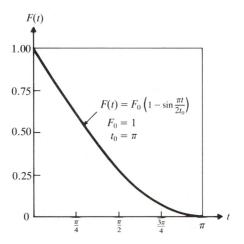

**Figure 4.17**

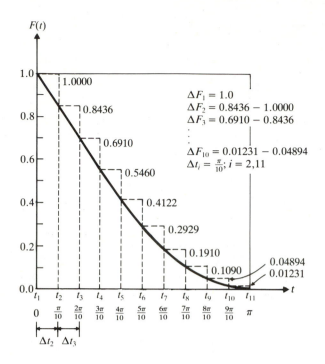

**Figure 4.18**

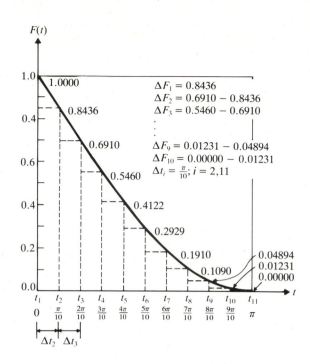

**Figure 4.19**

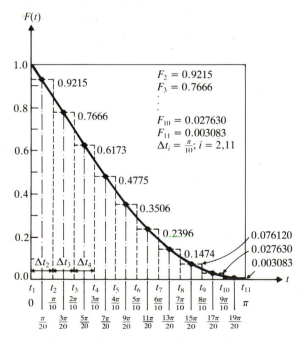

**Figure 4.20**

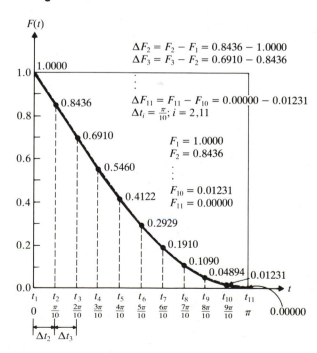

**Figure 4.21**

**TABLE 4.1**

| $i$ | $t_i$ | $x(t_i)$ Obtained According to | | | |
|---|---|---|---|---|---|
| | | Fig. 4.18 (Idealization 1) | Fig. 4.19 (Idealization 2) | Fig. 4.20 (Idealization 3) | Fig. 4.21 (Idealization 4) |
| 1 | 0 | 0.00000 | 0.00000 | 0.00000 | 0.00000 |
| 2 | $0.1\pi$ | 0.04794 | 0.04044 | 0.04417 | 0.04541 |
| 3 | $0.2\pi$ | 0.17578 | 0.14729 | 0.16147 | 0.16377 |
| 4 | $0.3\pi$ | 0.35188 | 0.29228 | 0.32190 | 0.32499 |
| 5 | $0.4\pi$ | 0.54248 | 0.44609 | 0.49392 | 0.49746 |
| 6 | $0.5\pi$ | 0.71540 | 0.58160 | 0.64790 | 0.65151 |
| 7 | $0.6\pi$ | 0.84330 | 0.67659 | 0.75906 | 0.76238 |
| 8 | $0.7\pi$ | 0.90630 | 0.71578 | 0.80986 | 0.81255 |
| 9 | $0.8\pi$ | 0.89367 | 0.69214 | 0.79142 | 0.79323 |
| 10 | $0.9\pi$ | 0.80449 | 0.60717 | 0.70403 | 0.70482 |
| 11 | $\pi$ | 0.64730 | 0.47152 | 0.55672 | 0.55647 |

overestimate and underestimate the true response, respectively. The results given by idealizations 3 and 4 are expected to lie between those given by idealizations 1 and 2. Further, the results obtained from idealization 4 will be the most accurate ones.

## 4.9 COMPUTER PROGRAMS

### 4.9.1 Response under an Arbitrary Periodic Forcing Function

A Fortran computer program, in the form of subroutine PERIOD, is given for finding the dynamic response of a damped oscillator excited by any periodic external force applied to the mass. The arguments of the subroutine are as follows:

| | | |
|---|---|---|
| XM | = | Mass of the system. Input data. |
| XK | = | Stiffness of the spring. Input data. |
| XAI | = | Damping ratio $\zeta$. Input data. |
| N | = | Number of equidistant points at which the values of the force $F(t)$ are known. Input data. |
| M | = | Number of Fourier coefficients to be considered in the solution. Input data. |
| TIME | = | Time period of the function $F(t)$. Input data. |
| F | = | Array of dimension $N$ which contains the known values of $F(t)$. $F(I) = F(t_i)$. Input data. |
| T | = | Array of dimension $N$ which contains the known values of time $t$. $T(I) = t_i$. Input data. |

FZERO = $F_0$. Output.

FC = Array of dimension $M$. FC(J) = $F_j$. Output.

X = Array of dimension $N$ which contains the computed response at time $t$. X(I) = $x_i$. Output.

A sample problem and the listing of the program are given below.

```
C =================================================================
C
C PROGRAM 4
C MAIN PROGRAM WHICH CALLS PERIOD
C
C =================================================================
C FOLLOWING 10 LINES CONTAIN PROBLEM-DEPENDENT DATA
      DIMENSION F(24),T(24),XSIN(20),XCOS(20),PSI(20),PHI(20),FC(20),
     2 X(24),XPC(20),XPS(20)
      DATA XM,XK,XAI /100.0,100000.0,0.1/
      DATA N,M,TIME /24,20,0.12/
      DATA F/24000.0,48000.0,72000.0,96000.0,120000.0,96000.0,72000.0,
     2 48000.0,24000.0,0.0,0.0,0.0,0.0,0.0,0.0,0.0,0.0,0.0,0.0,
     3 0.0,0.0,0.0,0.0/
      DATA T/0.005,0.010,0.015,0.020,0.025,0.030,0.035,0.040,0.045,
     2 0.050,0.055,0.060,0.065,0.070,0.075,0.080,0.085,0.090,0.095,
     3 0.100,0.105,0.110,0.115,0.120/
C END OF PROBLEM-DEPENDENT DATA
      CALL PERIOD (XM,XK,XAI,N,M,TIME,F,T,XSIN,XCOS,PSI,PHI,FZERO,FC,
     2 X,XPC,XPS)
      PRINT 100, XM,XK,XAI,N,M,TIME
100   FORMAT (/,56H RESPONSE OF A SINGLE D.O.F. SYSTEM UNDER PERIODIC FO
     2RCE,//,6H XM  =,E15.6,/,6H XK  =,E15.6,/,6H XAI =,E15.6,/,
     36H N  =,I3,/,6H M   =,I3,/,6H TIME=,E15.6,/)
      PRINT 200
200   FORMAT (/,27H APPLIED FORCE AND RESPONSE,//,3H  I,3X,5H T(I),10X,
     2 5H F(I),10X,5H X(I),/)
      DO 400 I=1,N
400   PRINT 500, I,T(I),F(I),X(I)
500   FORMAT (I3,3E15.6)
      STOP
      END
C =================================================================
C
C SUBROUTINE PERIOD
C
C =================================================================
      SUBROUTINE PERIOD (XM,XK,XAI,N,M,TIME,F,T,XSIN,XCOS,PSI,PHI,
     2 FZERO,FC,X,XPC,XPS)
      DIMENSION F(N),T(N),XSIN(M),XCOS(M),PSI(M),PHI(M),FC(M),X(N)
     2 ,XPC(M),XPS(M)
      OMEG=2.0*3.1416/TIME
      OMEGN=SQRT(XK/XM)
      SUMZ=0.0
      DO 100 I=1,N
100   SUMZ=SUMZ+F(I)
      FZERO=2.0*SUMZ/REAL(N)
```

```
              DO 300 J=1,M
              SUMS=0.0
              SUMC=0.0
              DO 200 I=1,N
              THETA=REAL(J)*OMEG*T(I)
              FSIN=F(I)*SIN(THETA)
              FCOS=F(I)*COS(THETA)
              SUMS=SUMS+FSIN
              SUMC=SUMC+FCOS
      200     CONTINUE
              AJ=2.0*SUMC/REAL(N)
              BJ=2.0*SUMS/REAL(N)
              R=OMEG/OMEGN
              PHI(J)=ATAN(2.0*XAI*REAL(J)*R/(1.0-(REAL(J)*R)**2))
              CON=SQRT((1.0-(REAL(J)*R)**2)**2+(2.0*XAI*REAL(J)*R)**2)
              XPC(J)=(AJ/XK)/CON
              XPS(J)=(BJ/XK)/CON
      300     CONTINUE
              DO 400 I=1,N
              X(I)=FZERO/(2.0*XK)
              DO 500 J=1,M
      500     X(I)=X(I)+XPC(J)*COS(REAL(J)*OMEG*T(I)-PHI(J))
             2 +XPS(J)*SIN(REAL(J)*OMEG*T(I)-PHI(J))
      400     CONTINUE
              RETURN
              END
```

RESPONSE OF A SINGLE D.O.F. SYSTEM UNDER PERIODIC FORCE

```
XM   =    0.100000E+03
XK   =    0.100000E+06
XAI  =    0.100000E+00
N    =       24
M    =       20
TIME=    0.120000E+00
```

APPLIED FORCE AND RESPONSE

| I | T(I) | F(I) | X(I) |
|---|------|------|------|
| 1 | 0.500000E-02 | 0.240000E+05 | 0.393120E+00 |
| 2 | 0.100000E-01 | 0.480000E+05 | 0.451156E+00 |
| 3 | 0.150000E-01 | 0.720000E+05 | 0.496755E+00 |
| 4 | 0.200000E-01 | 0.960000E+05 | 0.523365E+00 |
| 5 | 0.250000E-01 | 0.120000E+06 | 0.525113E+00 |
| 6 | 0.300000E-01 | 0.960000E+05 | 0.497451E+00 |
| 7 | 0.350000E-01 | 0.720000E+05 | 0.447280E+00 |
| 8 | 0.400000E-01 | 0.480000E+05 | 0.382350E+00 |
| 9 | 0.450000E-01 | 0.240000E+05 | 0.310534E+00 |
| 10 | 0.500000E-01 | 0.000000E+00 | 0.239646E+00 |
| 11 | 0.550000E-01 | 0.000000E+00 | 0.176984E+00 |
| 12 | 0.600000E-01 | 0.000000E+00 | 0.124139E+00 |
| 13 | 0.650000E-01 | 0.000000E+00 | 0.821526E-01 |
| 14 | 0.700000E-01 | 0.000000E+00 | 0.517498E-01 |

| | | | |
|---|---|---|---|
| 15 | 0.750000E-01 | 0.000000E+00 | 0.333252E-01 |
| 16 | 0.800000E-01 | 0.000000E+00 | 0.269447E-01 |
| 17 | 0.850000E-01 | 0.000000E+00 | 0.323697E-01 |
| 18 | 0.900000E-01 | 0.000000E+00 | 0.490896E-01 |
| 19 | 0.950000E-01 | 0.000000E+00 | 0.763507E-01 |
| 20 | 0.100000E+00 | 0.000000E+00 | 0.113176E+00 |
| 21 | 0.105000E+00 | 0.000000E+00 | 0.158378E+00 |
| 22 | 0.110000E+00 | 0.000000E+00 | 0.210580E+00 |
| 23 | 0.115000E+00 | 0.000000E+00 | 0.268249E+00 |
| 24 | 0.120000E+00 | 0.000000E+00 | 0.329747E+00 |

### 4.9.2    Response under Arbitrary Forcing Function
### Using the Methods of Section 4.8

A Fortran computer program is given for finding the response of a viscously damped single degree of freedom system under arbitrary forcing function using the methods outlined in Sec. 4.8. For illustration, the data of Example 4.11 is used. The following input data is required for this program.

F    =    Array containing the values of the forcing function at various time stations according to the idealization of Fig. 4.14 (Fig. 4.18 or 4.19 for Example 4.11).

FF    =    Array containing the values of the forcing function at various time stations according to the idealization of Fig. 4.16 (Fig. 4.20 or 4.21 for Example 4.11).

XAI    =    Damping factor.

OMN    =    Undamped natural frequency of the system.

DELT    =    Incremental time between consecutive time stations.

XK    =    Spring stiffness.

The program prints the values of $x(t_i)$ and $\dot{x}(t_i)$ given by four different methods at time stations $t_2, t_3, \ldots, t_{11}$. Although the program uses the data of Example 4.11 directly, it can be generalized to find the response under any arbitrary forcing function of any single degree of freedom system.

```
C =============================================================
C
C PROGRAM 5
C RESPONSE OF A SINGLE D.O.F. SYSTEM UNDER ARBITRARY FORCING FUNCTION
C USING THE METHODS OF SECTION 4.8
C
C =============================================================
C FOLLOWING 10 LINES CONTAIN PROBLEM-DEPENDENT DATA
      DIMENSION F(11),FF(11),DELF(11),T(11),X(11),XD(11),X1(11),
     2    XD1(11),X2(11),XD2(11),X3(11),XD3(11),X4(11),XD4(11)
      DATA F/0.0,1.0,0.84356554,0.69098301,0.54600950,0.41221475,
     2    0.29289322,0.19098301,0.10899348,0.04894348,0.01231166/
```

```
      DATA FF/1.0,0.92154090,0.76655464,0.61731657,0.47750144,
     2   0.35055195,0.23959404,0.14735984,0.07612047,0.02763008,
     3   0.00308267/
      DATA XAI,OMN,XK /0.1,1.0,1.0/
      DELT=3.14159265/10.0
      DATA NP,NP1,NP2 /11,10,9/
C NP = NUMBER OF POINTS AT WHICH VALUE OF F IS KNOWN, NP1=NP-1, NP2=NP-2
C END OF PROBLEM-DEPENDENT DATA
      XN=XAI*OMN
      PD=OMN*SQRT(1.0-XAI**2)
C SOLUTION ACCORDING TO METHOD 1 USING THE IDEALIZATION OF FIG. 4.18
      T(1)=0.0
      DO 10 I=2,NP
 10   T(I)=T(I-1)+DELT
      DO 20 I=1,NP1
 20   DELF(I)=F(I+1)-F(I)
      DO 40 J=2,NP
      X(J)=0.0
      XD(J)=0.0
      JM1=J-1
      DO 30 I=1,JM1
      X(J)=X(J)+(DELF(I)/XK)*(1.0-EXP(-XN*(T(J)-T(I)))*(COS(PD*(T(J)-
     2   T(I)))+(XN/PD)*SIN(PD*(T(J)-T(I)))))
C XD(J) OBTAINED BY DIFFERENTIATING EQ.(4.64)
      XD(J)=XD(J)+(DELF(I)/XK)*EXP(-XN*(T(J)-T(I)))*SIN(PD*(T(J)-
     2   T(I)))
 30      CONTINUE
 40      CONTINUE
      DO 50 I=2,NP
      X1(I)=X(I)
 50   XD1(I)=XD(I)
C SOLUTION ACCORDING TO METHOD 1 USING THE IDEALIZATION OF FIG. 4.19
      DO 60 K=2,NP2
 60   DELF(K)=DELF(K+1)
      DELF(1)=F(3)
      DELF(NP)=F(NP)
      DO 80 J=2,NP
      X(J)=0.0
      XD(J)=0.0
      JM1=J-1
      DO 70 I=1,JM1
      X(J)=X(J)+(DELF(I)/XK)*(1.0-EXP(-XN*(T(J)-T(I)))*(COS(PD*(T(J)-
     2   T(I)))+(XN/PD)*SIN(PD*(T(J)-T(I)))))
      XD(J)=XD(J)+(DELF(I)/XK)*EXP(-XN*(T(J)-T(I)))*SIN(PD*(T(J)-
     2   T(I)))
 70      CONTINUE
 80      CONTINUE
      DO 90 I=2,NP
      X2(I)=X(I)
 90   XD2(I)=XD(I)
C SOLUTION ACCORDING TO METHOD 2 USING THE IDEALIZATION OF FIG. 4.20
      X(1)=0.0
      XD(1)=0.0
      DO 100 J=2,NP
```

```
      DEL=DELT
      X(J)=(FF(J)/XK)*(1.0-EXP(-XN*DEL)*(COS(PD*DEL)+(XN/PD)*
     2   SIN(PD*DEL)))+EXP(-XN*DEL)*(X(J-1)*COS(PD*DEL)+((XD(J-1)
     3   +XN*X(J-1))/PD)*SIN(PD*DEL))
      XD(J)=(FF(J)*PD/XK)*EXP(-XN*DEL)*(1.0+XN**2/(PD**2))*SIN(PD*DEL)
     2   +PD*EXP(-XN*DEL)*(-X(J-1)*SIN(PD*DEL)+((XD(J-1)+XN*X(J-1))/PD)*
     3   COS(PD*DEL)-XN*(X(J-1)*COS(PD*DEL)+((XD(J-1)+XN*X(J-1))/PD)*
     4 SIN(PD*DEL))/PD)
100   CONTINUE
      DO 110 I=2,NP
      X3(I)=X(I)
110   XD3(I)=XD(I)
C SOLUTION ACCORDING TO METHOD 3 USING THE IDEALIZATION OF FIG. 4.21
      X(1)=0.0
      XD(1)=0.0
      DO 120 J=1,NP1
120   F(J)=F(J+1)
      F(NP)=0.0
      DO 130 J=2,NP
      DELF(J)=F(J)-F(J-1)
      X(J)=(DELF(J)/(XK*DEL))*(DEL-(2.0*XAI/OMN)+EXP(-XN*DEL)*
     2   ((2.0*XAI/OMN)*COS(PD*DEL)-((PD**2-XN**2)/(OMN*OMN*PD))*
     3   SIN(PD*DEL)))+(F(J-1)/XK)*(1.0-EXP(-XN*DEL)*(COS(PD*DEL)
     4   +(XN/PD)*SIN(PD*DEL)))+EXP(-XN*DEL)*(X(J-1)*COS(PD*DEL)
     5   +((XD(J-1)+XN*X(J-1))/PD)*SIN(PD*DEL))
      XD(J)=(DELF(J)/(XK*DEL))*(1.0-EXP(-XN*DEL)*(((XN**2+PD**2)/
     2   (OMN**2))*COS(PD*DEL)+((XN**3+XN*PD*PD)/(PD*(OMN**2)))*
     3   SIN(PD*DEL)))+(F(J-1)/XK)*EXP(-XN*DEL)*((XN**2/PD)+PD)*
     4   SIN(PD*DEL)+EXP(-XN*DEL)*(XD(J-1)*COS(PD*DEL)-((XN*XD(J-1)
     5   +XN*XN*X(J-1)+PD*PD*X(J-1))/PD)*SIN(PD*DEL))
130   CONTINUE
      DO 140 I=2,NP
      X4(I)=X(I)
140   XD4(I)=XD(I)
      PRINT 150
150   FORMAT (//,6H VALUE,6X,10H METHOD #1,7X,10H METHOD #1,7X,
     2   10H METHOD #2,7X,10H METHOD #3)
      PRINT 160
160   FORMAT (3X,3H OF,5X,11H (FIG.4.18),6X,11H (FIG.4.19),6X,
     2   11H (FIG.4.20),6X,11H (FIG.4.21),/)
      PRINT 170
170   FORMAT (3X,2H I,7X,5H X(I),12X,5H X(I),12X,5H X(I),12X,5H X(I),
     2   /)
      DO 180 I=2,NP
180   PRINT 190,I,X1(I),X2(I),X3(I),X4(I)
190   FORMAT (I5,2X,E15.6,2X,E15.6,2X,E15.6,2X,E15.6)
      PRINT 200
200   FORMAT (//,3X,2H I,6X,6H XD(I),11X,6H XD(I),11X,6H XD(I),11X,
     2   6H XD(I),/)
      DO 210 I=2,NP
210   PRINT 190,I,XD1(I),XD2(I),XD3(I),XD4(I)
      STOP
      END
```

| VALUE OF | METHOD #1 (FIG.4.18) | METHOD #1 (FIG.4.19) | METHOD #2 (FIG.4.20) | METHOD #3 (FIG.4.21) |
|---|---|---|---|---|
| I | X(I) | X(I) | X(I) | X(I) |
| 2 | 0.479360E-01 | 0.404372E-01 | 0.441750E-01 | 0.454151E-01 |
| 3 | 0.175781E+00 | 0.147294E+00 | 0.161471E+00 | 0.163773E+00 |
| 4 | 0.351883E+00 | 0.292277E+00 | 0.321877E+00 | 0.324989E+00 |
| 5 | 0.542483E+00 | 0.446091E+00 | 0.493842E+00 | 0.497464E+00 |
| 6 | 0.715396E+00 | 0.581603E+00 | 0.647699E+00 | 0.651514E+00 |
| 7 | 0.843296E+00 | 0.676586E+00 | 0.758676E+00 | 0.762379E+00 |
| 8 | 0.906301E+00 | 0.715783E+00 | 0.809225E+00 | 0.812552E+00 |
| 9 | 0.893674E+00 | 0.692145E+00 | 0.790486E+00 | 0.793231E+00 |
| 10 | 0.804490E+00 | 0.607167E+00 | 0.702788E+00 | 0.704820E+00 |
| 11 | 0.647299E+00 | 0.469170E+00 | 0.555198E+00 | 0.556465E+00 |
| I | XD(I) | XD(I) | XD(I) | XD(I) |
| 2 | 0.298008E+00 | 0.251389E+00 | 0.276010E+00 | 0.275640E+00 |
| 3 | 0.502976E+00 | 0.418148E+00 | 0.462605E+00 | 0.461687E+00 |
| 4 | 0.602270E+00 | 0.491876E+00 | 0.549249E+00 | 0.547683E+00 |
| 5 | 0.595174E+00 | 0.474576E+00 | 0.536630E+00 | 0.534405E+00 |
| 6 | 0.492171E+00 | 0.377744E+00 | 0.435845E+00 | 0.433036E+00 |
| 7 | 0.313187E+00 | 0.220601E+00 | 0.266613E+00 | 0.263378E+00 |
| 8 | 0.850188E-01 | 0.276649E-01 | 0.547668E-01 | 0.513272E-01 |
| 9 | -0.161754E+00 | -0.174042E+00 | -0.170711E+00 | -0.174093E+00 |
| 10 | -0.396047E+00 | -0.357870E+00 | -0.380784E+00 | -0.383830E+00 |
| 11 | -0.589414E+00 | -0.507466E+00 | -0.549289E+00 | -0.551730E+00 |

## REFERENCES

**4.1.** M. J. Maron, *Applied Numerical Analysis*, Macmillan, New York, 1982.

**4.2.** M. Paz, *Structural Dynamics: Theory and Computation*, Van Nostrand Reinhold, New York, 1980.

**4.3.** E. Kreyszig, *Advanced Engineering Mathematics*, (5th Ed.), John Wiley, New York, 1983.

**4.4.** F. Oberhettinger and L. Badii, *Tables of Laplace Transforms*, Springer Verlag, New York, 1973.

**4.5.** G. M. Hieber et al., "Understanding and measuring the shock response spectrum, Part I," *Sound and Vibration*, Vol. 8, March 1974, pp. 42–49.

**4.6.** R. E. D. Bishop, A. G. Parkinson, and J. W. Pendered, "Linear analysis of transient vibration," *Journal of Sound and Vibration*, Vol. 9, 1969, pp. 313–337.

**4.7.** Y. Matsuzaki and S. Kibe, "Shock and seismic response spectra in design problems," *Shock and Vibration Digest*, Vol. 15, October 1983, pp. 3–10.

**4.8.** F. E. Udwadia and S. Tabaie, "Pulse control of single-degree-of-freedom system," *Journal of the Engineering Mechanics Division*, *ASCE*, Vol. 107, No. EM 6, 1981, pp. 997–1010.

**4.9.** R. A. Spinelli, "Numerical inversion of a Laplace transform," *SIAM Journal of Numerical Analysis*, Vol. 3, 1966, pp. 636–649.

**4.10.** R. Bellman, R. E. Kalaba, and J. A. Lockett, *Numerical Inversion of the Laplace Transform*, American Elsevier, New York, 1966.

**4.11.** J. Inoue and S. Miyaura, "Dynamic stability of a vibrating hammer," *Journal of Sound and Vibration*, Vol. 105, 1983, pp. 321–325.

## REVIEW QUESTIONS

**4.1.** What is the basis for expressing the response of a system under periodic excitation as a summation of several harmonic responses?

**4.2.** Indicate some methods for finding the response of a system under nonperiodic forces.

**4.3.** What is Duhamel integral? What is its use?

**4.4.** How are the initial conditions determined for a single degree of freedom system subjected to an impulse at $t = 0$?

**4.5.** Derive the equation of motion of a system subjected to base excitation.

**4.6.** What is a response spectrum?

**4.7.** What are the advantages of the Laplace transformation method?

**4.8.** What is the use of the pseudo spectrum?

**4.9.** How is the Laplace transform of a function $x(t)$ defined?

**4.10.** Define these terms: generalized impedance and admittance of a system.

**4.11.** State the interpolation models that can be used for approximating an arbitrary forcing function.

**4.12.** How many resonant conditions are there when the external force is not harmonic?

**4.13.** How do you compute the frequency of the first harmonic of a periodic force?

**4.14.** What is the relation between the frequencies of higher harmonics and the frequency of the first harmonic for a periodic excitation?

## PROBLEMS

The problem assignments are organized as follows:

| Problems | Section covered | Topic covered |
|----------|-----------------|---------------|
| 4.1–4.8   | 4.2 | Response under general periodic force |
| 4.9–4.11  | 4.3 | Periodic force of irregular form |
| 4.12–4.27 | 4.5 | Convolution integral |
| 4.28–4.32 | 4.6 | Response spectrum |
| 4.33–4.35 | 4.7 | Laplace transformation |
| 4.36–4.39 | 4.8 | Irregular forcing conditions using numerical methods |
| 4.40–4.45 | 4.9 | Computer program |

**4.1.** A plane milling cutter is cutting a horizontal surface with a speed of 0.1 m/s. When passing over a groove of width $w = 0.02$ m, the cutting force drops from 1000 N to zero linearly. If the cutting head can be idealized as a spring-mass system with a mass of 20 kg and a stiffness of 250 kN/m, determine the maximum cutting error after the cutter passes over the groove. Assume the damping to be negligible and the cutting force to be independent of the depth of cut.

**4.2–4.6.** Find the steady-state response of a viscously damped system to the forcing functions obtained by replacing $x(t)$ with $F(t)$ and $A$ with $F_0$ in Figs. 1.53–1.57.

**4.7.** Find the steady-state response of a viscously damped system to the forcing function obtained by replacing $x(t)$ and $A$ with $F(t)$ and $F_0$, respectively, in Fig. 1.32($a$).

**4.8.** The rotor of a single-stage steam turbine rotates at a constant speed of 3000 rpm. The turbine stage has 11 buckets and the 12th is missing (see Fig. 4.22). The power generated at the stage is 500 kW. The shaft is made of steel and has a diameter of 10 cm and a length of 1 m. If the mass moment of inertia of the turbine stage is 1 N-m-$s^2$, find the torsional vibrations of the system due to the discontinuity of the power generation because of the missing bucket. Assume that the rotor is mounted at the mid-span of the shaft.

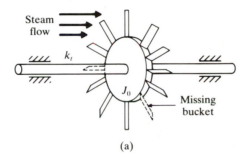

(a)

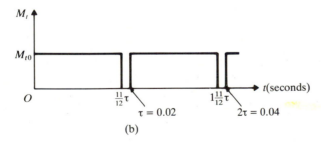

(b)

**Figure 4.22**

**4.9.** Find the response of a damped system with $m = 1$ kg, $k = 15$ kN/m, and $\zeta = 0.1$ under the action of a periodic forcing function, as shown in Fig. 1.58.

**4.10.** Find the response of a viscously damped system under the periodic force whose values are given in Problem 1.41. Assume that $x_i$ denotes the value of the force in Newtons at time $t_i$ seconds. Use $m = 0.5$ kg, $k = 8000$ N/m, and $\zeta = 0.06$.

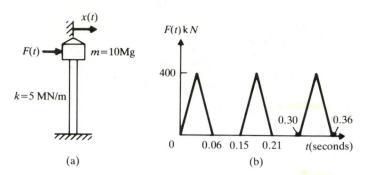

(a)                                    (b)

**Figure 4.23**

**4.11.** Find the displacement of the water tank shown in Fig. 4.23(a) under the periodic force shown in Fig. 4.23(b) by treating it as an undamped single degree of freedom system. Use the numerical procedure described in Sec. 4.3.

**4.12.** Show that the response to a unit step function $h(t)$ ($F_0 = 1$ in Fig. 4.6(a)) is related to the impulsive response $g(t)$, Eq. (4.27), as follows:

$$g(t) = \frac{dh(t)}{dt}$$

**4.13.** Show that the convolution integral, Eq. (4.33), can also be expressed in terms of the response to a unit step function $h(t)$ as

$$x(t) = F(0)h(t) + \int_0^t \frac{dF(\tau)}{d\tau} h(t - \tau) \, d\tau$$

**4.14.** Find the displacement of a damped single degree of freedom system under the forcing function $F(t) = F_0 e^{-\alpha t}$ where $\alpha$ is a constant.

**4.15.** Find the transient response of an undamped spring-mass system for $t > \pi/\omega$ when the mass is subjected to a force

$$F(t) = \begin{cases} \dfrac{F_0}{2}(1 - \cos \omega t) & \text{for } 0 \le t \le \dfrac{\pi}{\omega} \\[2mm] F_0 & \text{for } t > \dfrac{\pi}{\omega} \end{cases}$$

Assume that the displacement and velocity of the mass are zero at $t = 0$.

**4.16.** A piston of mass 10 kg slides in a smooth cylinder which is supported on a spring, as shown in Fig. 4.24(a). Due to leakage in the valve, the gas leaks into the cylinder from the gas reservoir and the pressure in the cylinder, $p(t)$, builds up, as shown in Fig. 4.24(b). Find the transient response of the piston and also its steady-state amplitude. The stiffness of the spring is 1000 N/m and the internal diameter of the cylinder is 100 mm.

**4.17–4.19.** Use the Duhamel integral method to derive expressions for the response of an undamped system subjected to the forcing functions shown in Figs. 4.25(a) to (c).

**4.20.** Find the time $t_m$ at which the response of a damped single degree of freedom system attains its maximum value under impulsive excitation.

**4.21.** Show that the time $t_m$ at which a damped spring-mass system reaches its maximum value under a step force $F_0$ is given by $t_m = \pi/\omega_d$.

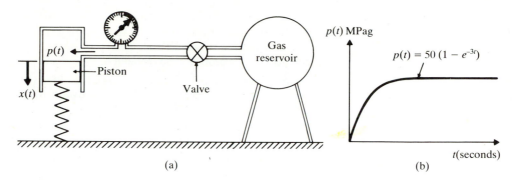

**Figure 4.24**

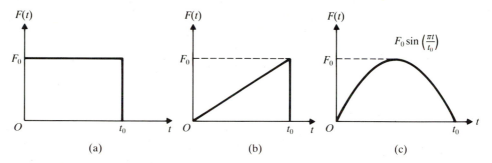

**Figure 4.25**

**4.22.** An automobile, having a mass of 1000 kg, runs over a road bump of the shape shown in Fig. 4.26. The speed of the automobile is 50 km/hr. If the undamped natural period of vibration in the vertical direction is 1.0 second, find the response of the car by assuming it as a single degree of freedom undamped system vibrating in the vertical direction.

**4.23.** To study the landing impact phenomenon of a light aircraft, the airplane is modeled as shown in Fig. 4.27 where $m$ denotes the lumped mass of the airplane and $k$ represents the stiffness of the landing gear. The airplane has a vertical descent velocity of $v$ when

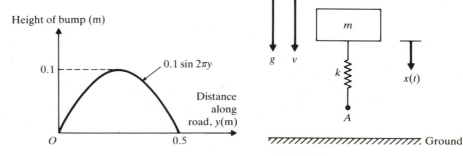

**Figure 4.26**                                      **Figure 4.27**

the landing gear (point $A$) touches the ground. Let $t = 0$ at the time of contact and let $x(0) = 0$. Find (1) the vertical position of the mass as a function of time during the time that the spring remains in contact with the ground, and (2) the time at which the spring loses contact with the ground upon rebound.

**4.24.** An undamped single degree of freedom system is subjected to a base excitation of $\dot{y}(t) = 50(1 - 4t)$. Determine the maximum relative displacement if the natural time period of the system is $\tau_n = 0.5$ sec.

**4.25.** An oscillatory system consists of a mass $m$, a spring of stiffness $k$, and a Coulomb damper that exerts a constant friction force $F$. If the base of the system is subjected to a velocity excitation as shown in Fig. 4.28, show that the relative displacement of the mass, $z$, is given by

$$z(t) = \frac{\dot{x}_0}{\omega_n^2 t_0}\left(1 - \frac{Ft_0}{m\dot{x}_0}\right)(1 - \cos \omega_n t) - \frac{\dot{x}_0}{\omega_n}\sin \omega_n t$$

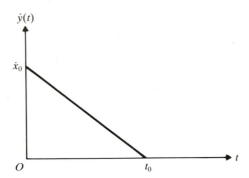

**Figure 4.28**

**4.26.** Derive Eq. (E.1) of Example 4.6.

**4.27.** In a static firing test of a rocket, the rocket is anchored to a rigid wall by a spring-damper system, as shown in Fig. 4.29(a). The thrust acting on the rocket reaches its maximum value $F$ in a negligibly short time and remains constant until the burnout

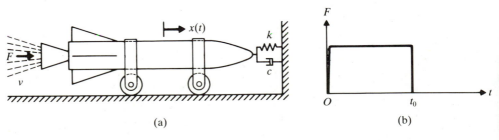

(a)                                         (b)

**Figure 4.29**

time $t_0$, as indicated in Fig. 4.29(b). The thrust acting on the rocket is given by $F = m_0 v$ where $m_0$ is the constant rate at which fuel is burnt and $v$ is the velocity of the jet stream. The initial mass of the rocket is $M$, so that its mass at any time $t$ is given by $m = M - m_0 t$, $0 \leq t \leq t_0$. If the data is $k = 7.5 \times 10^6$ N/m, $c = 0.1 \times 10^6$ N-s/m, $m_0 = 10$ kg/s, $v = 2000$ m/s, $M = 2000$ kg, and $t_0 = 100$ s, (1) derive the equation of motion of the rocket, and (2) find the maximum steady-state displacement of the rocket by assuming an average (constant) mass of $(M - \frac{1}{2} m_0 t_0)$.

**4.28.** Show that the response spectrum of an undamped system for the rectangular pulse shown in Fig. 4.25(a) is given by

$$\left( \frac{xk}{F_0} \right)_{max} = \begin{cases} 2\sin\dfrac{\pi t_0}{\tau_n} & \text{for } \dfrac{t_0}{\tau_n} < \dfrac{1}{2} \\[4mm] 2 & \text{for } \dfrac{t_0}{\tau_n} > \dfrac{1}{2} \end{cases}$$

**4.29.** Find the displacement response spectrum of an undamped system for the sine pulse shown in Fig. 4.25(c).

**4.30.** The base of an undamped spring-mass system is subjected to a velocity excitation, as shown in Fig. 4.28. Find the response spectrum of the relative displacement of the mass $z$.

**4.31.** Find the response spectrum of the system considered in Example 4.7. Plot $\left( \dfrac{kx}{F_0} \right)_{max}$ versus $\omega_n t_0$ in the range $0 \leq \omega_n t_0 \leq 15$.

**4.32.** A building frame is subjected to a blast load and the idealization of the frame and the load are shown in Fig 4.10. If $m = 5000$ kg, $k = 7.5$ GN/m, and $t_0 = 0.4$ s, determine the maximum blast force that can be sustained if the displacement is to be limited to 10 mm.

**4.33.** Find the steady state response of an undamped single degree of freedom system subjected to the force $F(t) = F_0 e^{i\omega t}$ by using the method of Laplace transformation.

**4.34.** Find the response of a damped spring-mass system subjected to a step function of magnitude $F_0$ by using the method of Laplace transformation.

**4.35.** Find the response of an undamped system subjected to a square pulse $F(t) = F_0$ for $0 \leq t \leq t_0$ and 0 for $t > t_0$ by using the Laplace transformation method. Assume all initial conditions as zero.

**4.36.** Determine the expression for the velocity $\dot{x}_j$ for the damped response represented by Eq. (4.64).

**4.37.** Derive Eqs. (4.68) and (4.71).

**4.38.** Compare the values of $\dot{x}_j$ given by Eqs. (4.68) and (4.71) in the case of Example 4.11.

**4.39.** Derive the expressions for $x_j$ and $\dot{x}_j$ according to the three interpolation functions considered in Sec. 4.8 for the undamped case. Using these expressions, find the solution of Example 4.11 by assuming the damping to be zero.

**4.40.** A damped single degree of freedom system has a mass $m = 2$, a spring of stiffness $k = 50$, and a damper with $c = 2$. A forcing function $F(t)$, whose magnitude is indicated in the following table, acts on the mass for one second. Find the response of the system by using the piecewise linear interpolation method described in Sec. 4.8.

| Time ($t_i$) | $F(t_i)$ |
|---|---|
| 0.0 | $-8.0$ |
| 0.1 | $-12.0$ |
| 0.2 | $-15.0$ |
| 0.3 | $-13.0$ |
| 0.4 | $-11.0$ |
| 0.5 | $-7.0$ |
| 0.6 | $-4.0$ |
| 0.7 | 3.0 |
| 0.8 | 10.0 |
| 0.9 | 15.0 |
| 1.0 | 18.0 |

**4.41.** The equation of motion of an undamped system is given by $2\ddot{x} + 1500x = F(t)$ where the forcing function is defined by the curve shown in Fig. 4.30. Find the response of the system numerically for $0 \le t \le 0.5$. Assume the initial conditions as $x_0 = \dot{x}_0 = 0$ and the step size as $\Delta t = 0.01$.

**4.42.** Solve Problem 4.41 if the system is viscously damped so that the equation of motion is $2\ddot{x} + 10\dot{x} + 1500x = F(t)$.

**4.43.** Find the relative displacement of the water tank shown in Fig. 4.23(a) when its base is subjected to the earthquake acceleration record shown in Fig. 1.59, by assuming the ordinate to represent acceleration in g's. Use the program of Problem 4.45.

**4.44.** The differential equation of motion of an undamped system is given by $2\ddot{x} + 150x = F(t)$ with the initial conditions $x_0 = \dot{x}_0 = 0$. If $F(t)$ is as shown in Fig. 4.31, find the response of the problem using the computer program of Problem 4.45.

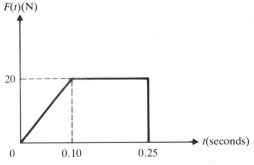

**Figure 4.30**

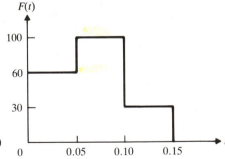

**Figure 4.31**

**4.45.** Write a computer program for finding the steady-state response of a single degree of freedom system subjected to an arbitrary force, by numerically evaluating the Duhamel integral. Using this program, solve Example 4.11.

# CHAPTER 5

# Two Degree of Freedom Systems

Daniel Bernoulli (1700–1782) was a Swiss who became a professor of mathematics at St. Petersburg in 1725 after receiving his doctorate in medicine for his thesis on the action of lungs. He later became professor of anatomy and botany at Basel. He developed the theory of hydrostatics and hydrodynamics and ''Bernoulli's theorem'' is well known to engineers. He derived the equation of motion for the vibration of beams (the Euler–Bernoulli theory) and studied the problem of vibrating strings. Bernoulli was the first person to propose the principle of superposition of harmonics in free vibration.

Courtesy Culver Pictures

## 5.1 INTRODUCTION

Systems that require two independent coordinates to describe their motion are called *two degree of freedom systems*. Some examples of systems having two degrees of freedom were shown in Fig. 1.7. We shall consider only two degree of freedom systems in this chapter, so as to provide a simple introduction to the behavior of systems with an arbitrarily large number of degrees of freedom, which is the subject of Chap. 6.

Consider the system shown in Fig. 5.1, in which a mass $m$ is supported on two equal springs. Assuming that the mass is constrained to move in a vertical plane, we find that the position of the mass $m$ at any time can be specified by a linear coordinate $x(t)$, indicating the vertical displacement of the center of gravity (C.G.) of the mass, and an angular coordinate $\theta(t)$, denoting the rotation of the mass $m$ about its C.G. Instead of $x(t)$ and $\theta(t)$, we can also use $x_1(t)$ and $x_2(t)$ as independent coordinates to specify the motion of the system. Thus the system has two degrees of freedom. It is important to note that in this case the mass $m$ is not treated as a point mass, but as a rigid body having two possible types of motion. (If it is a particle, there is no need to specify the rotation of the mass about its C.G.) The system shown in Fig. 5.2 does have one point mass $m$ but is a two degree of freedom system, because the mass has two possible types of motion (translations along the $x$ and $y$ directions). The general rule for the computation of the number of degrees of freedom can be stated as follows:

$$\text{Number of degrees of freedom of the system} = \frac{\text{Number of masses in the system}}{} \times \begin{array}{l}\text{number of possible types of motion} \\ \text{of each mass}\end{array}$$

There are two equations of motion for a two degree of freedom system, one for each mass (more precisely, for each degree of freedom). They are generally in the form of *coupled differential equations* — that is, each equation involves all the coordinates. If a harmonic solution is assumed for each coordinate, the equations of motion lead to a frequency equation that gives two natural frequencies for the system. If we give suitable initial excitation, the system vibrates at one of these natural frequencies.

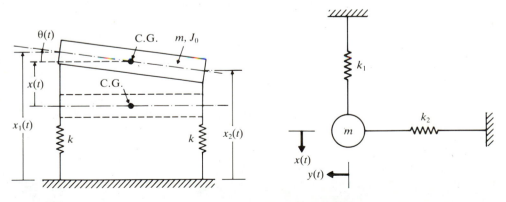

**Figure 5.1**                    **Figure 5.2**

When free vibration takes place at one of the natural frequencies, a definite relationship exists between the amplitudes of the two coordinates, and the configuration is called a *normal mode*, *principal mode*, or *natural mode* of vibration. Thus a two degree of freedom system has two normal modes of vibration corresponding to the two natural frequencies.

If we give an arbitrary initial excitation to the system, the resulting free vibration is the superposition of the two normal modes of vibration. However, if the system vibrates under the action of an external harmonic force, the resulting forced harmonic vibration takes place at the frequency of the exciting force and the amplitude of the two coordinates tends to a maximum at the two natural frequencies.

As is evident from the systems shown in Figs. 5.1 and 5.2, the configuration of a system can be specified by a set of independent coordinates such as length, angle, or some other physical parameters. Any such set of coordinates is called *generalized coordinates*. Although the equations of motion of a two degree of freedom system are generally coupled so that each equation involves all the coordinates, it is always possible to find a particular set of coordinates such that each equation of motion contains only one coordinate. The equations of motion are then *uncoupled* and can be solved independently of each other. Such a set of coordinates, which lead to an uncoupled system of equations, is called *principal coordinates*.

## 5.2 EQUATIONS OF MOTION FOR FORCED VIBRATION

Consider a viscously damped two degree of freedom spring-mass system, shown in Fig. 5.3(a). The motion of the system is completely described by the coordinates $x_1(t)$ and $x_2(t)$, which define the positions of the masses $m_1$ and $m_2$ at any time $t$

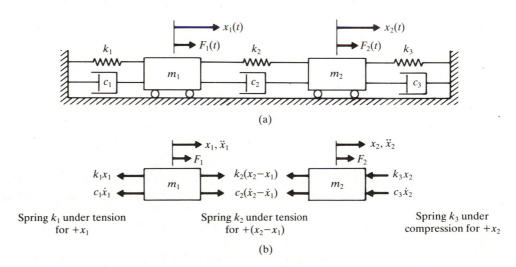

(a)

Spring $k_1$ under tension for $+x_1$ 

Spring $k_2$ under tension for $+(x_2-x_1)$ 

Spring $k_3$ under compression for $+x_2$

(b)

**Figure 5.3** A two degree of freedom spring-mass-damper system.

from the respective equilibrium positions. The external forces $F_1(t)$ and $F_2(t)$ act on the masses $m_1$ and $m_2$ respectively. The free-body diagrams of the masses $m_1$ and $m_2$ are shown in Fig. 5.3(b). The application of Newton's second law of motion to each of the masses gives the equations of motion:

coupled diff. eqs.
since each eq.
has $X_1$ and $X_2$

$$m_1\ddot{x}_1 + (c_1 + c_2)\dot{x}_1 - c_2\dot{x}_2 + (k_1 + k_2)x_1 - k_2x_2 = F_1 \tag{5.1}$$

$$m_2\ddot{x}_2 - c_2\dot{x}_1 + (c_2 + c_3)\dot{x}_2 - k_2x_1 + (k_2 + k_3)x_2 = F_2 \tag{5.2}$$

It can be seen that Eq. (5.1) contains terms involving $x_2$ (namely, $-c_2\dot{x}_2$ and $-k_2x_2$), whereas Eq. (5.2) contains terms involving $x_1$ (namely, $-c_2\dot{x}_1$ and $-k_2x_1$). Hence they represent a system of two coupled differential equations. We can therefore expect that the motion of the mass $m_1$ will influence the motion of the mass $m_2$, and vice versa. Equations (5.1) and (5.2) can be written in matrix form as

$$[m]\ddot{\vec{x}}(t) + [c]\dot{\vec{x}}(t) + [k]\vec{x}(t) = \vec{F}(t) \tag{5.3}$$

where $[m]$, $[c]$, and $[k]$ are called the *mass*, *damping*, and *stiffness matrices*, respectively, and are given by

$$[m] = \begin{bmatrix} m_1 & 0 \\ 0 & m_2 \end{bmatrix}$$

$$[c] = \begin{bmatrix} c_1 + c_2 & -c_2 \\ -c_2 & c_2 + c_3 \end{bmatrix}$$

$$[k] = \begin{bmatrix} k_1 + k_2 & -k_2 \\ -k_2 & k_2 + k_3 \end{bmatrix}$$

and $\vec{x}(t)$ and $\vec{F}(t)$ are called the *displacement* and *force vectors*, respectively, and are given by

$$\vec{x}(t) = \begin{Bmatrix} x_1(t) \\ x_2(t) \end{Bmatrix}$$

and

$$\vec{F}(t) = \begin{Bmatrix} F_1(t) \\ F_2(t) \end{Bmatrix}$$

It can be seen that the matrices $[m]$, $[c]$, and $[k]$ are all $2 \times 2$ matrices whose elements are the known masses, damping coefficients, and stiffnesses of the system, respectively. Further, these matrices can be seen to be symmetric, so that

$$[m]^T = [m], \qquad [c]^T = [c], \qquad [k]^T = [k]$$

where the superscript $T$ denotes the transpose of the matrix.

Notice that the equations of motion (5.1) and (5.2) become uncoupled (independent of one another) only when $c_2 = k_2 = 0$, which implies that the two masses $m_1$ and $m_2$ are not physically connected. In such a case, the matrices $[m]$, $[c]$, and $[k]$ become diagonal. The solution of the equations of motion (5.1) and (5.2) for any arbitrary forces $F_1(t)$ and $F_2(t)$ is difficult to obtain, mainly due to the coupling of the variables $x_1(t)$ and $x_2(t)$. We shall first consider the free vibration solution of Eqs. (5.1) and (5.2).

## 5.3 FREE VIBRATION ANALYSIS OF AN UNDAMPED SYSTEM

For the free vibration analysis of the system shown in Fig. 5.3(a), we set $F_1(t) = F_2(t) = 0$. Further, if damping is disregarded, $c_1 = c_2 = c_3 = 0$, and the equations of motion (5.1) and (5.2) reduce to

$$m_1\ddot{x}_1(t) + (k_1 + k_2)x_1(t) - k_2 x_2(t) = 0 \tag{5.4}$$

$$m_2\ddot{x}_2(t) - k_2 x_1(t) + (k_2 + k_3)x_2(t) = 0 \tag{5.5}$$

We are interested in knowing whether $m_1$ and $m_2$ can oscillate harmonically with the same frequency and phase angle but with different amplitudes. Assuming that it is possible to have harmonic motion of $m_1$ and $m_2$ at the same frequency $\omega$ and the same phase angle $\phi$, we take the solutions of Eqs. (5.4) and (5.5) as

$$x_1(t) = X_1\cos(\omega t + \phi)$$

$$x_2(t) = X_2\cos(\omega t + \phi) \tag{5.6}$$

where $X_1$ and $X_2$ are constants that denote the maximum amplitudes of $x_1(t)$ and $x_2(t)$, and $\phi$ is the phase angle. Substituting Eq. (5.6) into Eqs. (5.4) and (5.5), we obtain

$$\left[\{-m_1\omega^2 + (k_1 + k_2)\}X_1 - k_2 X_2\right]\cos(\omega t + \phi) = 0$$

$$\left[-k_2 X_1 + \{-m_2\omega^2 + (k_2 + k_3)\}X_2\right]\cos(\omega t + \phi) = 0 \tag{5.7}$$

Since Eqs. (5.7) must be satisfied for all values of the time $t$, the terms between brackets must be zero. This yields

homogeneous eqs.
in $X_1$ & $X_2$

$$\begin{cases} \{-m_1\omega^2 + (k_1 + k_2)\}X_1 - k_2 X_2 = 0 \\ -k_2 X_1 + \{-m_2\omega^2 + (k_2 + k_3)\}X_2 = 0 \end{cases} \tag{5.8}$$

which represent two simultaneous homogeneous algebraic equations in the unknowns $X_1$ and $X_2$. It can be seen that Eqs. (5.8) are satisfied by the trivial solution $X_1 = X_2 = 0$, which implies that there is no vibration. For a nontrivial solution of $X_1$ and $X_2$, the determinant of the coefficients of $X_1$ and $X_2$ must be zero:

$$\det\begin{bmatrix} \{-m_1\omega^2 + (k_1 + k_2)\} & -k_2 \\ -k_2 & \{-m_2\omega^2 + (k_2 + k_3)\} \end{bmatrix} = 0$$

or

*Frequency eqn. yields the Freq. of the system*

$$(m_1 m_2)\omega^4 - \{(k_1 + k_2)m_2 + (k_2 + k_3)m_1\}\omega^2 + \{(k_1 + k_2)(k_2 + k_3) - k_2^2\} = 0$$

$$\overset{a}{} \qquad \overset{b}{} \qquad \overset{c}{}$$

(5.9)

Equation (5.9) is called the *frequency* or *characteristic equation* because solution of this equation yields the frequencies or the characteristic values of the system. The roots of Eq. (5.9) are given by

*freq. roots :*

$$\omega_1^2, \omega_2^2 = \frac{1}{2}\left\{\frac{(k_1 + k_2)m_2 + (k_2 + k_3)m_1}{m_1 m_2}\right\}$$

$$\mp \frac{1}{2}\left[\left\{\frac{(k_1 + k_2)m_2 + (k_2 + k_3)m_1}{m_1 m_2}\right\}^2 - 4\left\{\frac{(k_1 + k_2)(k_2 + k_3) - k_2^2}{m_1 m_2}\right\}\right]^{1/2}$$

(5.10)

This shows that it is possible for the system to have a nontrivial harmonic solution of the form of Eqs. (5.6) when $\omega$ is equal to $\omega_1$ or $\omega_2$ given by Eq. (5.10). We call $\omega_1$ and $\omega_2$ the *natural frequencies* of the system.

The values of $X_1$ and $X_2$ remain to be determined. These values depend on the natural frequencies $\omega_1$ and $\omega_2$. We shall denote the values of $X_1$ and $X_2$ corresponding to $\omega_1$ as $X_1^{(1)}$ and $X_2^{(1)}$ and those corresponding to $\omega_2$ as $X_1^{(2)}$ and $X_2^{(2)}$. Further, since the Eqs. (5.8) are homogeneous, only the ratios $r_1 = \{X_2^{(1)}/X_1^{(1)}\}$ and $r_2 = \{X_2^{(2)}/X_1^{(2)}\}$ can be found. For $\omega^2 = \omega_1^2$ and $\omega^2 = \omega_2^2$, Eqs. (5.8) give

*ratios of amplitude for $\omega_1$ and $\omega_2$*

$$r_1 = \frac{X_2^{(1)}}{X_1^{(1)}} = \frac{-m_1\omega_1^2 + (k_1 + k_2)}{k_2} = \frac{k_2}{-m_2\omega_1^2 + (k_2 + k_3)}$$

(5.11)

$$r_2 = \frac{X_2^{(2)}}{X_1^{(2)}} = \frac{-m_1\omega_2^2 + (k_1 + k_2)}{k_2} = \frac{k_2}{-m_2\omega_2^2 + (k_2 + k_3)}$$

(5.12)

Notice that the two ratios given for each $r_i$ ($i = 1, 2$) in Eqs. (5.11) and (5.12) are identical. The normal modes of vibration corresponding to $\omega_1^2$ and $\omega_2^2$ can be expressed, respectively, as

$$\vec{X}^{(1)} = \begin{Bmatrix} X_1^{(1)} \\ X_2^{(1)} \end{Bmatrix} = \begin{Bmatrix} X_1^{(1)} \\ r_1 X_1^{(1)} \end{Bmatrix}$$

*Modal Vectors denote normal modes of vibration*

and

$$\vec{X}^{(2)} = \begin{Bmatrix} X_1^{(2)} \\ X_2^{(2)} \end{Bmatrix} = \begin{Bmatrix} X_1^{(2)} \\ r_2 X_1^{(2)} \end{Bmatrix}$$

The vectors $\vec{X}^{(1)}$ and $\vec{X}^{(2)}$, which denote the normal modes of vibration, are known as the *modal vectors* of the system.

The free vibration solution or the motion in time can be expressed as

$$\vec{x}^{(1)}(t) = \begin{Bmatrix} x_1^{(1)}(t) \\ x_2^{(1)}(t) \end{Bmatrix} = \begin{Bmatrix} X_1^{(1)}\cos(\omega_1 t + \phi_1) \\ r_1 X_1^{(1)}\cos(\omega_1 t + \phi_1) \end{Bmatrix} = \text{first mode} \qquad (5.13)$$

$$\vec{x}^{(2)}(t) = \begin{Bmatrix} x_1^{(2)}(t) \\ x_2^{(2)}(t) \end{Bmatrix} = \begin{Bmatrix} X_1^{(2)}\cos(\omega_2 t + \phi_2) \\ r_2 X_1^{(2)}\cos(\omega_2 t + \phi_2) \end{Bmatrix} = \text{second mode} \qquad (5.14)$$

where the constants $X_1^{(1)}$, $X_1^{(2)}$, $\phi_1$, and $\phi_2$ are determined by the initial conditions.

**Initial Conditions.** Since each of the two equations of motion, Eqs. (5.1) and (5.2), involves second-order time derivatives, we need to specify two initial conditions for each mass. As stated in Sec. 5.1, the system can be made to vibrate in its $i$th normal mode ($i = 1, 2$) by subjecting it to the specific initial conditions

$$x_1(t = 0) = X_1^{(i)} = \text{some constant}, \qquad \dot{x}_1(t = 0) = 0,$$
$$x_2(t = 0) = r_i X_1^{(i)}, \qquad \dot{x}_2(t = 0) = 0$$

However, for any other general initial conditions, both modes will be excited. The resulting motion, which is given by the general solution of Eqs. (5.4) and (5.5), can be obtained by superposing the two normal modes, Eqs. (5.13) and (5.14):

$$\vec{x}(t) = \vec{x}^{(1)}(t) + \vec{x}^{(2)}(t)$$

that is,

$$x_1(t) = x_1^{(1)}(t) + x_1^{(2)}(t) = X_1^{(1)}\cos(\omega_1 t + \phi_1) + X_1^{(2)}\cos(\omega_2 t + \phi_2)$$
$$x_2(t) = x_2^{(1)}(t) + x_2^{(2)}(t) = r_1 X_1^{(1)}\cos(\omega_1 t + \phi_1) + r_2 X_1^{(2)}\cos(\omega_2 t + \phi_2) \qquad (5.15)$$

Thus if the initial conditions are given by

$$x_1(t = 0) = x_1(0), \quad \dot{x}_1(t = 0) = \dot{x}_1(0), \quad x_2(t = 0) = x_2(0), \quad \dot{x}_2(t = 0) = \dot{x}_2(0) \qquad (5.16)$$

the constants $X_1^{(1)}$, $X_1^{(2)}$, $\phi_1$, and $\phi_2$ can be found by solving the following equations (obtained by substituting Eqs. (5.16) into Eqs. (5.15)):

$$x_1(0) = X_1^{(1)}\cos\phi_1 + X_1^{(2)}\cos\phi_2$$
$$\dot{x}_1(0) = -\omega_1 X_1^{(1)}\sin\phi_1 - \omega_2 X_1^{(2)}\sin\phi_2$$
$$x_2(0) = r_1 X_1^{(1)}\cos\phi_1 + r_2 X_1^{(2)}\cos\phi_2$$
$$\dot{x}_2(0) = -\omega_1 r_1 X_1^{(1)}\sin\phi_1 - \omega_2 r_2 X_1^{(2)}\sin\phi_2 \qquad (5.17)$$

Equations (5.17) can be regarded as four algebraic equations in the unknowns $X_1^{(1)}\cos\phi_1$, $X_1^{(2)}\cos\phi_2$, $X_1^{(1)}\sin\phi_1$, and $X_1^{(2)}\sin\phi_2$. The solution of Eqs. (5.17) can be expressed as

$$X_1^{(1)}\cos\phi_1 = \left\{ \frac{r_2 x_1(0) - x_2(0)}{r_2 - r_1} \right\}, \qquad X_1^{(2)}\cos\phi_2 = \left\{ \frac{-r_1 x_1(0) + x_2(0)}{r_2 - r_1} \right\}$$

$$X_1^{(1)}\sin\phi_1 = \left\{ \frac{-r_2 \dot{x}_1(0) + \dot{x}_2(0)}{\omega_1(r_2 - r_1)} \right\}, \qquad X_1^{(2)}\sin\phi_2 = \left\{ \frac{r_1 \dot{x}_1(0) - \dot{x}_2(0)}{\omega_2(r_2 - r_1)} \right\}$$

from which we obtain the desired solution

*Constants determined by I.C.'s*

$$X_1^{(1)} = \left[ \left\{ X_1^{(1)}\cos\phi_1 \right\}^2 + \left\{ X_1^{(1)}\sin\phi_1 \right\}^2 \right]^{1/2}$$

$$= \frac{1}{(r_2 - r_1)} \left[ \left\{ r_2 x_1(0) - x_2(0) \right\}^2 + \frac{\left\{ -r_2 \dot{x}_1(0) + \dot{x}_2(0) \right\}^2}{\omega_1^2} \right]^{1/2}$$

$$X_1^{(2)} = \left[ \left\{ X_1^{(2)}\cos\phi_2 \right\}^2 + \left\{ X_1^{(2)}\sin\phi_2 \right\}^2 \right]^{1/2}$$

$$= \frac{1}{(r_2 - r_1)} \left[ \left\{ -r_1 x_1(0) + x_2(0) \right\}^2 + \frac{\left\{ r_1 \dot{x}_1(0) - \dot{x}_2(0) \right\}^2}{\omega_2^2} \right]^{1/2}$$

$$\phi_1 = \tan^{-1} \left\{ \frac{X_1^{(1)}\sin\phi_1}{X_1^{(1)}\cos\phi_1} \right\} = \tan^{-1} \left\{ \frac{-r_2 \dot{x}_1(0) + \dot{x}_2(0)}{\omega_1[r_2 x_1(0) - x_2(0)]} \right\}$$

$$\phi_2 = \tan^{-1} \left\{ \frac{X_1^{(2)}\sin\phi_2}{X_1^{(2)}\cos\phi_2} \right\} = \tan^{-1} \left\{ \frac{r_1 \dot{x}_1(0) - \dot{x}_2(0)}{\omega_2[-r_1 x_1(0) + x_2(0)]} \right\} \qquad (5.18)$$

---

## EXAMPLE 5.1

Find the natural frequencies and mode shapes of a spring-mass system, shown in Fig. 5.4, which is constrained to move in the vertical direction only. Take $n = 1$.

*Solution.* If we measure $x_1$ and $x_2$ from the static equilibrium positions of the masses $m_1$ and $m_2$ respectively, the equations of motion and the solution obtained for the system of Fig. 5.3(a) are also applicable to this case if we substitute $m_1 = m_2 = m$ and $k_1 = k_2 = k_3 = k$. Thus the equations of motion, Eqs. (5.4) and (5.5), are given by

$$m\ddot{x}_1 + 2kx_1 - kx_2 = 0$$
$$m\ddot{x}_2 - kx_1 + 2kx_2 = 0 \qquad (E.1)$$

By assuming harmonic solution as

$$x_i(t) = X_i \cos(\omega t + \phi); \qquad i = 1, 2 \qquad (E.2)$$

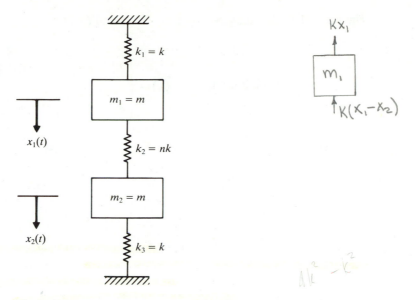

**Figure 5.4**

the frequency equation can be obtained by substituting Eq. (E.2) into Eq. (E.1):

$$\begin{vmatrix} (-m\omega^2 + 2k) & (-k) \\ (-k) & (-m\omega^2 + 2k) \end{vmatrix} = 0$$

or

$$m^2\omega^4 - 4km\omega^2 + 3k^2 = 0 \tag{E.3}$$

$(-m\omega^2 + 3k)(-m\omega^2 + k) = 0 \quad \text{quadratic formula}$

The solution of Eq. (E.3) gives the natural frequencies

$$\omega_1 = \left\{ \frac{4km - [16k^2m^2 - 12m^2k^2]^{1/2}}{2m^2} \right\}^{1/2} = \sqrt{\frac{k}{m}} \tag{E.4}$$

$$\omega_2 = \left\{ \frac{4km + [16k^2m^2 - 12m^2k^2]^{1/2}}{2m^2} \right\}^{1/2} = \sqrt{\frac{3k}{m}} \tag{E.5}$$

From Eqs. (5.11) and (5.12), the amplitude ratios are given by

$$r_1 = \frac{X_2^{(1)}}{X_1^{(1)}} = \frac{-m\omega_1^2 + 2k}{k} = \frac{k}{-m\omega_1^2 + 2k} = 1 \tag{E.6}$$

$$r_2 = \frac{X_2^{(2)}}{X_1^{(2)}} = \frac{-m\omega_2^2 + 2k}{k} = \frac{k}{-m\omega_2^2 + 2k} = -1 \tag{E.7}$$

The natural modes are given by Eqs. (5.13) and (5.14):

$$\text{First mode} = \vec{x}^{(1)}(t) = \left\{ \begin{array}{c} X_1^{(1)}\cos\left(\sqrt{\dfrac{k}{m}}\, t + \phi_1\right) \\[2ex] X_1^{(1)}\cos\left(\sqrt{\dfrac{k}{m}}\, t + \phi_1\right) \end{array} \right\} \tag{E.8}$$

$$\text{Second mode} = \vec{x}^{(2)}(t) = \left\{ \begin{array}{c} X_1^{(2)}\cos\left(\sqrt{\dfrac{3k}{m}}\, t + \phi_2\right) \\[2ex] - X_1^{(2)}\cos\left(\sqrt{\dfrac{3k}{m}}\, t + \phi_2\right) \end{array} \right\} \tag{E.9}$$

It can be seen from Eq. (E.8) that when the system vibrates in its first mode, the amplitudes of the two masses remain the same. This implies that the length of the middle spring remains constant. Thus the motions of $m_1$ and $m_2$ are in phase (see Fig. 5.5(a)). When the system vibrates in its second mode, Eq. (E.9) shows that the displacements of the two masses have the same magnitude with opposite signs. Thus the motions of $m_1$ and $m_2$ are 180° out of phase (see Fig. 5.5(b)). In this case the midpoint of the middle spring remains stationary for all time $t$. Such a point is called a *node*. Using Eqs. (5.15), the motion (general solution) of the system can be expressed as

$$x_1(t) = X_1^{(1)}\cos\left(\sqrt{\frac{k}{m}}\, t + \phi_1\right) + X_1^{(2)}\cos\left(\sqrt{\frac{3k}{m}}\, t + \phi_2\right)$$

$$x_2(t) = X_1^{(1)}\cos\left(\sqrt{\frac{k}{m}}\, t + \phi_1\right) - X_1^{(2)}\cos\left(\sqrt{\frac{3k}{m}}\, t + \phi_2\right) \tag{E.10}$$

**Figure 5.5**

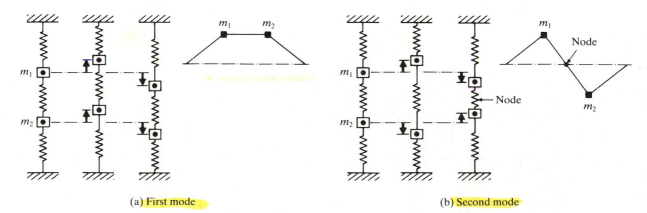

(a) First mode                    (b) Second mode

## EXAMPLE 5.2

Find the initial conditions that need to be applied to the system shown in Fig. 5.4 so as to make it vibrate in (1) the first mode, and (2) the second mode.

**Solution.** For arbitrary initial conditions, the motion of the masses is described by Eqs. (5.15). In the present case, $r_1 = 1$ and $r_2 = -1$, so Eqs. (5.15) reduce to

$$x_1(t) = X_1^{(1)}\cos\left(\sqrt{\frac{k}{m}}\, t + \phi_1\right) + X_1^{(2)}\cos\left(\sqrt{\frac{3k}{m}}\, t + \phi_2\right)$$

$$x_2(t) = X_1^{(1)}\cos\left(\sqrt{\frac{k}{m}}\, t + \phi_1\right) - X_1^{(2)}\cos\left(\sqrt{\frac{3k}{m}}\, t + \phi_2\right) \tag{E.1}$$

Assuming the initial conditions as in Eq. (5.16), the constants $X_1^{(1)}$, $X_1^{(2)}$, $\phi_1$, and $\phi_2$ can be obtained from Eqs. (5.18), using $r_1 = 1$ and $r_2 = -1$:

$$X_1^{(1)} = -\frac{1}{2}\left\{[x_1(0) + x_2(0)]^2 + \frac{m}{k}[\dot{x}_1(0) + \dot{x}_2(0)]^2\right\}^{1/2} \tag{E.2}$$

$$X_1^{(2)} = -\frac{1}{2}\left\{[-x_1(0) + x_2(0)]^2 + \frac{m}{3k}[\dot{x}_1(0) - \dot{x}_2(0)]^2\right\}^{1/2} \tag{E.3}$$

$$\phi_1 = \tan^{-1}\left\{\frac{-\sqrt{m}\,[\dot{x}_1(0) + \dot{x}_2(0)]}{\sqrt{k}\,[x_1(0) + x_2(0)]}\right\} \tag{E.4}$$

$$\phi_2 = \tan^{-1}\left\{\frac{\sqrt{m}\,[\dot{x}_1(0) - \dot{x}_2(0)]}{\sqrt{3k}\,[-x_1(0) + x_2(0)]}\right\} \tag{E.5}$$

(1) The first normal mode of the system is given by Eq. (E.8) of Example 5.1:

$$\vec{x}^{(1)}(t) = \left\{\begin{array}{c} X_1^{(1)}\cos\left(\sqrt{\dfrac{k}{m}}\, t + \phi_1\right) \\[2ex] X_1^{(1)}\cos\left(\sqrt{\dfrac{k}{m}}\, t + \phi_1\right) \end{array}\right\} \tag{E.6}$$

Comparison of Eqs. (E.1) and (E.6) shows that the motion of the system is identical with the first normal mode only if $X_1^{(2)} = 0$. This requires that (from Eq. (E.3))

$$x_1(0) = x_2(0) \qquad \text{and} \qquad \dot{x}_1(0) = \dot{x}_2(0) \tag{E.7}$$

(2) The second normal mode of the system is given by Eq. (E.9) of Example 5.1:

$$\vec{x}^{(2)}(t) = \left\{\begin{array}{c} X_1^{(2)}\cos\left(\sqrt{\dfrac{3k}{m}}\, t + \phi_2\right) \\[2ex] -X_1^{(2)}\cos\left(\sqrt{\dfrac{3k}{m}}\, t + \phi_2\right) \end{array}\right\} \tag{E.8}$$

Comparison of Eqs. (E.1) and (E.8) shows that the motion of the system coincides with the second normal mode only if $X_1^{(1)} = 0$. This implies that (from Eq. (E.2))

$$x_1(0) = -x_2(0) \qquad \text{and} \qquad \dot{x}_1(0) = -\dot{x}_2(0) \tag{E.9}$$

## 5.4 TORSIONAL SYSTEM

Consider a torsional system consisting of two discs mounted on a shaft, as shown in Fig. 5.6. The three segments of the shaft have rotational spring constants $k_{t1}$, $k_{t2}$, and $k_{t3}$, as indicated in the figure. Also shown are the discs of mass moments of inertia $J_1$ and $J_2$, the applied torques $M_{t1}$ and $M_{t2}$, and the rotational degrees of freedom $\theta_1$ and $\theta_2$. The differential equations of rotational motion for the discs $J_1$ and $J_2$ can be derived as follows:

$$J_1\ddot{\theta}_1 = -k_{t1}\theta_1 + k_{t2}(\theta_2 - \theta_1) + M_{t1}$$
$$J_2\ddot{\theta}_2 = -k_{t2}(\theta_2 - \theta_1) - k_{t3}\theta_2 + M_{t2} \tag{5.19}$$

which upon rearrangement become

$$J_1\ddot{\theta}_1 + (k_{t1} + k_{t2})\theta_1 - k_{t2}\theta_2 = M_{t1}$$
$$J_2\ddot{\theta}_2 - k_{t2}\theta_1 + (k_{t2} + k_{t3})\theta_2 = M_{t2} \tag{5.20}$$

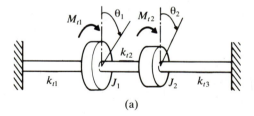

(a)

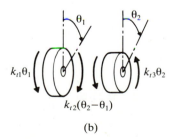

(b)

**Figure 5.6**

**EXAMPLE 5.3**

Find the natural frequencies and mode shapes for the torsional system shown in Fig. 5.7 for $J_1 = J_0$, $J_2 = 2J_0$, and $k_{t1} = k_{t2} = k_t$.

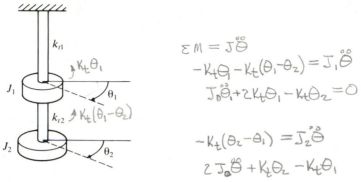

Figure 5.7

*Solution.* The differential equations of motion, Eq. (5.20), reduce to (with $M_{t1} = M_{t2} = k_{t3} = 0$, $k_{t1} = k_{t2} = k_t$, $J_1 = J_0$, and $J_2 = 2J_0$):

$$J_0 \ddot{\theta}_1 + 2k_t \theta_1 - k_t \theta_2 = 0$$
$$2J_0 \ddot{\theta}_2 - k_t \theta_1 + k_t \theta_2 = 0 \qquad \text{(E.1)}$$

Rearranging and substituting ==the harmonic solution==

$$\theta_i(t) = \Theta_i \cos(\omega t + \phi); \qquad i = 1, 2 \qquad \text{(E.2)}$$

gives the frequency equation:

$$2\omega^4 J_0^2 - 5\omega^2 J_0 k_t + k_t^2 = 0 \qquad \text{(E.3)}$$

The solution of Eq. (E.3) gives the natural frequencies:

$$\omega_1 = \sqrt{\frac{k_t}{4J_0}(5 - \sqrt{17})} \qquad \text{and} \qquad \omega_2 = \sqrt{\frac{k_t}{4J_0}(5 + \sqrt{17})} \qquad \text{(E.4)}$$

The amplitude ratios are given by

$$r_1 = \frac{\Theta_2^{(1)}}{\Theta_1^{(1)}} = 2 - \frac{(5 - \sqrt{17})}{4} = 1.781$$

$$r_2 = \frac{\Theta_2^{(2)}}{\Theta_1^{(2)}} = 2 - \frac{(5 + \sqrt{17})}{4} = -0.281 \qquad \text{(E.5)}$$

## 5.5 COORDINATE COUPLING AND PRINCIPAL COORDINATES

As stated earlier, an $n$ degree of freedom system requires $n$ independent coordinates to describe its configuration. Usually, these coordinates are independent geometrical quantities measured from the equilibrium position of the vibrating body. However, it is possible to select some other set of $n$ coordinates to describe the configuration of the system. The latter set may be, for example, different from the first set in that the coordinates may have their origin away from the equilibrium position of the body. There could be still other sets of coordinates to describe the configuration of the system. Each of these sets of $n$ coordinates is called the *generalized coordinates*.

As an example, consider the lathe shown in Fig. 5.8. An accurate model of this machine tool would involve the consideration of the lathe bed as an elastic beam with lumped masses attached to it [5.1–5.3]. However, for simplified vibration analysis, the lathe bed can be considered as a rigid body having mass and inertia, and the headstock and tailstock can each be replaced by lumped masses. The bed can be assumed to be supported on springs at the ends. Thus the final model will be a rigid body of total mass $m$ and mass moment of inertia $J_0$ about its C.G., resting on springs of stiffnesses $k_1$ and $k_2$, as shown in Fig. 5.9(a). For this two degree of freedom system, any of the following sets of coordinates may be used to describe the motion:

1. Deflections $x_1(t)$ and $x_2(t)$ of the two ends of the lathe $AB$
2. Deflection $x(t)$ of the C.G. and rotation $\theta(t)$
3. Deflection $x_1(t)$ of the end $A$ and rotation $\theta(t)$
4. Deflection $y(t)$ of point $P$ located at a distance $e$ to the left of the C.G. and rotation $\theta(t)$, as indicated in Fig. 5.9(b).

Thus any set of these coordinates—$(x_1, x_2)$, $(x, \theta)$, $(x_1, \theta)$, and $(y, \theta)$—represents the generalized coordinates of the system. Now we shall derive the equations of motion of the lathe using two different sets of coordinates to illustrate the concept of coordinate coupling.

**Equations of Motion Using $x(t)$ and $\theta(t)$.** From the free-body diagram shown in Fig. 5.9(a), with the positive values of the motion variables as indicated, the force equilibrium equation in the vertical direction can be written as

$$m\ddot{x} = -k_1(x - l_1\theta) - k_2(x + l_2\theta) \tag{5.21}$$

and the moment equation about the C.G. can be expressed as

$$J_0\ddot{\theta} = k_1(x - l_1\theta)l_1 - k_2(x + l_2\theta)l_2 \tag{5.22}$$

Equations (5.21) and (5.22) can be rearranged and written in matrix form as

$$\begin{bmatrix} m & 0 \\ 0 & J_0 \end{bmatrix} \begin{Bmatrix} \ddot{x} \\ \ddot{\theta} \end{Bmatrix} + \begin{bmatrix} (k_1 + k_2) & -(k_1 l_1 - k_2 l_2) \\ -(k_1 l_1 - k_2 l_2) & (k_1 l_1^2 + k_2 l_2^2) \end{bmatrix} \begin{Bmatrix} x \\ \theta \end{Bmatrix} = \begin{Bmatrix} 0 \\ 0 \end{Bmatrix} \tag{5.23}$$

It can be seen that each of the Eqs. (5.23) contain $x$ and $\theta$. They become

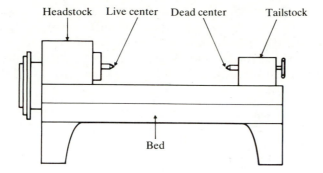

**Figure 5.8**

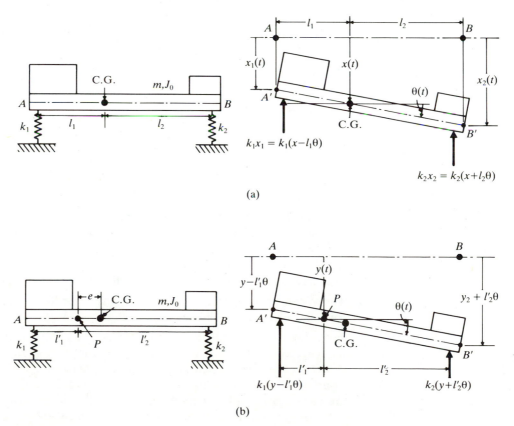

(a)

(b)

**Figure 5.9**

independent of each other if the coupling term $(k_1 l_1 - k_2 l_2)$ is equal to zero—that is, if $k_1 l_1 = k_2 l_2$. If $k_1 l_1 \neq k_2 l_2$, the resultant motion of the lathe $AB$ is both translational and rotational when either a displacement or torque is applied through the C.G. of the body as an initial condition. In other words, the lathe rotates in the vertical plane and has vertical motion as well, unless $k_1 l_1 = k_2 l_2$. This is known as *elastic* or *static coupling*.

**Equations of Motion using $y(t)$ and $\theta(t)$.** From Fig. 5.9(b), where $y(t)$ and $\theta(t)$ are used as the generalized coordinates of the system, the equations of motion for translation and rotation can be written as

$$m\ddot{y} = -k_1(y - l_1'\theta) - k_2(y + l_2'\theta) - me\ddot{\theta}$$
$$J_p\ddot{\theta} = k_1(y - l_1'\theta)l_1' - k_2(y + l_2'\theta)l_2' - me\ddot{y} \qquad (5.24)$$

These equations can be rearranged and written in matrix form as

$$\begin{bmatrix} m & me \\ me & J_p \end{bmatrix} \begin{Bmatrix} \ddot{y} \\ \ddot{\theta} \end{Bmatrix} + \begin{bmatrix} (k_1 + k_2) & (k_2 l_2' - k_1 l_1') \\ (-k_1 l_1' + k_2 l_2') & (k_1 l_1'^2 + k_2 l_2'^2) \end{bmatrix} \begin{Bmatrix} y \\ \theta \end{Bmatrix} = \begin{Bmatrix} 0 \\ 0 \end{Bmatrix} \qquad (5.25)$$

Both the equations of motion represented by Eq. (5.25) contain $y$ and $\theta$, so they are coupled equations. They contain static (or elastic) as well as dynamic (or mass) coupling terms. If $k_1 l_1' = k_2 l_2'$, the system will have dynamic or inertia coupling only. In this case, if the lathe moves up and down in the $y$ direction, the inertia force $m\ddot{y}$, which acts through the center of gravity of the body, induces a motion in the $\theta$ direction, by virtue of the moment $m\ddot{y}e$. Similarly, a motion in the $\theta$ direction induces a motion of the lathe in the $y$ direction due to the force $me\ddot{\theta}$.

Note the following characteristics of these systems:

1. In the most general case, a viscously damped two degree of freedom system has equations of motion in the following form:

$$\begin{bmatrix} m_{11} & m_{12} \\ m_{12} & m_{22} \end{bmatrix} \begin{Bmatrix} \ddot{x}_1 \\ \ddot{x}_2 \end{Bmatrix} + \begin{bmatrix} c_{11} & c_{12} \\ c_{12} & c_{22} \end{bmatrix} \begin{Bmatrix} \dot{x}_1 \\ \dot{x}_2 \end{Bmatrix} + \begin{bmatrix} k_{11} & k_{12} \\ k_{12} & k_{22} \end{bmatrix} \begin{Bmatrix} x_1 \\ x_2 \end{Bmatrix} = \begin{Bmatrix} 0 \\ 0 \end{Bmatrix} \qquad (5.26)$$

This equation reveals the type of coupling present. If the stiffness matrix is not diagonal, the system has elastic or static coupling. If the damping matrix is not diagonal, the system has damping or velocity coupling. Finally, if the mass matrix is not diagonal, the system has mass or inertial coupling. Both velocity and mass coupling come under the heading of dynamic coupling.

2. The system vibrates in its own natural way regardless of the coordinates used. The choice of the coordinates is a mere convenience.

3. From Eqs. (5.23) and (5.25), it is clear that the nature of the coupling depends on the coordinates used and is not an inherent property of the system. It is possible to choose a system of coordinates $q_1(t)$ and $q_2(t)$ which give equations of motion that are uncoupled both statically and dynamically. Such coordinates are called *principal* or *natural coordinates*. The main advantage of using principal coordinates is that the resulting uncoupled equations of motion can be solved independently of one another.

The following example illustrates the method of finding the principal coordinates in terms of the geometrical coordinates.

---

**EXAMPLE 5.4**

Determine the principal coordinates for the system shown in Fig. 5.4.

**Solution.** The general motion of the system shown in Fig. 5.4 is given by Eqs. (E.10) of Example 5.1:

$$x_1(t) = B_1 \cos\left(\sqrt{\frac{k}{m}}\, t + \phi_1\right) + B_2 \cos\left(\sqrt{\frac{3k}{m}}\, t + \phi_2\right)$$

$$x_2(t) = B_1 \cos\left(\sqrt{\frac{k}{m}}\, t + \phi_1\right) - B_2 \cos\left(\sqrt{\frac{3k}{m}}\, t + \phi_2\right) \tag{E.1}$$

where $B_1 = X_1^{(1)}$, $B_2 = X_1^{(2)}$, $\phi_1$ and $\phi_2$ are constants. We define a new set of coordinates $q_1(t)$ and $q_2(t)$ such that

$$q_1(t) = B_1 \cos\left(\sqrt{\frac{k}{m}}\, t + \phi_1\right)$$

$$q_2(t) = B_2 \cos\left(\sqrt{\frac{3k}{m}}\, t + \phi_2\right) \tag{E.2}$$

Since $q_1(t)$ and $q_2(t)$ are harmonic functions, their corresponding equations of motion can be written as*

$$\ddot{q}_1 + \left(\frac{k}{m}\right) q_1 = 0$$

$$\ddot{q}_2 + \left(\frac{3k}{m}\right) q_2 = 0 \tag{E.3}$$

These equations represent a two degree of freedom system whose natural frequencies are $\omega_1 = \sqrt{k/m}$ and $\omega_2 = \sqrt{3k/m}$. Because there is neither static nor dynamic coupling in the equations of motion (E.3), $q_1(t)$ and $q_2(t)$ are principal coordinates. From Eqs. (E.1) and (E.2), we can write

$$x_1(t) = q_1(t) + q_2(t)$$
$$x_2(t) = q_1(t) - q_2(t) \tag{E.4}$$

The solution of Eqs. (E.4) gives the principal coordinates:

$$q_1(t) = \frac{1}{2}\left[x_1(t) + x_2(t)\right]$$

$$q_2(t) = \frac{1}{2}\left[x_1(t) - x_2(t)\right] \tag{E.5}$$

---

*Note that the equation of motion corresponding to the solution $q = B \cos(\omega t + \phi)$ is given by $\ddot{q} + \omega^2 q = 0$.

### EXAMPLE 5.5

Determine the pitch (angular motion) and bounce (up and down linear motion) frequencies and the location of oscillation centers (nodes) of an automobile with the following data (see Fig. 5.10):

mass $= m = 1500$ kg

radius of gyration $= r = 1.1$ m

distance between front axle and C.G. $= l_1 = 1.5$ m

distance between rear axle and C.G. $= l_2 = 1.6$ m

front spring stiffness $= k_f = 36$ kN/m

rear spring stiffness $= k_r = 39$ kN/m

**Solution.** If $x$ and $\theta$ are used as independent coordinates, the equations of motion are given by Eq. (5.23) with $k_1 = k_f$, $k_2 = k_r$, and $J_0 = mr^2$. For free vibration, we assume a harmonic solution:

$$x(t) = X\cos(\omega t + \phi), \qquad \theta(t) = \Theta\cos(\omega t + \phi) \tag{E.1}$$

Using Eqs. (E.1) and (5.23), we obtain

$$\begin{bmatrix} (-m\omega^2 + k_1 + k_2) & (-k_1 l_1 + k_2 l_2) \\ (-k_1 l_1 + k_2 l_2) & (-J_0\omega^2 + k_1 l_1^2 + k_2 l_2^2) \end{bmatrix} \begin{Bmatrix} X \\ \Theta \end{Bmatrix} = \begin{Bmatrix} 0 \\ 0 \end{Bmatrix} \tag{E.2}$$

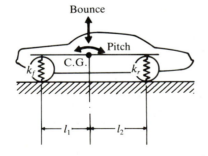

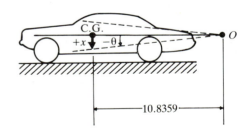

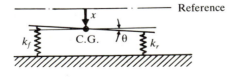

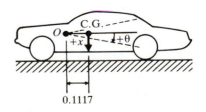

**Figure 5.10**

**Figure 5.11**

For the known data, Eq. (E.2) becomes

$$\begin{bmatrix} (-1500\omega^2 + 75\,000) & 8400 \\ 8400 & (-1815\omega^2 + 180\,840) \end{bmatrix} \begin{Bmatrix} X \\ \Theta \end{Bmatrix} = \begin{Bmatrix} 0 \\ 0 \end{Bmatrix} \tag{E.3}$$

from which the frequency equation can be derived:

$$2.7225\omega^4 - 407.385\omega^2 + 13\,492.44 = 0 \tag{E.4}$$

The natural frequencies can be found from Eq. (E.4):

$$\omega_1 = 7.0344 \text{ rad/sec}, \qquad \omega_2 = 10.0077 \text{ rad/sec} \tag{E.5}$$

With these values, the ratio of amplitudes can be found from Eq. (E.3):

$$\frac{X^{(1)}}{\Theta^{(1)}} = -10.8359, \qquad \frac{X^{(2)}}{\Theta^{(2)}} = 0.1117 \tag{E.6}$$

The node locations can be obtained by noting that the tangent of a small angle is approximately equal to the angle itself. Thus, from Fig. 5.11, we find the distance between the C.G. and the node as $-10.8359$ m for $\omega_1$ and 0.1117 m for $\omega_2$. The mode shapes are shown by dotted lines in Fig. 5.11.

## 5.6   FORCED VIBRATION ANALYSIS

The equations of motion of a general two degree of freedom system under external forces can be written as

$$\begin{bmatrix} m_{11} & m_{12} \\ m_{12} & m_{22} \end{bmatrix} \begin{Bmatrix} \ddot{x}_1 \\ \ddot{x}_2 \end{Bmatrix} + \begin{bmatrix} c_{11} & c_{12} \\ c_{12} & c_{22} \end{bmatrix} \begin{Bmatrix} \dot{x}_1 \\ \dot{x}_2 \end{Bmatrix} + \begin{bmatrix} k_{11} & k_{12} \\ k_{12} & k_{22} \end{bmatrix} \begin{Bmatrix} x_1 \\ x_2 \end{Bmatrix} = \begin{Bmatrix} F_1 \\ F_2 \end{Bmatrix} \tag{5.27}$$

Equations (5.1) and (5.2) can be seen to be special cases of Eq. (5.27), with $m_{11} = m_1$, $m_{22} = m_2$, and $m_{12} = 0$. We shall consider the external forces to be harmonic:

$$F_j(t) = F_{j0}e^{i\omega t}; \qquad j = 1, 2 \tag{5.28}$$

where $\omega$ is the forcing frequency. We can write the steady-state solutions as

$$x_j(t) = X_j e^{i\omega t}; \qquad j = 1, 2 \tag{5.29}$$

where $X_1$ and $X_2$ are, in general, complex quantities which depend on $\omega$ and the system parameters. Substitution of Eqs. (5.28) and (5.29) into Eq. (5.27) leads to

$$\begin{bmatrix} (-\omega^2 m_{11} + i\omega c_{11} + k_{11}) & (-\omega^2 m_{12} + i\omega c_{12} + k_{12}) \\ (-\omega^2 m_{12} + i\omega c_{12} + k_{12}) & (-\omega^2 m_{22} + i\omega c_{22} + k_{22}) \end{bmatrix} \begin{Bmatrix} X_1 \\ X_2 \end{Bmatrix} = \begin{Bmatrix} F_{10} \\ F_{20} \end{Bmatrix} \tag{5.30}$$

As in Sec. 3.5, we define the mechanical impedance $Z_{rs}(i\omega)$ as

$$Z_{rs}(i\omega) = -\omega^2 m_{rs} + i\omega c_{rs} + k_{rs}; \qquad r, s = 1, 2 \tag{5.31}$$

and write Eq. (5.30) as

$$[Z(i\omega)] \vec{X} = \vec{F}_0 \qquad (5.32)$$

where

$$[Z(i\omega)] = \begin{bmatrix} Z_{11}(i\omega) & Z_{12}(i\omega) \\ Z_{12}(i\omega) & Z_{22}(i\omega) \end{bmatrix} = \text{impedance matrix}$$

$$\vec{X} = \begin{Bmatrix} X_1 \\ X_2 \end{Bmatrix}$$

and

$$\vec{F}_0 = \begin{Bmatrix} F_{10} \\ F_{20} \end{Bmatrix}$$

Equation (5.32) can be solved to obtain

$$\vec{X} = [Z(i\omega)]^{-1} \vec{F}_0 \qquad (5.33)$$

where the inverse of the impedance matrix is given by

$$[Z(i\omega)]^{-1} = \frac{1}{Z_{11}(i\omega)Z_{22}(i\omega) - Z_{12}^2(i\omega)} \begin{bmatrix} Z_{22}(i\omega) & -Z_{12}(i\omega) \\ -Z_{12}(i\omega) & Z_{11}(i\omega) \end{bmatrix} \qquad (5.34)$$

Equations (5.33) and (5.34) lead to the solution

$$X_1(i\omega) = \frac{Z_{22}(i\omega)F_{10} - Z_{12}(i\omega)F_{20}}{Z_{11}(i\omega)Z_{22}(i\omega) - Z_{12}^2(i\omega)}$$

$$X_2(i\omega) = \frac{-Z_{12}(i\omega)F_{10} + Z_{11}(i\omega)F_{20}}{Z_{11}(i\omega)Z_{22}(i\omega) - Z_{12}^2(i\omega)} \qquad (5.35)$$

By substituting Eqs. (5.35) into Eqs. (5.29) we can find the complete solution, $x_1(t)$ and $x_2(t)$.

Reference [5.4] deals with the impact response of a two degree of freedom system, while Ref. [5.5] considers the steady-state response under harmonic excitation.

---

### EXAMPLE 5.6

Find the steady-state response of the system shown in Fig. 5.12 when the mass $m_1$ is excited by the force $F_1(t) = F_{10}\cos \omega t$. Also, plot its frequency response curve.

*Solution.* The equations of motion of the system can be expressed as

$$\begin{bmatrix} m & 0 \\ 0 & m \end{bmatrix} \begin{Bmatrix} \ddot{x}_1 \\ \ddot{x}_2 \end{Bmatrix} + \begin{bmatrix} 2k & -k \\ -k & 2k \end{bmatrix} \begin{Bmatrix} x_1 \\ x_2 \end{Bmatrix} = \begin{Bmatrix} F_{10}\cos \omega t \\ 0 \end{Bmatrix} \qquad (E.1)$$

Comparison of Eq. (E.1) with Eq. (5.27) shows that

$$m_{11} = m_{22} = m, \quad m_{12} = 0, \quad c_{11} = c_{12} = c_{22} = 0,$$
$$k_{11} = k_{22} = 2k, \quad k_{12} = -k, \quad F_1 = F_{10}\cos \omega t, \quad F_2 = 0$$

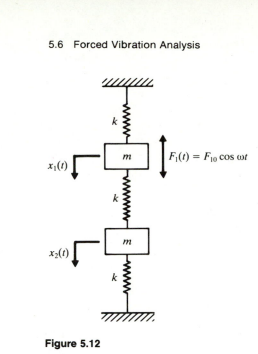

**Figure 5.12**

We assume the solution to be:*

$$x_j(t) = X_j \cos \omega t; \qquad j = 1, 2 \tag{E.2}$$

Equation (5.31) gives

$$Z_{11}(\omega) = Z_{22}(\omega) = -m\omega^2 + 2k, \qquad Z_{12}(\omega) = -k \tag{E.3}$$

Hence $X_1$ and $X_2$ are given by Eqs. (5.35):

$$X_1(\omega) = \frac{(-\omega^2 m + 2k) F_{10}}{(-\omega^2 m + 2k)^2 - k^2} = \frac{(-\omega^2 m + 2k) F_{10}}{(-m\omega^2 + 3k)(-m\omega^2 + k)} \tag{E.4}$$

$$X_2(\omega) = \frac{kF_{10}}{(-m\omega^2 + 2k)^2 - k^2} = \frac{kF_{10}}{(-m\omega^2 + 3k)(-m\omega^2 + k)} \tag{E.5}$$

By defining $\omega_1^2 = k/m$ and $\omega_2^2 = 3k/m$, Eqs. (E.4) and (E.5) can be expressed as

$$X_1(\omega) = \frac{\left\{2 - \left(\dfrac{\omega}{\omega_1}\right)^2\right\} F_{10}}{k\left[\left(\dfrac{\omega_2}{\omega_1}\right)^2 - \left(\dfrac{\omega}{\omega_1}\right)^2\right]\left[1 - \left(\dfrac{\omega}{\omega_1}\right)^2\right]} \tag{E.6}$$

$$X_2(\omega) = \frac{F_{10}}{k\left[\left(\dfrac{\omega_2}{\omega_1}\right)^2 - \left(\dfrac{\omega}{\omega_1}\right)^2\right]\left[1 - \left(\dfrac{\omega}{\omega_1}\right)^2\right]} \tag{E.7}$$

---

*Since $F_{10}\cos \omega t = \mathrm{Real}(F_{10}e^{i\omega t})$, we shall assume the solution also to be $x_j = \mathrm{Real}(X_j e^{i\omega t}) = X_j \cos \omega t$, $j = 1, 2$. It can be verified that $X_j$ are real for an undamped system.

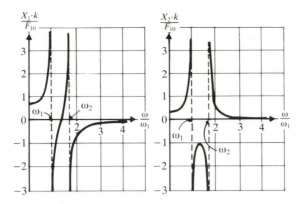

**Figure 5.13**  Frequency response curves of Example 5.6.

The responses $X_1$ and $X_2$ are shown in Fig. 5.13 in terms of the dimensionless parameter $\omega/\omega_1$. In the dimensionless parameter $\omega/\omega_1$, $\omega_1$ was selected arbitrarily; $\omega_2$ could have been selected just as easily. It can be seen that the amplitudes $X_1$ and $X_2$ become infinite when $\omega^2 = \omega_1^2$ or $\omega^2 = \omega_2^2$. Thus there are two resonance conditions for the system: one at $\omega_1$ and another at $\omega_2$. At all other values of $\omega$, the amplitudes of vibration are finite. It can be noted from Fig. 5.13 that there is a particular value of the frequency $\omega$ at which the vibration of the first mass $m_1$, to which the force $F_1(t)$ is applied, is reduced to zero. This characteristic forms the basis of the dynamic vibration absorber discussed in Chap. 9.

## 5.7  SEMI-DEFINITE SYSTEMS

*Semi-definite systems* are also known as *unrestrained* or *degenerate systems*. An example of such a system is shown in Fig. 5.14. This arrangement may be considered to represent two railway cars of masses $m_1$ and $m_2$ with a coupling spring $k$. The equations of motion of the system can be written as

$$m_1\ddot{x}_1 + k(x_1 - x_2) = 0$$
$$m_2\ddot{x}_2 + k(x_2 - x_1) = 0 \tag{5.36}$$

For free vibration, we assume the motion to be harmonic:

$$x_j(t) = X_j\cos(\omega t + \phi_1); \qquad j = 1,2 \tag{5.37}$$

Substitution of Eq. (5.37) into Eq. (5.36) gives

$$\left(-m_1\omega^2 + k\right)X_1 - kX_2 = 0$$
$$-kX_1 + \left(-m_2\omega^2 + k\right)X_2 = 0 \tag{5.38}$$

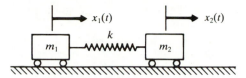

**Figure 5.14**

By equating the determinant of the coefficients of $X_1$ and $X_2$ to zero, we obtain the frequency equation as

$$\omega^2 \left[ m_1 m_2 \omega^2 - k(m_1 + m_2) \right] = 0 \qquad (5.39)$$

from which the natural frequencies can be obtained:

$$\omega_1 = 0 \qquad \text{and} \qquad \omega_2 = \sqrt{\frac{k(m_1 + m_2)}{m_1 m_2}} \qquad (5.40)$$

It can be seen that one of the natural frequencies of the system is zero, which means that the system is not oscillating. In other words, the system moves as a whole with-out any relative motion between the two masses (rigid body translation). Such systems, which have one of the natural frequencies equal to zero, are called *semi-definite systems*. We can verify, by substituting $\omega_2$ into Eq. (5.38), that $X_1^{(2)}$ and $X_2^{(2)}$ are opposite in phase. There would thus be a node at the middle of the spring.

## 5.8 SELF-EXCITATION AND STABILITY

In the case of a single degree of freedom system, it is simple to establish the stability conditions in terms of the physical constants of the system. However, the problem becomes more involved with a greater number of degrees of freedom. In general, we can determine the stability of a system by examining the values and signs of the coefficients of the frequency equation [5.6, 5.7]. We shall consider a damped two degree of freedom system in which the forcing functions are linear functions of the velocities. Thus the system is subjected to self-exciting functions, which, when combined with the damping terms, give the equations of motion:

$$\begin{bmatrix} 1 & 0 \\ 0 & 1 \end{bmatrix} \begin{Bmatrix} \ddot{x}_1 \\ \ddot{x}_2 \end{Bmatrix} + \begin{bmatrix} a_{11} & a_{12} \\ a_{21} & a_{22} \end{bmatrix} \begin{Bmatrix} \dot{x}_1 \\ \dot{x}_2 \end{Bmatrix} + \begin{bmatrix} b_{11} & b_{12} \\ b_{21} & b_{22} \end{bmatrix} \begin{Bmatrix} x_1 \\ x_2 \end{Bmatrix} = \begin{Bmatrix} 0 \\ 0 \end{Bmatrix} \qquad (5.41)$$

By substituting the solution

$$x_j(t) = X_j e^{st}; \qquad j = 1, 2 \qquad (5.42)$$

in Eq. (5.41) and setting the determinant of the coefficient matrix to zero, we obtain the characteristic equation of the form

$$s^4 + A_3 s^3 + A_2 s^2 + A_1 s + A_0 = 0 \qquad (5.43)$$

The coefficients $A_0$, $A_1$, $A_2$, and $A_3$ are real, since they are derived from the physical parameters of the system.

First we assume the system to be semi-definite or degenerate so that Eq. (5.43) reduces to a cubic equation:

$$s^3 + A_2 s^2 + A_1 s + A_0 = 0 \qquad (5.44)$$

If the roots of Eq. (5.44) are denoted as $s_1$, $s_2$, and $s_3$, Eq. (5.44) can be expressed as

$$(s - s_1)(s - s_2)(s - s_3) = s^3 - (s_1 + s_2 + s_3)s^2 + (s_1 s_2 + s_2 s_3 + s_3 s_1)s - s_1 s_2 s_3 = 0 \qquad (5.45)$$

A comparison of Eqs. (5.44) and (5.45) shows that

$$A_0 = -s_1 s_2 s_3$$
$$A_1 = s_1 s_2 + s_2 s_3 + s_3 s_1$$
$$A_2 = -(s_1 + s_2 + s_3) \qquad (5.46)$$

One of the three roots must always be real, and the other two can be real or complex conjugates. Thus the roots can be expressed in general form as

$$s_1 = p_1, \qquad s_2 = p_2 + i q_2, \qquad s_3 = p_2 - i q_2 \qquad (5.47)$$

The use of Eq. (5.47) in Eq. (5.46) gives

$$A_0 = -p_1(p_2^2 + q_2^2)$$
$$A_1 = 2 p_1 p_2 + p_2^2 + q_2^2$$
$$A_2 = -(p_1 + 2 p_2) \qquad (5.48)$$

The criterion of stability is that the real parts of $s_1$, $s_2$, and $s_3$ must be negative (to avoid increasing exponentials in Eq. (5.42)). In the first place, we note that all the coefficients $A_0$, $A_1$, and $A_2$ must be positive, because, if any one of them were negative, Eq. (5.48) requires that either $p_1$ or $p_2$ or both $p_1$ and $p_2$ would be positive. Assuming that $A_0$, $A_1$, and $A_2$ are all positive, the first equation (5.48) requires that $p_1$ be negative. No information is available as yet about $p_2$. But on the borderline between stability and instability, $p_2$ must pass from a positive to a negative value through zero. Taking $p_2 = 0$, Eq. (5.48) gives

$$A_0 = -p_1 q_2^2, \qquad A_1 = q_2^2, \qquad A_2 = -p_1 \qquad (5.49)$$

Since these relations must be satisfied on the boundary of stability, we have

$$A_0 = A_1 A_2 \qquad (5.50)$$

by eliminating $p_1$ and $p_2$ in Eq. (5.49). Since we do not yet know on which side of Eq. (5.50) stability exists, we test one particular case. Let $s_1 = -1$, $s_2 = -1 + i$, and $s_3 = -1 - i$; this is obviously a stable solution. By substituting this in Eq. (5.48), we find $A_0 = 2$, $A_1 = 4$, and $A_2 = 3$, so that $A_0 < A_1 A_2$.

Thus the complete criterion for dynamic stability is that all coefficients $A_0$, $A_1$, and $A_2$ are positive and that $A_0 < A_1 A_2$. A similar procedure can be used for the stability analysis of Eq. (5.43). For this equation, the complete criterion for dynamic stability is that all coefficients $A_0$, $A_1$, $A_2$, and $A_3$ are positive and that $A_1^2 + A_0 A_3^2 < A_1 A_2 A_3$ [5.8].

## **5.9**  COMPUTER PROGRAMS

The determination of the natural frequencies of a damped two degree of freedom system involves the solution of a fourth order polynomial equation. Similarly, a damped degenerate two degree of freedom system requires the determination of the roots of a cubic equation. An undamped system, on the other hand, requires the solution of a quadratic equation. This section presents three Fortran subroutines (QUADRA, CUBIC, and QUART) for the solution of quadratic, cubic, and quartic equations, respectively. The listing of these subroutines and typical main programs for calling them are given below. The input data required and the output of the programs are explained in the comment lines of the programs.

### **5.9.1**  **Roots of a Quadratic Equation**

```
C =======================================================================
C
C PROGRAM 6
C MAIN PROGRAM WHICH CALLS QUADRA
C
C =======================================================================
C FOLLOWING 2 LINES CONTAIN PROBLEM-DEPENDENT DATA
C EXAMPLE X**2 - 2.0*X + 5.0 = 0.0
      DATA A1,A2,A3/1.0,-2.0,5.0/
C END OF PROBLEM-DEPENDENT DATA
      CALL QUADRA (A1,A2,A3,RR1,RR2,RI1,RI2)
      PRINT 10, A1,A2,A3
   10 FORMAT (/,2X,28H POLYNOMIAL COEFFICIENTS ARE,/,3E15.6,//,
     2   2X,10H ROOTS ARE,//,4X,5H REAL,14X,10H IMAGINARY)
      PRINT 20, RR1,RI1
      PRINT 20, RR2,RI2
   20 FORMAT (4X,E15.8,4X,E15.8)
      STOP
      END
C =======================================================================
C
C SUBROUTINE QUADRA
C
C =======================================================================
C     SOLUTION OF QUADRATIC EQUATION  A1*(X**2)+A2*(X)+A3 = 0
C     A1,A2,A3 ARE INPUT, (RR1,RI1) AND (RR2,RI2) ARE ROOTS (OUTPUT)
C     A1 MUST NOT BE EQUAL TO ZERO
      SUBROUTINE QUADRA (A1,A2,A3,RR1,RR2,RI1,RI2)
      RAD=A2**2-4.0*A1*A3
      IF (RAD) 20,10,10
   10 SRAD=SQRT(RAD)
      RR1=(-A2-SRAD)/(2.0*A1)
```

```
          RR2=(-A2+SRAD)/(2.0*A1)
          RI1=0.0
          RI2=0.0
          RETURN
   20     SRAD=SQRT(-RAD)
          RR1=-A2/(2.0*A1)
          RR2=RR1
          RI1=SRAD/(2.0*A1)
          RI2=-RI1
          RETURN
          END
```

```
POLYNOMIAL COEFFICIENTS ARE
0.100000E+01   -0.200000E+01    0.500000E+01

ROOTS ARE

     REAL              IMAGINARY
  0.10000000E+01      0.20000000E+01
  0.10000000E+01     -0.20000000E+01
```

## 5.9.2   Roots of a Cubic Equation

```
C ================================================================
C
C PROGRAM 7
C MAIN PROGRAM FOR CALLING THE SUBROUTINE CUBIC
C
C ================================================================
          DIMENSION A(4),RR(3),RI(3)
          DATA A/1.0,0.0,6.0,20.0/
          PRINT 10
   10     FORMAT (//,24H ROOTS OF CUBIC EQUATION,//,
          2    51H GIVEN POLYNOMIAL COEFFICIENTS A(1),A(2),A(3),A(4):,/)
          PRINT 20,(A(I),I=1,4)
   20     FORMAT (4E15.6)
          CALL CUBIC (A,RR,RI)
          PRINT 30
   30     FORMAT (//,38H ROOTS (REAL PART AND IMAGINARY PART):,/)
          DO 40 I=1,3
   40     PRINT 50,RR(I),RI(I)
   50     FORMAT (2E15.6)
          STOP
          END
C ================================================================
C
C SUBROUTINE CUBIC
C
C ================================================================
C     ROOTS OF CUBIC EQUATION   A(1)*(X**3)+A(2)*(X**2)+A(3)*X+A(4)=0
          SUBROUTINE CUBIC (A,RR,RI)
          DIMENSION A(4),RR(3),RI(3)
          DO 10 I=1,3
          RR(I)=0.0
```

```
10     RI(I)=0.0
       A0=A(1)
       A1=A(2)/3.0
       A2=A(3)/3.0
       A3=A(4)
       G=(A0**2)*A3-3.0*A0*A1*A2+2.0*(A1**3)
       H=A0*A2-A1**2
       Y1=G**2+4.0*(H**3)
       IF (Y1 .LT. 0.0) GO TO 100
       Y2=SQRT(Y1)
       Z1=(G+Y2)/2.0
       Z2=(G-Y2)/2.0
       IF(Z1 .LT. 0.0) GO TO 21
       Z3=Z1**(1.0/3.0)
       GO TO 22
21     Z3=(-Z1)**(1.0/3.0)
       Z3=-Z3
22     IF(Z2 .LT. 0.0) GO TO 23
       Z4=Z2**(1.0/3.0)
       GO TO 24
23     Z4=(-Z2)**(1.0/3.0)
       Z4=-Z4
24     CONTINUE
       RR(1)=-(A1+Z3+Z4)/A0
       RR(2)=(-2.0*A1+Z3+Z4)/(2.0*A0)
       RI(2)=SQRT(3.0)*(Z4-Z3)/(2.0*A0)
       RR(3)=RR(2)
       RI(3)=-RI(2)
       GO TO 200
100    SH=SQRT(-H)
       XK=2.0*SH
       THETA=ACOS(G/(2.0*H*SH))/3.0
       XY1=2.0*SH*COS(THETA)
       PI=3.1416
       XY2=2.0*SH*COS(THETA+(2.0*PI/3.0))
       XY3=2.0*SH*COS(THETA+(4.0*PI/3.0))
       RR(1)=(XY1-A1)/A0
       RR(2)=(XY2-A1)/A0
       RR(3)=(XY3-A1)/A0
200    RETURN
       END
```

ROOTS OF CUBIC EQUATION

GIVEN POLYNOMIAL COEFFICIENTS A(1),A(2),A(3),A(4):

```
 0.100000E+01    0.000000E+00    0.600000E+01    0.200000E+02
```

ROOTS (REAL PART AND IMAGINARY PART):

```
-0.200000E+01    0.000000E+00
 0.100000E+01   -0.300000E+01
 0.100000E+01    0.300000E+01
```

### 5.9.3    Roots of a Quartic Equation

```
C =================================================================
C
C PROGRAM 8
C MAIN PROGRAM FOR CALLING THE SUBROUTINE QUART
C
C =================================================================
C SOLUTION OF:  A(1)*(X**4)+A(2)*(X**3)+A(3)*(X**2)+A(4)*X+A(5)=0
      DIMENSION A(5),RR(4),RI(4)
C FOLLOWING LINE CONTAINS PROBLEM-DEPENDENT DATA
      DATA A/1.0,0.0,0.0,-8.0,12.0/
C END OF PROBLEM-DEPENDENT DATA
      PRINT 10,(A(I),I=1,5)
  10  FORMAT (//,31H SOLUTION OF A QUARTIC EQUATION,//,6H DATA:,/,
     2    7H A(1) =,E15.6,/,7H A(2) =,E15.6,/,7H A(3) =,E15.6,/,
     3    7H A(4) =,E15.6,/,7H A(5) =,E15.6,/)
      CALL QUART (A,RR,RI)
      PRINT 20
  20  FORMAT (/,7H ROOTS:,//,9H ROOT NO.,3X,10H REAL PART,5X,
     2    15H IMAGINARY PART,/)
      DO 30 I=1,4
  30  PRINT 40,I,RR(I),RI(I)
  40  FORMAT (I5,3X,E15.6,3X,E15.6)
      STOP
      END
C =================================================================
C
C SUBROUTINE QUART
C
C =================================================================
      SUBROUTINE QUART (A,RR,RI)
      DIMENSION A(5),RR(4),RI(4),B(4),RRC(3),RIC(3)
      DO 10 I=2,5
  10  A(I)=A(I)/A(1)
      B(1)=1.0
      B(2)=-A(3)
      B(3)=A(4)*A(2)-4.0*A(5)
      B(4)=A(5)*(4.0*A(3)-A(2)**2)-A(4)**2
      CALL CUBIC (B,RRC,RIC)
      IF (RIC(2) .NE. 0.0) GO TO 20
      X=AMAX1(RRC(1),RRC(2),RRC(3))
      RRC(1)=X
  20  X=RRC(1)/2.0
      IF ((X**2-A(5)) .GT. 0.0) GO TO 30
      Y=0.0
      Z=SQRT((A(2)/2.0)**2+2.0*X-A(3))
C ADD TO ABOVE EQUATION
      GO TO 40
  30  Y=SQRT(X**2-A(5))
      Z=-(A(4)-A(2)*X)/(2.0*Y)
  40  C1=1.0
      C2=A(2)/2.0+Z
      C3=X+Y
      CALL QUADRA (C1,C2,C3,QR1,QR2,QI1,QI2)
      RR(1)=QR1
```

```
      RR(2)=QR2
      RI(1)=QI1
      RI(2)=QI2
      C1=1.0
      C2=A(2)/2.0-Z
      C3=X-Y
      CALL QUADRA (C1,C2,C3,QR1,QR2,QI1,QI2)
      RR(3)=QR1
      RR(4)=QR2
      RI(3)=QI1
      RI(4)=QI2
      RETURN
      END
```

SOLUTION OF A QUARTIC EQUATION

DATA:
A(1) =     0.100000E+01
A(2) =     0.000000E+00
A(3) =     0.000000E+00
A(4) =    -0.800000E+01
A(5) =     0.120000E+02

ROOTS:

| ROOT NO. | REAL PART | IMAGINARY PART |
|---|---|---|
| 1 | -0.137091E+01 | 0.182709E+01 |
| 2 | -0.137091E+01 | -0.182709E+01 |
| 3 | 0.137091E+01 | 0.648457E+00 |
| 4 | 0.137091E+01 | -0.648457E+00 |

## REFERENCES

**5.1.** H. Sato, Y. Kuroda, and M. Sagara, "Development of the finite element method for vibration analysis of machine tool structure and its application," *Proceedings of the Fourteenth International Machine Tool Design and Research Conference*, Macmillan, London, 1974, pp. 545–552.

**5.2.** F. Koenigsberger and J. Tlusty, *Machine Tool Structures*, Pergamon Press, Oxford, 1970.

**5.3.** C. P. Reddy and S. S. Rao, "Automated optimum design of machine tool structures for static rigidity, natural frequencies and regenerative chatter stability," *Journal of Engineering for Industry*, Vol. 100, 1978, pp. 137–146.

**5.4.** M. S. Hundal, "Effect of damping on impact response of a two degree of freedom system," *Journal of Sound and Vibration*, Vol. 68, 1980, pp. 407–412.

**5.5.** J. A. Linnett, "The effect of rotation on the steady state response of a spring-mass system under harmonic excitation," *Journal of Sound and Vibration*, Vol. 35, 1974, pp. 1–11.

**5.6.** A. Hurwitz, "On the conditions under which an equation has only roots with negative real parts," in *Selected Papers on Mathematical Trends in Control Theory*, Dover Publications, New York, 1964, pp. 70–82.

**5.7.** R. C. Dorf, *Modern Control Systems* (3rd Ed.), Addison-Wesley, Reading, Mass., 1980.

**5.8.** J. P. Den Hartog, "*Mechanical Vibrations* (4th Ed.), McGraw-Hill, New York, 1956.

## REVIEW QUESTIONS

**5.1.** How do you determine the number of degrees of freedom of a lumped-mass system?

**5.2.** Define mass coupling, velocity coupling, and elastic coupling.

**5.3.** Is the nature of the coupling dependent on the coordinates used?

**5.4.** How many degrees of freedom does an airplane in flight have if it is treated as (a) a rigid body, and (b) an elastic body?

**5.5.** What are principal coordinates? What is their use?

**5.6.** Why are the mass, damping, and stiffness matrices symmetrical?

**5.7.** What is a node?

**5.8.** What is meant by static and dynamic coupling? How can you eliminate coupling of the equations of motion?

**5.9.** Define the impedance matrix.

**5.10.** How can we make a system vibrate in one of its natural modes?

**5.11.** What is a degenerate system? Give two examples of physical systems that are degenerate.

**5.12.** How many degenerate modes can a vibrating system have?

## PROBLEMS

The problem assignments are organized as follows:

| Problems | Section covered | Topic covered |
|---|---|---|
| 5.1–5.18 | 5.3 | Free vibration of undamped systems |
| 5.19–5.24 | 5.4 | Torsional systems |
| 5.25–5.32 | 5.5 | Coordinate coupling |
| 5.33–5.40 | 5.6 | Forced vibrations |
| 5.41–5.46 | 5.7 | Semi-definite systems |
| 5.47–5.49 | 5.9 | Computer programs |

**5.1.** Find the natural frequencies of the system shown in Fig. 5.15, with $m_1 = m$, $m_2 = 2m$, $k_1 = k$, and $k_2 = 2k$. Determine the response of the system when $k = 1000$ N/m, $m = 20$ kg, and the initial values of the displacements of the masses $m_1$ and $m_2$ are 1 and $-1$, respectively.

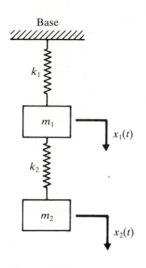

Figure 5.15                                    Figure 5.16

**5.2.** Set up the differential equations of motion for the double pendulum shown in Fig. 5.16, using the coordinates $x_1$ and $x_2$ and assuming small amplitudes. Find the natural frequencies, the ratios of amplitudes, and the locations of nodes for the two modes of vibration when $m_1 = m_2 = m$ and $l_1 = l_2 = l$.

**5.3.** Determine the natural modes of the system shown in Fig. 5.17 when $k_1 = k_2 = k_3 = k$.

**5.4.** The following data has been obtained for the automobile shown in Fig. 5.18:

Body mass $(M) = 1000$ kg

Axle mass $(m) = 200$ kg

Stiffness of springs $(k_1) = 50$ N/mm

Stiffness of tires $(k_2) = 500$ N/mm, Damping of tires $(c_2) = 2$ N-s/mm

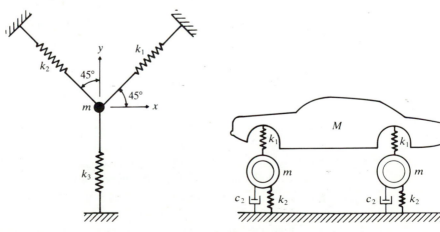

Figure 5.17                                    Figure 5.18

By idealizing the automobile as a two degree of freedom system, determine the natural frequencies and mode shapes of the system.

**5.5.** An overhead traveling crane can be modeled as shown in Fig. 5.19. The beam has an area moment of inertia ($I$) of 0.02 $m^4$ and modulus of elasticity ($E$) of $2.06 \times 10^{11}$ $N/m^2$, the truck has a mass ($m_1$) of 1000 kg, the load being lifted has a mass ($m_2$) of 5000 kg, and the cable through which the mass ($m_2$) is lifted has a stiffness ($k$) of $3.0 \times 10^5$ N/m. Determine the natural frequencies and mode shapes of the system. Assume the span of the beam as 40 m.

**5.6.** One means of packaging fragile objects for shipping consists of supporting the fragile object inside the shipping container by springs under tension, as shown in Fig. 5.20. To simplify vibration analysis, the system is idealized as two masses and three springs. If $M = 2m = 5$ kg and $k_2 = 2k_1 = 4000$ N/m determine (a) the natural frequencies and the natural modes of the system, and (b) the maximum deflection of the inner mass if it is dropped from a height of 1 m.

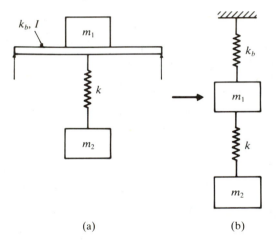

**Figure 5.19**

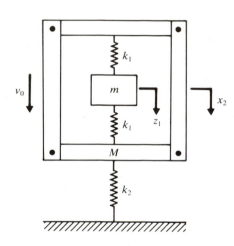

**Figure 5.20**

**5.7.** Derive the equations of motion of the double pendulum shown in Fig. 5.16, using the coordinates $\theta_1$ and $\theta_2$. Also find the natural frequencies and mode shapes of the system for $m_1 = m_2 = m$ and $l_1 = l_2 = l$.

**5.8.** Find the natural frequencies and mode shapes of the system shown in Fig. 5.15 for $m_1 = m_2 = m$ and $k_1 = k_2 = k$.

**5.9.** The normal modes of a two degree of freedom system are orthogonal if $X^{(1)^T}[m]\vec{X}^{(2)} = 0$. Prove that the mode shapes of the system shown in Fig. 5.3(a) are orthogonal.

**5.10.** Find the natural frequencies of the system shown in Fig. 5.4 for $k_1 = 300$ N/m, $k_2 = 500$ N/m, $k_3 = 200$ N/m, $m_1 = 2$ kg, and $m_2 = 1$ kg.

**5.11.** Find the natural frequencies and mode shapes of the system shown in Fig. 5.15 for $m_1 = m_2 = 1$ kg, $k_1 = 2000$ N/m, and $k_2 = 6000$ N/m.

**5.12.** If $m_1 = m_2 = 5$ kg and $k_1 = k_2 = k_3 = 10$ kN/m, determine the general motion of the system shown in Fig. 5.4.

**5.13.** For the system shown in Fig. 5.4, $m_1 = 1$ kg, $m_2 = 2$ kg, $k_1 = 2000$ N/m, $k_2 = 1000$ N/m, $k_3 = 3000$ N/m, and an initial velocity of 20 m/s is imparted to mass $m_1$. Find the resulting motion of the two masses.

**5.14.** For Problem 5.11, calculate $x_1(t)$ and $x_2(t)$ for the following initial conditions: (a) $x_1(0) = 0.2$, $\dot{x}_1(0) = x_2(0) = \dot{x}_2(0) = 0$; and (b) $x_1(0) = 0.2$, $\dot{x}_1(0) = x_2(0) = 0$, $\dot{x}_2(0) = 5.0$.

**5.15.** A two-story building frame is modeled as shown in Fig. 5.21. The girders are assumed to be rigid, and the columns have flexural rigidities $EI_1$ and $EI_2$, with negligible masses. The stiffness of each column can be computed as

$$\frac{24 EI_i}{h_i^3}; \qquad i = 1, 2$$

For $m_1 = 2m$, $m_2 = m$, $h_1 = h_2 = h$, and $EI_1 = EI_2 = EI$, determine the natural frequencies and mode shapes of the frame.

**5.16.** Figure 5.22 shows a system of two masses attached to a tightly stretched string, fixed at both ends. Determine the natural frequencies and mode shapes of the system for $m_1 = m_2 = m$ and $l_1 = l_2 = l_3 = l$.

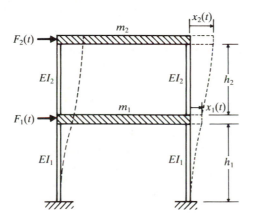

**Figure 5.21**

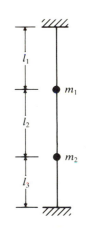

**Figure 5.22**

**5.17.** Find the normal modes of the two-story building shown in Fig. 5.21 when $m_1 = 3m$, $m_2 = m$, $k_1 = 3k$, and $k_2 = k$, where $k_1$ and $k_2$ represent the total equivalent stiffnesses of the lower and upper columns, respectively.

**5.18.** A forging hammer is mounted on a large heavy reinforced concrete block, which is supported by a nest of isolating springs. A linearly elastic pad is placed between the forging hammer and the concrete block, as shown in Fig. 5.23. The mass of the hammer is 10000 kg, and the mass of the block is 100000 kg. The static deflection of the isolation springs under the weight of the block and the hammer is 6 cm, and the static

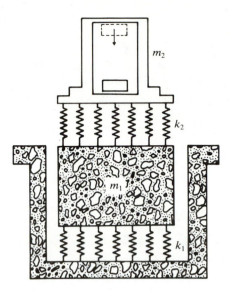

**Figure 5.23**

deflection of the elastic pad due to the weight of the hammer is 0.3 cm. Determine the natural frequencies and amplitude ratios for the vertical vibration of the system.

**5.19.** A pulley 250 mm in diameter drives a second pulley 1000 mm in diameter by means of a belt (see Fig. 5.24). The moment of inertia of the driving pulley is 0.1 kg-m$^2$ and of the driven pulley is 0.2 kg-m$^2$. The belt connecting these pulleys is represented by two springs, each of stiffness $k = 5000$ N/m. Find the two natural frequencies of the system.

**5.20.** Determine the natural frequencies and normal modes of the torsional system shown in Fig. 5.25 for $k_{t2} = 2k_{t1}$ and $J_2 = 2J_1$.

**5.21.** Determine the natural frequencies of the system shown in Fig. 5.26 by assuming that the rope passing over the cylinder does not slip.

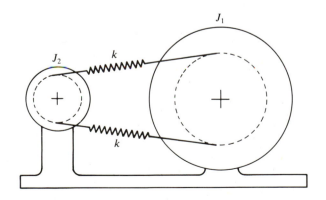

**Figure 5.24**

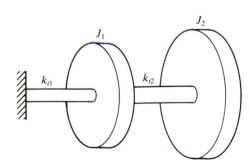

**Figure 5.25**

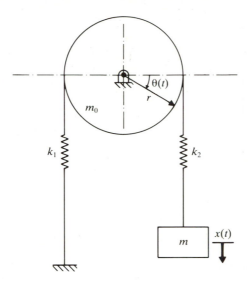

**Figure 5.26**

**5.22.** Find the natural frequencies and mode shapes of the system shown in Fig. 5.6(a) by assuming that $J_1 = J_0$, $J_2 = 2J_0$, and $k_{t1} = k_{t2} = k_{t3} = k_t$.

**5.23.** Determine the normal modes of the torsional system shown in Fig. 5.7 when $k_{t1} = k_t$, $k_{t2} = 5k_t$, $J_1 = J_0$, and $J_2 = 5J_0$.

**5.24.** A simplified ride model of a military vehicle is shown in Fig. 5.27(b). This model can be used to obtain information about the bounce and pitch modes of the vehicle. If the total mass of the vehicle is $m$ and the mass moment of inertia about its C.G. is $J_0$, derive the equations of motion of the vehicle using two different sets of coordinates, as indicated in Sec. 5.5.

**5.25.** Find the natural frequencies and the amplitude ratios of the system shown in Fig. 5.28.

**5.26.** A simple means of studying the effect of an earthquake on a building involves assuming the building to be rigid and the base of the building to be connected to the ground

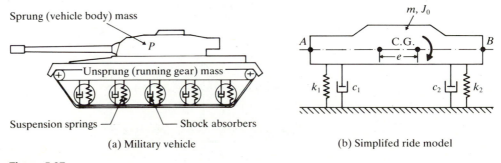

(a) Military vehicle

(b) Simplifed ride model

**Figure 5.27**

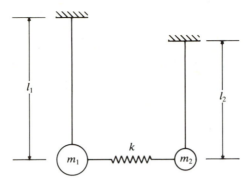

**Figure 5.28**

through two springs and two dampers, as shown in Fig. 5.29. If the horizontal ground motion is given by $x_g = X_g \cos \omega t$, derive the equations of motion in terms of the coordinates $x$ and $\theta$. Assume the mass and moment of inertia about the C.G. of the building to be $M$ and $J_0$, respectively.

**5.27.** Figure 5.30 shows a spring-restrained piston of mass $m$ moving in a smooth cylinder. A simple pendulum of length $l$ and mass $m_0$ is connected to the piston. Using $x$ and $\theta$ as coordinates, determine the natural frequencies of the system for $m = m_0 = 1$ kg, $k = 500$ N/m, and $l = 0.2$ m.

**5.28.** A rigid rod of negligible mass and length $2l$ is pivoted at the middle point and is constrained to move in the vertical plane by springs and masses, as shown in Fig. 5.31. Find the natural frequencies and mode shapes of the system.

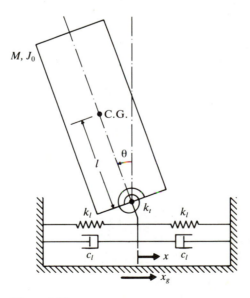

**Figure 5.29**

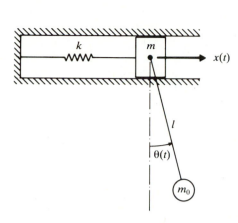

**Figure 5.30**

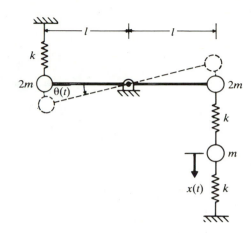

**Figure 5.31**

**5.29.** The equivalent spring constants of the nose and main landing gears of an airplane are shown in Fig. 5.32. The mass of the airplane is $m$ and the mass moment of inertia about its centroid is $J_0$. Assuming that $l_1 = l$, $l_2 = 2l$, $k_1 = k$, $k_2 = 5k$, and the radius of gyration is $5l$, find the natural frequencies and mode shapes of the system by treating it as a two degree of freedom system.

**5.30.** An airfoil of mass $m$ is suspended by a linear spring of stiffness $k$ and a torsional spring of stiffness $k_t$ in a wind tunnel, as shown in Fig. 5.33. The C.G. is located at a distance of $e$ from point $O$. The mass moment of inertia of the airfoil about an axis passing through point $O$ is $J_0$. Find the natural frequencies of the airfoil.

**5.31.** The expansion joints of a concrete highway, which are located at 15 m intervals, cause a series of impulses to affect cars running at a constant speed. Determine the speeds at which bounce motion and pitch motion are most likely to arise for the automobile of Example 5.5.

**5.32.** The rotor shown in Fig. 5.34 is free to move in the vertical direction and rotate in the plane of the paper. The rotor is mounted on bearings with the indicated stiffness and damping constant. If there is a small unbalance $me$ at a distance $a$ from the C.G. of the rotor, find the equations of motion of the rotor for a rotational speed of $\omega$.

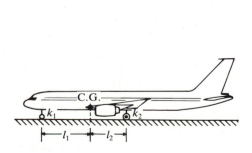

**Figure 5.32**

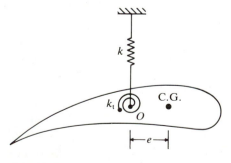

**Figure 5.33**

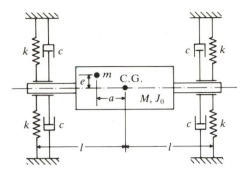

**Figure 5.34**

**5.33.** Find the steady-state response of the system shown in Fig. 5.15 by using the mechanical impedance method, when the mass $m_1$ is subjected to the force $F(t) = F_0 \sin \omega t$ in the direction of $x_1(t)$.

**5.34.** Find the steady-state response of the system shown in Fig. 5.15 when the base is subjected to a displacement $y(t) = Y_0 \cos \omega t$.

**5.35.** The mass $m_1$ of the two degree of freedom system shown in Fig. 5.15 is subjected to a force $F_0 \cos \omega t$. Assuming that the surrounding air damping is equivalent to $c = 200$ N·s/m, find the steady-state response of the two masses. Assume $m_1 = m_2 = 1$ kg, $k_1 = k_2 = 500$ N/m, and $\omega = 1s^{-1}$.

**5.36.** Determine the steady-state vibration of the system shown in Fig. 5.3(a), assuming that $c_1 = c_2 = c_3 = 0$, $F_1(t) = F_{10} \cos \omega t$, and $F_2(t) = F_{20} \cos \omega t$.

**5.37.** In the system shown in Fig. 5.15, the mass $m_1$ is excited by a harmonic force having a maximum value of 50 N and a frequency of 2 Hz. Find the forced amplitude of each mass for $m_1 = 10$ kg, $m_2 = 5$ kg, $k_1 = 8000$ N/m, and $k_2 = 2000$ N/m.

**5.38.** Find the response of the two masses of the two-story frame shown in Fig. 5.21 under the ground displacement $y(t) = 0.2 \sin \pi t$ m. Assume the equivalent stiffness of the lower and upper columns to be 800 N/m and 600 N/m, respectively, and $m_1 = m_2 = 50$ kg.

**5.39.** Find the forced vibration response of the system shown in Fig. 5.12 when $F_1(t)$ is a step force of magnitude 5 N using the Laplace transformation method. Assume $x_1(0) = \dot{x}_1(0) = x_2(0) = \dot{x}_2(0) = 0$, $m = 1$ kg and $k = 100$ N/m.

**5.40.** A punch press and its foundation are schematically shown in Fig. 5.35(a). The punch press and the upper slab on which it is mounted are represented by $m_1$ and the base or foundation by $m_2$ in Fig. 5.35(b). $k_1$ is the stiffness of the supports between the upper and the foundation slabs, and $k_2$ is the stiffness of the soil on which the foundation slab rests. If $m_1 = 200$ Mg, $m_2 = 250$ Mg, $k_1 = 150$ MN/m, and $k_2 = 75$ MN/m, determine (a) the natural frequencies of the system, and (b) the response of the masses $m_1$ and $m_2$ when the time history of the punch force is as indicated in Fig. 5.35(c). Assume $F_0 = 10^5$ N and $T = 0.5$ s.

**5.41.** Determine the equations of motion and the natural frequencies of the system shown in Fig. 5.36.

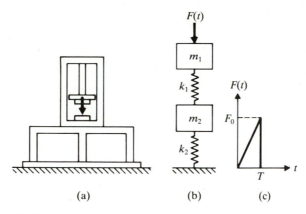

(a)                     (b)                 (c)

**Figure 5.35**

**5.42.** An automobile of mass 2000 kg pulls a trailer of mass 500 kg (Fig. 5.37). If the flexibility of the hitch is 150 N/mm, determine the natural frequencies and natural modes of the system.

**5.43.** Two identical circular cylinders, of radius $r$ and mass $m$ each, are connected by a spring as shown in Fig. 5.38. Determine the natural frequencies of oscillation of the system.

**5.44.** The differential equations of motion for a two degree of freedom system are given by

$$a_1 \ddot{x}_1 + b_1 x_1 + c_1 x_2 = 0$$
$$a_2 \ddot{x}_2 + b_2 x_1 + c_2 x_2 = 0$$

Derive the condition to be satisfied for the system to be degenerate.

**5.45.** Find the angular displacements $\theta_1(t)$ and $\theta_2(t)$ of the system shown in Fig. 5.39 for the initial conditions $\theta_1(t=0) = \theta_1(0)$, $\theta_2(t=0) = \theta_2(0)$, and $\dot{\theta}_1(t=0) = \dot{\theta}_2(t=0) = 0$.

**5.46.** Determine the normal modes of the system shown in Fig. 5.7 with $k_{t1} = 0$. Show that the system with $k_{t1} = 0$ can be treated as a single degree of freedom system by using the coordinate $\alpha = \theta_1 - \theta_2$.

**5.47.** Find the response of the system shown in Fig. 5.3(a) using a numerical procedure when $k_1 = k$, $k_2 = 2k$, $k_3 = k$, $m_1 = 2m$, $m_2 = m$, $F_2(t) = 0$, and $F_1(t)$ is a rectangular pulse of magnitude 500 N and duration 0.5 sec. Assume $m = 10$ kg, $c_1 = c_2 = c_3 = 0$, and $k = 2000$ N/m.

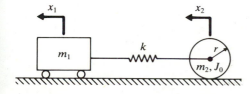

**Figure 5.36**

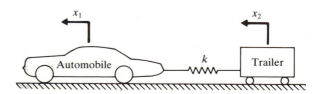

**Figure 5.37**

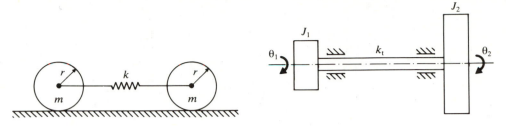

**Figure 5.38**                                        **Figure 5.39**

**5.48.** (a) Find the roots of the frequency equation of the system shown in Fig. 5.3 using subroutine QUART with the following data: $m_1 = m_2 = 0.2$ lb-s$^2$/in., $k_1 = k_2 = 18$ lb/in., $k_3 = 0$, $c_1 = c_2 = c_3 = 0$. (b) If the initial conditions are $x_1(0) = x_2(0) = 2$ in., $\dot{x}_1(0) = \dot{x}_2(0) = 0$, determine the displacements $x_1(t)$ and $x_2(t)$ of the masses.

**5.49.** Write a computer program for finding the steady-state response of a two degree of freedom system under the harmonic excitation $F_j(t) = F_{j0}e^{i\omega t}$; $j = 1, 2$ using Eqs. (5.29) and (5.35). Use this program to find the response of a system with $m_{11} = m_{22} = 0.1$ lb-s$^2$/in., $m_{12} = 0$, $c_{11} = 1.0$ lb-s/in., $c_{12} = c_{22} = 0$, $k_{11} = 40$ lb/in., $k_{22} = 20$ lb/in., $k_{12} = -20$ lb/in., $F_{10} = 1$ lb, $F_{20} = 2$ lb, and $\omega = 5$ rad/s.

# CHAPTER 6

# Multidegree of Freedom Systems

Joseph Louis Lagrange (1736–1813) was an Italian-born mathematician famous for his work on theoretical mechanics. He was made professor of mathematics in 1755 at the Artillery School at Turin. Lagrange's masterpiece, his *Mèchanique*, contains what are now known as ''Lagrange's equations,'' which are very useful in the study of vibrations. His work on elasticity and strength of materials, where he considered the strength and deflection of struts, is less well-known.

Courtesy Brown Brothers

## 6.1 INTRODUCTION

All the concepts introduced in the preceding chapter can be directly extended to the case of multidegree of freedom systems. For example, there is one equation of motion for each degree of freedom; if generalized coordinates are used, there is one generalized coordinate for each degree of freedom. The equations of motion can be obtained from Newton's second law of motion or by using the influence coefficients defined in Sec. 6.3. However, it is often more convenient to derive the equations of motion of a multidegree of freedom system by using Lagrange's equations.

There are $n$ natural frequencies, each associated with its own mode shape, for a system having $n$ degrees of freedom. The method of determining the natural frequencies from the characteristic equation obtained by equating the determinant to zero also applies to these systems. However, as the number of degrees of freedom increases, the solution of the characteristic equation becomes more complex. The mode shapes exhibit a property known as *orthogonality*, which often enables us to simplify the analysis of multidegree of freedom systems.

## 6.2 MULTIDEGREE OF FREEDOM SPRING-MASS SYSTEM

Consider a simple $n$ degree of freedom system, as shown in Fig. 6.1(a). With reference to the free-body diagram of a typical interior mass $m_i$, the equation of motion can be derived:

$$m_i \ddot{x}_i = -k_i(x_i - x_{i-1}) + k_{i+1}(x_{i+1} - x_i) + F_i; \qquad i = 2, 3, \ldots, n-1$$

or

$$m_i \ddot{x}_i - k_i x_{i-1} + (k_i + k_{i+1}) x_i - k_{i+1} x_{i+1} = F_i; \qquad i = 2, 3, \ldots, n-1 \qquad (6.1)$$

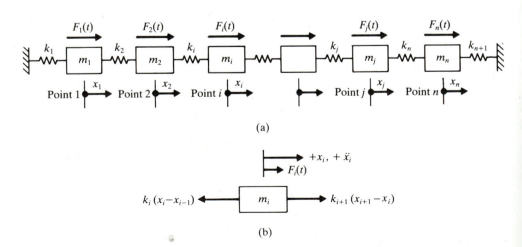

(a)

(b)

**Figure 6.1**

The equations of motion of the masses $m_1$ and $m_n$ can be derived from Eq. (6.1) by setting $i = 1$ along with $x_0 = 0$ and $i = n$ along with $x_{n+1} = 0$, respectively:

$$m_1 \ddot{x}_1 + (k_1 + k_2)x_1 - k_2 x_2 = F_1 \tag{6.2}$$

$$m_n \ddot{x}_n - k_n x_{n-1} + (k_n + k_{n+1})x_n = F_n \tag{6.3}$$

Equations (6.1) to (6.3) can be expressed in matrix form as

$$[m]\ddot{\vec{x}} + [k]\vec{x} = \vec{F} \tag{6.4}$$

where $[m]$ and $[k]$ are called the *mass matrix* and the *stiffness matrix*, respectively, and are given by

$$[m] = \begin{bmatrix} m_1 & 0 & 0 & \cdots & 0 & 0 \\ 0 & m_2 & 0 & \cdots & 0 & 0 \\ 0 & 0 & m_3 & \cdots & 0 & 0 \\ \vdots & & & & & \\ 0 & 0 & 0 & \cdots & 0 & m_n \end{bmatrix} \tag{6.5}$$

$$[k] = \begin{bmatrix} (k_1 + k_2) & -k_2 & 0 & \cdots & 0 & 0 \\ -k_2 & (k_2 + k_3) & -k_3 & \cdots & 0 & 0 \\ 0 & -k_3 & (k_3 + k_4) & \cdots & 0 & 0 \\ \vdots & & & & & \\ 0 & 0 & 0 & \cdots & -k_n & (k_n + k_{n+1}) \end{bmatrix} \tag{6.6}$$

and $\vec{x}$, $\ddot{\vec{x}}$, and $\vec{F}$ are the displacement, acceleration, and force vectors, given by

$$\vec{x} = \begin{Bmatrix} x_1(t) \\ x_2(t) \\ \vdots \\ x_n(t) \end{Bmatrix}, \qquad \ddot{\vec{x}} = \begin{Bmatrix} \ddot{x}_1(t) \\ \ddot{x}_2(t) \\ \vdots \\ \ddot{x}_n(t) \end{Bmatrix}, \qquad \vec{F} = \begin{Bmatrix} F_1(t) \\ F_2(t) \\ \vdots \\ F_n(t) \end{Bmatrix} \tag{6.7}$$

The spring-mass system considered above is a particular case of a general $n$ degree of freedom spring-mass system. In their most general form, the mass and stiffness matrices are given by

$$[m] = \begin{bmatrix} m_{11} & m_{12} & m_{13} & \cdots & m_{1n} \\ m_{12} & m_{22} & m_{23} & \cdots & m_{2n} \\ \vdots & & & & \\ m_{1n} & m_{2n} & m_{3n} & \cdots & m_{nn} \end{bmatrix} \tag{6.8}$$

and

$$[k] = \begin{bmatrix} k_{11} & k_{12} & k_{13} & \cdots & k_{1n} \\ k_{12} & k_{22} & k_{23} & \cdots & k_{2n} \\ \vdots & & & & \\ k_{1n} & k_{2n} & k_{3n} & \cdots & k_{nn} \end{bmatrix} \tag{6.9}$$

## 6.3 INFLUENCE COEFFICIENTS

The equations of motion of a multidegree of freedom system can also be written in terms of influence coefficients, which are extensively used in structural engineering. For a linear spring, the force necessary to cause a unit elongation is called the *spring constant*. In more complex systems, we can express the relation between the displacement at a point and the forces acting at various other points of the system by means of influence coefficients. There are two types of influence coefficients: flexibility influence coefficients and stiffness influence coefficients. To illustrate the concept of an influence coefficient, consider the multidegree of freedom spring-mass system shown in Fig. 6.1.

Let the system be acted on by just one force $F_j$, and let the displacement at point $i$ (i.e., mass $m_i$) due to $F_j$ be $x_{ij}$. The flexibility influence coefficient, denoted by $a_{ij}$, is defined as the deflection at point $i$ due to a unit load at point $j$. Since the deflection increases proportionately with the load for a linear system, we have

$$x_{ij} = a_{ij}F_j \tag{6.10}$$

If several forces $F_j$ ($j = 1, 2, \ldots, n$) act at different points of the system, the total deflection at any point $i$ can be found by summing up the contributions of all forces $F_j$:

$$x_i = \sum_{j=1}^{n} x_{ij} = \sum_{j=1}^{n} a_{ij}F_j, \qquad i = 1, 2, \ldots, n \tag{6.11}$$

Equation (6.11) can be expressed in matrix form as

$$\vec{x} = [a]\vec{F} \tag{6.12}$$

where $\vec{x}$ and $\vec{F}$ are the displacement and force vectors defined in Eq. (6.7) and $[a]$ is the flexibility matrix given by

$$[a] = \begin{bmatrix} a_{11} & a_{12} & \cdots & a_{1n} \\ a_{21} & a_{22} & \cdots & a_{2n} \\ \vdots & & & \\ a_{n1} & a_{n2} & \cdots & a_{nn} \end{bmatrix} \tag{6.13}$$

The stiffness influence coefficient, denoted by $k_{ij}$, is defined as the force at point $i$ due to a unit displacement at point $j$ when all the points other than the point $j$ are fixed. The total force at point $i$, $F_i$, can be obtained by summing up the forces due to all displacements $x_j$ ($j = 1, 2, \ldots, n$):

$$F_i = \sum_{j=1}^{n} k_{ij}x_j, \qquad i = 1, 2, \ldots, n \tag{6.14}$$

Equation (6.14) can be stated in matrix form as

$$\vec{F} = [k]\vec{x} \tag{6.15}$$

where $[k]$ is the stiffness matrix given by

$$[k] = \begin{bmatrix} k_{11} & k_{12} & \cdots & k_{1n} \\ k_{21} & k_{22} & \cdots & k_{2n} \\ \vdots & & & \\ k_{n1} & k_{n2} & \cdots & k_{nn} \end{bmatrix} \tag{6.16}$$

An examination of Eqs. (6.12) and (6.15) indicates that the flexibility and stiffness matrices are related. If we substitute Eq. (6.15) into Eq. (6.12), we obtain

$$\vec{x} = [a]\vec{F} = [a][k]\vec{x} \tag{6.17}$$

from which we can obtain the relation

$$[a][k] = [I] \tag{6.18}$$

where $[I]$ denotes the unit matrix. Equation (6.18) is equivalent to

$$[k] = [a]^{-1}, \qquad [a] = [k]^{-1} \tag{6.19}$$

That is, the stiffness and flexibility matrices are the inverse of one another. The use of dynamic stiffness influence coefficients in the vibration of non-uniform beams is discussed in Ref. [6.10].

Note the following aspects of influence coefficients:

1. Since the deflection at point $i$ due to a unit load at point $j$ is the same as the deflection at point $j$ due to a unit load at point $i$ for a linear system (Maxwell's reciprocity theorem [6.1]), we have $a_{ij} = a_{ji}$. By a similar reasoning, we have $k_{ij} = k_{ji}$.

2. The flexibility and stiffness influence coefficients can be calculated from the principles of solid mechanics.

3. The influence coefficients for torsional systems can be defined in terms of unit torque and the angular deflection it causes. For example, in a multirotor torsional system, $a_{ij}$ can be defined as the angular displacement of point $i$ (rotor $i$) due to a unit torque at point $j$.

---

**EXAMPLE 6.1**

Find the flexibility influence coefficients of the system shown in Fig. 6.2(a).

**Solution.** Let $x_1$, $x_2$, and $x_3$ denote the displacements of the masses $m_1$, $m_2$, and $m_3$, respectively. The flexibility influence coefficients $a_{ij}$ of the system can be determined in terms of the spring stiffnesses $k_1$, $k_2$, and $k_3$ as follows. If we apply a unit force at mass $m_1$ and no force at the other masses ($F_1 = 1$, $F_2 = F_3 = 0$), as shown in Fig. 6.2(b), the deflection of the mass $m_1$ is equal to $\delta_1 = 1/k_1 = a_{11}$. Since the other two masses $m_2$ and $m_3$ move (undergo rigid body translation) by the same amount of deflection $\delta_1$, we have, by definition,

$$a_{21} = a_{31} = \delta_1 = \frac{1}{k_1}$$

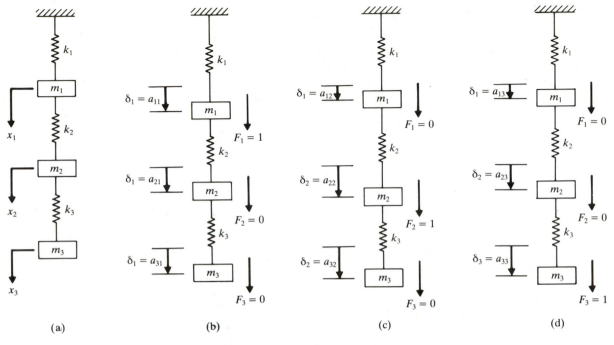

**Figure 6.2**

Next, we apply a unit force at mass $m_2$ and no force at masses $m_1$ and $m_3$, as shown in Fig. 6.2(c). Since the two springs $k_1$ and $k_2$ offer resistance, the deflection of mass $m_2$ is given by

$$\delta_2 = \frac{1}{k_{eq}} = \frac{1}{k_1} + \frac{1}{k_2} = \frac{k_1 + k_2}{k_1 k_2} = a_{22}$$

The mass $m_3$ undergoes the same displacement $\delta_2$ (rigid body translation) while the mass $m_1$ moves through a smaller distance given by $\delta_1 = 1/k_1$. Hence

$$a_{32} = \delta_2 = \frac{k_1 + k_2}{k_1 k_2} \qquad \text{and} \qquad a_{12} = \delta_1 = \frac{1}{k_1}$$

Finally, when we apply a unit force to mass $m_3$ and no force to masses $m_1$ and $m_2$, as shown in Fig. 6.2(d), the displacement of mass $m_3$ is given by

$$\delta_3 = \frac{1}{k_{eq}} = \frac{1}{k_1} + \frac{1}{k_2} + \frac{1}{k_3} = \frac{k_1 k_2 + k_2 k_3 + k_3 k_1}{k_1 k_2 k_3} = a_{33}$$

while the displacements of masses $m_2$ and $m_1$ are given by

$$\delta_2 = \frac{1}{k_1} + \frac{1}{k_2} = \frac{k_1 + k_2}{k_1 k_2} = a_{23}$$

and

$$\delta_1 = \frac{1}{k_1} = a_{13}$$

According to Maxwell's reciprocity theorem, we have

$$a_{ij} = a_{ji}$$

Thus the flexibility matrix of the system is given by

$$[a] = \begin{bmatrix} \dfrac{1}{k_1} & \dfrac{1}{k_1} & \dfrac{1}{k_1} \\[2mm] \dfrac{1}{k_1} & \left(\dfrac{1}{k_1} + \dfrac{1}{k_2}\right) & \left(\dfrac{1}{k_1} + \dfrac{1}{k_2}\right) \\[2mm] \dfrac{1}{k_1} & \left(\dfrac{1}{k_1} + \dfrac{1}{k_2}\right) & \left(\dfrac{1}{k_1} + \dfrac{1}{k_2} + \dfrac{1}{k_3}\right) \end{bmatrix} \quad\text{(E.1)}$$

The stiffness matrix of the system can be found from the relation $[k] = [a]^{-1}$ or can be derived by using the definition of $k_{ij}$ (see Problem 6.8):

$$[k] = \begin{bmatrix} (k_1 + k_2) & -k_2 & 0 \\ -k_2 & (k_2 + k_3) & -k_3 \\ 0 & -k_3 & k_3 \end{bmatrix} \quad\text{(E.2)}$$

## EXAMPLE 6.2

Derive the flexibility matrix of the weightless beam shown in Fig. 6.3(a). The beam is simply supported at both ends, and the three masses are placed at equal intervals. Assume the beam to be uniform with stiffness $EI$.

**Figure 6.3**

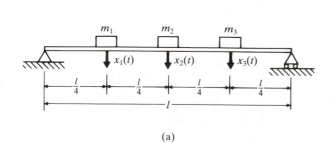

(a)

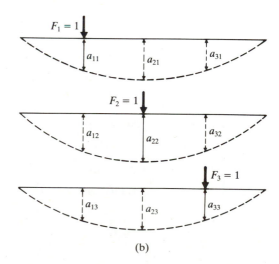

(b)

**Solution.** Let $x_1$, $x_2$, and $x_3$ denote the total transverse deflection of the masses $m_1$, $m_2$, and $m_3$, respectively. From the known formula for the deflection of a pinned-pinned beam [6.2], the influence coefficients $a_{1j}$ ($j=1,2,3$) can be found by applying a unit load at the location of $m_1$ (see Fig. 6.3(b)):

$$a_{11} = \frac{9}{768} \frac{l^3}{EI}, \qquad a_{12} = \frac{11}{768} \frac{l^3}{EI}, \qquad a_{13} = \frac{7}{768} \frac{l^3}{EI} \tag{E.1}$$

Similarly, by applying a unit load at the locations of $m_2$ and $m_3$ separately, we obtain

$$a_{21} = a_{12} = \frac{11}{768} \frac{l^3}{EI}, \qquad a_{22} = \frac{1}{48} \frac{l^3}{EI}, \qquad a_{23} = \frac{11}{768} \frac{l^3}{EI} \tag{E.2}$$

and

$$a_{31} = a_{13} = \frac{7}{768} \frac{l^3}{EI}, \qquad a_{32} = a_{23} = \frac{11}{768} \frac{l^3}{EI}, \qquad a_{33} = \frac{9}{768} \frac{l^3}{EI} \tag{E.3}$$

Thus the flexibility matrix of the system is given by

$$[a] = \frac{l^3}{768EI} \begin{bmatrix} 9 & 11 & 7 \\ 11 & 16 & 11 \\ 7 & 11 & 9 \end{bmatrix} \tag{E.4}$$

## 6.4 POTENTIAL AND KINETIC ENERGY EXPRESSIONS IN MATRIX FORM

Let $x_i$ denote the displacement of mass $m_i$ and $F_i$ the force applied in the direction of $x_i$ at mass $m_i$ in an $n$ degree of freedom system similar to the one shown in Fig. 6.1. The elastic potential energy (also known as *strain energy* or *energy of deformation*) of the $i$th spring is given by

$$V_i = \frac{1}{2} F_i x_i \tag{6.20}$$

The total potential energy can be expressed as

$$V = \sum_{i=1}^{n} V_i = \frac{1}{2} \sum_{i=1}^{n} F_i x_i \tag{6.21}$$

Since

$$F_i = \sum_{j=1}^{n} k_{ij} x_j \tag{6.22}$$

Eq. (6.21) becomes

$$V = \frac{1}{2} \sum_{i=1}^{n} \left( \sum_{j=1}^{n} k_{ij} x_j \right) x_i = \frac{1}{2} \sum_{i=1}^{n} \sum_{j=1}^{n} k_{ij} x_i x_j \tag{6.23}$$

Equation (6.23) can also be written in matrix form as*

$$V = \frac{1}{2} \vec{x}^T [k] \vec{x}$$ (6.24)

where the displacement vector is given by Eq. (6.7) and the stiffness matrix is given by

$$[k] = \begin{bmatrix} k_{11} & k_{12} & \cdots & k_{1n} \\ k_{21} & k_{22} & \cdots & k_{2n} \\ \vdots & & & \\ k_{n1} & k_{n2} & \cdots & k_{nn} \end{bmatrix}$$ (6.25)

The kinetic energy associated with mass $m_i$ is, by definition, equal to

$$T_i = \frac{1}{2} m_i \dot{x}_i^2$$ (6.26)

The total kinetic energy of the system can be expressed as

$$T = \sum_{i=1}^{n} T_i = \frac{1}{2} \sum_{i=1}^{n} m_i \dot{x}_i^2$$ (6.27)

which can be written in matrix form as

$$T = \frac{1}{2} \dot{\vec{x}}^T [m] \dot{\vec{x}}$$ (6.28)

where the velocity vector $\dot{\vec{x}}$ is given by

$$\dot{\vec{x}} = \begin{Bmatrix} \dot{x}_1 \\ \dot{x}_2 \\ \vdots \\ \dot{x}_n \end{Bmatrix}$$

and the mass matrix $[m]$ is a diagonal matrix given by

$$[m] = \begin{bmatrix} m_1 & & & 0 \\ & m_2 & & \\ & & \ddots & \\ 0 & & & m_n \end{bmatrix}$$ (6.29)

If generalized coordinates $(q_i)$, discussed in Sec. 6.5, are used instead of the physical displacements $(x_i)$, the kinetic energy can be expressed as

$$T = \frac{1}{2} \dot{\vec{q}}^T [m] \dot{\vec{q}}$$ (6.30)

---

*Since the indices $i$ and $j$ can be interchanged in Eq. (6.23), we have the relation $k_{ij} = k_{ji}$.

where $\dot{\vec{q}}$ is the vector of generalized velocities, given by

$$\dot{\vec{q}} = \begin{Bmatrix} \dot{q}_1 \\ \dot{q}_2 \\ \vdots \\ \dot{q}_n \end{Bmatrix} \tag{6.31}$$

and $[m]$ is called the *generalized mass matrix*, given by

$$[m] = \begin{bmatrix} m_{11} & m_{12} & \cdots & m_{1n} \\ m_{21} & m_{22} & \cdots & m_{2n} \\ \vdots & & & \\ m_{n1} & m_{n2} & \cdots & m_{nn} \end{bmatrix} \tag{6.32}$$

with $m_{ij} = m_{ji}$. The generalized mass matrix given by Eq. (6.32) is full, as opposed to the diagonal mass matrix of Eq. (6.29).

It can be seen that the potential energy is a quadratic function of the displacements, and the kinetic energy is a quadratic function of the velocities. Hence they are said to be in quadratic form. Since kinetic energy, by definition, cannot be negative and vanishes only when all the velocities vanish, Eqs. (6.28) and (6.30) are called *positive definite quadratic forms* and the mass matrix $[m]$ is called a *positive definite matrix*. On the other hand, the potential energy expression, Eq. (6.24), is a positive definite quadratic form, but the matrix $[k]$ is positive definite only if the system is a stable one. There are systems for which the potential energy is zero without the displacements or coordinates $x_1, x_2, \ldots, x_n$ being zero. In these cases the potential energy will be a positive quadratic function rather than positive definite; correspondingly, the matrix $[k]$ is said to be positive. A system for which $[k]$ is positive and $[m]$ is positive definite is called a semi-definite system (see Sec. 6.11).

## 6.5  GENERALIZED COORDINATES AND GENERALIZED FORCES

The equations of motion of a vibrating system can be formulated in a number of different coordinate systems. As stated earlier, $n$ independent coordinates are necessary to describe the motion of a system having $n$ degrees of freedom. Any set of $n$ independent coordinates is called generalized coordinates, usually designated by $q_1, q_2, \ldots, q_n$. The generalized coordinates may be lengths, angles, or any other set of numbers that define uniquely the dynamic configuration of the system. They are also independent of the conditions of constraint.

To illustrate the concept of generalized coordinates, consider the triple pendulum shown in Fig. 6.4. The configuration of the system can be specified by the six coordinates $(x_j, y_j)$, $j = 1, 2, 3$. However, these coordinates are not independent but

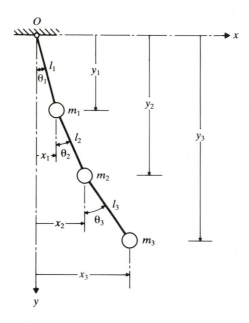

**Figure 6.4**

are constrained by the relations

$$x_1^2 + y_1^2 = l_1^2$$
$$(x_2 - x_1)^2 + (y_2 - y_1)^2 = l_2^2 \qquad (6.33)$$
$$(x_3 - x_2)^2 + (y_3 - y_2)^2 = l_3^2$$

Since the coordinates $(x_j, y_j)$, $j = 1, 2, 3$ are not independent, they cannot be called generalized coordinates. It can be seen that the number of coordinates (6) minus the number of constraints (3) gives the number of degrees of freedom (3). If there had been no constraints—that is, if the pendulums were removed so that $m_1$, $m_2$, and $m_3$ were free to move anywhere in the $xy$ plane—then there would be six degrees of freedom requiring six coordinates such as $(x_j, y_j)$, $j = 1, 2, 3$, which would then be generalized coordinates.

For the triple pendulum, the coordinates shown $(\theta_j, j = 1, 2, 3)$ would properly define the configuration of the system. Since they are independent of each other and of the constraints, they form a set of generalized coordinates and are denoted as $q_j = \theta_j$, $j = 1, 2, 3$. The number of these generalized coordinates (3) is equal to the number of degrees of freedom (3) in this case.

In general, certain forces act on the system. When the generalized coordinate $q_j$ is changed by a small amount $\delta q_j$, the work done can be denoted as $U_j$. Then the generalized force $Q_j$ corresponding to $q_j$ can be defined as

$$Q_j = \frac{U_j}{\delta q_j}, \qquad j = 1, 2, \dots, n \qquad (6.34)$$

Note that if $q_j$ is a linear displacement, then $Q_j$ is a force; when $q_j$ is an angular displacement, then $Q_j$ is a moment.

## 6.6 LAGRANGE'S EQUATIONS

The equations of motion of a vibrating system can often be derived in a simple manner in terms of generalized coordinates by the use of Lagrange's equations [6.3]. Lagrange's equations can be stated as

$$\frac{d}{dt}\left(\frac{\partial T}{\partial \dot{q}_j}\right) - \frac{\partial T}{\partial q_j} + \frac{\partial V}{\partial q_j} = Q_j^{(n)}, \qquad j = 1, 2, \dots, n \tag{6.35}$$

where $\dot{q}_j = \partial q_j / \partial t$ is the generalized velocity and $Q_j^{(n)}$ is the nonconservative generalized force corresponding to the generalized coordinate $q_j$. The forces represented by $Q_j^{(n)}$ may be dissipative (damping) forces or other external forces which are not derivable from a potential function. For example, if $F_{xk}$, $F_{yk}$, and $F_{zk}$ represent the external forces acting on the $k$th mass of the system in the $x$, $y$, and $z$ directions, respectively, then the generalized force $Q_j^{(n)}$ can be computed as follows:

$$Q_j^{(n)} = \sum_k \left( F_{xk} \frac{\partial x_k}{\partial q_j} + F_{yk} \frac{\partial y_k}{\partial q_j} + F_{zk} \frac{\partial z_k}{\partial q_j} \right) \tag{6.36}$$

where $x_k$, $y_k$, and $z_k$ are the displacements of the $k$th mass in the $x$, $y$, and $z$ directions, respectively. For a conservative system, $Q_j^{(n)} = 0$, so Eqs. (6.35) take the form

$$\frac{d}{dt}\left(\frac{\partial T}{\partial \dot{q}_j}\right) - \frac{\partial T}{\partial q_j} + \frac{\partial V}{\partial q_j} = 0, \qquad j = 1, 2, \dots, n \tag{6.37}$$

Equations (6.35) or (6.37) represent a system of $n$ differential equations, one corresponding to each of the $n$ generalized coordinates. Thus the equations of motion of the vibrating system can be derived, provided the energy expressions are available.

### EXAMPLE 6.3

In the torsional system shown in Fig. 6.5, $J_j$ denote the mass moments of inertia of the discs, $M_{tj}$ indicate the external moments acting on the discs, and $k_{tj}$ represent the torsional spring constants of the various shaft portions. Using the angular displacements of the discs $\theta_j$ as the generalized coordinates $q_j$, write the expressions for the kinetic and potential energy functions and derive the equations of motion using Lagrange's equations.

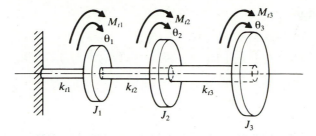

**Figure 6.5**

***Solution.*** Here $q_1 = \theta_1$, $q_2 = \theta_2$, and $q_3 = \theta_3$, and the kinetic energy of the system is given by

$$T = \frac{1}{2} J_1 \dot{\theta}_1^2 + \frac{1}{2} J_2 \dot{\theta}_2^2 + \frac{1}{2} J_3 \dot{\theta}_3^2 \tag{E.1}$$

For the shaft, the potential energy is equal to the work done by the shaft as it returns from the dynamic configuration to the reference equilibrium position. Thus if $\theta$ denotes the angular displacement,

$$V = \int_\theta^0 (-k_t \theta)\, d\theta = \frac{1}{2} k_t \theta^2 \tag{E.2}$$

The application of Eq. (E.2) to the present example gives

$$V = \frac{1}{2} k_{t1} \theta_1^2 + \frac{1}{2} k_{t2} (\theta_2 - \theta_1)^2 + \frac{1}{2} k_{t3} (\theta_3 - \theta_2)^2 \tag{E.3}$$

There are external moments applied to the discs, so Eq. (6.36) gives

$$Q_j^{(n)} = \sum_{k=1}^{3} M_{tk} \frac{\partial \theta_k}{\partial q_j} = \sum_{k=1}^{3} M_{tk} \frac{\partial \theta_k}{\partial \theta_j} \tag{E.4}$$

from which we can obtain

$$Q_1^{(n)} = M_{t1} \frac{\partial \theta_1}{\partial \theta_1} + M_{t2} \frac{\partial \theta_2}{\partial \theta_1} + M_{t3} \frac{\partial \theta_3}{\partial \theta_1} = M_{t1}$$

$$Q_2^{(n)} = M_{t1} \frac{\partial \theta_1}{\partial \theta_2} + M_{t2} \frac{\partial \theta_2}{\partial \theta_2} + M_{t3} \frac{\partial \theta_3}{\partial \theta_2} = M_{t2}$$

$$Q_3^{(n)} = M_{t1} \frac{\partial \theta_1}{\partial \theta_3} + M_{t2} \frac{\partial \theta_2}{\partial \theta_3} + M_{t3} \frac{\partial \theta_3}{\partial \theta_3} = M_{t3} \tag{E.5}$$

Substituting Eqs. (E.1), (E.3), and (E.5) in Lagrange's equations, Eq. (6.35), we obtain for $j = 1, 2, 3$ the equations of motion:

$$J_1 \ddot{\theta}_1 + (k_{t1} + k_{t2}) \theta_1 - k_{t2} \theta_2 = M_{t1}$$
$$J_2 \ddot{\theta}_2 + (k_{t2} + k_{t3}) \theta_2 - k_{t2} \theta_1 - k_{t3} \theta_3 = M_{t2}$$
$$J_3 \ddot{\theta}_3 + k_{t3} \theta_3 - k_{t3} \theta_2 = M_{t3} \tag{E.6}$$

which can be expressed in matrix form as

$$\begin{bmatrix} J_1 & 0 & 0 \\ 0 & J_2 & 0 \\ 0 & 0 & J_3 \end{bmatrix}\begin{Bmatrix} \ddot{\theta}_1 \\ \ddot{\theta}_2 \\ \ddot{\theta}_3 \end{Bmatrix} + \begin{bmatrix} (k_{t1}+k_{t2}) & -k_{t2} & 0 \\ -k_{t2} & (k_{t2}+k_{t3}) & -k_{t3} \\ 0 & -k_{t3} & k_{t3} \end{bmatrix}\begin{Bmatrix} \theta_1 \\ \theta_2 \\ \theta_3 \end{Bmatrix} = \begin{Bmatrix} M_{t1} \\ M_{t2} \\ M_{t3} \end{Bmatrix}$$

$$(\text{E.7})$$

## EXAMPLE 6.4

A simplified model of a three-story building whose foundation is subjected to translation and rotation is shown in Fig. 6.6. Here $x_0$ and $\theta$ denote the translation and rotation of the foundation, and $x_j$ indicate the elastic displacements of the $j$th floor. Further, $m_0$ and $m_j$ represent the mass of the foundation and the $j$th floor, while $J_0$ and $J_i$ denote the mass moment of inertia of the foundation and the $j$th floor, respectively. Derive the equations of motion using Lagrange's equations.

***Solution.*** By choosing $x_0$, $\theta$, $x_1$, $x_2$, and $x_3$ as the generalized coordinates $q_1, q_2, \ldots, q_5$, the kinetic and potential energies of the system can be expressed as

$$T = \frac{1}{2}m_0\dot{x}_0^2 + \frac{1}{2}J_0\dot{\theta}^2 + \frac{1}{2}m_1(\dot{x}_0 + h_1\dot{\theta} + \dot{x}_1)^2 + \frac{1}{2}J_1\dot{\theta}^2$$

$$+ \frac{1}{2}m_2[\dot{x}_0 + (h_1 + h_2)\dot{\theta} + \dot{x}_2]^2 + \frac{1}{2}J_2\dot{\theta}^2$$

$$+ \frac{1}{2}m_3[\dot{x}_0 + (h_1 + h_2 + h_3)\dot{\theta} + \dot{x}_3]^2 + \frac{1}{2}J_3\dot{\theta}^2 \qquad (\text{E.1})$$

$$V = \frac{1}{2}k_0x_0^2 + \frac{1}{2}k_{t0}\theta^2 + \frac{1}{2}k_1x_1^2 + \frac{1}{2}k_2(x_2 - x_1)^2 + \frac{1}{2}k_3(x_3 - x_2)^2 \qquad (\text{E.2})$$

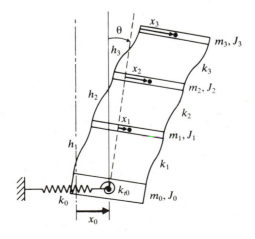

**Figure 6.6**

Since there are no external forces, substitution of Eqs. (E.1) and (E.2) in Lagrange's equations, Eq. (6.37), gives the equations of motion:

$$(m_0 + m_1 + m_2 + m_3)\ddot{x}_0 + [m_1 h_1 + m_2(h_1 + h_2) + m_3(h_1 + h_2 + h_3)]\ddot{\theta}$$
$$+ m_1 \ddot{x}_1 + m_2 \ddot{x}_2 + m_3 \ddot{x}_3 + k_0 x_0 = 0 \tag{E.3}$$

$$[m_1 h_1 + m_2(h_1 + h_2) + m_3(h_1 + h_2 + h_3)]\ddot{x}_0$$
$$+ \left[ J_0 + J_1 + J_2 + J_3 + m_1 h_1^2 + m_2(h_1 + h_2)^2 + m_3(h_1 + h_2 + h_3)^2 \right]\ddot{\theta}$$
$$+ m_1 h_1 \ddot{x}_1 + m_2(h_1 + h_2)\ddot{x}_2 + m_3(h_1 + h_2 + h_3)\ddot{x}_3 + k_{t0}\theta = 0 \tag{E.4}$$

$$m_1 \ddot{x}_0 + m_1 h_1 \ddot{\theta} + m_1 \ddot{x}_1 + (k_1 + k_2)x_1 - k_2 x_2 = 0 \tag{E.5}$$

$$m_2 \ddot{x}_0 + m_2(h_1 + h_2)\ddot{\theta} + m_2 \ddot{x}_2 - k_2 x_1 + (k_2 + k_3)x_2 - k_3 x_3 = 0 \tag{E.6}$$

$$m_3 \ddot{x}_0 + m_3(h_1 + h_2 + h_3)\ddot{\theta} + m_3 \ddot{x}_3 - k_3 x_2 + k_3 x_3 = 0 \tag{E.7}$$

## 6.7 GENERAL EQUATIONS OF MOTION IN MATRIX FORM

We can derive the equations of motion of a multidegree of freedom system in matrix form from Lagrange's equations,*

$$\frac{d}{dt}\left(\frac{\partial T}{\partial \dot{x}_i}\right) - \frac{\partial T}{\partial x_i} + \frac{\partial V}{\partial x_i} = F_i, \quad i = 1, 2, \ldots, n \tag{6.38}$$

where $F_i$ is the nonconservative generalized force corresponding to the $i$th generalized coordinate $x_i$ and $\dot{x}_i$ is the time derivative of $x_i$ (generalized velocity). The kinetic and potential energies of a multidegree of freedom system can be expressed in matrix form as indicated in Sec. 6.4:

$$T = \frac{1}{2}\dot{\vec{x}}^T[m]\dot{\vec{x}} \tag{6.39}$$

$$V = \frac{1}{2}\vec{x}^T[k]\vec{x} \tag{6.40}$$

where $\vec{x}$ is the column vector of the generalized coordinates

$$\vec{x} = \begin{Bmatrix} x_1 \\ x_2 \\ \vdots \\ x_n \end{Bmatrix} \tag{6.41}$$

---

*The generalized coordinates are denoted as $x_i$ instead of $q_i$ and the generalized forces as $F_i$ instead of $Q_i^{(n)}$ in Eq. (6.38).

From the theory of matrices, we obtain, by taking note of the symmetry of $[m]$,

$$\frac{\partial T}{\partial \dot{x}_i} = \frac{1}{2}\vec{\delta}^T[m]\dot{\vec{x}} + \frac{1}{2}\dot{\vec{x}}^T[m]\vec{\delta} = \vec{\delta}^T[m]\dot{\vec{x}}$$

$$= \vec{m}_i^T\dot{\vec{x}}, \qquad i=1,2,\ldots,n \tag{6.42}$$

where $\delta_{ji}$ is the Kronecker delta ($\delta_{ji}=1$ if $j=i$ and $=0$ if $j \ne i$), $\vec{\delta}$ is the column vector of Kronecker deltas whose elements in the rows for which $j \ne i$ are equal to zero and whose element in the row $i=j$ is equal to 1, and $\vec{m}_i^T$ is a row vector which is identical to the $i$th row of the matrix $[m]$. All the relations represented by Eq. (6.42) can be expressed as

$$\frac{\partial T}{\partial \dot{x}_i} = \vec{m}_i^T\dot{\vec{x}} \tag{6.43}$$

Differentiation of Eq. (6.43) with respect to time gives

$$\frac{d}{dt}\left(\frac{\partial T}{\partial \dot{x}_i}\right) = \vec{m}_i^T\ddot{\vec{x}}, \qquad i=1,2,\ldots,n \tag{6.44}$$

since the mass matrix is not a function of time. Further, the kinetic energy is a function of only the velocities $\dot{x}_i$, and so

$$\frac{\partial T}{\partial x_i} = 0, \qquad i=1,2,\ldots,n \tag{6.45}$$

Similarly, we can differentiate Eq. (6.40), taking note of the symmetry of $[k]$:

$$\frac{\partial V}{\partial x_i} = \frac{1}{2}\vec{\delta}^T[k]\vec{x} + \frac{1}{2}\vec{x}^T[k]\vec{\delta} = \vec{\delta}^T[k]\vec{x}$$

$$= \vec{k}_i^T\vec{x}, \qquad i=1,2,\ldots,n \tag{6.46}$$

where $\vec{k}_i^T$ is a row vector identical to the $i$th row of the matrix $[k]$. By substituting Eqs. (6.44) to (6.46) into Eq. (6.38), we obtain the desired equations of motion in matrix form:

$$[m]\ddot{\vec{x}} + [k]\vec{x} = \vec{F} \tag{6.47}$$

where

$$\vec{F} = \begin{Bmatrix} F_1 \\ F_2 \\ \vdots \\ F_n \end{Bmatrix} \tag{6.48}$$

Note that if the system is conservative, there are no nonconservative forces $F_i$, so the equations of motion become

$$[m]\ddot{\vec{x}} + [k]\vec{x} = \vec{0} \tag{6.49}$$

Note also that if the generalized coordinates $x_i$ are same as the actual (physical) displacements, the mass matrix $[m]$ is a diagonal matrix.

## **6.8** EIGENVALUE PROBLEM

The solution of Eq. (6.49) corresponds to the undamped free vibration of the system. In this case, if the system is given some energy in the form of initial displacements or initial velocities or both, it vibrates indefinitely because there is no dissipation of energy. We can find the solution of Eq. (6.49) by assuming a solution of the form

$$x_i(t) = X_i T(t), \qquad i = 1, 2, \dots, n \tag{6.50}$$

where $X_i$ is a constant and $T$ is a function of time $t$. Equation (6.50) shows that the amplitude ratio of two coordinates

$$\left\{ \frac{x_i(t)}{x_j(t)} \right\}$$

is independent of time. Physically, this means that all coordinates have synchronous motions. The configuration of the system does not change its shape during motion, but its amplitude does. The configuration of the system, given by the vector

$$\vec{X} = \begin{Bmatrix} X_1 \\ X_2 \\ \vdots \\ X_n \end{Bmatrix}$$

is known as the *mode shape* of the system. Substituting Eq. (6.50) into Eq. (6.49), we obtain

$$[m]\vec{X}\ddot{T}(t) + [k]\vec{X}T(t) = \vec{0} \tag{6.51}$$

Equation (6.51) can be written in scalar form as $n$ separate equations:

$$\left( \sum_{j=1}^{n} m_{ij} X_j \right) \ddot{T}(t) + \left( \sum_{j=1}^{n} k_{ij} X_j \right) T(t) = 0, \qquad i = 1, 2, \dots, n \tag{6.52}$$

from which we can obtain the relations

$$-\frac{\ddot{T}(t)}{T(t)} = \frac{\left( \displaystyle\sum_{j=1}^{n} k_{ij} X_j \right)}{\left( \displaystyle\sum_{j=1}^{n} m_{ij} X_j \right)}, \qquad i = 1, 2, \dots, n \tag{6.53}$$

Since the left-hand side of Eq. (6.53) is independent of the index $i$, and the right-hand side is independent of $t$, both sides must be equal to a constant. By assuming this constant* as $\omega^2$, we can write Eqs. (6.53) as

$$\ddot{T}(t) + \omega^2 T(t) = 0 \tag{6.54}$$

$$\sum_{j=1}^{n} \left( k_{ij} - \omega^2 m_{ij} \right) X_j = 0, \qquad i = 1, 2, \ldots, n$$

or

$$\left[ [k] - \omega^2 [m] \right] \vec{X} = \vec{0} \tag{6.55}$$

The solution of Eq. (6.54) can be expressed as

$$T(t) = C_1 \cos(\omega t + \phi) \tag{6.56}$$

where $C_1$ and $\phi$ are constants, known as the *amplitude* and the *phase angle*, respectively. Equation (6.56) shows that all the coordinates can perform a harmonic motion with the same frequency $\omega$ and the same phase angle $\phi$. However, the frequency $\omega$ cannot take any arbitrary value; it has to satisfy Eq. (6.55). Since Eq. (6.55) represents a set of $n$ linear homogeneous equations in the unknowns $X_i$ ($i = 1, 2, \ldots, n$), the trivial solution is $X_1 = X_2 = \cdots = X_n = 0$. For a nontrivial solution of Eq. (6.55), the determinant $\Delta$ of the coefficient matrix must be zero. That is,

$$\Delta = \left| k_{ij} - \omega^2 m_{ij} \right| = \left| [k] - \omega^2 [m] \right| = 0 \tag{6.57}$$

Equation (6.55) represents what is known as the *eigenvalue* or *characteristic value* problem, Eq. (6.57) is called the *characteristic equation*, $\omega^2$ is known as the *eigenvalue* or the *characteristic value*, and $\omega$ is called the *natural frequency* of the system.

The expansion of Eq. (6.57) leads to an $n$th order polynomial equation in $\omega^2$. The solution (roots) of this polynomial or characteristic equation gives $n$ values of $\omega^2$. It can be shown that all the $n$ roots are real and positive when the matrices $[k]$ and $[m]$ are symmetric and positive definite [6.4], as in the present case. If $\omega_1^2, \omega_2^2, \ldots, \omega_n^2$ denote the $n$ roots in ascending order of magnitude, their positive square roots give the $n$ natural frequencies of the system $\omega_1 \leq \omega_2 \leq \cdots \leq \omega_n$. The lowest value ($\omega_1$) is called the *fundamental or first natural frequency*. In general, all the natural frequencies $\omega_i$ are distinct, although in some cases two natural frequencies might possess the same value. A method of reducing the size of the eigenvalue problem was presented by Guyan [6.17] and Irons [6.18]. We shall consider a simple method of solving the eigenvalue problem, Eq. (6.55), in the following section.

## 6.9  SOLUTION OF THE EIGENVALUE PROBLEM

Several methods are available to solve an eigenvalue problem. We shall consider an elementary method in this section.

---

*The constant is assumed to be a positive number, $\omega^2$, so as to obtain a harmonic solution to the resulting Eq. (6.54). Otherwise, the solution of $T(t)$ and hence that of $x(t)$ become exponential, which violates the physical limitations of finite total energy.

### 6.9.1    Solution of the Characteristic (Polynomial) Equation

Equation (6.55) can also be expressed as

$$[\lambda[k]-[m]]\,\vec{X} = \vec{0} \tag{6.58}$$

where

$$\lambda = \frac{1}{\omega^2} \tag{6.59}$$

By premultiplying Eq. (6.58) by $[k]^{-1}$, we obtain

$$[\lambda[I]-[D]]\,\vec{X} = \vec{0}$$

or

$$\lambda[I]\,\vec{X} = [D]\,\vec{X} \tag{6.60}$$

where $[I]$ is the identity matrix and

$$[D] = [k]^{-1}[m] \tag{6.61}$$

is called the *dynamical matrix*. The eigenvalue problem of Eq. (6.60) is known as the *standard eigenvalue problem*. For a nontrivial solution of $\vec{X}$, the characteristic determinant must be zero—that is,

$$\Delta = |\lambda[I]-[D]| = 0 \tag{6.62}$$

On expansion, Eq. (6.62) gives an $n$th degree polynomial in $\lambda$, known as the *characteristic* or *frequency equation*. If the degree of freedom of the system ($n$) is large, the solution of this polynomial equation becomes quite tedious. We must use some numerical method, several of which are available to find the roots of a polynomial equation [6.5].

### EXAMPLE 6.5

Find the natural frequencies and mode shapes of the system shown in Fig. 6.2 for $k_1 = k_2 = k_3 = k$ and $m_1 = m_2 = m_3 = m$.

**Solution.** The dynamical matrix is given by

$$[D] = [k]^{-1}[m] \equiv [a][m] \tag{E.1}$$

where the flexibility and mass matrices can be obtained from Example 6.1:

$$[a] = \frac{1}{k}\begin{bmatrix} 1 & 1 & 1 \\ 1 & 2 & 2 \\ 1 & 2 & 3 \end{bmatrix} \tag{E.2}$$

and

$$[m] = m\begin{bmatrix} 1 & 0 & 0 \\ 0 & 1 & 0 \\ 0 & 0 & 1 \end{bmatrix} \tag{E.3}$$

Thus

$$[D] = \frac{m}{k} \begin{bmatrix} 1 & 1 & 1 \\ 1 & 2 & 2 \\ 1 & 2 & 3 \end{bmatrix} \tag{E.4}$$

By setting the characteristic determinant equal to zero, we obtain the frequency equation:

$$\Delta = |\lambda[I] - [D]| = \left\| \begin{bmatrix} \lambda & 0 & 0 \\ 0 & \lambda & 0 \\ 0 & 0 & \lambda \end{bmatrix} - \frac{m}{k} \begin{bmatrix} 1 & 1 & 1 \\ 1 & 2 & 2 \\ 1 & 2 & 3 \end{bmatrix} \right\| = 0 \tag{E.5}$$

where

$$\lambda = \frac{1}{\omega^2} \tag{E.6}$$

By dividing throughout by $\lambda$, Eq. (E.5) gives

$$\begin{vmatrix} 1 - \alpha & -\alpha & -\alpha \\ -\alpha & 1 - 2\alpha & -2\alpha \\ -\alpha & -2\alpha & 1 - 3\alpha \end{vmatrix} = \alpha^3 - 5\alpha^2 + 6\alpha - 1 = 0 \tag{E.7}$$

where

$$\alpha = \frac{m}{k\lambda} = \frac{m\omega^2}{k} \tag{E.8}$$

The roots of the cubic equation (E.7) are given by

$$\alpha_1 = \frac{m\omega_1^2}{k} = 0.19806, \qquad \omega_1 = 0.44504\sqrt{\frac{k}{m}} \tag{E.9}$$

$$\alpha_2 = \frac{m\omega_2^2}{k} = 1.5553, \qquad \omega_2 = 1.2471\sqrt{\frac{k}{m}} \tag{E.10}$$

$$\alpha_3 = \frac{m\omega_3^2}{k} = 3.2490, \qquad \omega_3 = 1.8025\sqrt{\frac{k}{m}} \tag{E.11}$$

Once the natural frequencies are known, the mode shapes or *eigenvectors* can be calculated, using Eq. (6.60):

$$[\lambda_i[I] - [D]] \vec{X}^{(i)} = \vec{0}, \qquad i = 1, 2, 3 \tag{E.12}$$

where

$$\vec{X}^{(i)} = \begin{Bmatrix} X_1^{(i)} \\ X_2^{(i)} \\ X_3^{(i)} \end{Bmatrix}$$

denotes the $i$th mode shape. The procedure is outlined below.

*First mode:* By substituting the value of $\omega_1$ $\left(i.e., \lambda_1 = 5.0489 \dfrac{m}{k}\right)$ in Eq. (E.12), we obtain

$$\left[ 5.0489 \frac{m}{k} \begin{bmatrix} 1 & 0 & 0 \\ 0 & 1 & 0 \\ 0 & 0 & 1 \end{bmatrix} - \frac{m}{k} \begin{bmatrix} 1 & 1 & 1 \\ 1 & 2 & 2 \\ 1 & 2 & 3 \end{bmatrix} \right] \begin{Bmatrix} X_1^{(1)} \\ X_2^{(1)} \\ X_3^{(1)} \end{Bmatrix} = \begin{Bmatrix} 0 \\ 0 \\ 0 \end{Bmatrix}$$

That is,

$$\begin{bmatrix} 4.0489 & -1.0 & -1.0 \\ -1.0 & 3.0489 & -2.0 \\ -1.0 & -2.0 & 2.0489 \end{bmatrix} \begin{Bmatrix} X_1^{(1)} \\ X_2^{(1)} \\ X_3^{(1)} \end{Bmatrix} = \begin{Bmatrix} 0 \\ 0 \\ 0 \end{Bmatrix} \tag{E.13}$$

Equation (E.13) denotes a system of three homogeneous linear equations in the three unknowns $X_1^{(1)}$, $X_2^{(1)}$, and $X_3^{(1)}$. Any two of these unknowns can be expressed in terms of the remaining one. If we choose, arbitrarily, to express $X_2^{(1)}$ and $X_3^{(1)}$ in terms of $X_1^{(1)}$, we obtain from the first two rows of Eq. (E.13)

$$X_2^{(1)} + X_3^{(1)} = 4.0489 X_1^{(1)}$$

$$3.0489 X_2^{(1)} - 2.0 X_3^{(1)} = X_1^{(1)} \tag{E.14}$$

Once Eqs. (E.14) are satisfied, the third row of Eq. (E.13) is satisfied automatically. The solution of Eqs. (E.14) can be obtained:

$$X_2^{(1)} = 1.8019 X_1^{(1)} \qquad \text{and} \qquad X_3^{(1)} = 2.2470 X_1^{(1)} \tag{E.15}$$

Thus the first mode shape is given by

$$\vec{X}^{(1)} = X_1^{(1)} \begin{Bmatrix} 1.0 \\ 1.8019 \\ 2.2470 \end{Bmatrix} \tag{E.16}$$

where the value of $X_1^{(1)}$ can be chosen arbitrarily.

*Second mode:* The substitution of the value of $\omega_2$ $\left(i.e., \lambda_2 = 0.6430 \dfrac{m}{k}\right)$ in Eq. (E.12) leads to

$$\left[ 0.6430 \frac{m}{k} \begin{bmatrix} 1 & 0 & 0 \\ 0 & 1 & 0 \\ 0 & 0 & 1 \end{bmatrix} - \frac{m}{k} \begin{bmatrix} 1 & 1 & 1 \\ 1 & 2 & 2 \\ 1 & 2 & 3 \end{bmatrix} \right] \begin{Bmatrix} X_1^{(2)} \\ X_2^{(2)} \\ X_3^{(2)} \end{Bmatrix} = \begin{Bmatrix} 0 \\ 0 \\ 0 \end{Bmatrix}$$

that is,

$$\begin{bmatrix} -0.3570 & -1.0 & -1.0 \\ -1.0 & -1.3570 & -2.0 \\ -1.0 & -2.0 & -2.3570 \end{bmatrix} \begin{Bmatrix} X_1^{(2)} \\ X_2^{(2)} \\ X_3^{(2)} \end{Bmatrix} = \begin{Bmatrix} 0 \\ 0 \\ 0 \end{Bmatrix} \tag{E.17}$$

As before, the first two rows of Eq. (E.17) can be used to obtain

$$- X_2^{(2)} - X_3^{(2)} = 0.3570 X_1^{(2)}$$
$$-1.3570 X_2^{(2)} - 2.0 X_3^{(2)} = X_1^{(2)} \tag{E.18}$$

The solution of Eqs. (E.18) leads to

$$X_2^{(2)} = 0.4450 X_1^{(2)} \qquad \text{and} \qquad X_3^{(2)} = -0.8020 X_1^{(2)} \tag{E.19}$$

Thus the second mode shape can be expressed as

$$\vec{X}^{(2)} = X_1^{(2)} \begin{Bmatrix} 1.0 \\ 0.4450 \\ -0.8020 \end{Bmatrix} \tag{E.20}$$

where the value of $X_1^{(2)}$ can be chosen arbitrarily.

*Third mode:* To find the third mode, we substitute the value of $\omega_3$ $\left( \text{i.e., } \lambda_3 = 0.3078 \dfrac{m}{k} \right)$ in Eq. (E.12) and obtain

$$\left[ 0.3078 \frac{m}{k} \begin{bmatrix} 1 & 0 & 0 \\ 0 & 1 & 0 \\ 0 & 0 & 1 \end{bmatrix} - \frac{m}{k} \begin{bmatrix} 1 & 1 & 1 \\ 1 & 2 & 2 \\ 1 & 2 & 3 \end{bmatrix} \right] \begin{Bmatrix} X_1^{(3)} \\ X_2^{(3)} \\ X_3^{(3)} \end{Bmatrix} = \begin{Bmatrix} 0 \\ 0 \\ 0 \end{Bmatrix}$$

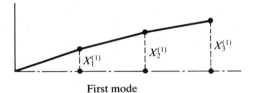

First mode

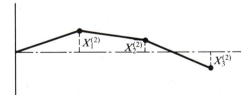

Second mode

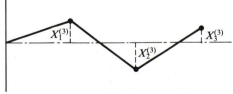

Third mode

**Figure 6.7**

that is,

$$\begin{bmatrix} -0.6922 & -1.0 & -1.0 \\ -1.0 & -1.6922 & -2.0 \\ -1.0 & -2.0 & -2.6922 \end{bmatrix} \begin{Bmatrix} X_1^{(3)} \\ X_2^{(3)} \\ X_3^{(3)} \end{Bmatrix} = \begin{Bmatrix} 0 \\ 0 \\ 0 \end{Bmatrix} \qquad \text{(E.21)}$$

The first two rows of Eq. (E.21) can be written as

$$- X_2^{(3)} - X_3^{(3)} = 0.6922\, X_1^{(3)}$$

$$-1.6922\, X_2^{(3)} - 2.0\, X_3^{(3)} = X_1^{(3)} \qquad \text{(E.22)}$$

Equations (E.22) give

$$X_2^{(3)} = -1.2468\, X_1^{(3)} \qquad \text{and} \qquad X_3^{(3)} = 0.5544\, X_1^{(3)} \qquad \text{(E.23)}$$

Hence the third mode shape can be written as

$$\vec{X}^{(3)} = X_1^{(3)} \begin{Bmatrix} 1.0 \\ -1.2468 \\ 0.5544 \end{Bmatrix} \qquad \text{(E.24)}$$

where the value of $X_1^{(3)}$ is arbitrary. The values of $X_1^{(1)}$, $X_1^{(2)}$, and $X_1^{(3)}$ are usually taken as 1, and the mode shapes are shown in Fig. 6.7.

### 6.9.2   Orthogonality of Normal Modes

In the previous section we considered a method of finding the $n$ natural frequencies $\omega_i$ and the corresponding normal modes or modal vectors $\vec{X}^{(i)}$. We shall now see an important property of the normal modes— *orthogonality*. The natural frequency $\omega_i$ and the corresponding modal vector $\vec{X}^{(i)}$ satisfy Eq. (6.55) so that

$$\omega_i^2 [m]\vec{X}^{(i)} = [k]\vec{X}^{(i)} \qquad \text{(6.63)}$$

If we consider another natural frequency $\omega_j$ and the corresponding modal vector $\vec{X}^{(j)}$, they also satisfy Eq. (6.55) so that

$$\omega_j^2 [m]\vec{X}^{(j)} = [k]\vec{X}^{(j)} \qquad \text{(6.64)}$$

By premultiplying Eqs. (6.63) and (6.64) by $\vec{X}^{(j)^T}$ and $\vec{X}^{(i)^T}$ respectively, we obtain, by considering the symmetry of the matrices $[k]$ and $[m]$,

$$\omega_i^2 \vec{X}^{(j)^T}[m]\vec{X}^{(i)} = \vec{X}^{(j)^T}[k]\vec{X}^{(i)} \equiv \vec{X}^{(i)^T}[k]\vec{X}^{(j)} \qquad \text{(6.65)}$$

$$\omega_j^2 \vec{X}^{(i)^T}[m]\vec{X}^{(j)} \equiv \omega_j^2 \vec{X}^{(j)^T}[m]\vec{X}^{(i)} = \vec{X}^{(i)^T}[k]\vec{X}^{(j)} \qquad \text{(6.66)}$$

By subtracting Eq. (6.66) from Eq. (6.65), we obtain

$$\left(\omega_i^2 - \omega_j^2\right) \vec{X}^{(j)^T}[m]\,\vec{X}^{(i)} = 0 \tag{6.67}$$

In general, $\omega_i^2 \neq \omega_j^2$, so Eq. (6.67) leads to*

$$\vec{X}^{(j)^T}[m]\,\vec{X}^{(i)} = 0, \qquad i \neq j \tag{6.68}$$

From Eqs. (6.65) and (6.66), we obtain, in view of Eq. (6.68),

$$\vec{X}^{(j)^T}[k]\,\vec{X}^{(i)} = 0, \qquad i \neq j \tag{6.69}$$

Equations (6.68) and (6.69) indicate that the modal vectors $\vec{X}^{(i)}$ and $\vec{X}^{(j)}$ are orthogonal with respect to both mass and stiffness matrices.

When $i = j$, the left-hand sides of Eqs. (6.68) and (6.69) are not equal to zero, but yield the generalized mass and stiffness coefficients of the $i$th mode:

$$M_{ii} = \vec{X}^{(i)^T}[m]\,\vec{X}^{(i)}, \qquad i = 1, 2, \ldots, n \tag{6.70}$$

$$K_{ii} = \vec{X}^{(i)^T}[k]\,\vec{X}^{(i)}, \qquad i = 1, 2, \ldots, n \tag{6.71}$$

Equations (6.70) and (6.71) can be written in matrix form as

$$[M] = \begin{bmatrix} M_{11} & & & 0 \\ & M_{22} & & \\ & & \ddots & \\ 0 & & & M_{nn} \end{bmatrix} = [X]^T[m][X] \tag{6.72}$$

$$[K] = \begin{bmatrix} K_{11} & & & 0 \\ & K_{22} & & \\ & & \ddots & \\ 0 & & & K_{nn} \end{bmatrix} = [X]^T[k][X] \tag{6.73}$$

where $[X]$ is called the *modal matrix*, in which the $i$th column corresponds to the $i$th modal vector:

$$[X] = [\vec{X}^{(1)} \quad \vec{X}^{(2)} \quad \cdots \quad \vec{X}^{(n)}] \tag{6.74}$$

In many cases, we normalize the modal vectors $\vec{X}^{(i)}$ such that $[M] = [I]$—that is,

$$\vec{X}^{(i)^T}[m]\,\vec{X}^{(i)} = 1, \qquad i = 1, 2, \ldots, n \tag{6.75}$$

In this case the matrix $[K]$ reduces to

$$[K] = [\omega_i^2] = \begin{bmatrix} \omega_1^2 & & & 0 \\ & \omega_2^2 & & \\ & & \ddots & \\ 0 & & & \omega_n^2 \end{bmatrix} \tag{6.76}$$

---

*In the case of repeated eigenvalues, $\omega_i = \omega_j$, the associated modal vectors are orthogonal to all the remaining modal vectors, but are not usually orthogonal to each other.

EXAMPLE 6.6

Orthonormalize the eigenvectors of Example 6.5 with respect to the mass matrix.

*Solution.* The eigenvectors of Example 6.5 are given by

$$\vec{X}^{(1)} = X_1^{(1)} \begin{Bmatrix} 1.0 \\ 1.8019 \\ 2.2470 \end{Bmatrix}$$

$$\vec{X}^{(2)} = X_1^{(2)} \begin{Bmatrix} 1.0 \\ 0.4450 \\ -0.8020 \end{Bmatrix}$$

$$\vec{X}^{(3)} = X_1^{(3)} \begin{Bmatrix} 1.0 \\ -1.2468 \\ 0.5544 \end{Bmatrix}$$

The mass matrix is given by

$$[m] = m \begin{bmatrix} 1 & 0 & 0 \\ 0 & 1 & 0 \\ 0 & 0 & 1 \end{bmatrix}$$

The eigenvector $\vec{X}^{(i)}$ is said to be $[m]$-orthonormal if the following condition is satisfied:

$$\vec{X}^{(i)^T}[m]\vec{X}^{(i)} = 1 \qquad (E.1)$$

Thus for $i = 1$, Eq. (E.1) leads to

$$m\left(X_1^{(1)}\right)^2 (1.0^2 + 1.8019^2 + 2.2470^2) = 1$$

or

$$X_1^{(1)} = \frac{1}{\sqrt{m(9.2959)}} = \frac{0.3280}{\sqrt{m}}$$

Similarly, for $i = 2$ and $i = 3$, Eq. (E.1) gives

$$m\left(X_1^{(2)}\right)^2 (1.0^2 + 0.4450^2 + \{-0.8020\}^2) = 1 \qquad \text{or} \qquad X_1^{(2)} = \frac{0.7370}{\sqrt{m}}$$

and

$$m\left(X_1^{(3)}\right)^2 (1.0^2 + \{-1.2468\}^2 + 0.5544^2) = 1 \qquad \text{or} \qquad X_1^{(3)} = \frac{0.5911}{\sqrt{m}}$$

### 6.9.3 Repeated Eigenvalues

When the characteristic equation possesses repeated roots, the corresponding mode shapes are not unique. To see this, let $\vec{X}^{(1)}$ and $\vec{X}^{(2)}$ be the mode shapes corresponding to the repeated eigenvalue $\lambda_1 = \lambda_2 = \lambda$ and $\vec{X}^{(3)}$ be the mode shape corresponding to a different eigenvalue $\lambda_3$. Equation (6.60) can be written as

$$[D]\vec{X}^{(1)} = \lambda\vec{X}^{(1)} \tag{6.77}$$

$$[D]\vec{X}^{(2)} = \lambda\vec{X}^{(2)} \tag{6.78}$$

$$[D]\vec{X}^{(3)} = \lambda_3\vec{X}^{(3)} \tag{6.79}$$

By multiplying Eq. (6.77) by a constant $p$ and adding to Eq. (6.78), we obtain

$$[D](p\vec{X}^{(1)} + \vec{X}^{(2)}) = \lambda(p\vec{X}^{(1)} + \vec{X}^{(2)}) \tag{6.80}$$

This shows that the new mode shape, $(p\vec{X}^{(1)} + \vec{X}^{(2)})$, which is a linear combination of the first two, also satisfies Eq. (6.60), so the mode shape corresponding to $\lambda$ is not unique. Any $\vec{X}$ corresponding to $\lambda$ must be orthogonal to $\vec{X}^{(3)}$ if it is to be a normal mode. If all three modes are orthogonal, they will be linearly independent and can be used to describe the free vibration resulting from any initial conditions.

The response of a multidegree of freedom system with repeated natural frequencies to force and displacement excitation was presented by Mahalingam and Bishop [6.16].

---

### EXAMPLE 6.7

Determine the eigenvalues and eigenvectors of a vibrating system for which

$$[m] = \begin{bmatrix} 1 & 0 & 0 \\ 0 & 2 & 0 \\ 0 & 0 & 1 \end{bmatrix} \quad \text{and} \quad [k] = \begin{bmatrix} 1 & -2 & 1 \\ -2 & 4 & -2 \\ 1 & -2 & 1 \end{bmatrix}$$

**Solution.** The eigenvalue equation $[[k] - \lambda[m]]\vec{X} = \vec{0}$ can be written in the form

$$\begin{bmatrix} (1-\lambda) & -2 & 1 \\ -2 & 2(2-\lambda) & -2 \\ 1 & -2 & (1-\lambda) \end{bmatrix} \begin{Bmatrix} X_1 \\ X_2 \\ X_3 \end{Bmatrix} = \begin{Bmatrix} 0 \\ 0 \\ 0 \end{Bmatrix} \tag{E.1}$$

where $\lambda = \omega^2$. The characteristic equation gives

$$|[k] - \lambda[m]| = \lambda^2(\lambda - 4) = 0$$

so

$$\lambda_1 = 0, \quad \lambda_2 = 0, \quad \lambda_3 = 4 \tag{E.2}$$

*Eigenvector for $\lambda_3 = 4$:* Using $\lambda_3 = 4$, Eq. (E.1) gives

$$-3X_1^{(3)} - 2X_2^{(3)} + X_3^{(3)} = 0$$

$$-2X_1^{(3)} - 4X_2^{(3)} - 2X_3^{(3)} = 0$$

$$X_1^{(3)} - 2X_2^{(3)} - 3X_3^{(3)} = 0 \tag{E.3}$$

If $X_1^{(3)}$ is set equal to 1, Eqs. (E.3) give the eigenvector $\vec{X}^{(3)}$:

$$\vec{X}^{(3)} = \begin{Bmatrix} 1 \\ -1 \\ 1 \end{Bmatrix} \tag{E.4}$$

*Eigenvector for* $\lambda_1 = \lambda_2 = 0$: The value $\lambda_1 = 0$ or $\lambda_2 = 0$ indicates that the system is degenerate (see Sec. 6.11). Using $\lambda_1 = 0$ in Eq. (E.1), we obtain

$$X_1^{(1)} - 2X_2^{(1)} + X_3^{(1)} = 0$$

$$-2X_1^{(1)} + 4X_2^{(1)} - 2X_3^{(1)} = 0$$

$$X_1^{(1)} - 2X_2^{(1)} + X_3^{(1)} = 0 \qquad \text{(E.5)}$$

All these equations are of the form

$$X_1^{(1)} = 2X_2^{(1)} - X_3^{(1)}$$

Thus the eigenvector corresponding to $\lambda_1 = \lambda_2 = 0$ can be written as

$$\vec{X}^{(1)} = \begin{Bmatrix} 2X_2^{(1)} - X_3^{(1)} \\ X_2^{(1)} \\ X_3^{(1)} \end{Bmatrix} \qquad \text{(E.6)}$$

If we choose $X_2^{(1)} = 1$ and $X_3^{(1)} = 1$, we obtain

$$\vec{X}^{(1)} = \begin{Bmatrix} 1 \\ 1 \\ 1 \end{Bmatrix} \qquad \text{(E.7)}$$

If we select $X_2^{(1)} = 1$ and $X_3^{(1)} = -1$, Eq. (E.6) gives

$$\vec{X}^{(1)} = \begin{Bmatrix} 3 \\ 1 \\ -1 \end{Bmatrix} \qquad \text{(E.8)}$$

As shown earlier in Eq. (6.80), $\vec{X}^{(1)}$ and $\vec{X}^{(2)}$ are not unique; any linear combination of $\vec{X}^{(1)}$ and $\vec{X}^{(2)}$ will also satisfy the original Eq. (E.1). Note that $\vec{X}^{(1)}$ given by Eq. (E.6) is orthogonal to $\vec{X}^{(3)}$ of Eq. (E.4) for all values of $X_2^{(1)}$ and $X_3^{(1)}$, since

$$\vec{X}^{(3)^T}[m]\vec{X}^{(1)} = (1 \quad -1 \quad 1)\begin{bmatrix} 1 & 0 & 0 \\ 0 & 2 & 0 \\ 0 & 0 & 1 \end{bmatrix}\begin{Bmatrix} 2X_2^{(1)} - X_3^{(1)} \\ X_2^{(1)} \\ X_3^{(1)} \end{Bmatrix} = 0$$

## 6.10 EXPANSION THEOREM

The eigenvectors, due to their property of orthogonality, are linearly independent*. Hence they form a basis in the $n$-dimensional space[†]. This means that any vector in the $n$-dimensional space can be expressed by a linear combination of the $n$ linearly independent vectors. If $\vec{x}$ is an arbitrary vector in $n$-dimensional space, it can be

---

*A set of vectors is called linearly independent if no vector in the set can be obtained by a linear combination of the remaining ones.

[†]Any set of $n$ linearly independent vectors in an $n$-dimensional space is called a *basis* in that space.

expressed as

$$\vec{x} = \sum_{i=1}^{n} c_i \vec{X}^{(i)} \tag{6.81}$$

where $c_i$ are constants. By premultiplying Eq. (6.81) throughout by $\vec{X}^{(i)^T}[m]$, the value of the constant $c_i$ can be determined as

$$c_i = \frac{\vec{X}^{(i)^T}[m]\vec{x}}{\vec{X}^{(i)^T}[m]\vec{X}^{(i)}} = \frac{\vec{X}^{(i)^T}[m]\vec{x}}{M_{ii}}, \qquad i = 1, 2, \ldots, n \tag{6.82}$$

where $M_{ii}$ is the generalized mass in the $i$th normal mode. If the modal vectors $\vec{X}^{(i)}$ are normalized according to Eq. (6.75), $c_i$ is given by

$$c_i = \vec{X}^{(i)^T}[m]\vec{x}, \qquad i = 1, 2, \ldots, n \tag{6.83}$$

Equation (6.83) represents what is known as the *expansion theorem* [6.6]. It is very useful in finding the response of multidegree of freedom systems subjected to arbitrary forcing conditions according to a procedure called *modal analysis*.

## 6.11  UNRESTRAINED SYSTEMS

As stated in Sec. 5.7, an unrestrained system is one which has no restraints or supports and which can move as a rigid body. It is not uncommon to see, in practice, systems which are not attached to any stationary frame. A common example is the motion of two railway cars with masses $m_1$ and $m_2$ and a coupling spring $k$. Such systems are capable of moving as rigid bodies, which can be considered as modes of oscillation with zero frequency. For a conservative system, the kinetic and potential energies are given by Eqs. (6.27) and (6.24), respectively. By definition, the kinetic energy is always positive, so the mass matrix $[m]$ is a positive definite matrix. However, the stiffness matrix $[k]$ is a semi-definite matrix; $V$ is zero without the displacement vector $\vec{x}$ being zero for unrestrained systems. To see this, consider the equation of motion for free vibration in normal coordinates:

$$\ddot{q}(t) + \omega^2 q(t) = 0 \tag{6.84}$$

For $\omega = 0$, the solution of Eq. (6.84) can be expressed as

$$q(t) = \alpha + \beta t \tag{6.85}$$

where $\alpha$ and $\beta$ are constants. Equation (6.85) represents a rigid body translation. Let the modal vector of a multidegree of freedom system corresponding to the rigid body mode be denoted by $\vec{X}^{(0)}$. The eigenvalue problem, Eq. (6.58), can be expressed as

$$\omega^2[m]\vec{X}^{(0)} = [k]\vec{X}^{(0)} \tag{6.86}$$

With $\omega = 0$, Eq. (6.86) gives

$$[k]\vec{X}^{(0)} = \vec{0}$$

That is,

$$k_{11}X_1^{(0)} + k_{12}X_2^{(0)} + \cdots + k_{1n}X_n^{(0)} = 0$$
$$k_{21}X_1^{(0)} + k_{22}X_2^{(0)} + \cdots + k_{2n}X_n^{(0)} = 0$$
$$\vdots$$
$$k_{n1}X_1^{(0)} + k_{n2}X_2^{(0)} + \cdots + k_{nn}X_n^{(0)} = 0 \qquad (6.87)$$

If the system undergoes rigid body translation, not all the components $X_i^{(0)}$, $i = 1, 2, \ldots, n$ are zero—that is, the vector $\vec{X}^{(0)}$ is not equal to $\vec{0}$. Hence, in order to satisfy Eq. (6.87), the determinant of $[k]$ must be zero. Thus the stiffness matrix of an unrestrained system (having zero natural frequency) is singular. If $[k]$ is singular, the potential energy is given by

$$V = \frac{1}{2}\vec{X}^{(0)^T}[k]\vec{X}^{(0)} \qquad (6.88)$$

by virtue of Eq. (6.87). The mode $\vec{X}^{(0)}$ is called a *zero mode* or *rigid body mode*. If we substitute any vector $\vec{X}$ other than $\vec{X}^{(0)}$ and $\vec{0}$ for $\vec{x}$ in Eq. (6.24), the potential energy $V$ becomes a positive quantity. The matrix $[k]$ is then a positive semi-definite matrix. This is why an unrestrained system is also called a *semi-definite system*.

Note that a multidegree of freedom system can have at most six rigid body modes with the corresponding frequencies equal to zero. There can be three modes for rigid body translation, one for translation along each of the three Cartesian coordinates, and three modes for rigid body rotation, one for rotation about each of the three Cartesian coordinates. We can determine the mode shapes and natural frequencies of a semi-definite system by the procedures outlined in Sec. 6.9.

### EXAMPLE 6.8

Find the natural frequencies and mode shapes of the system shown in Fig. 6.8 for $m_1 = m_2 = m_3 = m$ and $k_1 = k_2 = k$.

***Solution.*** The kinetic energy of the system can be written as

$$T = \frac{1}{2}\left(m_1\dot{x}_1^2 + m_2\dot{x}_2^2 + m_3\dot{x}_3^2\right) = \frac{1}{2}\dot{\vec{x}}^T[m]\dot{\vec{x}} \qquad (E.1)$$

where

$$\vec{x} = \begin{Bmatrix} x_1 \\ x_2 \\ x_3 \end{Bmatrix}, \qquad \dot{\vec{x}} = \begin{Bmatrix} \dot{x}_1 \\ \dot{x}_2 \\ \dot{x}_3 \end{Bmatrix}$$

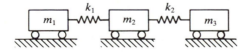

**Figure 6.8**

and

$$[m] = \begin{bmatrix} m_1 & 0 & 0 \\ 0 & m_2 & 0 \\ 0 & 0 & m_3 \end{bmatrix} \tag{E.2}$$

The elongations of the springs $k_1$ and $k_2$ are $(x_2 - x_1)$ and $(x_3 - x_2)$, respectively, so the potential energy of the system is given by

$$V = \frac{1}{2}\left\{ k_1(x_2 - x_1)^2 + k_2(x_3 - x_2)^2 \right\} = \frac{1}{2}\vec{x}^T[k]\vec{x} \tag{E.3}$$

where

$$[k] = \begin{bmatrix} k_1 & -k_1 & 0 \\ -k_1 & k_1 + k_2 & -k_2 \\ 0 & -k_2 & k_2 \end{bmatrix} \tag{E.4}$$

It can be verified that the stiffness matrix $[k]$ is singular. Furthermore, if we take all the displacement components to be the same as $x_1 = x_2 = x_3 = c$ (rigid body motion), the potential energy $V$ can be seen to be equal to zero.

To find the natural frequencies and the mode shapes of the system, we express the eigenvalue problem as

$$\left[[k] - \omega^2[m]\right]\vec{X} = \vec{0} \tag{E.5}$$

Since $[k]$ is singular, we cannot find its inverse $[k]^{-1}$, so the dynamical matrix $[D] = [k]^{-1}[m]$. Hence we set the determinant of the coefficient matrix of $\vec{X}$ in Eq. (E.5) equal to zero. For $k_1 = k_2 = k$ and $m_1 = m_2 = m_3 = m$, this yields

$$\begin{vmatrix} (k - \omega^2 m) & -k & 0 \\ -k & (2k - \omega^2 m) & -k \\ 0 & -k & (k - \omega^2 m) \end{vmatrix} = 0 \tag{E.6}$$

The expansion of the determinant in Eq. (E.6) leads to

$$m^3\omega^6 - 4m^2 k\omega^4 + 3mk^2\omega^2 = 0 \tag{E.7}$$

By setting

$$\lambda = \omega^2 \tag{E.8}$$

Eq. (E.7) can be rewritten as

$$m\lambda\left(\lambda - \frac{k}{m}\right)\left(\lambda - \frac{3k}{m}\right) = 0 \tag{E.9}$$

As $m \neq 0$, the roots of Eq. (E.9) are

$$\lambda_1 = \omega_1^2 = 0$$

$$\lambda_2 = \omega_2^2 = \frac{k}{m}$$

$$\lambda_3 = \omega_3^2 = \frac{3k}{m} \tag{E.10}$$

The first natural frequency $\omega_1$ can be observed to be zero in Eq. (E.10). To find the mode shapes, we substitute the values of $\omega_1$, $\omega_2$, and $\omega_3$ into Eq. (E.5) and solve for

$\vec{X}^{(1)}$, $\vec{X}^{(2)}$, and $\vec{X}^{(3)}$, respectively. For $\omega_1 = 0$, Eq. (E.5) gives

$$kX_1^{(1)} - kX_2^{(1)} = 0$$

$$-kX_1^{(1)} + 2kX_2^{(1)} - kX_3^{(1)} = 0$$

$$-kX_2^{(1)} + kX_3^{(1)} = 0 \qquad \text{(E.11)}$$

By fixing the value of one component of $\vec{X}^{(1)}$—say, $X_1^{(1)}$ as 1—Eqs. (E.11) can be solved to obtain

$$X_2^{(1)} = X_1^{(1)} = 1 \qquad \text{and} \qquad X_3^{(1)} = X_2^{(1)} = 1$$

Thus the first (rigid body) mode $\vec{X}^{(1)}$ corresponding to $\omega_1 = 0$ is given by

$$\vec{X}^{(1)} = \begin{Bmatrix} 1 \\ 1 \\ 1 \end{Bmatrix} \qquad \text{(E.12)}$$

For $\omega_2 = (k/m)^{1/2}$, Eq. (E.5) yields

$$-kX_2^{(2)} = 0$$

$$-kX_1^{(2)} + kX_2^{(2)} - kX_3^{(2)} = 0$$

$$-kX_2^{(2)} = 0 \qquad \text{(E.13)}$$

By fixing the value of one component of $\vec{X}^{(2)}$—say, $X_1^{(2)}$ as 1—Eqs. (E.13) can be solved to obtain

$$X_2^{(2)} = 0 \qquad \text{and} \qquad X_3^{(2)} = -X_1^{(2)} = -1$$

Thus the second mode $\vec{X}^{(2)}$ corresponding to $\omega_2 = (k/m)^{1/2}$ is given by

$$\vec{X}^{(2)} = \begin{Bmatrix} 1 \\ 0 \\ -1 \end{Bmatrix} \qquad \text{(E.14)}$$

For $\omega_3 = (3k/m)^{1/2}$, Eq. (E.5) gives

$$-2kX_1^{(3)} - kX_2^{(3)} = 0$$

$$-kX_1^{(3)} - kX_2^{(3)} - kX_3^{(3)} = 0$$

$$-kX_2^{(3)} - 2kX_3^{(3)} = 0 \qquad \text{(E.15)}$$

By fixing the value of one component of $\vec{X}^{(3)}$—say, $X_1^{(3)}$ as 1—Eqs. (E.15) can be solved to obtain

$$X_2^{(3)} = -2X_1^{(3)} = -2 \qquad \text{and} \qquad X_3^{(3)} = -\frac{1}{2}X_2^{(3)} = 1$$

Thus the third mode $\vec{X}^{(3)}$ corresponding to $\omega_3 = (3k/m)^{1/2}$ is given by

$$\vec{X}^{(3)} = \begin{Bmatrix} 1 \\ -2 \\ 1 \end{Bmatrix} \qquad \text{(E.16)}$$

## 6.12 FORCED VIBRATION

When external forces act on a multidegree of freedom system, the system undergoes forced vibration. For a system with $n$ coordinates or degrees of freedom, the governing equations of motion are a set of $n$ coupled ordinary differential equations of second order. The solution of these equations becomes more complex when the degree of freedom of the system ($n$) is large and/or when the forcing functions are nonperiodic.* In such cases, a more convenient method known as *modal analysis* can be used to solve the problem. In this method, the expansion theorem is used, and the displacements of the masses are expressed as a linear combination of the normal modes of the system. This linear transformation uncouples the equations of motion so that we obtain a set of $n$ uncoupled differential equations of second order. The solution of these equations, which is equivalent to the solution of the equations of $n$ single degree of freedom systems, can be readily obtained. We shall now consider the procedure of modal analysis.

**Modal Analysis.** The equations of motion of a multidegree of freedom system under external forces are given by

$$[m]\ddot{\vec{x}} + [k]\vec{x} = \vec{F} \tag{6.89}$$

where $\vec{F}$ is the vector of arbitrary external forces. To solve Eq. (6.89) by modal analysis, it is necessary first to solve the eigenvalue problem,

$$\omega^2[m]\vec{X} = [k]\vec{X} \tag{6.90}$$

and find the natural frequencies $\omega_1, \omega_2, \ldots, \omega_n$ and the corresponding normal modes $\vec{X}^{(1)}, \vec{X}^{(2)}, \ldots, \vec{X}^{(n)}$. According to the expansion theorem, the solution vector of Eq. (6.89) can be expressed by a linear combination of the normal modes:

$$\vec{x}(t) = q_1(t)\vec{X}^{(1)} + q_2(t)\vec{X}^{(2)} + \cdots + q_n(t)\vec{X}^{(n)} \tag{6.91}$$

where $q_1(t), q_2(t), \ldots, q_n(t)$ are time-dependent generalized coordinates, also known as the *modal participation coefficients*. By defining a modal matrix $[X]$ in which the $j$th column is the vector $\vec{X}^{(j)}$—that is,

$$[X] = [\vec{X}^{(1)} \quad \vec{X}^{(2)} \quad \cdots \quad \vec{X}^{(n)}] \tag{6.92}$$

Eq. (6.91) can be rewritten as

$$\vec{x}(t) = [X]\vec{q}(t) \tag{6.93}$$

where

$$\vec{q}(t) = \begin{Bmatrix} q_1(t) \\ q_2(t) \\ \vdots \\ q_n(t) \end{Bmatrix} \tag{6.94}$$

---

*The dynamic response of multidegree of freedom systems with statistical properties was considered in Ref. [6.15].

Since $[X]$ is not a function of time, we obtain from Eq. (6.93)

$$\ddot{\vec{x}}(t) = [X]\ddot{\vec{q}}(t) \tag{6.95}$$

Using Eqs. (6.93) and (6.95), Eq. (6.89) can be written as

$$[m][X]\ddot{\vec{q}} + [k][X]\vec{q} = \vec{F} \tag{6.96}$$

Premultiplying Eq. (6.96) throughout by $[X]^T$, we obtain

$$[X]^T[m][X]\ddot{\vec{q}} + [X]^T[k][X]\vec{q} = [X]\vec{F} \tag{6.97}$$

If the normal modes are normalized according to Eqs. (6.68) and (6.69), we have

$$[X]^T[m][X] = [I] \tag{6.98}$$

$$[X]^T[k][X] = \lceil\omega^2\rfloor \tag{6.99}$$

By defining the vector of generalized forces $\vec{Q}(t)$ associated with the generalized coordinates $\vec{q}(t)$ as

$$\vec{Q}(t) = [X]^T\vec{F}(t) \tag{6.100}$$

Eq. (6.97) can be expressed, using Eqs. (6.98) and (6.99), as

$$\ddot{\vec{q}}(t) + \lceil\omega^2\rfloor\vec{q}(t) = \vec{Q}(t) \tag{6.101}$$

Equation (6.101) denotes a set of $n$ uncoupled differential equations of second order*

$$\ddot{q}_i(t) + \omega_i^2 q_i(t) = Q_i(t), \qquad i = 1, 2, \ldots, n \tag{6.102}$$

It can be seen that Eqs. (6.102) have precisely the form of the differential equation describing the motion of an undamped single degree of freedom system. The

---

*It is possible to approximate the solution vector $\vec{x}(t)$ by only the first $r(r < n)$ modal vectors (instead of $n$ vectors as in Eq. (6.91)):

$$\vec{x}(t) = [X]\vec{q}(t)$$
$$n \times 1 \quad n \times r \quad r \times 1$$

where

$$[X] = [\vec{X}^{(1)} \quad \vec{X}^{(2)} \quad \cdots \quad \vec{X}^{(r)}] \quad \text{and} \quad \vec{q}(t) = \begin{Bmatrix} q_1(t) \\ q_2(t) \\ \vdots \\ q_r(t) \end{Bmatrix}$$

This leads to only $r$ uncoupled differential equations,

$$\ddot{q}_i(t) + \omega_i^2 q_i(t) = Q_i(t), \qquad i = 1, 2, \ldots, r$$

instead of $n$ equations. The resulting solution $\vec{x}(t)$ will be an approximate solution. This procedure is called the mode displacement method. An alternate procedure, known as the mode acceleration method, for finding an approximate solution is indicated in Problem 6.40.

solution of Eqs. (6.102) can be expressed (see Eq. (4.33)) as:

$$q_i(t) = q_i(0)\cos \omega_i t + \left( \frac{\dot{q}_i(0)}{\omega_i} \right) \sin \omega_i t + \frac{1}{\omega_i} \int_0^t Q_i(\tau) \sin \omega_i(t - \tau) \, d\tau,$$

$$i = 1, 2, \ldots, n \qquad (6.103)$$

The initial generalized displacements $q_i(0)$ and the initial generalized velocities $\dot{q}_i(0)$ can be obtained from the initial values of the physical displacements $x_i(0)$ and physical velocities $\dot{x}_0(0)$:

$$\vec{q}(0) = [X]^T [m] \vec{x}(0) \qquad (6.104)$$

$$\dot{\vec{q}}(0) = [X]^T [m] \dot{\vec{x}}(0) \qquad (6.105)$$

where

$$\vec{q}(0) = \begin{Bmatrix} q_1(0) \\ q_2(0) \\ \vdots \\ q_n(0) \end{Bmatrix}, \quad \dot{\vec{q}}(0) = \begin{Bmatrix} \dot{q}_1(0) \\ \dot{q}_2(0) \\ \vdots \\ \dot{q}_n(0) \end{Bmatrix}, \quad \vec{x}(0) = \begin{Bmatrix} x_1(0) \\ x_2(0) \\ \vdots \\ x_n(0) \end{Bmatrix}, \quad \dot{\vec{x}}(0) = \begin{Bmatrix} \dot{x}_1(0) \\ \dot{x}_2(0) \\ \vdots \\ \dot{x}_n(0) \end{Bmatrix}$$

Once the generalized displacements $q_i(t)$ are found, using Eqs. (6.103) to (6.105), the physical displacements $x_i(t)$ can be found with the help of Eq. (6.93).

## 6.13 VISCOUSLY DAMPED SYSTEMS

Modal analysis, as presented in Sec. 6.12, applies only to undamped systems. In many cases, the influence of damping upon the response of a vibratory system is minor and can be disregarded. However, the effect of damping must be considered if the response of the system is required for a relatively long period of time compared to the natural periods of the system. Further, if the frequency of excitation (in the case of a periodic force) is at or near one of the natural frequencies of the system, damping is of primary importance and must be taken into account. In general, since the effects are not known in advance, damping must be considered in the vibration analysis of any system. In this section, we shall consider the equations of motion of a damped multidegree of freedom system and their solution using Lagrange's equations. If the system has viscous damping, its motion will be resisted by a force whose magnitude is proportional to that of the velocity but in the opposite direction. It is convenient to introduce a function $R$, known as Rayleigh's dissipation function, in deriving the equations of motion by means of Lagrange's equations [6.7]. This function is defined as

$$R = \frac{1}{2} \dot{x}^T [c] \dot{x} \qquad (6.106)$$

where the matrix $[c]$ is called the *damping matrix* and is positive definite, like the mass and stiffness matrices. Lagrange's equations, in this case [6.8], can be written as

$$\frac{d}{dt}\left(\frac{\partial T}{\partial \dot{x}_i}\right) - \frac{\partial T}{\partial x_i} + \frac{\partial R}{\partial \dot{x}_i} + \frac{\partial V}{\partial x_i} = F_i, \qquad i = 1, 2, \ldots, n \tag{6.107}$$

where $F_i$ is the force applied to mass $m_i$. By substituting Eqs. (6.24), (6.28), and (6.106) into Eq. (6.107), we obtain the equations of motion of a damped multidegree of freedom system in matrix form:

$$[m]\ddot{\vec{x}} + [c]\dot{\vec{x}} + [k]\vec{x} = \vec{F} \tag{6.108}$$

For simplicity, we shall consider a special system for which the damping matrix can be expressed as a linear combination of the mass and stiffness matrices:

$$[c] = \alpha[m] + \beta[k] \tag{6.109}$$

where $\alpha$ and $\beta$ are constants. This type of damping is known as *proportional damping* because $[c]$ is proportional to a linear combination of $[m]$ and $[k]$. By substituting Eq. (6.109) into Eq. (6.108) we obtain

$$[m]\ddot{\vec{x}} + [\alpha[m] + \beta[k]]\dot{\vec{x}} + [k]\vec{x} = \vec{F} \tag{6.110}$$

By expressing the solution vector $\vec{x}$ as a linear combination of the natural modes of the undamped system, as in the case of Eq. (6.93),

$$\vec{x}(t) = [X]\vec{q}(t) \tag{6.111}$$

Eq. (6.110) can be rewritten as

$$[m][X]\ddot{\vec{q}}(t) + [\alpha[m] + \beta[k]][X]\dot{\vec{q}}(t) + [k][X]\vec{q}(t) = \vec{F}(t) \tag{6.112}$$

Premultiplication of Eq. (6.112) by $[X]^T$ leads to

$$[X]^T[m][X]\ddot{\vec{q}} + \left[\alpha[X]^T[m][X] + \beta[X]^T[k][X]\right]\dot{\vec{q}}$$
$$+ [X]^T[k][X]\vec{q} = [X]^T\vec{F} \tag{6.113}$$

If the eigenvectors $\vec{X}^{(j)}$ are normalized according to Eqs. (6.68) and (6.69), Eq. (6.113) reduces to

$$[I]\ddot{\vec{q}}(t) + \left[\alpha[I] + \beta\lceil\omega^2\rfloor\right]\dot{\vec{q}}(t) + \lceil\omega^2\rfloor\vec{q}(t) = \vec{Q}(t)$$

that is,

$$\ddot{q}_i(t) + \left(\alpha + \omega_i^2\beta\right)\dot{q}_i(t) + \omega_i^2 q_i(t) = Q_i(t), \qquad i = 1, 2, \ldots, n \tag{6.114}$$

where $\omega_i$ is the $i$th natural frequency of the undamped system and

$$\vec{Q}(t) = [X]^T\vec{F}(t) \tag{6.115}$$

By writing

$$\alpha + \omega_i^2\beta = 2\zeta_i\omega_i \tag{6.116}$$

where $\zeta_i$ is called the *modal damping ratio* for the $i$th normal mode, Eqs. (6.114) can be rewritten as

$$\ddot{q}_i(t) + 2\zeta_i\omega_i\dot{q}_i(t) + \omega_i^2 q_i(t) = Q_i(t), \qquad i = 1, 2, \ldots, n \tag{6.117}$$

It can be seen that each of the $n$ equations represented by this expression is uncoupled from all of the others. Hence we can find the response of the $i$th mode in the same manner as that of a viscously damped single degree of freedom system. The solution of Eqs. (6.117), when $\zeta_i < 1$, can be expressed as

$$q_i(t) = e^{-\zeta_i \omega_i t} \left\{ \cos \omega_{di} t + \frac{\zeta_i}{\sqrt{1 - \zeta_i^2}} \sin \omega_{di} t \right\} q_i(0)$$

$$+ \left\{ \frac{1}{\omega_{di}} e^{-\zeta_i \omega_i t} \sin \omega_{di} t \right\} \dot{q}_i(0) + \frac{1}{\omega_{di}} \int_0^t Q_i(\tau) e^{-\zeta_i \omega_i (t - \tau)} \sin \omega_{di}(t - \tau) \, d\tau,$$

$$i = 1, 2, \ldots, n \qquad (6.118)$$

where

$$\omega_{di} = \omega_i \sqrt{1 - \zeta_i^2} \qquad (6.119)$$

Note the following aspects of these systems:

1. It has been shown by Caughey [6.9] that the condition given by Eq. (6.109) is sufficient but not necessary for the existence of normal modes in damped systems. The necessary condition is that the transformation that diagonalizes the damping matrix also uncouples the coupled equations of motion. This condition is less restrictive than Eq. (6.109) and covers more possibilities.

2. In the general case of damping, the damping matrix cannot be diagonalized simultaneously with the mass and stiffness matrices. In this case, the eigenvalues of the system are either real and negative or complex with negative real parts. The complex eigenvalues exist as conjugate pairs; the associated eigenvectors also consist of complex conjugate pairs. A common procedure for finding the solution of the eigenvalue problem of a damped system involves the transformation of the $n$ coupled second order equations of motion into $2n$ uncoupled first order equations [6.6].

3. The error bounds and numerical methods in the modal analysis of dynamic systems were discussed in Refs. [6.11, 6.12].

## EXAMPLE 6.9

Derive the equations of motion of the system shown in Fig. 6.9.

*Solution.* The kinetic energy of the system is

$$T = \tfrac{1}{2} \left( m_1 \dot{x}_1^2 + m_2 \dot{x}_2^2 + m_3 \dot{x}_3^2 \right) \qquad (E.1)$$

The potential energy has the form

$$V = \tfrac{1}{2} \left[ k_1 x_1^2 + k_2 (x_2 - x_1)^2 + k_3 (x_3 - x_2)^2 \right] \qquad (E.2)$$

and Rayleigh's dissipation function is

$$R = \tfrac{1}{2} \left[ c_1 \dot{x}_1^2 + c_2 (\dot{x}_2 - \dot{x}_1)^2 + c_3 (\dot{x}_3 - \dot{x}_2)^2 + c_4 \dot{x}_2^2 + c_5 (\dot{x}_3 - \dot{x}_1)^2 \right] \qquad (E.3)$$

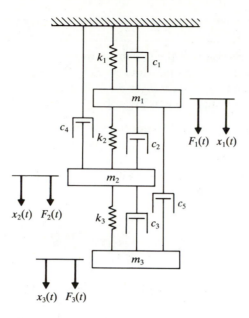

**Figure 6.9**

Lagrange's equations can be written as

$$\frac{d}{dt}\left(\frac{\partial T}{\partial \dot{x}_i}\right) - \frac{\partial T}{\partial x_i} + \frac{\partial R}{\partial \dot{x}_i} + \frac{\partial V}{\partial x_i} = F_i, \qquad i = 1,2,3 \tag{E.4}$$

By substituting Eqs. (E.1) to (E.3) into Eq. (E.4), we obtain the differential equations of motion:

$$[m]\ddot{\vec{x}} + [c]\dot{\vec{x}} + [k]\vec{x} = \vec{F} \tag{E.5}$$

where

$$[m] = \begin{bmatrix} m_1 & 0 & 0 \\ 0 & m_2 & 0 \\ 0 & 0 & m_3 \end{bmatrix} \tag{E.6}$$

$$[c] = \begin{bmatrix} c_1 + c_2 + c_5 & -c_2 & -c_5 \\ -c_2 & c_2 + c_3 + c_4 & -c_3 \\ -c_5 & -c_3 & c_3 + c_5 \end{bmatrix} \tag{E.7}$$

$$[k] = \begin{bmatrix} k_1 + k_2 & -k_2 & 0 \\ -k_2 & k_2 + k_3 & -k_3 \\ 0 & -k_3 & k_3 \end{bmatrix} \tag{E.8}$$

$$\vec{x} = \begin{Bmatrix} x_1(t) \\ x_2(t) \\ x_3(t) \end{Bmatrix} \qquad \text{and} \qquad \vec{F} = \begin{Bmatrix} F_1(t) \\ F_2(t) \\ F_3(t) \end{Bmatrix} \tag{E.9}$$

---

## EXAMPLE 6.10

Find the steady-state response of the system shown in Fig. 6.9 when the masses are subjected to the simple harmonic forces $F_1 = F_2 = F_3 = F_0 \cos \omega t$, where $\omega = 1.75\sqrt{k/m}$. Assume that $m_1 = m_2 = m_3 = m$, $k_1 = k_2 = k_3 = k$, $c_4 = c_5 = 0$, and the damping ratio in each normal mode is given by $\zeta_i = 0.01$, $i = 1, 2, 3$.

***Solution.*** The (undamped) natural frequencies of the system (see Example 6.5) are given by

$$\omega_1 = 0.44504\sqrt{\frac{k}{m}}$$

$$\omega_2 = 1.2471\sqrt{\frac{k}{m}}$$

$$\omega_3 = 1.8025\sqrt{\frac{k}{m}} \tag{E.1}$$

and the corresponding $[m]$-orthonormal mode shapes (see Example 6.6) are given by

$$\vec{X}^{(1)} = \frac{0.3280}{\sqrt{m}}\begin{Bmatrix} 1.0 \\ 1.8019 \\ 2.2470 \end{Bmatrix}, \qquad \vec{X}^{(2)} = \frac{0.7370}{\sqrt{m}}\begin{Bmatrix} 1.0 \\ 0.4450 \\ -0.8020 \end{Bmatrix},$$

$$\vec{X}^{(3)} = \frac{0.5911}{\sqrt{m}}\begin{Bmatrix} 1.0 \\ -1.2468 \\ 0.5544 \end{Bmatrix} \tag{E.2}$$

Thus the modal vector can be expressed as

$$[X] = [\vec{X}^{(1)} \vec{X}^{(2)} \vec{X}^{(3)}] = \frac{1}{\sqrt{m}}\begin{bmatrix} 0.3280 & 0.7370 & 0.5911 \\ 0.5911 & 0.3280 & -0.7370 \\ 0.7370 & -0.5911 & 0.3280 \end{bmatrix} \tag{E.3}$$

The generalized force vector $\vec{Q}(t)$ can be obtained:

$$\vec{Q}(t) = [X]^T \vec{F}(t) = \frac{1}{\sqrt{m}}\begin{bmatrix} 0.3280 & 0.5911 & 0.7370 \\ 0.7370 & 0.3280 & -0.5911 \\ 0.5911 & -0.7370 & 0.3280 \end{bmatrix}\begin{Bmatrix} F_0 \cos \omega t \\ F_0 \cos \omega t \\ F_0 \cos \omega t \end{Bmatrix}$$

$$= \begin{Bmatrix} Q_{10} \\ Q_{20} \\ Q_{30} \end{Bmatrix}\cos \omega t \tag{E.4}$$

where

$$Q_{10} = 1.6561\frac{F_0}{\sqrt{m}}, \qquad Q_{20} = 0.4739\frac{F_0}{\sqrt{m}}, \qquad Q_{30} = 0.1821\frac{F_0}{\sqrt{m}} \tag{E.5}$$

If the generalized coordinates or the modal participation factors for the three principal modes are denoted as $q_1(t)$, $q_2(t)$, and $q_3(t)$, the equations of motion can be expressed as

$$\ddot{q}_i(t) + 2\zeta_i\omega_i\dot{q}_i(t) + \omega_i^2 q_i(t) = Q_i(t), \qquad i = 1, 2, 3 \tag{E.6}$$

The steady-state solution of Eqs. (E.6) can be written as

$$q_i(t) = q_{i0}\cos(\omega t - \phi_i), \qquad i = 1, 2, 3 \tag{E.7}$$

where

$$q_{i0} = \frac{Q_{i0}}{\omega_i^2} \frac{1}{\left[\left\{1 - \left(\frac{\omega}{\omega_i}\right)^2\right\}^2 + \left(2\zeta_i\frac{\omega}{\omega_i}\right)^2\right]^{1/2}} \tag{E.8}$$

and

$$\phi_i = \tan^{-1}\left\{\frac{2\zeta_i\dfrac{\omega}{\omega_i}}{1 - \left(\dfrac{\omega}{\omega_i}\right)^2}\right\} \tag{E.9}$$

By substituting the values given in Eqs. (E.5) and (E.1) into Eqs. (E.8) and (E.9), we obtain

$$q_{10} = 0.57815\frac{F_0\sqrt{m}}{k}, \qquad \phi_1 = \tan^{-1}(-0.00544)$$

$$q_{20} = 0.31429\frac{F_0\sqrt{m}}{k}, \qquad \phi_2 = \tan^{-1}(-0.02988)$$

$$q_{30} = 0.92493\frac{F_0\sqrt{m}}{k}, \qquad \phi_3 = \tan^{-1}(0.33827) \tag{E.10}$$

Finally the steady-state response can be found using Eq. (6.111).

## 6.14 SELF-EXCITATION AND STABILITY ANALYSIS

The stability of a multidegree of freedom system involving damping and forcing conditions can be analyzed by extending the methods of Sec. 5.8, although this becomes more difficult as the number of degrees of freedom increases [6.13, 6.14]. Let us consider the viscously damped $n$ degree of freedom system shown in Fig. 6.10, with self-exciting forcing conditions defined by

$$F_j(t) = F_{j0}\dot{x}_j, \qquad j = 1, 2, \ldots \tag{6.120}$$

This forcing term can be combined with the damping term while writing the differential equations of motion of the system. By substituting the solution

$$x_j(t) = C_j e^{st}, \qquad j = 1, 2, \ldots \tag{6.121}$$

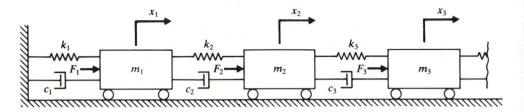

**Figure 6.10**

in the differential equations of motion and expanding the determinant of the resulting auxiliary equations, we obtain the frequency equation in general form:

$$a_0 s^m + a_1 s^{m-1} + a_2 s^{m-2} + \cdots + a_{m-1} s + a_m = 0 \qquad \text{with } a_0 > 0 \qquad (6.122)$$

where $m = 2n$. For dynamic stability, Eq. (6.122) should not contain any positive real root. This requires that all the coefficients $a_0, a_1, a_2, \ldots$ be positive. Now consider the case in which all the roots of Eq. (6.122) are complex. The $m$ complex roots can be expressed as

$$s_1 = p_1 + iq_1, \quad s_2 = p_1 - iq_1, \quad s_3 = p_2 + iq_2, \quad s_4 = p_2 - iq_2, \ldots \qquad (6.123)$$

The real parts of the roots ($p_1, p_2, \ldots$) must be negative for the system to be stable. The criteria for this can be expressed in terms of the $m$th order determinant $H_m$, defined as

$$H_m = \begin{vmatrix} a_1 & a_3 & a_5 & a_7 & \cdots & a_{2m-1} \\ a_0 & a_2 & a_4 & a_6 & \cdots & a_{2m-2} \\ 0 & a_1 & a_3 & a_5 & \cdots & a_{2m-3} \\ 0 & a_0 & a_2 & a_4 & \cdots & a_{2m-4} \\ 0 & 0 & a_1 & a_3 & \cdots & a_{2m-5} \\ \vdots & & & & & \\ \vdots & \cdot & \cdot & \cdot & \cdots & a_m \end{vmatrix} \qquad (6.124)$$

The value of $a_j$ is to be taken as zero if its subscript $j$ is less than zero or greater than $m$ in Eq. (6.124). Then the following minors, indicated by the dashed lines in Eq. (6.124), are formulated:

$$H_1 = a_1 \qquad (6.125)$$

$$H_2 = \begin{vmatrix} a_1 & a_3 \\ a_0 & a_2 \end{vmatrix} \qquad (6.126)$$

$$H_3 = \begin{vmatrix} a_1 & a_3 & a_5 \\ a_0 & a_2 & a_4 \\ 0 & a_1 & a_3 \end{vmatrix} \qquad (6.127)$$

$$\vdots$$

Then the criterion for the dynamic stability of the system is that $a_0$, as well as all the determinants $H_1, H_2, H_3, \ldots, H_m$, must be positive. This will ensure that all the roots of Eq. (6.122) have negative real parts.

## 6.15 COMPUTER PROGRAMS

### 6.15.1 Generating the Characteristic Polynomial from the Matrix

A Fortran program, in the form of subroutine POLCOF, is given for expanding a determinantal equation to a polynomial form. Thus the equation

$$|[A] - x[I]| = \left| \begin{bmatrix} a_{11} & a_{12} & \cdots & a_{1n} \\ a_{21} & a_{22} & \cdots & a_{2n} \\ \vdots & & & \\ a_{n1} & a_{n2} & \cdots & a_{nn} \end{bmatrix} - x \begin{bmatrix} 1 & 0 & \cdots & 0 \\ 0 & 1 & \cdots & 0 \\ \vdots & & & \\ 0 & 0 & \cdots & 1 \end{bmatrix} \right| = 0$$

is expanded:

$$c_{n+1}x^n + c_n x^{n-1} + \cdots + c_2 x + c_1 = 0$$

The arguments of the subroutine are as follows:

A = Array of dimension $N \times N$. Contains the matrix $[A]$. Input data.

B, C = Arrays of dimension $N \times N$ each.

P, S = Arrays of dimension $N$ each.

PCF = Array of dimension $NP$. Contains the polynomial coefficients in the order $c_1, c_2, \ldots, c_{n+1}$. Thus the coefficient of the highest order term is stored as the last number. Output.

N = Order of the matrix $[A]$. Input data.

NP = Number of polynomial coefficients $= N + 1$. Input data.

A main program for calling subroutine POLCOF is written with the data:

$$N = 3, \, NP = 4, [A] = \begin{bmatrix} 2 & -1 & 0 \\ -1 & 2 & -1 \\ 0 & -1 & 2 \end{bmatrix}$$

The program listing and the output are given below.

```
C ===============================================================
C
C PROGRAM 9
C MAIN PROGRAM WHICH CALLS POLCOF
C
C ===============================================================
C FOLLOWING 4 LINES CONTAIN PROBLEM-DEPENDENT DATA
      DIMENSION A(3,3),B(3,3),C(3,3),P(3),S(3),PCF(4)
      N=3
      NP=4
      DATA A/2.0,-1.0,0.0,-1.0,2.0,-1.0,0.0,-1.0,2.0/
```

```
C END OF PROBLEM-DEPENDENT DATA
      CALL POLCOF (A,B,C,P,S,PCF,N,NP)
      PRINT 10
   10 FORMAT (//,49H POLYNOMIAL EXPANSION OF A DETERMINANTAL EQUATION)
      PRINT 20
   20 FORMAT (//,21H DATA: DETERMINANT A:,/)
      DO 30 I=1,N
   30 PRINT 40,(A(I,J),J=1,N)
   40 FORMAT (4E15.6)
      PRINT 50
   50 FORMAT (/,35H RESULT: POLYNOMIAL COEFFICIENTS IN,/,
     2    53H PCF(NP)*(X**N)+PCF(N)*(X**N-1)+...+PCF(2)*X+PCF(1)=0,/)
      PRINT 60,(PCF(I),I=1,NP)
   60 FORMAT (4E15.6)
      STOP
      END
C ==============================================================
C
C SUBROUTINE POLCOF
C
C ==============================================================
      SUBROUTINE POLCOF (A,B,C,P,S,PCF,N,NP)
      DIMENSION A(N,N),B(N,N),C(N,N),P(N),S(N),PCF(NP)
      DO 10 I=1,N
      P(I)=0.0
      DO 10 J=1,N
   10 B(I,J)=0.0
      DO 20 J=1,N
   20 B(J,J)=1.0
      DO 60 K=1,N
      S(K)=0.0
      DO 30 I=1,N
      DO 30 J=1,N
   30 C(I,J)=B(I,J)
      CALL MATMUL (B,C,A,N,N,N)
      DO 40 J=1,N
   40 S(K)=S(K)+B(J,J)
      P(1)=-S(1)
      IF (K .EQ. 1) GO TO 60
      KM=K-1
      DO 50 I=1,KM
   50 P(K)=P(K)-(1.0/FLOAT(K))*(S(I)*P(K-I))
      P(K)=P(K)-S(K)/FLOAT(K)
   60 CONTINUE
      DO 70 I=1,N
   70 PCF(I)=P(NP-I)
      PCF(NP)=1.0
      RETURN
      END
C ==============================================================
C
C SUBROUTINE MATMUL
C
C ==============================================================
```

```
C MATRIX MULTIPLICATION SUBROUTINE:  A = B * C
C B(L,M) AND C(M,N) ARE INPUT MATRICES, A(L,N) IS OUTPUT MATRIX
      SUBROUTINE MATMUL (A,B,C,L,M,N)
      DIMENSION A(L,N),B(L,M),C(M,N)
      DO 10 I=1,L
      DO 10 J=1,N
      A(I,J)=0.0
      DO 10 K=1,M
   10 A(I,J)=A(I,J)+B(I,K)*C(K,J)
      RETURN
      END
```

POLYNOMIAL EXPANSION OF A DETERMINANTAL EQUATION


DATA: DETERMINANT A:

```
  0.200000E+01   -0.100000E+01    0.000000E+00
 -0.100000E+01    0.200000E+01   -0.100000E+01
  0.000000E+00   -0.100000E+01    0.200000E+01
```

RESULT: POLYNOMIAL COEFFICIENTS IN
PCF(NP)*(X**N)+PCF(N)*(X**N-1)+...+PCF(2)*X+PCF(1)=0

```
 -0.400000E+01    0.100000E+02   -0.600000E+01    0.100000E+01
```


## 6.15.2 Roots of an nth Order Polynomial Equation with Complex Coefficients

A Fortran program, in the form of subroutine CROOTS, is given to find the roots of the polynomial equation

$$a_1 x^n + a_2 x^{n-1} + \cdots + a_n x + a_{n+1} = 0$$

The program takes the polynomial coefficients $a_1, a_2, \ldots, a_n, a_{n+1}$ in complex form as input data and gives the roots in the complex form. If the coefficients of the polynomial are real numbers, they can always be given in complex form by assuming the imaginary parts to be zero. The arguments of subroutine CROOTS are as follows:

A   = Complex array of dimension *NPS* containing the complex polynomial coefficients in the order $a_1, a_2, \ldots, a_{n+1}$. Input data.

N   = Order of the polynomial $n$. Input data.

NPS = Number of polynomial coefficients = $N + 1$. Input data.

EPS = Convergence requirement. A small value such as $10^{-6}$ is to be used. Input data.

KMAX = Maximum number of iterations to be used in finding the roots. A number in the range 50 to 100 is to be used. Input data.

| X | = | Complex array of dimension $N$ containing the computed complex roots. Input data. |
|---|---|---|
| J | = | Number of roots actually computed. Output. |
| ID | = | A value of ID equal to zero indicates that the program could not find all the roots. Output. |

This program is used to compute the roots of the equation

$$x^3 - 6x^2 + 11x - 6 = 0$$

The main program for this problem, subroutine CROOTS, and the output of the program are given below.

```
C =========================================================================
C
C PROGRAM 10
C MAIN PROGRAM FOR CALLING CROOTS
C
C =========================================================================
C FOLLOWING 6 LINES CONTAIN PROBLEM-DEPENDENT DATA
C A(NP1) DENOTES THE COEFFICIENT OF (X**N) IN THE POLYNOMIAL
      COMPLEX A(4),X(3)
      DATA N,NP1,IMAX,EPS/3,4,50,1.0E-06/
      A(4)=(1.0,0.0)
      A(3)=(-6.0,0.0)
      A(2)=(11.0,0.0)
      A(1)=(-6.0,0.0)
C END OF PROBLEM-DEPENDENT DATA
      CALL CROOTS (A,N,NP1,EPS,IMAX,X,J,ID)
      PRINT 10
   10 FORMAT (//,28H COMPLEX ROOTS OF POLYNOMIAL,/)
      DO 20 I=1,N
   20 PRINT 30,I,X(I)
   30 FORMAT ((I5,2E15.6))
      STOP
      END
C =========================================================================
C
C SUBROUTINE CROOTS
C
C =========================================================================
      SUBROUTINE CROOTS (A,N,NP1,EPS,IMAX,X,J,ID)
      COMPLEX A(NP1),X(N)
      COMPLEX XOLD,XNEW,ZX0,ZX1,ZX2,ALN,ALO,HN,HO,DELTA,ZG,ZR,DEN
     2    ,AZ1,AZ2
      DO 10 J=1,N
   10 X(J)=(0.0,0.0)
      J=0
      IF (N .EQ. 1) GO TO 100
   20 J=J+1
      XOLD=(0.0,0.0)
      ZX0=A(J)-A(J+1)+A(J+2)
      ZX1=A(J)+A(J+1)+A(J+2)
      ZX2=A(J)
      ALN=(-0.5,0.0)
```

```
          HN=(-1.0,0.0)
          DO 70 K=1,IMAX
          ALO=ALN
          HO=HN
          DELTA=(1.0,0.0)+ALN
          ZG=ZX0*(ALO**2)-ZX1*(DELTA**2)+ZX2*(ALO+DELTA)
          ZR=CSQRT(ZG**2-4.0*ZX2*DELTA*ALO*(ZX0*ALO-ZX1*DELTA+ZX2))
          AZ3=REAL(ZR)
          AZ4=AIMAG(ZR)
          AZ1=CMPLX(AZ3,-AZ4)
          AZ2=ZG*AZ1
          ZZZSS=REAL (AZ2)
          IF (ZZZSS .LT. 0.0) GO TO 30
          DEN=ZG+ZR
          GO TO 40
   30     DEN=ZG-ZR
   40     ALN=-2.0*ZX2*DELTA
          ALN=ALN/DEN
          IF (REAL(ALN) .GT. 1.0E25 .OR. AIMAG(ALN) .GT. 1.0E25) ALN=
         2    (1.0,0.0)
   50     HN=ALN*HO
          XNEW=XOLD+HN
          IF (CABS((XNEW-XOLD)/XNEW) .LT. EPS) GO TO 80
          ZX0=ZX1
          ZX1=ZX2
          ZS=CABS(ZX2)
          ZX2=A(NP1)
          NJ1=N-J+1
          DO 60 II=1,NJ1
          I=N-II+1
   60     ZX2=ZX2*XNEW+A(I)
          IF (CABS(ZX2/ZS) .LT. 10.0) GO TO 70
          ALN=0.5*ALN
          GO TO 50
   70     XOLD=XNEW
          X(J)=XNEW
          ID=0
          RETURN
   80     X(J)=XNEW
          JP1=J+1
          NJ1=N-JP1+1
          DO 90 II=1,NJ1
          L=N-II+1
   90     A(L)=A(L+1)*X(J)+A(L)
          IF (JP1 .LT. N) GO TO 20
  100     X(N)=-A(N)/A(NP1)
          J=N
          RETURN
          END
```

COMPLEX ROOTS OF POLYNOMIAL

```
     1    0.100000E+01    0.157652E-13
     2    0.200000E+01   -0.315303E-13
     3    0.300000E+01    0.157652E-13
```

### 6.15.3  Modal Analysis of a Multidegree of Freedom System

Subroutine MODAL is given for the modal analysis of a multidegree of freedom system. The arguments of this subroutine are as follows:

| | | |
|---|---|---|
| XM | = | Array of size $N \times N$, in which the mass matrix $[m]$ is stored. Input data. |
| OM | = | Array of size $NVEC$, in which the natural frequencies are stored. Input data. |
| T | = | Array of size $NSTEP$, in which the times $t_1, t_2, \ldots, t_{NSTEP}$ are stored. |
| Z | = | Array of size $NVEC$, in which the modal damping ratios of various modes are stored. Input data. |
| X0 | = | Array of size $N$, in which the initial values $x_1(0), x_2(0), \ldots, x_n(0)$ are stored. Input data. |
| XD0 | = | Array of size $N$, in which the initial values $\dot{x}_1(0), \dot{x}_2(0), \ldots, \dot{x}_n(0)$ are stored. Input data. |
| Y0, YD0 | = | Arrays of size $NVEC$. |
| Q | = | Array of size $NVEC \times NSTEP$. |
| F | = | Array of size $N \times NSTEP$, in which the magnitudes of the forces applied to the different masses at times $t_1, t_2, \ldots, t_{NSTEP}$ are stored. Input data. |
| DELT | = | Time step used for calculation, $\Delta t = t_{i+1} - t_i$. Input data. |
| EV | = | Array of size $N \times NVEC$, in which the normal modes are stored columnwise. Input data. |
| EVT | = | Array of size $NVEC \times N$ = transpose of the matrix $EV$. |
| XMX | = | Array of size $N \times NVEC$. |
| XTMX | = | Array of size $NVEC \times NVEC$. |
| X | = | Array of size $N \times NSTEP$, in which the displacements of the masses $m_1, m_2, \ldots, m_n$ at various time stations $t_1, t_2, \ldots, t_{NSTEP}$ are stored. Output. |
| U, V | = | Arrays of size $NVEC \times NSTEP$. |
| NSTEP | = | Number of time stations or integration points $t_1, t_2, \ldots, t_{NSTEP}$. Input data. |
| N | = | Number of degrees of freedom of the system. Input data. |
| NVEC | = | Number of modes used in the analysis. Input data. |

To illustrate the use of the program, the solution of Example 6.10 is considered:

$$[m]\ddot{\vec{x}} + [c]\dot{\vec{x}} + [k]\vec{x} = \vec{f} \quad \text{with} \quad x(0) = x_0 \quad \text{and} \quad \dot{x}(0) = \dot{x}_0$$

Data:

$$[m] = \begin{bmatrix} 1 & 0 & 0 \\ 0 & 1 & 0 \\ 0 & 0 & 1 \end{bmatrix}, \qquad [EV] = \begin{bmatrix} 1.0000 & 1.0000 & 1.0000 \\ 1.8019 & 0.4450 & -1.2468 \\ 2.2470 & -0.8020 & 0.5544 \end{bmatrix}$$

$$n = 3, \qquad NVEC = 3, \qquad \zeta_i = 0.01 \qquad \text{for } i = 1, 2, 3$$

$$\omega_1 = 0.89008, \qquad \omega_2 = 1.4942, \qquad \omega_3 = 3.6050$$

$$X0 = \begin{Bmatrix} 0 \\ 0 \\ 0 \end{Bmatrix}, \qquad XD0 = \begin{Bmatrix} 0 \\ 0 \\ 0 \end{Bmatrix}, \qquad \vec{f} = \begin{Bmatrix} F_0 \\ F_0 \\ F_0 \end{Bmatrix} \cos \omega t, \qquad \omega = 3.5, \qquad F_0 = 2.0$$

$$NSTEP = 20, \qquad DELT = 0.1$$

The force array $F$ is generated in the main program that calls subroutine MODAL. The program listing and the output are given below.

```
C ===========================================================================
C
C PROGRAM 11
C MAIN PROGRAM FOR CALLING THE SUBROUTINE MODAL
C
C ===========================================================================
      DIMENSION XM(3,3),OM(3),Z(3),X0(3),XD0(3),Y0(3),YD0(3),EV(3,3),
     2    EVT(3,3),XMX(3,3),XTMX(3,3),T(20),F(3,20),X(3,20),U(3,20),
     3    V(3,20),Q(3,20)
      DATA N,NVEC,NSTEP,DELT/3,3,20,0.1/
      DATA XM/1.0,0.0,0.0,0.0,1.0,0.0,0.0,0.0,1.0/
      OMF=3.5
      DATA OM/0.89008,1.4942,3.6050/
      DATA Z/0.01,0.01,0.01/
      DATA X0/0.0,0.0,0.0/
      DATA XD0/0.0,0.0,0.0/
      DATA (EV(I,1),I=1,3)/1.0,1.8019,2.2470/
      DATA (EV(I,2),I=1,3)/1.0,0.4450,-0.8020/
      DATA (EV(I,3),I=1,3)/1.0,-1.2468,0.5544/
      DO 5 I=1,NSTEP
      TIME=REAL(I)*DELT
5     F(1,I)=2.0*COS(3.5*TIME)
      DO 10 I=1,20
      F(2,I)=F(1,I)
10    F(3,I)=F(1,I)
      DO 20 I=1,NVEC
      DO 20 J=1,N
20    EVT(I,J)=EV(J,I)
      CALL MODAL (XM,OM,OMF,T,Z,X0,XD0,Y0,YD0,Q,F,DELT,EV,EVT,XMX,
     2    XTMX,X,U,V,NSTEP,N,NVEC)
      PRINT 30
30    FORMAT (//,40H RESPONSE OF SYSTEM USING MODAL ANALYSIS,/)
      DO 40 I=1,N
40    PRINT 50,I,(X(I,J),J=1,NSTEP)
50    FORMAT (/,11H COORDINATE,I5,/,(1X,5E14.6))
      STOP
      END
```

```
C ================================================================
C
C SUBROUTINE MODAL
C
C ================================================================
      SUBROUTINE MODAL (XM,OM,OMF,T,Z,X0,XD0,Y0,YD0,Q,F,DELT,EV,EVT,
     2    XMX,XTMX,X,U,V,NSTEP,N,NVEC)
      DIMENSION XM(N,N),OM(NVEC),T(NSTEP),Z(NVEC),X0(N),XD0(N),
     2    Y0(NVEC),YD0(NVEC),Q(NVEC,NSTEP),F(N,NSTEP),EV(N,NVEC),
     3    EVT(NVEC,N),XMX(N,NVEC),XTMX(NVEC,NVEC),X(N,NSTEP),U(NVEC,
     4    NSTEP),V(NVEC,NSTEP)
      T(1)=DELT
      DO 10 I=2,NSTEP
  10  T(I)=T(I-1)+DELT
C NORMALIZATION OF MODAL MATRIX WITH RESPECT TO THE MASS MATRIX
      CALL MATMUL (XMX,XM,EV,N,N,NVEC)
      CALL MATMUL (XTMX,EVT,XMX,NVEC,N,NVEC)
      DO 30 I=1,NVEC
      DO 20 J=1,N
  20  EV(J,I)=EV(J,I)/SQRT(XTMX(I,I))
  30  CONTINUE
C CONVERTION OF INFORMATION TO NORMAL COORDINATES
      DO 40 I=1,NVEC
      Y0(I)=0.0
  40  YD0(I)=0.0
      DO 60 I=1,NVEC
      DO 50 J=1,N
      Y0(I)=Y0(I)+EV(J,I)*X0(J)
  50  YD0(I)=YD0(I)+EV(J,I)*XD0(J)
  60  CONTINUE
      DO 70 I=1,NVEC
      DO 70 J=1,N
  70  EVT(I,J)=EV(J,I)
      CALL MATMUL (Q,EVT,F,NVEC,N,NSTEP)
      DO 100 I=1,NVEC
      R=OMF/OM(I)
      PP=Y0(I)
      QQ=YD0(I)
      ZI=Z(I)
      OMEG=OM(I)
      OMD=OMEG*SQRT(1.0-ZI**2)
      DO 90 J=1,NSTEP
      IF (J .EQ. 1) GO TO 80
      PP=U(I,J-1)
      QQ=V(I,J-1)
  80  C1=EXP(-ZI*OMEG*DELT)
      C2=COS(OMD*DELT)
      C3=SIN(OMD*DELT)
      C4=(QQ+OMEG*ZI*PP)/OMD
      C5=OMEG*ZI/OMD
      C6=Q(I,J)/(OMEG**2)
      U(I,J)=C1*(PP*C2+C3*C4)+C6*(1.0-C1*(C2+C3*C5))
      V(I,J)=OMD*C1*(-PP*C3+C2*C4-C5*(PP*C2+C3*C4))+C6*OMD*C1*C3*
     2    (1.0+C5**2)
```

```
 90    CONTINUE
100    CONTINUE
C FINDING THE SOLUTION IN THE ORIGINAL COORDINATES
       CALL MATMUL (X,EV,U,N,NVEC,NSTEP)
       RETURN
       END
```

RESPONSE OF SYSTEM USING MODAL ANALYSIS

```
COORDINATE    1
   0.936322E-02   0.354378E-01   0.731183E-01   0.115538E+00   0.155010E+00
   0.184085E+00   0.196572E+00   0.188401E+00   0.158198E+00   0.107508E+00
   0.406290E-01  -0.359245E-01  -0.114247E+00  -0.186060E+00  -0.243782E+00
  -0.281509E+00  -0.295776E+00  -0.286009E+00  -0.254587E+00  -0.206525E+00

COORDINATE    2
   0.939528E-02   0.358739E-01   0.751277E-01   0.121321E+00   0.167711E+00
   0.207370E+00   0.233936E+00   0.242311E+00   0.229239E+00   0.193704E+00
   0.137094E+00   0.631125E-01  -0.225565E-01  -0.112803E+00  -0.199814E+00
  -0.275895E+00  -0.334293E+00  -0.369942E+00  -0.380052E+00  -0.364457E+00

COORDINATE    3
   0.937892E-02   0.356481E-01   0.740876E-01   0.118353E+00   0.161294E+00
   0.195874E+00   0.216076E+00   0.217673E+00   0.198772E+00   0.160053E+00
   0.104682E+00   0.379048E-01  -0.336272E-01  -0.102705E+00  -0.162484E+00
  -0.207353E+00  -0.233644E+00  -0.240091E+00  -0.227975E+00  -0.200947E+00
```

### 6.15.4   Solution of Simultaneous Linear Equations

Subroutine SIMUL is given for solving a system of $N$ linear equations of the form $[A]\vec{X} = \vec{B}$. The following arguments are used:

A      =   Array of size $N \times N$. It is used to store the matrix $[A]$ in the beginning (Input) and contains the inverse of the matrix $[A]$ upon return from the subroutine SIMUL (Output).

B      =   Array of size $N$. It is used to store the vector $\vec{B}$ in the beginning (Input) and contains the solution vector $\vec{X}$ upon return from subroutine SIMUL (Output).

N      =   Number of equations to be solved. Input data.

IND    =   Zero if only the inverse $[A]^{-1}$ is required and = any non-zero integer if $\vec{X}$ is needed. Input data.

LA,S   =   Arrays of size $N$ each.

LB     =   Array of size $N \times 2$.

The program listing and typical results are given below.

```
C ================================================================
C
C PROGRAM 12
C MAIN PROGRAM WHICH CALLS SIMUL
C
C ================================================================
C FOLLOWING 4 LINES CONTAIN PROBLEM-DEPEMDENT DATA
      DIMENSION A(3,3),B(3),LA(3),LB(3,2),S(3)
      DATA A/1.0,2.0,3.0,10.0,0.0,3.0,1.0,1.0,2.0/
      DATA B/7.0,0.0,14.0/
      DATA N,IND/3,1/
C END OF PROBLEM-DEPENDENT DATA
      PRINT 100
  100 FORMAT (//,42H SOLUTION OF SIMULTANEOUS LINEAR EQUATIONS)
      PRINT 200, ((A(I,J),J=1,N),I=1,N)
  200 FORMAT (//,2X,28H ORIGINAL COEFFICIENT MATRIX,//,3(E12.4,1X))
      PRINT 300, (B(I),I=1,N)
  300 FORMAT (//,2X,23H RIGHT HAND SIDE VECTOR,//,3(E12.4,1X))
      CALL SIMUL (A,B,N,IND,LA,LB,S)
      PRINT 400, ((A(I,J),J=1,N),I=1,N)
  400 FORMAT`(//,2X,30H INVERSE OF COEFFICIENT MATRIX,//,3(E12.4,1X))
      PRINT 500, (B(I),I=1,N)
  500 FORMAT (//,2X,16H SOLUTION VECTOR,//,3(E12.4,1X))
      STOP
      END
C ================================================================
C
C SUBROUTINE SIMUL
C
C ================================================================
      SUBROUTINE SIMUL (A,B,N,IND,LA,LB,S)
      DIMENSION A(N,N),B(N),LA(N),LB(N,2),S(N)
      DO 100 I=1,N
  100 LA(I)=0
      DO 250 K=1,N
      Z=0.0
      DO 150 I=1,N
      IF (LA(I) .EQ. 1) GO TO 150
      DO 140 J=1,N
      IF (LA(J)-1) 130,140,300
  130 IF (ABS(Z) .GE. ABS(A(I,J))) GO TO 140
      IA=I
      IB=J
      Z=A(I,J)
  140 CONTINUE
  150 CONTINUE
      LA(IB)=LA(IB)+1
      IF (IA .EQ. IB) GO TO 190
      DO 160 I=1,N
      Z=A(IA,I)
      A(IA,I)=A(IB,I)
  160 A(IB,I)=Z
      IF (IND .EQ. 0) GO TO 190
      Z=B(IA)
```

```
              B(IA)=B(IB)
              B(IB)=Z
    190   LB(K,1)=IA
              LB(K,2)=IB
              S(K)=A(IB,IB)
              A(IB,IB)=1.0
              DO 200 I=1,N
    200   A(IB,I)=A(IB,I)/S(K)
              IF (IND .EQ. 0) GO TO 220
              B(IB)=B(IB)/S(K)
    220   DO 250 I=1,N
              IF(I .EQ. IB) GO TO 250
              Z=A(I,IB)
              A(I,IB)=0.0
              DO 230 J=1,N
    230   A(I,J)=A(I,J)-A(IB,J)*Z
              IF (IND .EQ. 0) GO TO 250
              B(I)=B(I)-B(IB)*Z
    250   CONTINUE
              DO 270 I=1,N
              J=N-I+1
              IF (LB(J,1) .EQ. LB(J,2)) GO TO 270
              IA=LB(J,1)
              IB=LB(J,2)
              DO 260 K=1,N
              Z=A(K,IA)
              A(K,IA)=A(K,IB)
              A(K,IB)=Z
    260   CONTINUE
    270   CONTINUE
    300   RETURN
              END
```

SOLUTION OF SIMULTANEOUS LINEAR EQUATIONS

ORIGINAL COEFFICIENT MATRIX

```
0.1000E+01    0.1000E+02    0.1000E+01
0.2000E+01    0.0000E+00    0.1000E+01
0.3000E+01    0.3000E+01    0.2000E+01
```

RIGHT HAND SIDE VECTOR

```
0.7000E+01    0.0000E+00    0.1400E+02
```

INVERSE OF COEFFICIENT MATRIX

```
 0.4286E+00    0.2429E+01   -0.1429E+01
 0.1429E+00    0.1429E+00   -0.1429E+00
-0.8571E+00   -0.3857E+01    0.2857E+01
```

SOLUTION VECTOR

```
-0.1700E+02   -0.1000E+01    0.3400E+02
```

## REFERENCES

**6.1.** F. W. Beaufait, *Basic Concepts of Structural Analysis*, Prentice-Hall, Englewood Cliffs, N.J., 1977.

**6.2.** R. J. Roark and W. C. Young, *Formulas for Stress and Strain* (5th Ed.), McGraw-Hill, New York, 1975.

**6.3.** D. A. Wells, *Theory and Problems of Lagrangian Dynamics*, Schaum's Outline Series, McGraw-Hill, New York, 1967.

**6.4.** J. H. Wilkinson, *The Algebraic Eigenvalue Problem*, Clarendon Press, Oxford, 1965.

**6.5.** A. Ralston, *A First Course in Numerical Analysis*, McGraw-Hill, New York, 1965.

**6.6.** L. Meirovitch, *Analytical Methods in Vibrations*, Macmillan, New York, 1967.

**6.7.** J. W. Strutt, Lord Rayleigh, *The Theory of Sound*, Macmillan, London, 1877 (reprinted by Dover Publications, New York in 1945).

**6.8.** W. C. Hurty and M. F. Rubinstein, *Dynamics of Structures*, Prentice-Hall, Englewood Cliffs, N.J., 1964.

**6.9.** T. K. Caughey, "Classical normal modes in damped linear dynamic systems," *Journal of Applied Mechanics*, Vol. 27, 1960, pp. 269–271.

**6.10.** A. Avakian and D. E. Beskos, "Use of dynamic stiffness influence coefficients in vibrations of non-uniform beams," letter to the editor, *Journal of Sound and Vibration*, Vol. 47, 1976, pp. 292–295.

**6.11.** I. Gladwell and P. M. Hanson, "Some error bounds and numerical experiments in modal methods for dynamics of systems," *Earthquake Engineering and Structural Dynamics*, Vol. 12, 1984, pp. 9–36.

**6.12.** R. Bajan, A. R. Kukreti, and C. C. Feng, "Method for improving incomplete modal coupling," *Journal of Engineering Mechanics*, Vol. 109, 1983, pp. 937–949.

**6.13.** D. W. Nicholson and D. J. Inman, "Stable response of damped linear systems," *Shock and Vibration Digest*, Vol. 15, November 1983, pp. 19–25.

**6.14.** W. W. Walter and G. L. Anderson, "Stability of a system of three degrees of freedom subjected to a circulatory force," *Journal of Sound and Vibration*, Vol. 45, 1976, pp. 105–114.

**6.15.** P. C. Chen and W. W. Soroka, "Multidegree dynamic response of a system with statistical properties," *Journal of Sound and Vibration*, Vol. 37, 1974, pp. 547–556.

**6.16.** S. Mahalingam and R. E. D. Bishop, "The response of a system with repeated natural frequencies to force and displacement excitation," *Journal of Sound and Vibration*, Vol. 36, 1974, pp. 285–295.

**6.17.** R. J. Guyan, "Reduction of stiffness and mass matrices," *AIAA Journal*, Vol. 3, 1965, p. 380.

**6.18.** B. M. Irons, "Structural eigenvalue problems—Elimination of unwanted variables," *AIAA Journal*, Vol. 3, 1965, pp. 961–962.

## REVIEW QUESTIONS

**6.1.** Define the flexibility and stiffness influence coefficients. What is the relation between them?

**6.2.** Write the equations of motion of a multidegree of freedom system in matrix form using (1) the flexibility matrix and (2) the stiffness matrix.

**6.3.** Express the potential and kinetic energy of an $n$ degree of freedom system, using matrix notation.

**6.4.** What is a generalized mass matrix?

**6.5.** Why is the mass matrix $[m]$ always positive definite?

**6.6.** Is the stiffness matrix $[k]$ always positive definite? Why?

**6.7.** What is the difference between generalized coordinates and Cartesian coordinates?

**6.8.** State Lagrange's equations.

**6.9.** What is an eigenvalue problem?

**6.10.** What is a mode shape? How is it computed?

**6.11.** How many distinct natural frequencies can exist for an $n$ degree of freedom system?

**6.12.** What is a dynamical matrix? What is its use?

**6.13.** How is the frequency equation derived for a multidegree of freedom system?

**6.14.** What is meant by the orthogonality of normal modes? What are orthonormal modal vectors?

**6.15.** What is a basis in $n$-dimensional space?

**6.16.** What is the expansion theorem? What is its importance?

**6.17.** Explain the modal analysis procedure.

**6.18.** What is a rigid body mode? How is it determined?

**6.19.** What is a degenerate system?

**6.20.** How can we find the response of a multidegree of freedom system using the first few modes only?

**6.21.** Define Rayleigh's dissipation function.

**6.22.** Define these terms: proportional damping, modal damping ratio, and modal participation factor.

**6.23.** When do we get complex eigenvalues?

**6.24.** What is the reason for the occurrence of an irregular mode instead of a principal mode of vibration?

---

## PROBLEMS

The problem assignments are organized as follows:

| Problems | Section covered | Topic covered |
|---|---|---|
| 6.1–6.9 | 6.3 | Influence coefficients |
| 6.10 | 6.4 | Potential and kinetic energies |
| 6.11 | 6.5 | Generalized coordinates |
| 6.12–6.16 | 6.6 | Lagrange's equations |
| 6.17–6.18 | 6.8 | Eigenvalue problem |

**6.1.** Find the flexibility and stiffness influence coefficients of the torsional system shown in Fig. 6.11. Also write the equations of motion of the system.

**6.2.** Find the flexibility and stiffness influence coefficients of the system shown in Fig. 6.12.

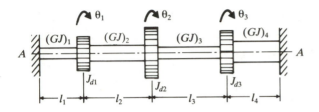

**Figure 6.11**

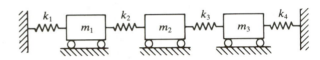

**Figure 6.12**

**6.3.** Derive the flexibility matrix and the equations of motion of a cantilever beam with concentrated masses (see Fig. 6.13) by idealizing it as a three degree of freedom system. Assume that all $A_i = A$, $(EI)_i = EI$, and $l_i = l$.

**6.4.** Determine the flexibility matrix of the uniform beam shown in Fig. 6.14. Disregard the mass of the beam compared to the concentrated masses placed on the beam and assume all $l_i = l$.

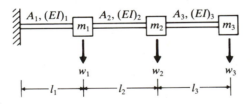

**Figure 6.13**

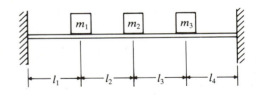

**Figure 6.14**

**6.5.** Derive the flexibility and stiffness matrices of the spring-mass system shown in Fig. 6.15, assuming that all the contacting surfaces are frictionless.

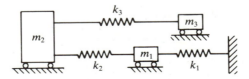

**Figure 6.15**

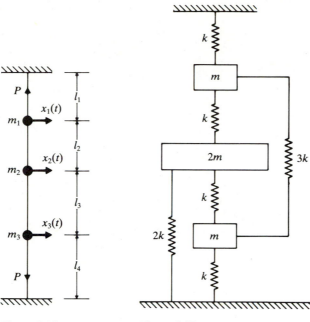

**Figure 6.16**       **Figure 6.17**

**6.6.** Derive the equations of motion for the tightly stretched string carrying three masses, as shown in Fig. 6.16. Assume the ends of the string to be fixed.

**6.7.** Derive the equations of motion of the system shown in Fig. 6.17.

**6.8.** Find the stiffness influence coefficients for the spring-mass system shown in Fig. 6.2(a).

**6.9.** Four identical springs, each having a stiffness $k$, are arranged symmetrically at 90° from each other, as shown in Fig. 2.33. Find the influence coefficient of the junction point in an arbitrary direction.

**6.10.** Show that the stiffness matrix of the spring-mass system shown in Fig. 6.1 is a band matrix along the diagonal.

**6.11.** Derive the constraint equation for the spherical pendulum shown in Fig. 6.18.

**6.12.** Derive the equations of motion of the system shown in Fig. 6.19 by using Lagrange's equations with $x$ and $\theta$ as generalized coordinates.

**6.13.** Derive the equations of motion of the system shown in Fig. 5.9(a), using Lagrange's equations with (1) $x_1$ and $x_2$ as generalized coordinates and (2) $x$ and $\theta$ as generalized coordinates.

**6.14.** Derive the equations of motion of the system shown in Fig. 6.12, using Lagrange's equations.

**6.15.** Derive the equations of motion of the triple pendulum shown in Fig. 6.4, using Lagrange's equations.

**6.16.** To study the symmetric vibrations of an airplane, as in Fig. 6.20(a), the fuselage can be idealized as a central mass $M_0$ and the wings can be modeled as rigid bars carrying end masses $M$, as shown in Fig. 6.20(b). The flexural behavior of the wings can be

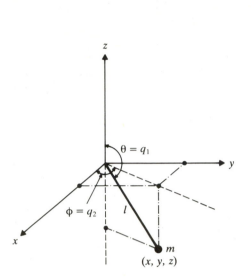

**Figure 6.18**

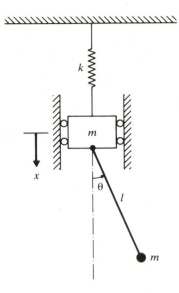

**Figure 6.19**

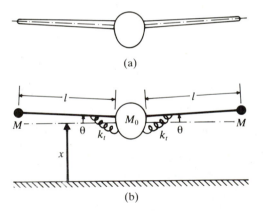

**Figure 6.20**

represented by torsional springs of stiffness $k_t$ between the fuselage and the wings. Derive the equations of motion of the airplane, using Lagrange's equations with $x$ and $\theta$ as generalized coordinates. Assume that $\theta$ is small.

**6.17.** Set up the eigenvalue problem of Example 6.5 in terms of the coordinates $q_1 = x_1$, $q_2 = x_2 - x_1$ and $q_3 = x_3 - x_2$, and solve the resulting problem. Compare the results obtained with those of Example 6.5 and draw conclusions.

**6.18.** Derive the frequency equation of the system shown in Fig. 6.12.

**6.19.** Find the natural frequencies and mode shapes of the system shown in Fig. 6.2 when $k_1 = k$, $k_2 = 2k$, $k_3 = 3k$, $m_1 = m$, $m_2 = 2m$, and $m_3 = 3m$. Plot the mode shapes.

**6.20.** Set up the matrix equation of motion and determine the three principal modes of vibration for the system shown in Fig. 6.2 with $k_1 = 3k$, $k_2 = k_3 = k$, $m_1 = 3m$, and $m_2 = m_3 = m$. Check the orthogonality of the modes found.

**6.21.** Find the natural frequencies of the system shown in Fig. 6.4 with $l_1 = 20$ cm, $l_2 = 30$ cm, $l_3 = 40$ cm, $m_1 = 1$ kg, $m_2 = 2$ kg, and $m_3 = 3$ kg.

**6.22.** Find the natural frequencies of the system shown in Fig. 6.14 with $m_1 = m_2 = m_3 = m$ and $l_1 = l_2 = l_3 = l_4 = l/4$.

**6.23.** The frequency equation of a three degree of freedom system is given by

$$\begin{vmatrix} \lambda - 5 & -3 & -2 \\ -3 & \lambda - 6 & -4 \\ -1 & -2 & \lambda - 6 \end{vmatrix} = 0$$

Find the roots of this equation.

**6.24.** Determine the eigenvalues and eigenvectors of the system shown in Fig. 6.12, taking $k_1 = k_2 = k_3 = k_4 = k$ and $m_1 = m_2 = m_3 = m$.

**6.25.** Find the natural frequencies and mode shapes of the system shown in Fig. 6.12 for $k_1 = k_2 = k_3 = k_4 = k$, $m_1 = 2m$, $m_2 = 3m$, and $m_3 = 2m$.

**6.26.** Find the natural frequencies and principal modes of the triple pendulum shown in Fig. 6.4, assuming that $l_1 = l_2 = l_3 = l$ and $m_1 = m_2 = m_3 = m$.

**6.27.** Find the natural frequencies and mode shapes of the system considered in Problem 6.5 with $m_1 = m$, $m_2 = 2m$, $m_3 = m$, $k_1 = k_2 = k$, and $k_3 = 2k$.

**6.28.** Show that the natural frequencies of the system shown in Fig. 6.2(a), with $k_1 = 3k$, $k_2 = k_3 = k$, $m_1 = 4m$, $m_2 = 2m$, and $m_3 = m$, are given by $\omega_1 = 0.46\sqrt{k/m}$, $\omega_2 = \sqrt{k/m}$, and $\omega_3 = 1.34\sqrt{k/m}$. Find the eigenvectors of the system.

**6.29.** Find the natural frequencies of the system considered in Problem 6.6 with $m_1 = 2m$, $m_2 = m$, $m_3 = 3m$, and $l_1 = l_2 = l_3 = l_4 = l$.

**6.30.** Find the natural frequencies and principal modes of the torsional system shown in Fig. 6.11 for $(GJ)_i = GJ$, $i = 1,2,3,4$, $J_{d1} = J_{d2} = J_{d3} = J_0$, and $l_1 = l_2 = l_3 = l_4 = l$.

**6.31.** The mass matrix $[m]$ and the stiffness matrix $[k]$ of a uniform bar are

$$[m] = \frac{\rho A l}{4}\begin{bmatrix} 1 & 0 & 0 \\ 0 & 2 & 0 \\ 0 & 0 & 1 \end{bmatrix} \quad \text{and} \quad [k] = \frac{2AE}{l}\begin{bmatrix} 1 & -1 & 0 \\ -1 & 2 & -1 \\ 0 & -1 & 1 \end{bmatrix}$$

where $\rho$ is the density, $A$ is the cross-sectional area, $E$ is Young's modulus, and $l$ is the length of the bar. Find the natural frequencies of the system by finding the roots of the characteristic equation. Also find the principal modes.

**6.32.** The mass matrix of a vibrating system is given by

$$[m] = \begin{bmatrix} 1 & 0 & 0 \\ 0 & 2 & 0 \\ 0 & 0 & 1 \end{bmatrix}$$

and the eigenvectors by

$$\begin{Bmatrix} 1 \\ -1 \\ 1 \end{Bmatrix}, \quad \begin{Bmatrix} 1 \\ 1 \\ 1 \end{Bmatrix} \quad \text{and} \quad \begin{Bmatrix} 0 \\ 1 \\ 2 \end{Bmatrix}$$

Find the $[m]$-orthonormal modal matrix of the system.

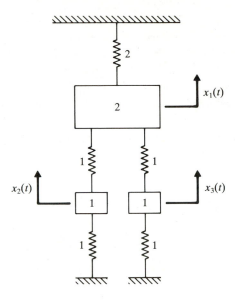

**Figure 6.21**

**6.33.** For the system shown in Fig. 6.21, (a) determine the characteristic polynomial $\Delta(\omega^2) = \det|[k] - \omega^2[m]|$, (b) plot $\Delta(\omega^2)$ from $\omega^2 = 0$ to $\omega^2 = 4.0$ (using increments $\Delta\omega^2 = 0.2$), and (c) find $\omega_1^2$, $\omega_2^2$, and $\omega_3^2$.

**6.34.** (a) Two of the eigenvectors of a vibrating system are known to be

$$\left\{ \begin{array}{c} 0.2754946 \\ 0.3994672 \\ 0.4490562 \end{array} \right\} \quad \text{and} \quad \left\{ \begin{array}{c} 0.6916979 \\ 0.2974301 \\ -0.3389320 \end{array} \right\}$$

Prove that these are orthogonal with respect to the mass matrix

$$[m] = \begin{bmatrix} 1 & 0 & 0 \\ 0 & 2 & 0 \\ 0 & 0 & 3 \end{bmatrix}$$

Find the remaining $[m]$-orthogonal eigenvector. (b) If the stiffness matrix of the system is given by

$$\begin{bmatrix} 6 & -4 & 0 \\ -4 & 10 & 0 \\ 0 & 0 & 6 \end{bmatrix}$$

determine all the natural frequencies of the system, using the eigenvectors of part (a).

**6.35.** Find the natural frequencies and mode shapes of the system shown in Fig. 6.8 with $m_1 = m$, $m_2 = 2m$, $m_3 = 3m$, and $k_1 = k_2 = k$.

**6.36.** Find the modal matrix for the semi-definite system shown in Fig. 6.22 for $J_1 = J_2 = J_3 = J_0$, $k_{t1} = k_t$, and $k_{t2} = 2k_t$.

**6.37.** Determine the amplitudes of motion of the three masses in Fig. 6.23 when a harmonic force $F(t) = F_0 \sin \omega t$ is applied to the lower left mass with $m = 1$ kg, $k = 1000$ N/m, $F_0 = 5$ N and $\omega = 10$ rad/sec using the mode superposition method.

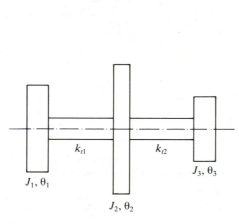

Figure 6.22

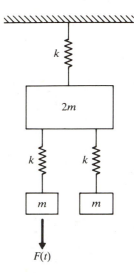

Figure 6.23

**6.38.** (a) Determine the natural frequencies and mode shapes of the three-disc system shown in Fig. 6.5 for $k_{t1} = k_{t2} = k_{t3} = k_t$ and $J_1 = J_2 = J_3 = J_0$. (b) If a torque $M_{t3}(t) = M_{t0}\cos \omega t$, with $M_{t0} = 500$ N-m and $\omega = 100$ rad/sec, acts on the disc $J_3$, find the amplitude of each disc. Assume $M_{t1} = M_{t2} = 0$, $k_{t1} = 100$ N-m/rad, and $J_0 = 1$ kg-m$^2$.

**6.39.** Using the results of Problems 6.2 and 6.24, determine the modal matrix $[X]$ of the system shown in Fig. 6.12 and derive the uncoupled equations of motion.

**6.40.** An approximate solution of a multidegree of freedom system can be obtained using the mode acceleration method. According to this method, the equations of motion of an undamped system, for example, are expressed as

$$\vec{x} = [k]^{-1}(\vec{F} - [m]\ddot{\vec{x}}) \qquad (E.1)$$

and $\ddot{\vec{x}}$ is approximated using the first $r$ modes $(r < n)$ as

$$\underset{n \times 1}{\ddot{\vec{x}}} = \underset{n \times r}{[X]} \underset{r \times 1}{\ddot{\vec{q}}} \qquad (E.2)$$

Since $([k] - \omega_i^2 [m]) \vec{X}^{(i)} = \vec{0}$, Eq. (E.1) can be written as

$$\vec{x}(t) = [k]^{-1}\vec{F}(t) - \sum_{i=1}^{r} \frac{1}{\omega_i^2} \vec{X}^{(i)} \ddot{q}_i(t) \qquad (E.3)$$

Find the approximate response of the system described in Example 6.10 (without damping), using the mode acceleration method with $r = 1$.

**6.41.** Determine the response of the system in Problem 6.19 to the initial conditions $x_1(0) = 1$, $\dot{x}_1(0) = 0$, $x_2(0) = 2$, $\dot{x}_2(0) = 1$, $x_3(0) = 1$, and $\dot{x}_3(0) = -1$. Assume $k/m = 1$.

**6.42.** Find the steady state response of the system shown in Fig. 6.10 with $k_1 = k_2 = k_3 = k_4 = 100$ N/m, $c_1 = c_2 = c_3 = c_4 = 1$ N-s/m, $m_1 = m_2 = m_3 = 1$ kg, $F_1(t) = F_0\cos \omega t$, $F_0 = 10$ N and $\omega = 1$ rad/sec. Assume that the spring $k_4$ and the damper $c_4$ are connected to a rigid wall at the right end. Use the mechanical impedance method described in Sec. 5.6 for solution.

**6.43.** A twelve-story building having equal values of floor masses and interfloor stiffnesses is modeled as a viscously damped multidegree of freedom system. The first three natural frequencies and normal modes of the undamped system are given below:

| Mode shape | Floor level | | | | | | | | | | | | |
|---|---|---|---|---|---|---|---|---|---|---|---|---|
| | **0** | **1** | **2** | **3** | **4** | **5** | **6** | **7** | **8** | **9** | **10** | **11** | **12** |
| $\vec{X}^{(1)}$ | 0.0 | 0.126 | 0.249 | 0.369 | 0.483 | 0.589 | 0.686 | 0.772 | 0.846 | 0.907 | 0.953 | 0.984 | 1.000 |
| $\vec{X}^{(2)}$ | 0.0 | −0.375 | −0.697 | −0.922 | −1.017 | −0.969 | −0.785 | −0.491 | −0.127 | 0.254 | 0.599 | 0.860 | 1.000 |
| $\vec{X}^{(3)}$ | 0.0 | 0.618 | 1.000 | 1.000 | 0.618 | 0.000 | −0.618 | −1.000 | −1.000 | −0.618 | 0.000 | 0.618 | 1.000 |

$\omega_1 = 225$ rad/sec, $\omega_2 = 660$ rad/sec, and $\omega_3 = 1100$ rad/sec. If the foundation of the building is subjected to a known horizontal displacement $x_0(t)$, derive the uncoupled equations for determining the response of the building by approximating it as a linear combination of the first three normal modes. Hint: The equation of motion of the building under ground motion is

$$[m]\ddot{\vec{x}} + [c]\dot{\vec{x}} + [k]\vec{x} = -[m]\vec{1}\ddot{x}_0(t)$$

where $\vec{1}$ is a unit vector.

**6.44.** Write a computer program for finding the eigenvectors using the known eigenvalues in Eq. (6.55). Find the mode shapes of Problem 6.25 using this program.

**6.45.** Write a computer program for generating the $[m]$-orthonormal modal matrix $[X]$. The program should accept the number of degrees of freedom, the normal modes, and the mass matrix as input. Solve Problem 6.32 using this program.

**6.46.** Generate the characteristic polynomial of Problem 6.23 using subroutine POLCOF.

**6.47.** Find the characteristic values of Problem 6.46 using subroutine CROOTS.

**6.48.** Write a computer program for finding the natural frequencies and mode shapes of a multidegree of freedom system when the mass and stiffness matrices are known, using the following steps:
1. Find the dynamical matrix using subroutines DECOMP and MATMUL.
2. Find the characteristic polynomial using subroutine POLCOF.
3. Find the natural frequencies using subroutine CROOTS.
4. Find the mode shapes using subroutine SIMUL.

Solve Problem 6.21 using this program.

**6.49.** The equations of motion of an undamped system in SI units are given by

$$\begin{bmatrix} 2 & 0 & 0 \\ 0 & 2 & 0 \\ 0 & 0 & 2 \end{bmatrix}\ddot{\vec{x}} + \begin{bmatrix} 16 & -8 & 0 \\ -8 & 16 & -8 \\ 0 & -8 & 16 \end{bmatrix}\vec{x} = \begin{Bmatrix} 10\sin\omega t \\ 0 \\ 0 \end{Bmatrix}$$

Find the steady-state response of the system when $\omega = 5$ rad/s, using subroutine MODAL.

**6.50.** Find the response of the system in Problem 6.49 by varying $\omega$ between 1 and 10 rad/sec in increments of 1 rad/sec. Plot the graphs showing the variations of magnitudes of the first peaks of $x_i(t)$, $i = 1, 2, 3$ with respect to $\omega$.

# CHAPTER 7

# Determination of Natural Frequencies and Mode Shapes

John William Strutt, Lord Rayleigh (1842–1919), was an English physicist who held the positions of professor of experimental physics at Cambridge University, professor of natural philosophy at the Royal Institution in London, president of the Royal Society, and chancellor of Cambridge University. His works in optics and acoustics are well known, with *Theory of Sound* (1877) considered as a standard reference even today. The method of computing approximate natural frequencies of vibrating bodies using an energy approach has become known as "Rayleigh's method."

Courtesy Brown Brothers

## 7.1  INTRODUCTION

In the preceding chapter, the natural frequencies (eigenvalues) and the natural modes (eigenvectors) of a multidegree of freedom system were found by setting the characteristic determinant equal to zero. Although this is an exact method, the expansion of the characteristic determinant and the solution of the resulting $n$th degree polynomial equation to obtain the natural frequencies can become quite tedious for large values of $n$. Several analytical and numerical methods have been developed to compute the natural frequencies and mode shapes of multidegree of freedom systems. In this chapter, we shall consider Dunkerley's formula, Rayleigh's method, Holzer's method, the matrix iteration method, and Jacobi's method. Dunkerley's formula and Rayleigh's method are useful for estimating the fundamental natural frequency only. Holzer's method is essentially a tabular method that can be used to find partial or full solutions to eigenvalue problems. The matrix iteration method finds one natural frequency at a time, usually starting from the lowest value. The method can thus be terminated after finding the required number of natural frequencies and mode shapes. When all the natural frequencies and mode shapes are required, Jacobi's method can be used; it finds all the eigenvalues and eigenvectors simultaneously.

## 7.2  DUNKERLEY'S FORMULA

Dunkerley's formula gives the approximate value of the fundamental frequency of a composite system in terms of the natural frequencies of its component parts. It is derived by making use of the fact that the higher natural frequencies of most vibratory systems are large compared to their fundamental frequencies [7.1–7.3]. To derive Dunkerley's formula, consider the three degree of freedom system shown in Fig. 6.2. The equations of motion of this system can be derived in terms of the flexibility influence coefficients found in Example 6.1. By assuming harmonic motion during free vibration, we can derive the frequency equation:

$$\left| [a][m] - \frac{1}{\omega^2}[I] \right| = \begin{vmatrix} \left( a_{11}m_1 - \dfrac{1}{\omega^2} \right) & a_{12}m_2 & a_{13}m_3 \\ a_{21}m_1 & \left( a_{22}m_2 - \dfrac{1}{\omega^2} \right) & a_{23}m_3 \\ a_{31}m_1 & a_{32}m_2 & \left( a_{33}m_3 - \dfrac{1}{\omega^2} \right) \end{vmatrix} = 0$$

$$(7.1)$$

By expanding the determinantal equation, we obtain

$$\left(\frac{1}{\omega^2}\right)^3 - (a_{11}m_1 + a_{22}m_2 + a_{33}m_3)\left(\frac{1}{\omega^2}\right)^2$$

$$+ (a_{11}a_{22}m_1m_2 + a_{22}a_{33}m_2m_3 + a_{33}a_{11}m_3m_1 - a_{12}a_{21}m_1m_2$$

$$- a_{23}a_{32}m_2m_3 - a_{31}a_{13}m_3m_1)\left(\frac{1}{\omega^2}\right) - m_1m_2m_3$$

$$\times (a_{11}a_{22}a_{33} + a_{12}a_{23}a_{31} + a_{21}a_{13}a_{32} - a_{11}a_{23}a_{32} - a_{22}a_{13}a_{31}$$

$$- a_{33}a_{12}a_{21}) = 0 \tag{7.2}$$

This is a polynomial equation of third degree in $(1/\omega^2)$. Let the roots of Eq. (7.2) be denoted as $1/\omega_1^2$, $1/\omega_2^2$, and $1/\omega_3^2$. From algebra we know that the sum of the roots of Eq. (7.2) must be equal to the negative of the coefficient of its second term:

$$\frac{1}{\omega_1^2} + \frac{1}{\omega_2^2} + \frac{1}{\omega_3^2} = a_{11}m_1 + a_{22}m_2 + a_{33}m_3 \tag{7.3}$$

This equation can be generalized to an $n$ degree of freedom system:

$$\frac{1}{\omega_1^2} + \frac{1}{\omega_2^2} + \cdots + \frac{1}{\omega_n^2} = a_{11}m_1 + a_{22}m_2 + \cdots + a_{nn}m_n \tag{7.4}$$

In most cases, the higher frequencies $\omega_2, \omega_3, \ldots, \omega_n$ are considerably larger than the fundamental frequency $\omega_1$, and so

$$\frac{1}{\omega_i^2} \ll \frac{1}{\omega_1^2}, \qquad i = 2, 3, \ldots, n \tag{7.5}$$

In view of Eq. (7.5), Eq. (7.4) can be approximately written as

$$\frac{1}{\omega_1^2} \simeq a_{11}m_1 + a_{22}m_2 + \cdots + a_{nn}m_n \tag{7.6}$$

This equation is known as *Dunkerley's formula*. The fundamental frequency given by Eq. (7.6) will always be smaller than the exact value. In some cases, it will be more convenient to rewrite Eq. (7.6) as

$$\frac{1}{\omega_1^2} \simeq \frac{1}{\omega_{1n}^2} + \frac{1}{\omega_{2n}^2} + \cdots + \frac{1}{\omega_{nn}^2} \tag{7.7}$$

where $\omega_{in} = (k_{ii}/m_i)^{1/2}$ denotes the natural frequency of a single degree of freedom system consisting of mass $m_i$ and spring of stiffness $k_{ii}$, $i = 1, 2, \ldots, n$. The use of Dunkerley's formula for finding the lowest frequency of elastic systems is presented in Refs. [7.4, 7.5].

---

**EXAMPLE 7.1**

Estimate the fundamental natural frequency of a simply supported beam carrying three identical equally spaced masses, as shown in Fig. 6.3(a).

*Solution.* The flexibility influence coefficients (see Example 6.2) required for the application of Dunkerley's formula are given by

$$a_{11} = a_{33} = \frac{3}{256} \frac{l^3}{EI}, \qquad a_{22} = \frac{1}{48} \frac{l^3}{EI} \tag{E.1}$$

Equation (7.6) thus gives, using $m_1 = m_2 = m_3 = m$,

$$\frac{1}{\omega_1^2} \simeq \left( \frac{3}{256} + \frac{1}{48} + \frac{3}{256} \right) \frac{ml^3}{EI} = 0.04427 \frac{ml^3}{EI}$$

$$\omega_1 \simeq 4.75375 \sqrt{\frac{EI}{ml^3}}$$

This value can be compared with the exact value of the fundamental frequency

$$4.934 \sqrt{\frac{EI}{ml^3}}$$

---

## 7.3 RAYLEIGH'S METHOD

Rayleigh's method was presented in Sec. 2.5 to find the natural frequencies of single degree of freedom systems. The method can be extended to find the approximate value of the fundamental natural frequency of a discrete system.* The method is based on *Rayleigh's principle*, which can be stated as follows [7.6]:

> *The frequency of vibration of a conservative system vibrating about an equilibrium position has a stationary value in the neighborhood of a natural mode. This stationary value, in fact, is a minimum value in the neighborhood of the fundamental natural mode.*

We shall now derive an expression for the approximate value of the first natural frequency of a multidegree of freedom system according to Rayleigh's method.

The kinetic and potential energies of an *n* degree of freedom discrete system can be expressed as

$$T = \frac{1}{2} \dot{\vec{x}}^T [m] \dot{\vec{x}} \tag{7.8}$$

$$V = \frac{1}{2} \vec{x}^T [k] \vec{x} \tag{7.9}$$

---

*Rayleigh's method for continuous systems is presented in Sec. 8.7.

To find the natural frequencies, we assume harmonic motion to be

$$\vec{x} = \vec{X} \cos \omega t \tag{7.10}$$

where $\vec{X}$ denotes the vector of amplitudes (mode shape) and $\omega$ represents the natural frequency of vibration. If the system is conservative, the maximum kinetic energy is equal to the maximum potential energy:

$$T_{max} = V_{max} \tag{7.11}$$

By substituting Eq. (7.10) into Eqs. (7.8) and (7.9), we find

$$T_{max} = \frac{1}{2} \vec{X}^T [m] \vec{X} \omega^2 \tag{7.12}$$

$$V_{max} = \frac{1}{2} \vec{X}^T [k] \vec{X} \tag{7.13}$$

By equating $T_{max}$ and $V_{max}$, we obtain*

$$\omega^2 = \frac{\vec{X}^T [k] \vec{X}}{\vec{X}^T [m] \vec{X}} \tag{7.14}$$

The right-hand side of Eq. (7.14) is known as *Rayleigh's quotient* and is denoted as $R(\vec{X})$.

### 7.3.1   Properties of Rayleigh's Quotient

As stated earlier, $R(\vec{X})$ has a stationary value when the arbitrary vector $\vec{X}$ is in the neighborhood of any eigenvector $\vec{X}^{(r)}$. To prove this, we express the arbitrary vector $\vec{X}$ in terms of the normal modes of the system, $\vec{X}^{(i)}$, as

$$\vec{X} = c_1 \vec{X}^{(1)} + c_2 \vec{X}^{(2)} + c_3 \vec{X}^{(3)} + \cdots \tag{7.15}$$

Then

$$\vec{X}^T [k] \vec{X} = c_1^2 \vec{X}^{(1)^T} [k] \vec{X}^{(1)} + c_2^2 \vec{X}^{(2)^T} [k] \vec{X}^{(2)} + c_3^2 \vec{X}^{(3)^T} [k] \vec{X}^{(3)} + \cdots \tag{7.16}$$

and

$$\vec{X}^T [m] \vec{X} = c_1^2 \vec{X}^{(1)^T} [m] \vec{X}^{(1)} + c_2^2 \vec{X}^{(2)^T} [m] \vec{X}^{(2)} + c_3^2 \vec{X}^{(3)^T} [m] \vec{X}^{(3)} + \cdots \tag{7.17}$$

as the cross terms of the form $c_i c_j \vec{X}^{(i)^T} [k] \vec{X}^{(j)}$ and $c_i c_j \vec{X}^{(i)^T} [m] \vec{X}^{(j)}$, $i \neq j$, are zero by the orthogonality property. Using Eqs. (7.16) and (7.17) and the relation

$$\vec{X}^{(i)^T} [k] \vec{X}^{(i)} = \omega_i^2 \vec{X}^{(i)^T} [m] \vec{X}^{(i)} \tag{7.18}$$

the Rayleigh's quotient of Eq. (7.14) can be expressed as

$$\omega^2 = R(\vec{X}) = \frac{c_1^2 \omega_1^2 \vec{X}^{(1)^T} [m] \vec{X}^{(1)} + c_2^2 \omega_2^2 \vec{X}^{(2)^T} [m] \vec{X}^{(2)} + \cdots}{c_1^2 \vec{X}^{(1)^T} [m] \vec{X}^{(1)} + c_2^2 \vec{X}^{(2)^T} [m] \vec{X}^{(2)} + \cdots} \tag{7.19}$$

---

*Equation (7.14) can also be obtained from the relation $[k] \vec{X} = \omega^2 [m] \vec{X}$. Premultiplying this equation by $\vec{X}^T$ and solving the resulting equation gives Eq. (7.14).

If the normal modes are normalized, this equation becomes

$$\omega^2 = R(\vec{X}) = \frac{c_1^2\omega_1^2 + c_2^2\omega_2^2 + \cdots}{c_1^2 + c_2^2 + \cdots} \tag{7.20}$$

If $\vec{X}$ differs little from the eigenvector $\vec{X}^{(r)}$, the coefficient $c_r$ will be much larger than the remaining coefficients $c_i$ $(i \neq r)$ and Eq. (7.20) can be written as

$$R(\vec{X}) = \frac{c_r^2\omega_r^2 + c_r^2 \sum_{\substack{i=1,2,\ldots \\ i \neq r}} \left(\frac{c_i}{c_r}\right)^2 \omega_i^2}{c_r^2 + c_r^2 \sum_{\substack{i=1,2,\ldots \\ i \neq r}} \left(\frac{c_i}{c_r}\right)^2} \tag{7.21}$$

Since $|c_i/c_r| = \varepsilon_i \ll 1$ where $\varepsilon_i$ is a small number for all $i \neq r$, Eq. (7.21) gives

$$R(\vec{X}) = \omega_r^2 \{1 + O(\varepsilon^2)\} \tag{7.22}$$

where $O(\varepsilon^2)$ represents an expression in $\varepsilon$ of the second order or higher. Equation (7.22) indicates that if the arbitrary vector $\vec{X}$ differs from the eigenvector $\vec{X}^{(r)}$ by a small quantity of the first order, $R(\vec{X})$ differs from the eigenvalue $\omega_r^2$ by a small quantity of the second order. This means that Rayleigh's quotient has a stationary value in the neighborhood of an eigenvector.

The stationary value is actually a minimum value in the neighborhood of the fundamental mode, $\vec{X}^{(1)}$. To see this, let $r = 1$ in Eq. (7.21) and write

$$R(\vec{X}) = \frac{\omega_1^2 + \sum_{i=2,3,\ldots} \left(\frac{c_i}{c_1}\right)^2 \omega_i^2}{\left\{1 + \sum_{i=2,3,\ldots} \left(\frac{c_i}{c_1}\right)^2\right\}}$$

$$\simeq \omega_1^2 + \sum_{i=2,3,\ldots} \varepsilon_i^2\omega_i^2 - \omega_1^2 \sum_{i=2,3,\ldots} \varepsilon_i^2$$

$$\simeq \omega_1^2 + \sum_{i=2,3,\ldots} \left(\omega_i^2 - \omega_1^2\right)\varepsilon_i^2 \tag{7.23}$$

Since, in general, $\omega_i^2 > \omega_1^2$ for $i = 2,3,\ldots$, Eq. (7.23) leads to

$$R(\vec{X}) \geq \omega_1^2 \tag{7.24}$$

which shows that Rayleigh's quotient is never lower than the first eigenvalue. By proceeding in a similar manner, we can show that

$$R(\vec{X}) \leq \omega_n^2 \tag{7.25}$$

which means that Rayleigh's quotient is never higher than the highest eigenvalue. Thus Rayleigh's quotient provides an upper bound for $\omega_1^2$ and a lower bound for $\omega_n^2$.

### 7.3.2 Computation of the Fundamental Natural Frequency

Equation (7.14) can be used to find an approximate value of the first natural frequency ($\omega_1$) of the system. For this, we select a trial vector $\vec{X}$ to represent the first natural mode $\vec{X}^{(1)}$ and substitute it on the right-hand side of Eq. (7.14). This yields the approximate value of $\omega_1^2$. Because Rayleigh's quotient is stationary, remarkably good estimates of $\omega_1^2$ can be obtained even if the trial vector $\vec{X}$ deviates greatly from the true natural mode $\vec{X}^{(1)}$. Obviously, the estimated value of the fundamental frequency $\omega_1$ is more accurate if the trial vector ($\vec{X}$) chosen resembles the true natural mode $\vec{X}^{(1)}$ closely. Rayleigh's method is compared with Dunkerley's and other methods in Refs. [7.7–7.9].

---

### EXAMPLE 7.2

Estimate the fundamental frequency of vibration of the system shown in Fig. 7.1. Assume that $m_1 = m_2 = m_3 = m$, $k_1 = k_2 = k_3 = k$, and the mode shape is

$$\vec{X} = \begin{Bmatrix} 1 \\ 2 \\ 3 \end{Bmatrix}$$

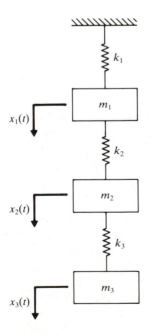

**Figure 7.1**

*Solution.* The stiffness and mass matrices of the system are:

$$[k] = k \begin{bmatrix} 2 & -1 & 0 \\ -1 & 2 & -1 \\ 0 & -1 & 1 \end{bmatrix} \tag{E.1}$$

$$[m] = m \begin{bmatrix} 1 & 0 & 0 \\ 0 & 1 & 0 \\ 0 & 0 & 1 \end{bmatrix} \tag{E.2}$$

By substituting the assumed mode shape in the expression for Rayleigh's quotient, we obtain

$$R(\vec{X}) = \omega^2 = \frac{\vec{X}^T [k] \vec{X}}{\vec{X}^T [m] \vec{X}} = \frac{(1 \quad 2 \quad 3)k \begin{bmatrix} 2 & -1 & 0 \\ -1 & 2 & -1 \\ 0 & -1 & 1 \end{bmatrix} \begin{Bmatrix} 1 \\ 2 \\ 3 \end{Bmatrix}}{(1 \quad 2 \quad 3)m \begin{bmatrix} 1 & 0 & 0 \\ 0 & 1 & 0 \\ 0 & 0 & 1 \end{bmatrix} \begin{Bmatrix} 1 \\ 2 \\ 3 \end{Bmatrix}}$$

$$= 0.2143 \frac{k}{m} \tag{E.3}$$

$$\omega_1 = 0.4629 \sqrt{\frac{k}{m}} \tag{E.4}$$

This value is 4.0225% larger than the exact value of $0.4450\sqrt{k/m}$. The exact fundamental mode shape (see Example 6.5) in this case is

$$\vec{X}^{(1)} = \begin{Bmatrix} 1.0000 \\ 1.8019 \\ 2.2470 \end{Bmatrix} \tag{E.5}$$

### 7.3.3    Fundamental Frequency of Beams and Shafts

Although the procedure outlined above is applicable to all discrete systems, a simpler equation can be derived for the fundamental frequency of the lateral vibration of a beam or a shaft carrying several masses such as pulleys, gears, or flywheels. In these cases, the static deflection curve is used as an approximation of the dynamic deflection curve.

Consider a shaft carrying several masses, as shown in Fig. 7.2. The shaft is assumed to have negligible mass. The potential energy of the system is the strain energy of the deflected shaft, which is equal to the work done by the static loads. Thus

$$V_{\max} = \frac{1}{2} (m_1 g w_1 + m_2 g w_2 + \cdots) \tag{7.26}$$

where $m_i g$ is the static load due to the mass $m_i$, and $w_i$ is the total static deflection of mass $m_i$ due to all the masses. For harmonic oscillation (free vibration), the

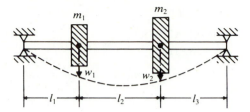

**Figure 7.2**

maximum kinetic energy due to the masses is

$$T_{max} = \frac{\omega^2}{2}\left(m_1 w_1^2 + m_2 w_2^2 + \cdots\right) \tag{7.27}$$

where $\omega$ is the frequency of oscillation. Equating $V_{max}$ and $T_{max}$, we obtain

$$\omega = \left\{\frac{g(m_1 w_1 + m_2 w_2 + \cdots)}{(m_1 w_1^2 + m_2 w_2^2 + \cdots)}\right\}^{1/2} \tag{7.28}$$

## EXAMPLE 7.3

Estimate the fundamental frequency of lateral vibration of the system shown in Fig. 7.2 for $m_1 = 20$ kg, $m_2 = 50$ kg, $l_1 = 1$ m, $l_2 = 3$ m, and $l_3 = 2$ m.

**Solution.** From strength of materials, the deflection of the beam shown in Fig. 7.3 due to a static load $P$ [7.10] is given by

$$w(x) = \begin{cases} \dfrac{Pbx}{6EIl}(l^2 - b^2 - x^2); & 0 \le x \le a \tag{E.1} \\[2ex] -\dfrac{Pa(l-x)}{6EIl}[a^2 + x^2 - 2lx]; & a \le x \le l \tag{E.2} \end{cases}$$

The deflection of mass $m_1$ due to the static load $m_1 g$ can be obtained from Eq. (E.1)

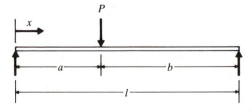

**Figure 7.3**

with $x = 1$ m, $b = 5$ m, and $l = 6$ m as

$$w_1' = \frac{(20 \times 9.81)(5)(1)}{6EI(6)}(36 - 25 - 1) = \frac{272.5}{EI}$$

The deflection of mass $m_2$ due to the load $m_1 g$ is given by Eq. (E.2) with $a = 1$ m, $x = 4$ m, and $l = 6$ m:

$$w_2' = -\frac{(20 \times 9.81)(1)(2)}{6EI(6)}(1 + 16 - 2 \times 6 \times 4) = \frac{337.9}{EI}$$

The deflection of $m_1$ due to the load $m_2 g$ can be found from Eq. (E.1) using $b = 2$ m, $x = 1$ m, and $l = 6$ m:

$$w_1'' = \frac{(50 \times 9.81)(2)(1)}{6EI(6)}(36 - 4 - 1) = \frac{844.75}{EI}$$

The deflection of $m_2$ due to the load $m_2 g$ is given by Eq. (E.1) with $b = 2$ m, $x = 4$ m, and $l = 6$ m:

$$w_2'' = \frac{(50 \times 9.81)(2)(4)}{6EI(6)}(36 - 4 - 16) = \frac{1744.0}{EI}$$

The total deflections of masses $m_1$ and $m_2$ are

$$w_1 = w_1' + w_1'' = \frac{1117.25}{EI} \qquad \text{and} \qquad w_2 = w_2' + w_2'' = \frac{2081.90}{EI}$$

Substituting into Eqs. (7.28), we find the fundamental natural frequency:

$$\omega = \left\{ \frac{9.81(20 \times 1117.25 + 50 \times 2081.90)EI}{(20 \times 1117.25^2 + 50 \times 2081.90^2)} \right\}^{1/2} = 0.07164\sqrt{EI} \text{ rad/sec.}$$

## 7.4  HOLZER'S METHOD

Holzer's method is basically a trial-and-error scheme to find the natural frequencies of undamped, damped, semi-definite, fixed, or branched vibrating systems involving linear and angular displacements [7.11, 7.12]. The method can also be programmed for computer applications. A trial frequency of the system is first assumed, and a solution is found when the assumed frequency satisfies the constraints of the system. This generally requires several trials. Depending on the trial frequency used, the fundamental as well as the higher frequencies of the system can be determined. The method also gives the mode shapes. We shall illustrate the method for the undamped semi-definite system shown in Fig. 7.4. The equations of motion of the discs can be derived as follows:

$$J_1 \ddot{\theta}_1 + k_{t1}(\theta_1 - \theta_2) = 0 \tag{7.29}$$

$$J_2 \ddot{\theta}_2 + k_{t1}(\theta_2 - \theta_1) + k_{t2}(\theta_2 - \theta_3) = 0 \tag{7.30}$$

$$J_3 \ddot{\theta}_3 + k_{t2}(\theta_3 - \theta_2) = 0 \tag{7.31}$$

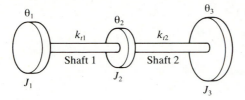

**Figure 7.4**

Since the motion is harmonic at a natural mode of vibration, we assume that $\theta_i = \Theta_i \cos(\omega t + \phi)$ in Eqs. (7.29) to (7.31) and obtain

$$\omega^2 J_1 \Theta_1 = k_{t1}(\Theta_1 - \Theta_2) \tag{7.32}$$

$$\omega^2 J_2 \Theta_2 = k_{t1}(\Theta_2 - \Theta_1) + k_{t2}(\Theta_2 - \Theta_3) \tag{7.33}$$

$$\omega^2 J_3 \Theta_3 = k_{t2}(\Theta_3 - \Theta_2) \tag{7.34}$$

Summing these equations gives

$$\sum_{i=1}^{3} \omega^2 J_i \Theta_i = 0 \tag{7.35}$$

Equation (7.35) states that the sum of the inertia torques of the semi-definite system must be zero. This equation can be treated as another form of the frequency equation, and the trial frequency must satisfy this requirement.

In Holzer's method, a trial frequency $\omega$ is assumed, and $\Theta_1$ is arbitrarily chosen as unity. Next, $\Theta_2$ is computed from Eq. (7.32), and then $\Theta_3$ is found from Eq.

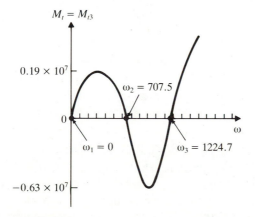

**Figure 7.5**

(7.33). Thus we obtain

$$\Theta_1 = 1 \qquad\qquad (7.36)$$

$$\Theta_2 = \Theta_1 - \frac{\omega^2 J_1 \Theta_1}{k_{t1}} \qquad\qquad (7.37)$$

$$\Theta_3 = \Theta_2 - \frac{\omega^2}{k_{t2}} (J_1 \Theta_1 + J_2 \Theta_2) \qquad\qquad (7.38)$$

These values are substituted in Eq. (7.35) to verify whether the constraint is satisfied. If Eq. (7.35) is not satisfied, a new trial value of $\omega$ is assumed and the process repeated. Equations (7.35), (7.37), and (7.38) can be generalized for an $n$ disc system as follows:

$$\sum_{i=1}^{n} \omega^2 J_i \Theta_i = 0 \qquad\qquad (7.39)$$

$$\Theta_i = \Theta_{i-1} - \frac{\omega^2}{k_{t_{i-1}}} \left( \sum_{k=1}^{i-1} J_k \Theta_k \right); \qquad i = 2, 3, \ldots, n \qquad\qquad (7.40)$$

Thus the method uses Eqs. (7.39) and (7.40) repeatedly for different trial frequencies. If the assumed trial frequency is not a natural frequency of the system, Eq. (7.39) is not satisfied. The resultant torque in Eq. (7.39) represents a torque applied at the last disc. This torque, $M_t$, is then plotted for the chosen $\omega$. When the calculation is repeated with other values of $\omega$, the resulting graph appears as shown in Fig. 7.5. From this graph, the natural frequencies of the system can be identified as the values of $\omega$ at which $M_t = 0$. The amplitudes $\Theta_i$ ($i = 1, 2, \ldots, n$) corresponding to the natural frequencies are the mode shapes of the system.

Holzer's method can also be applied to systems with fixed ends. At a fixed end, the amplitude of vibration must be zero. In this case, the natural frequencies can be found by plotting the resulting amplitude (instead of the resultant torque) against the assumed frequencies. For a system with one end free and the other end fixed, Eq. (7.40) can be used for checking the amplitude at the fixed end. An improvement of Holzer's method was presented in Refs. [7.13, 7.14].

---

### EXAMPLE 7.4

Figure 7.6 shows a gas turbine rotor arrangement. Find the natural frequencies and mode shapes of the system. Disregard the mass moment of inertia of the shafts and the couplings.

*Solution.* This is a free-free (unsupported) system; Table 7.1 shows its parameters and the sequence of computations. The calculations for the trial frequencies $\omega = 0$, 10, 20, 700, and 710 are shown in this table. The quantity $M_{t3}$ denotes the torque to the right of Station 3 (generator) which must be zero at the natural frequencies. Figure 7.5 shows the graph of $M_{t3}$ versus $\omega$. Closely spaced trial values of $\omega$ are

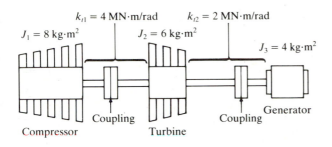

**Figure 7.6**

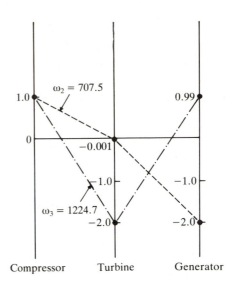

**Figure 7.7**

**TABLE 7.1**

| Parameters of the system | Quantity | Trial | | | | | |
|---|---|---|---|---|---|---|---|
| | | 1 | 2 | 3 | ... | 71 | 72 |
| | $\omega$ | 0 | 10 | 20 | | 700 | 710 |
| | $\omega^2$ | 0 | 100 | 400 | | 490000 | 504100 |
| Station 1: $J_1 = 8$ | $\Theta_1$ | 1.0 | 1.0 | 1.0 | | 1.0 | 1.0 |
| $k_{t1} = 4 \times 10^6$ | $M_{t1} = \omega^2 J_1 \Theta_1$ | 0 | 800 | 3200 | | 0.392E7 | 0.403E7 |
| Station 2: $J_2 = 6$ | $\Theta_2 = 1 - \dfrac{M_{t1}}{k_{t1}}$ | 1.0 | 0.9998 | 0.9992 | | 0.0200 | −0.0082 |
| $k_{t2} = 2 \times 10^6$ | $M_{t2} = M_{t1} + \omega^2 J_2 \Theta_2$ | 0 | 1400 | 5598 | | 0.398E7 | 0.401E7 |
| Station 3: $J_3 = 4$ | $\Theta_3 = \Theta_2 - \dfrac{M_{t2}}{k_{t2}}$ | 1.0 | 0.9991 | 0.9964 | | −1.9690 | −2.0120 |
| $k_{t3} = 0$ | $M_{t3} = M_{t2} + \omega^2 J_3 \Theta_3$ | 0 | 1800 | 7192 | | 0.119E6 | −0.494E5 |

used in the vicinity of $M_{t3} = 0$ to obtain accurate values of the first two flexible mode shapes, shown in Fig. 7.7. Note that the value $\omega = 0$ corresponds to the rigid body rotation.

### 7.4.1 Spring-Mass Systems

Although Holzer's method has been extensively applied to torsional systems, the procedure is equally applicable to the vibration analysis of spring-mass systems. The

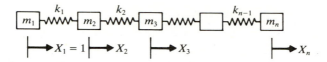

**Figure 7.8**

equations of motion of a spring-mass system (see Fig. 7.8) can be expressed as

$$m_1 \ddot{x}_1 + k_1(x_1 - x_2) = 0 \tag{7.41}$$

$$m_2 \ddot{x}_2 + k_1(x_2 - x_1) + k_2(x_2 - x_3) = 0 \tag{7.42}$$

$$\cdots\cdots$$

For harmonic motion, $x_i(t) = X_i \cos \omega t$, where $X_i$ is the amplitude of mass $m_i$, and Eqs. (7.41) and (7.42) can be rewritten as

$$\omega^2 m_1 X_1 = k_1(X_1 - X_2) \tag{7.43}$$

$$\omega^2 m_2 X_2 = k_1(X_2 - X_1) + k_2(X_2 - X_3) = -\omega^2 m_1 X_1 + k_2(X_2 - X_3) \tag{7.44}$$

$$\cdots\cdots$$

The procedure of Holzer's method starts with a trial frequency $\omega$ and the amplitude of mass $m_1$ as $X_1 = 1$. Equations (7.43) and (7.44) can then be used to obtain the amplitudes of the masses $m_2, m_3, \ldots, m_i$:

$$X_2 = X_1 - \frac{\omega^2 m_1 X_1}{k_1} \tag{7.45}$$

$$X_3 = X_2 - \frac{\omega^2}{k_2}(m_1 X_1 + m_2 X_2) \tag{7.46}$$

$$\cdots\cdots$$

$$X_i = X_{i-1} - \frac{\omega^2}{k_{i-1}} \left( \sum_{k=1}^{i-1} m_k X_k \right); \qquad i = 2, 3, \ldots, n \tag{7.47}$$

As in the case of torsional systems, the resultant force applied to the last ($n$th) mass can be computed as follows:

$$F = \sum_{i=1}^{n} \omega^2 m_i X_i \tag{7.48}$$

The calculations are repeated with several other trial frequencies $\omega$. The natural frequencies are identified as those values of $\omega$ that give $F = 0$ for a free-free system. For this, it is convenient to plot a graph between $F$ and $\omega$, using the same procedure for spring-mass systems as for torsional systems.

## 7.5  MATRIX ITERATION METHOD

The matrix iteration method assumes that the natural frequencies are distinct and well separated such that $\omega_1 < \omega_2 < \cdots < \omega_n$. The iteration is started by selecting a trial vector $\vec{X}_1$, which is then premultiplied by the dynamical matrix $[D]$. The resulting column vector is then normalized, usually by making one of its components to unity. The normalized column vector is premultiplied by $[D]$ to obtain a third column vector, which is normalized in the same way as before and becomes still another trial column vector. The process is repeated until the successive normalized column vectors converge to a common vector: the fundamental eigenvector. The normalizing factor gives the largest value of $\lambda = 1/\omega^2$ —that is, the smallest or the fundamental natural frequency [7.15]. The convergence of the process can be explained as follows.

According to the expansion theorem, any arbitrary $n$-dimensional vector $\vec{X}_1$ can be expressed as a linear combination of the $n$ orthogonal eigenvectors of the system $\vec{X}^{(i)}$, $i = 1, 2, \ldots, n$:

$$\vec{X}_1 = c_1 \vec{X}^{(1)} + c_2 \vec{X}^{(2)} + \cdots + c_n \vec{X}^{(n)} \tag{7.49}$$

where $c_1, c_2, \ldots, c_n$ are constants. In the iteration method, the trial vector $\vec{X}_1$ is selected arbitrarily and is therefore a known vector. The modal vectors $\vec{X}^{(i)}$, although unknown, are constant vectors because they depend upon the properties of the system. The constants $c_i$ are unknown numbers to be determined. According to the iteration method, we premultiply $\vec{X}_1$ by the matrix $[D]$. In view of Eq. (7.49), this gives

$$[D]\vec{X}_1 = c_1[D]\vec{X}^{(1)} + c_2[D]\vec{X}^{(2)} + \cdots + c_n[D]\vec{X}^{(n)} \tag{7.50}$$

Now, according to Eq. (6.60), we have

$$[D]\vec{X}^{(i)} = \lambda_i[I]\vec{X}^{(i)} = \frac{1}{\omega_i^2}\vec{X}^{(i)}; \qquad i = 1, 2, \ldots, n \tag{7.51}$$

Substitution of Eq. (7.51) into Eq. (7.50) yields

$$[D]\vec{X}_1 = \vec{X}_2$$

$$= \frac{c_1}{\omega_1^2}\vec{X}^{(1)} + \frac{c_2}{\omega_2^2}\vec{X}^{(2)} + \cdots + \frac{c_n}{\omega_n^2}\vec{X}^{(n)} \tag{7.52}$$

where $\vec{X}_2$ is the second trial vector. We now repeat the process and premultiply $\vec{X}_2$ by $[D]$ to obtain, by Eqs. (7.49) and (6.60),

$$[D]\vec{X}_2 = \vec{X}_3$$

$$= \frac{c_1}{\omega_1^4}\vec{X}^{(1)} + \frac{c_2}{\omega_2^4}\vec{X}^{(2)} + \cdots + \frac{c_n}{\omega_n^4}\vec{X}^{(n)} \tag{7.53}$$

By repeating the process we obtain, after the $r$th iteration,

$$[D]\vec{X}_r = \vec{X}_{r+1}$$

$$= \frac{c_1}{\omega_1^{2r}}\vec{X}^{(1)} + \frac{c_2}{\omega_2^{2r}}\vec{X}^{(2)} + \cdots + \frac{c_n}{\omega_n^{2r}}\vec{X}^{(n)} \qquad (7.54)$$

Since the natural frequencies are assumed to be $\omega_1 < \omega_2 < \cdots < \omega_n$, a sufficiently large value of $r$ yields

$$\frac{1}{\omega_1^{2r}} \gg \frac{1}{\omega_2^{2r}} \gg \cdots \gg \frac{1}{\omega_n^{2r}} \qquad (7.55)$$

Thus the first term on the right-hand side of Eq. (7.54) becomes the only significant one. Hence we have

$$\vec{X}_{r+1} = \frac{c_1}{\omega_1^{2r}}\vec{X}^{(1)} \qquad (7.56)$$

which means that the $(r+1)$th trial vector becomes identical to the fundamental modal vector to within a multiplicative constant. Since

$$\vec{X}_r = \frac{c_1}{\omega_1^{2(r-1)}}\vec{X}^{(1)} \qquad (7.57)$$

the fundamental natural frequency $\omega_1$ can be found by taking the ratio of any two corresponding components in the vectors $\vec{X}_r$ and $\vec{X}_{r+1}$:

$$\omega_1^2 \simeq \frac{X_{i,r}}{X_{i,r+1}} \qquad \text{for any } i = 1, 2, \ldots, n \qquad (7.58)$$

where $X_{i,r}$ and $X_{i,r+1}$ are the $i$th elements of the vectors $\vec{X}_r$ and $\vec{X}_{r+1}$, respectively.

**Discussion**

1. In the above proof, nothing has been said about the normalization of the successive trial vectors $\vec{X}_i$. Actually, it is not necessary to establish the proof of convergence of the method. The normalization amounts to a readjustment of the constants $c_1, c_2, \ldots, c_n$ in each iteration.

2. Although it is theoretically necessary to have $r \to \infty$ for the convergence of the method, in practice only a finite number of iterations suffices to obtain a reasonably good estimate of $\omega_1$.

3. The actual number of iterations necessary to find the value of $\omega_1$ to within a desired degree of accuracy depends on how closely the arbitrary trial vector $\vec{X}_1$ resembles the fundamental mode $\vec{X}^{(1)}$ and on how well $\omega_1$ and $\omega_2$ are separated. The required number of iterations is less if $\omega_2$ is very large compared to $\omega_1$.

4. The method has a distinct advantage in that any computational errors made do not yield incorrect results. Any error made in premultiplying $\vec{X}_i$ by $[D]$ results in a vector other than the desired one, $\vec{X}_{i+1}$. But this wrong vector can be

considered as a new trial vector. This may delay the convergence but does not produce wrong results.

5. One can take any set of $n$ numbers for the first trial vector $\vec{X}_1$ and still achieve convergence to the fundamental modal vector. Only in the unusual case in which the trial vector $\vec{X}_1$ is exactly proportional to one of the modes $\vec{X}^{(i)}$ ($i \neq 1$) does the method fail to converge to the first mode. In such a case, the premultiplication of $\vec{X}^{(i)}$ by $[D]$ results in a vector proportional to $\vec{X}^{(i)}$ itself.

### 7.5.1 Convergence to the Highest Natural Frequency

To obtain the highest natural frequency $\omega_n$ and the corresponding mode shape or eigenvector $\vec{X}^{(n)}$ by the matrix iteration method, we first rewrite Eq. (6.60) as

$$[D]^{-1}\vec{X} = \omega^2[I]\vec{X} = \omega^2\vec{X} \tag{7.59}$$

where $[D]^{-1}$ is the inverse of the dynamical matrix $[D]$ given by

$$[D]^{-1} = [m]^{-1}[k] \tag{7.60}$$

Now we select any arbitrary trial vector $\vec{X}_1$ and premultiply it by $[D]^{-1}$ to obtain an improved trial vector $\vec{X}_2$. The sequence of trial vectors $\vec{X}_{i+1}$ ($i = 1, 2, \ldots$) obtained by premultiplying by $[D]^{-1}$ converges to the highest normal mode $\vec{X}^{(n)}$. It can be seen that the procedure is similar to the one already described. The constant of proportionality in this case is $\omega^2$ instead of $1/\omega^2$.

### 7.5.2 Computation of Intermediate Natural Frequencies

Once the first natural frequency $\omega_1$ (or the largest eigenvalue $\lambda_1 = 1/\omega_1^2$) and the corresponding eigenvector $\vec{X}^{(1)}$ are determined, we can proceed to find the higher natural frequencies and the corresponding mode shapes by the matrix iteration method. Before we proceed, it should be remembered that any arbitrary trial vector premultiplied by $[D]$ would lead again to the largest eigenvalue. It is thus necessary to remove the largest eigenvalue from the matrix $[D]$. The succeeding eigenvalues and eigenvectors can be obtained by eliminating the root $\lambda_1$ from the characteristic or frequency equation

$$|[D] - \lambda[I]| = 0 \tag{7.61}$$

A procedure known as *matrix deflation* can be used for this purpose [7.16]. To find the eigenvector $\vec{X}^{(i)}$ by this procedure, the previous eigenvector $\vec{X}^{(i-1)}$ is normalized with respect to the mass matrix such that

$$\vec{X}^{(i-1)^T}[m]\vec{X}^{(i-1)} = 1 \tag{7.62}$$

Then the deflated matrix $[D_i]$ is constructed as

$$[D_i] = [D_{i-1}] - \lambda_{i-1}\vec{X}^{(i-1)}\vec{X}^{(i-1)^T}[m]; \qquad i = 2, 3, \ldots, n \tag{7.63}$$

where $[D_1] = [D]$. Once $[D_i]$ is constructed, the iterative scheme

$$\vec{X}_{r+1} = [D_i]\vec{X}_r \tag{7.64}$$

is used, where $\vec{X}_1$ is an arbitrary trial eigenvector.

**EXAMPLE 7.5**

Find the natural frequencies and mode shapes of the system shown in Fig. 7.1 for $k_1 = k_2 = k_3 = k$ and $m_1 = m_2 = m_3 = m$ by the matrix iteration method.

**Solution.** The mass and stiffness matrices of the system are given in Example 7.2. The flexibility matrix is

$$[a] = [k]^{-1} = \frac{1}{k}\begin{bmatrix} 1 & 1 & 1 \\ 1 & 2 & 2 \\ 1 & 2 & 3 \end{bmatrix} \tag{E.1}$$

and so the dynamical matrix is

$$[k]^{-1}[m] = \frac{m}{k}\begin{bmatrix} 1 & 1 & 1 \\ 1 & 2 & 2 \\ 1 & 2 & 3 \end{bmatrix} \tag{E.2}$$

The eigenvalue problem can be stated as

$$[D]\vec{X} = \lambda\vec{X} \tag{E.3}$$

where

$$[D] = \begin{bmatrix} 1 & 1 & 1 \\ 1 & 2 & 2 \\ 1 & 2 & 3 \end{bmatrix} \tag{E.4}$$

and

$$\lambda = \frac{k}{m} \cdot \frac{1}{\omega^2} \tag{E.5}$$

*First natural frequency:* By assuming the first trial eigenvector or mode shape to be

$$\vec{X}_1 = \begin{Bmatrix} 1 \\ 1 \\ 1 \end{Bmatrix} \tag{E.6}$$

the second trial eigenvector can be obtained:

$$\vec{X}_2 = [D]\vec{X}_1 = \begin{Bmatrix} 3 \\ 5 \\ 6 \end{Bmatrix} \tag{E.7}$$

By making the first element equal to unity, we obtain

$$\vec{X}_2 = 3.0\begin{Bmatrix} 1.0000 \\ 1.6667 \\ 2.0000 \end{Bmatrix} \tag{E.8}$$

and the corresponding eigenvalue is given by

$$\lambda_1 \simeq 3.0 \quad \text{or} \quad \omega_1 \simeq 0.5773\sqrt{\frac{k}{m}} \tag{E.9}$$

The subsequent trial eigenvector can be obtained from the relation

$$\vec{X}_{i+1} = [D]\vec{X}_i \tag{E.10}$$

and the corresponding eigenvalues are given by

$$\lambda_1 \simeq X_{1,i+1} \tag{E.11}$$

where $X_{1,i+1}$ is the first component of the vector $\vec{X}_{i+1}$ before normalization. The various trial eigenvectors and eigenvalues obtained by using Eqs. (E.10) and (E.11) are shown below:

| $i$ | $\vec{X}_i$ with $X_{1,i}=1$ | $\vec{X}_{i+1} = [D]\vec{X}_i$ | $\lambda_1 \simeq X_{1,i+1}$ | $\omega_1$ |
|---|---|---|---|---|
| 1 | $\begin{Bmatrix} 1 \\ 1 \\ 1 \end{Bmatrix}$ | $\begin{Bmatrix} 3 \\ 5 \\ 6 \end{Bmatrix}$ | 3.0 | $0.5773\sqrt{\dfrac{k}{m}}$ |
| 2 | $\begin{Bmatrix} 1.00000 \\ 1.66667 \\ 2.00000 \end{Bmatrix}$ | $\begin{Bmatrix} 4.66667 \\ 8.33333 \\ 10.33333 \end{Bmatrix}$ | 4.66667 | $0.4629\sqrt{\dfrac{k}{m}}$ |
| 3 | $\begin{Bmatrix} 1.0000 \\ 1.7857 \\ 2.2143 \end{Bmatrix}$ | $\begin{Bmatrix} 5.00000 \\ 9.00000 \\ 11.2143 \end{Bmatrix}$ | 5.00000 | $0.4472\sqrt{\dfrac{k}{m}}$ |
| $\vdots$ | | | | |
| 7 | $\begin{Bmatrix} 1.00000 \\ 1.80193 \\ 2.24697 \end{Bmatrix}$ | $\begin{Bmatrix} 5.04891 \\ 9.09781 \\ 11.34478 \end{Bmatrix}$ | 5.04891 | $0.44504\sqrt{\dfrac{k}{m}}$ |
| 8 | $\begin{Bmatrix} 1.00000 \\ 1.80194 \\ 2.24698 \end{Bmatrix}$ | $\begin{Bmatrix} 5.04892 \\ 9.09783 \\ 11.34481 \end{Bmatrix}$ | 5.04892 | $0.44504\sqrt{\dfrac{k}{m}}$ |

It can be seen that the mode shape and the natural frequency converged (to the fourth decimal place) in eight iterations. Thus the first eigenvalue and the corresponding natural frequency and mode shape are given by

$$\lambda_1 = 5.04892, \quad \omega_1 = 0.44504\sqrt{\frac{k}{m}}$$

$$\vec{X}^{(1)} = \begin{Bmatrix} 1.00000 \\ 1.80194 \\ 2.24698 \end{Bmatrix} \tag{E.12}$$

*Second natural frequency:* To compute the second eigenvalue and the eigenvector, we must first produce a deflated matrix:

$$[D_2] = [D_1] - \lambda_1 \vec{X}^{(1)} \vec{X}^{(1)^T} [m] \tag{E.13}$$

This equation, however, calls for a normalized vector $\vec{X}^{(1)}$ satisfying $\vec{X}^{(1)^T}[m]\vec{X}^{(1)} = 1$. Let the normalized vector be denoted as

$$\vec{X}^{(1)} = \alpha \begin{Bmatrix} 1.00000 \\ 1.80194 \\ 2.24698 \end{Bmatrix}$$

where $\alpha$ is a constant whose value must be such that

$$\vec{X}^{(1)^T}[m]\vec{X}^{(1)} = \alpha^2 m \begin{Bmatrix} 1.00000 \\ 1.80194 \\ 2.24698 \end{Bmatrix}^T \begin{bmatrix} 1 & 0 & 0 \\ 0 & 1 & 0 \\ 0 & 0 & 1 \end{bmatrix} \begin{Bmatrix} 1.00000 \\ 1.80194 \\ 2.24698 \end{Bmatrix}$$

$$= \alpha^2 m (9.29591) = 1 \tag{E.14}$$

from which we obtain $\alpha = 0.32799 m^{-1/2}$. Hence the first normalized eigenvector is

$$\vec{X}^{(1)} = m^{-1/2} \begin{Bmatrix} 0.32799 \\ 0.59102 \\ 0.73699 \end{Bmatrix} \tag{E.15}$$

Next we use Eq. (E.13) and form the first deflated matrix

$$[D_2] = \begin{bmatrix} 1 & 1 & 1 \\ 1 & 2 & 2 \\ 1 & 2 & 3 \end{bmatrix} - 5.04892 \begin{Bmatrix} 0.32799 \\ 0.59102 \\ 0.73699 \end{Bmatrix} \begin{Bmatrix} 0.32799 \\ 0.59102 \\ 0.73699 \end{Bmatrix}^T \begin{bmatrix} 1 & 0 & 0 \\ 0 & 1 & 0 \\ 0 & 0 & 1 \end{bmatrix}$$

$$= \begin{bmatrix} 0.45684 & 0.02127 & -0.22048 \\ 0.02127 & 0.23641 & -0.19921 \\ -0.22048 & -0.19921 & 0.25768 \end{bmatrix} \tag{E.16}$$

Since the trial vector can be chosen arbitrarily, we again take

$$\vec{X}_1 = \begin{Bmatrix} 1 \\ 1 \\ 1 \end{Bmatrix} \tag{E.17}$$

By using the iterative scheme

$$\vec{X}_{i+1} = [D_2]\vec{X}_i \tag{E.18}$$

we obtain $\vec{X}_2$:

$$\vec{X}_2 = \begin{Bmatrix} 0.25763 \\ 0.05847 \\ -0.16201 \end{Bmatrix} = 0.25763 \begin{Bmatrix} 1.00000 \\ 0.22695 \\ -0.62885 \end{Bmatrix} \tag{E.19}$$

Hence $\lambda_2$ can be found from the general relation

$$\lambda_2 \simeq X_{1,i+1} \tag{E.20}$$

as 0.25763. Continuation of this procedure gives the following results:

| $i$ | $\vec{X}_i$ with $X_{1,i}=1$ | $\vec{X}_{i+1}=[D_2]\vec{X}_i$ | $\lambda_2 \simeq X_{1,i+1}$ | $\omega_2$ |
|---|---|---|---|---|
| 1 | $\begin{Bmatrix} 1 \\ 1 \\ 1 \end{Bmatrix}$ | $\begin{Bmatrix} 0.25763 \\ 0.05847 \\ -0.16201 \end{Bmatrix}$ | 0.25763 | $1.97016\sqrt{\dfrac{k}{m}}$ |
| 2 | $\begin{Bmatrix} 1.00000 \\ 0.22695 \\ -0.62885 \end{Bmatrix}$ | $\begin{Bmatrix} 0.60032 \\ 0.20020 \\ -0.42773 \end{Bmatrix}$ | 0.60032 | $1.29065\sqrt{\dfrac{k}{m}}$ |
| $\vdots$ | | | | |
| 10 | $\begin{Bmatrix} 1.00000 \\ 0.44443 \\ -0.80149 \end{Bmatrix}$ | $\begin{Bmatrix} 0.64300 \\ 0.28600 \\ -0.51554 \end{Bmatrix}$ | 0.64300 | $1.24708\sqrt{\dfrac{k}{m}}$ |
| 11 | $\begin{Bmatrix} 1.00000 \\ 0.44479 \\ -0.80177 \end{Bmatrix}$ | $\begin{Bmatrix} 0.64307 \\ 0.28614 \\ -0.51569 \end{Bmatrix}$ | 0.64307 | $1.24701\sqrt{\dfrac{k}{m}}$ |

Thus the converged second eigenvalue and the eigenvector are

$$\lambda_2 = 0.64307, \qquad \omega_2 = 1.24701\sqrt{\frac{k}{m}}$$

$$\vec{X}^{(2)} = \begin{Bmatrix} 1.00000 \\ 0.44496 \\ -0.80192 \end{Bmatrix} \tag{E.21}$$

*Third natural frequency:* For the third eigenvalue and the eigenvector, we use a similar procedure. The detailed calculations are left as an exercise to the reader. Note that before computing the deflated matrix $[D_3]$, we need to normalize $\vec{X}^{(2)}$ by using Eq. (7.62), which gives

$$\vec{X}^{(2)} = m^{-1/2} \begin{Bmatrix} 0.73700 \\ 0.32794 \\ -0.59102 \end{Bmatrix} \tag{E.22}$$

## 7.6 JACOBI'S METHOD

The matrix iteration method described in the preceding section produces the eigenvalues and eigenvectors of a matrix $[D]$ one at a time. Jacobi's method is also an iterative method but produces all the eigenvalues and eigenvectors of $[D]$ simultaneously, where $[D]=[d_{ij}]$ is a real symmetric matrix of order $n \times n$. The method is based on a theorem in linear algebra that states that a real symmetric matrix $[D]$ has only real eigenvalues and that there exists a real orthogonal matrix

$[R]$ such that $[R]^T[D][R]$ is diagonal [7.17]. The diagonal elements are the eigenvalues, and the columns of the matrix $[R]$ are the eigenvectors. In Jacobi's method, the matrix $[R]$ is generated as a product of several rotation matrices [7.18] of the form

$$[R_1] = \begin{bmatrix} 1 & 0 & & & & \\ 0 & 1 & & & & \\ & & \ddots & & & \\ & & & \cos\theta & \cdots & -\sin\theta & \\ & & & & \ddots & \\ & & & \sin\theta & \cdots & \cos\theta & \\ & & & & & & \ddots \\ & & & & & & & 1 \end{bmatrix} \begin{matrix} \\ \\ \\ i\text{th row} \\ \\ j\text{th row} \\ \\ \\ \end{matrix} \quad (7.65)$$

$$\underset{n \times n}{}$$

where all elements other than those appearing in columns and rows $i$ and $j$ are identical with those of the identity matrix $[I]$. If the sine and cosine entries appear in positions $(i, i)$, $(i, j)$, $(j, i)$, and $(j, j)$, then the corresponding elements of $[R_1]^T[D][R_1]$ can be computed as follows:

$$\underline{d}_{ii} = d_{ii}\cos^2\theta + 2d_{ij}\sin\theta\cos\theta + d_{jj}\sin^2\theta \qquad (7.66)$$

$$\underline{d}_{ij} = \underline{d}_{ji} = (d_{jj} - d_{ii})\sin\theta\cos\theta + d_{ij}(\cos^2\theta - \sin^2\theta) \qquad (7.67)$$

$$\underline{d}_{jj} = d_{ii}\sin^2\theta - 2d_{ij}\sin\theta\cos\theta + d_{jj}\cos^2\theta \qquad (7.68)$$

If $\theta$ is chosen to be

$$\tan 2\theta = \left( \frac{2d_{ij}}{d_{ii} - d_{jj}} \right) \qquad (7.69)$$

then it makes $\underline{d}_{ij} = \underline{d}_{ji} = 0$. Thus each step of Jacobi's method reduces a pair of off-diagonal elements to zero. Unfortunately, in the next step, while the method reduces a new pair of zeros, it introduces nonzero contributions to formerly zero positions. However, successive matrices of the form

$$[R_2]^T[R_1]^T[D][R_1][R_2], \qquad [R_3]^T[R_2]^T[R_1]^T[D][R_1][R_2][R_3], \ldots$$

converge to the required diagonal form; the final matrix $[R]$, whose columns give the eigenvectors, then becomes

$$[R] = [R_1][R_2][R_3]\ldots \qquad (7.70)$$

## EXAMPLE 7.6

Find the eigenvalues and eigenvectors of the matrix

$$[D] = \begin{bmatrix} 1 & 1 & 1 \\ 1 & 2 & 2 \\ 1 & 2 & 3 \end{bmatrix}$$

using Jacobi's method.

*Solution.* We start with the largest off-diagonal term $d_{23} = 2$ in the matrix $[D]$ and try to reduce it to zero. From Eq. (7.69),

$$\theta_1 = \frac{1}{2}\tan^{-1}\left(\frac{2d_{23}}{d_{22}-d_{33}}\right) = \frac{1}{2}\tan^{-1}\left(\frac{4}{2-3}\right) = -37.981878°$$

$$[R_1] = \begin{bmatrix} 1.0 & 0.0 & 0.0 \\ 0.0 & 0.7882054 & 0.6154122 \\ 0.0 & -0.6154122 & 0.7882054 \end{bmatrix}$$

$$[D'] = [R_1]^T[D][R_1] = \begin{bmatrix} 1.0 & 0.1727932 & 1.4036176 \\ 0.1727932 & 0.4384472 & 0.0 \\ 1.4036176 & 0.0 & 4.5615525 \end{bmatrix}$$

Next we try to reduce the largest off-diagonal term of $[D']$, namely, $d'_{13} = 1.4036176$ to zero. Equation (7.69) gives

$$\theta_2 = \frac{1}{2}\tan^{-1}\left(\frac{2d'_{13}}{d'_{11}-d'_{33}}\right) = \frac{1}{2}\tan^{-1}\left(\frac{2.8072352}{1.0-4.5615525}\right) = -19.122686°$$

$$[R_2] = \begin{bmatrix} 0.9448193 & 0.0 & 0.3275920 \\ 0.0 & 1.0 & 0.0 \\ -0.3275920 & 0.0 & 0.9448193 \end{bmatrix}$$

$$[D''] = [R_2]^T[D'][R_2] = \begin{bmatrix} 0.5133313 & 0.1632584 & 0.0 \\ 0.1632584 & 0.4384472 & 0.0566057 \\ 0.0 & 0.0566057 & 5.0482211 \end{bmatrix}$$

The largest off-diagonal element in $[D'']$ is $d''_{12} = 0.1632584$. $\theta_3$ can be obtained from Eq. (7.69) as

$$\theta_3 = \frac{1}{2}\tan^{-1}\left(\frac{2d''_{12}}{d''_{11}-d''_{22}}\right) = \frac{1}{2}\tan^{-1}\left(\frac{0.3265167}{0.5133313-0.4384472}\right) = 38.541515°$$

$$[R_3] = \begin{bmatrix} 0.7821569 & -0.6230815 & 0.0 \\ 0.6230815 & 0.7821569 & 0.0 \\ 0.0 & 0.0 & 1.0 \end{bmatrix}$$

$$[D'''] = [R_3]^T[D''][R_3] = \begin{bmatrix} 0.6433861 & 0.0 & 0.0352699 \\ 0.0 & 0.3083924 & 0.0442745 \\ 0.0352699 & 0.0442745 & 5.0482211 \end{bmatrix}$$

Assuming that all the off-diagonal terms in $[D''']$ are close to zero, we can stop the process here. The diagonal elements of $[D''']$ give the eigenvalues (values of $1/\omega^2$) as 0.6433861, 0.3083924, and 5.0482211. The corresponding eigenvectors are given by the columns of the matrix $[R]$ where

$$[R] = [R_1][R_2][R_3] = \begin{bmatrix} 0.7389969 & -0.5886994 & 0.3275920 \\ 0.3334301 & 0.7421160 & 0.5814533 \\ -0.5854125 & -0.3204631 & 0.7447116 \end{bmatrix}$$

The iterative process can be continued for obtaining a more accurate solution. The present eigenvalues can be compared with the exact values: 0.6431041, 0.3079786, and 5.0489173.

## 7.7 STANDARD EIGENVALUE PROBLEM

In the preceding chapter, the eigenvalue problem was stated as

$$[k]\vec{X} = \omega^2[m]\vec{X} \tag{7.71}$$

which can be rewritten in the form of a standard eigenvalue problem [7.19] as

$$[D]\vec{X} = \lambda\vec{X} \tag{7.72}$$

where

$$[D] = [k]^{-1}[m] \tag{7.73}$$

and

$$\lambda = \frac{1}{\omega^2} \tag{7.74}$$

In general, the matrix $[D]$ is nonsymmetric, although the matrices $[k]$ and $[m]$ are both symmetric. Since Jacobi's method (described in Sec. 7.6) is applicable only to symmetric matrices $[D]$, we can adopt the following procedure [7.18] to derive a standard eigenvalue problem with a symmetric matrix $[D]$.

Assuming that the matrix $[k]$ is symmetric and positive definite, we can use Choleski decomposition (see Sec. 7.7.1) and express $[k]$ as

$$[k] = [U]^T[U] \tag{7.75}$$

where $[U]$ is an upper triangular matrix. Using this relation, the eigenvalue problem of Eq. (7.71) can be stated as

$$\lambda[U]^T[U]\vec{X} = [m]\vec{X} \tag{7.76}$$

Premultiplying this equation by $([U]^T)^{-1}$, we obtain

$$\lambda[U]\vec{X} = ([U]^T)^{-1}[m]\vec{X} = ([U]^T)^{-1}[m][U]^{-1}[U]\vec{X} \tag{7.77}$$

By defining a new vector $\vec{Y}$ as

$$\vec{Y} = [U]\vec{X} \tag{7.78}$$

Eq. (7.77) can be written as a standard eigenvalue problem:

$$[D]\vec{Y} = \lambda\vec{Y} \tag{7.79}$$

where

$$[D] = ([U]^T)^{-1}[m][U]^{-1} \tag{7.80}$$

Thus, to formulate $[D]$ according to Eq. (7.80), we first decompose the symmetric matrix $[k]$ as shown in Eq. (7.75), find $[U]^{-1}$ and $([U]^T)^{-1} = ([U]^{-1})^T$ as outlined in the next section, and then carry out the matrix multiplication as stated in Eq. (7.80). The solution of the eigenvalue problem stated in Eq. (7.79) yields $\lambda_i$ and $\vec{Y}^{(i)}$. Then we apply inverse transformation and find the desired eigenvectors:

$$\vec{X}^{(i)} = [U]^{-1}\vec{Y}^{(i)} \tag{7.81}$$

### 7.7.1 Choleski Decomposition

Any symmetric and positive definite matrix $[A]$ of order $n \times n$ can be decomposed uniquely [7.20]:

$$[A] = [U]^T[U] \tag{7.82}$$

where $[U]$ is an upper triangular matrix given by

$$[U] = \begin{bmatrix} u_{11} & u_{12} & u_{13} & \cdots & u_{1n} \\ 0 & u_{22} & u_{23} & \cdots & u_{2n} \\ 0 & 0 & u_{33} & \cdots & u_{3n} \\ \vdots & & & & \\ 0 & 0 & 0 & \cdots & u_{nn} \end{bmatrix} \tag{7.83}$$

with

$$u_{11} = (a_{11})^{1/2}$$

$$u_{1j} = \frac{a_{1j}}{u_{11}}; \qquad j = 2,3,\ldots,n$$

$$u_{ii} = \left( a_{ii} - \sum_{k=1}^{i-1} u_{ki}^2 \right)^{1/2}; \qquad i = 2,3,\ldots,n$$

$$u_{ij} = \frac{1}{u_{ii}} \left( a_{ij} - \sum_{k=1}^{i-1} u_{ki}u_{kj} \right); \qquad i = 2,3,\ldots,n \text{ and } j = i+1, i+2,\ldots,n$$

$$u_{ij} = 0; \qquad i > j \tag{7.84}$$

**Inverse of the Matrix [U].** If the inverse of the upper triangular matrix $[U]$ is denoted as $[\alpha_{ij}]$, the elements $\alpha_{ij}$ can be determined from the relation

$$[U][U]^{-1} = [I] \tag{7.85}$$

which gives

$$\alpha_{ii} = \frac{1}{u_{ii}}$$

$$\alpha_{ij} = \frac{-1}{u_{ii}} \left( \sum_{k=i+1}^{j} u_{ik}\alpha_{kj} \right); \qquad i < j$$

$$\alpha_{ij} = 0; \qquad i > j \tag{7.86}$$

Thus the inverse of $[U]$ is also an upper triangular matrix.

---

**EXAMPLE 7.7**

Decompose the matrix

$$[A] = \begin{bmatrix} 5 & 1 & 0 \\ 1 & 3 & 2 \\ 0 & 2 & 8 \end{bmatrix}$$

into the form of Eq. (7.82).

*Solution.* Equations (7.84) give

$$u_{11} = \sqrt{a_{11}} = \sqrt{5} = 2.2360680$$

$$u_{12} = a_{12}/u_{11} = 1/2.236068 = 0.4472136$$

$$u_{13} = a_{13}/u_{11} = 0$$

$$u_{22} = \left[a_{22} - u_{12}^2\right]^{1/2} = (3 - 0.4472136^2)^{1/2} = 1.6733201$$

$$u_{33} = \left[a_{33} - u_{13}^2 - u_{23}^2\right]^{1/2}$$

where

$$u_{23} = (a_{23} - u_{12}u_{13})/u_{22} = (2 - 0.4472136 \times 0)/1.6733201 = 1.1952286$$

$$u_{33} = (8 - 0^2 - 1.1952286^2)^{1/2} = 2.5634799$$

Since $u_{ij} = 0$ for $i > j$, we have

$$[U] = \begin{bmatrix} 2.2360680 & 0.4472136 & 0.0 \\ 0.0 & 1.6733201 & 1.1952286 \\ 0.0 & 0.0 & 2.5634799 \end{bmatrix}$$

---

### 7.7.2    Other Solution Methods

Several other methods have been developed for finding the numerical solution of an eigenvalue problem [7.18, 7.21]. Bathe and Wilson [7.22] have done a comparative study of some of these methods. Recent emphasis has been on the economical solution of large eigenproblems [7.23–7.27]. The estimation of natural frequencies by the use of Sturm sequences is presented in Refs. [7.28] and [7.29]. An alternative way to solve a class of lumped mechanical vibration problems using topological methods is presented in Ref. [7.30].

## 7.8  COMPUTER PROGRAMS

### 7.8.1    Jacobi's Method

A Fortran program, in the form of subroutine JACOBI, is given for finding the eigenvalues and eigenvectors of a real symmetric matrix $[D]$ according to Jacobi's method. The arguments of this subroutine are as follows:

D       =    Array of size $N \times N$, containing the elements of the matrix $[D]$. Input data. The diagonal elements of the array contain the eigenvalues upon return to the main program; $D(I, I) = 1/\omega_i^2$. Output.

N       =    Order of the matrix $[D]$. Input data.

E       =    Array of size $N \times N$ in which the eigenvectors are stored columnwise. Output.

EPS   =    Convergence specification. A small quantity on the order of $10^{-5}$ is to be used. Input data.

ITMAX  =    Maximum number of iterations or rotations permitted. Input data.

By way of illustration, Example 7.6 is solved using subroutine JACOBI. The main program, subroutine JACOBI, and the output are given below.

```
C ==================================================================
C
C PROGRAM 13
C MAIN PROGRAM FOR CALLING JACOBI
C
C ==================================================================
C FOLLOWING 3 LINES CONTAIN PROBLEM-DEPENDENT DATA
      DIMENSION D(3,3),E(3,3)
      DATA N,ITMAX,EPS/3,200,1.0E-05/
      DATA D/1.0,1.0,1.0,1.0,2.0,2.0,1.0,2.0,3.0/
C END OF PROBLEM-DEPENDENT DATA
      PRINT 50
   50 FORMAT (//,37H EIGENVALUE SOLUTION BY JACOBI METHOD)
      PRINT 40
   40 FORMAT (/,13H GIVEN MATRIX)
      DO 30 I=1,N
   30 PRINT 20,(D(I,J),J=1,N)
   20 FORMAT (3E15.6)
      CALL JACOBI (D,N,E,EPS,ITMAX)
      PRINT 10, (D(I,I),I=1,N), ((E(I,J),J=1,N),I=1,N)
   10 FORMAT (/,1X,17H EIGEN VALUES ARE,/,3E15.6,//,1X,
     2    14H EIGEN VECTORS,/,6X,5HFIRST,10X,6HSECOND,9X,5HTHIRD,/,
     3    (1X,3E15.6))
      STOP
      END
C ==================================================================
C
C SUBROUTINE JACOBI
C
C ==================================================================
      SUBROUTINE JACOBI (D,N,E,EPS,ITMAX)
      DIMENSION D(N,N),E(N,N)
      ITER=0
      DO 110 I=1,N
      DO 110 J=1,N
      E(I,J)=0.0
```

```
110  E(I,I)=1.0
120  ZZ=0.0
     NM1=N-1
     DO 130 I=1,NM1
     IP1=I+1
     DO 130 J=IP1,N
     IF (ABS(D(I,J)) .LE. ZZ) GO TO 130
     ZZ=ABS(D(I,J))
     IR=I
     IC=J
130  CONTINUE
     IF (ITER .EQ. 0) YY=ZZ*EPS
     IF (ZZ .LE. YY) GO TO 210
     DIF=D(IR,IR) - D(IC,IC)
     TANZ=(-DIF+SQRT(DIF**2+4.0*ZZ**2))/(2.0*D(IR,IC))
     COSZ=1.0/SQRT(1.0+TANZ**2)
     SINZ=COSZ*TANZ
     DO 140 I=1,N
     ZZZ=E(I,IR)
     E(I,IR)=COSZ*ZZZ+SINZ*E(I,IC)
140  E(I,IC)=COSZ*E(I,IC)-SINZ*ZZZ
     I=1
150  IF (I .EQ. IR) GO TO 160
     YYY=D(I,IR)
     D(I,IR)=COSZ*YYY+SINZ*D(I,IC)
     D(I,IC)=COSZ*D(I,IC)-SINZ*YYY
     I=I+1
     GO TO 150
160  I=IR+1
170  IF (I .EQ. IC) GO TO 180
     YYY=D(IR,I)
     D(IR,I)=COSZ*YYY+SINZ*D(I,IC)
     D(I,IC)=COSZ*D(I,IC)-SINZ*YYY
     I=I+1
     GO TO 170
180  I=IC+1
190  IF (I .GT. N) GO TO 200
     ZZZ=D(IR,I)
     D(IR,I)=COSZ*ZZZ+SINZ*D(IC,I)
     D(IC,I)=COSZ*D(IC,I)-SINZ*ZZZ
     I=I+1
     GO TO 190
200  YYY=D(IR,IR)
     D(IR,IR)=YYY*COSZ**2+D(IR,IC)*2.0*COSZ*SINZ+D(IC,IC)
    2    *SINZ**2
     D(IC,IC)=D(IC,IC)*COSZ**2+YYY*SINZ**2-D(IR,IC)*2.0*COSZ*SINZ
     D(IR,IC)=0.0
     ITER=ITER+1
     IF (ITER .LT. ITMAX) GO TO 120
210  RETURN
     END
```

EIGENVALUE SOLUTION BY JACOBI METHOD

GIVEN MATRIX
   0.100000E+01    0.100000E+01    0.100000E+01

```
0.100000E+01    0.200000E+01    0.200000E+01
0.100000E+01    0.200000E+01    0.300000E+01

EIGEN VALUES ARE
0.504892E+01    0.643104E+00    0.307979E+00

EIGEN VECTORS
      FIRST          SECOND           THIRD
   0.327985E+00   -0.736984E+00    0.590999E+00
   0.591007E+00   -0.327977E+00   -0.736981E+00
   0.736978E+00    0.591004E+00    0.327991E+00
```

### 7.8.2   Matrix Iteration Method

Subroutine MITER is given for implementing the matrix iteration method. This subroutine uses the following arguments:

D      =    Array of size $N \times N$, containing the matrix $[D]$. Input data.

X, XX    =    Arrays of size $N$ each.

XS     =    Array of size $N$. Contains the initial guess vector such as $\begin{Bmatrix} 1 \\ 1 \\ \cdot \\ \cdot \\ 1 \end{Bmatrix}$. Input data.

N      =    Order of the matrix $[D]$. Input data.

NVEC    =    Number of eigenvalues and eigenvectors to be found. Input data.

B, C     =    Arrays of size $N \times N$ each.

XM     =    Array of size $N \times N$, containing the mass matrix $[m]$. Input data.

EPS     =    Convergence requirement. A small number on the order of $10^{-5}$ is to be used. Input data.

FREQ    =    Array of size $NVEC$, containing the computed natural frequencies. Output.

EIG     =    Array of size $N \times NVEC$, containing the computed eigenvectors columnwise. Output.

Example 7.5 is solved using subroutine MITER. The main program that calls subroutine MITER, subroutine MITER, and the output of the program are given below.

```
C ================================================================
C
C PROGRAM 14
C MAIN PROGRAM FOR CALLING THE SUBROUTINE MITER
C
C ================================================================
C FOLLOWING 8 LINES CONTAIN PROBLEM-DEPENDENT DATA
      DIMENSION D(3,3),X(3),XS(3),B(3,3),C(3,3),XX(3),XM(3,3),FREQ(3),
     2    EIG(3,3)
      N=3
      NVEC=3
```

```
          DATA D/1.0,1.0,1.0,1.0,2.0,2.0,1.0,2.0,3.0/
          DATA XM/1.0,0.0,0.0,0.0,1.0,0.0,0.0,0.0,1.0/
          EPS=0.00001
          DATA XS/1.0,1.0,1.0/
C END OF PROBLEM-DEPENDENT DATA
          CALL MITER (D,X,XS,N,NVEC,B,C,XX,XM,EPS,FREQ,EIG)
          PRINT 10
   10     FORMAT (//,34H SOLUTION OF EIGENVALUE PROBLEM BY,/,
          2    24H MATRIX ITERATION METHOD)
          PRINT 20,(FREQ(I),I=1,NVEC)
   20     FORMAT (//,20H NATURAL FREQUENCIES,//,3(E15.8,1X))
          PRINT 30
   30     FORMAT (//,26H MODE SHAPES (COLUMNWISE):,/)
          DO 40 I=1,NVEC
   40     PRINT 50,(EIG(I,J),J=1,N)
   50     FORMAT (3(E15.8,1X),/)
          STOP
          END
C================================================================
C
C SUBROUTINE MITER
C
C================================================================
          SUBROUTINE MITER (D,X,XS,N,NVEC,B,C,XX,XM,EPS,FREQ,EIG)
          DIMENSION D(N,N),X(N),XS(N),B(N,N),C(N,N),XX(N),XM(N,N),
          2    FREQ(NVEC),EIG(N,NVEC)
          CON=XS(1)
          DO 10 I=1,N
   10     X(I)=XS(I)/CON
   50     ICON=0
          CALL MULT(D,X,N,XX)
          ALAM=XX(1)
          DO 20 I=1,N
   20     XX(I)=XX(I)/ALAM
          DO 30 I=1,N
   30     IF(ABS((XX(I)-X(I))/X(I)) .GT. EPS) ICON=1
          DO 40 I=1,N
   40     X(I)=XX(I)
          IF (ICON .EQ. 0) GO TO 60
          GO TO 50
   60     ICON=0
          FREQ(1)=SQRT(1.0/ALAM)
          DO 70 I=1,N
   70     EIG(I,1)=X(I)
          II=1
  100     II=II+1
          SUM=0.0
          DO 110 I=1,N
  110     SUM=SUM+X(I)**2
          ALP=SQRT(1.0/SUM)
          DO 120 I=1,N
  120     X(I)=X(I)*ALP
          DO 130 I=1,N
          DO 130 J=1,N
  130     C(I,J)=X(I)*X(J)
          CALL MATMUL (B,C,XM,N,N,N)
```

```
              DO 140 I=1,N
              DO 140 J=1,N
      140     D(I,J)=D(I,J)-ALAM*B(I,J)
              CON=XS(1)
              DO 150 I=1,N
      150     X(I)=XS(I)/CON
      250     ICON=0
              CALL MULT (D,X,N,XX)
              ALAM=XX(1)
              DO 220 I=1,N
      220     XX(I)=XX(I)/ALAM
              DO 230 I=1,N
      230     IF (ABS((XX(I)-X(I))/X(I)) .GT. EPS) ICON=1
              DO 240 I=1,N
      240     X(I)=XX(I)
              IF (ICON .EQ. 0) GO TO 260
              GO TO 250
      260     ICON=0
              FREQ(II)=SQRT(1.0/ALAM)
              DO 270 I=1,N
      270     EIG(I,II)=X(I)
              IF (II .EQ. NVEC) GO TO 300
              GO TO 100
      300     RETURN
              END
C ================================================================
C
C SUBROUTINE MULT
C
C ================================================================
              SUBROUTINE MULT (D,X,N,XX)
              DIMENSION D(N,N),X(N),XX(N)
              DO 20 I=1,N
              XX(I)=0.0
              DO 10 J=1,N
      10      XX(I)=XX(I)+D(I,J)*X(J)
      20      CONTINUE
              RETURN
              END
```

SOLUTION OF EIGENVALUE PROBLEM BY
MATRIX ITERATION METHOD


NATURAL FREQUENCIES

0.44504240E+00   0.12469811E+01   0.18019377E+01


MODE SHAPES (COLUMNWISE):

0.10000000E+01   0.10000000E+01   0.10000000E+01

0.18019372E+01   0.44503731E+00  -0.12469926E+01

0.22469788E+01  -0.80193609E+00   0.55496848E+00

### 7.8.3    Choleski Decomposition

Subroutine DECOMP is given for decomposing a matrix $[A]$ of order $N$ using Choleski decomposition. The program gives the upper triangular matrix $[U]$ as output. The listing of DECOMP is given below.

```
C  ==============================================================
C
C PROGRAM 15
C MAIN PROGRAM WHICH CALLS DECOMP
C
C  ==============================================================
       DIMENSION A(3,3),U(3,3)
       DATA A/5.0,1.0,0.0,1.0,3.0,2.0,0.0,2.0,8.0/
       N=3
       CALL DECOMP (A,U,N)
       PRINT 10
10     FORMAT (/,25H UPPER TRIANGULAR MATRIX:,/)
       DO 30 I=1,N
       PRINT 20, (U(I,J),J=1,N)
20     FORMAT (3E15.8)
30     CONTINUE
       STOP
       END
C  ==============================================================
C
C SUBROUTINE DECOMP
C
C  ==============================================================
       SUBROUTINE DECOMP (A,U,N)
       DIMENSION A(N,N),U(N,N)
       DO 10 I=1,N
       DO 10 J=1,N
10     U(I,J)=0.0
       U(1,1)=SQRT(A(1,1))
       DO 90 J=2,N
90     U(1,J)=A(1,J)/U(1,1)
       DO 40 I=2,N
       IM=I-1
       SUM=0.0
       DO 30 K=1,IM
30     SUM=SUM+U(K,I)**2
       U(I,I)=SQRT(A(I,I)-SUM)
       J=I+1
       SUM=0.0
       DO 50 K=1,IM
50     SUM=SUM+U(K,I)*U(K,J)
       U(I,J)=(A(I,J)-SUM)/U(I,I)
40     CONTINUE
       RETURN
       END
UPPER TRIANGULAR MATRIX:

0.22360680E+01  0.44721359E+00  0.00000000E+00
0.00000000E+00  0.16733201E+01  0.11952286E+01
0.00000000E+00  0.00000000E+00  0.25634799E+01
```

### 7.8.4   Eigenvalue Solution Using Choleski Decomposition

A Fortran program is written for solving the general eigenvalue problem

$$[k]\vec{X} = \omega^2[m]\vec{X}$$

The problem is converted into a special eigenvalue problem

$$[D]\vec{Y} = \frac{1}{\omega^2}[I]\vec{Y}$$

by generating the matrix $[D]$ using the relation

$$[D] = \left([U]^T\right)^{-1}[m][U]^{-1}$$

The following data is needed for this program:

BK = Array of size $N \times N$, containing the matrix $[k]$.

BM = Array of size $N \times N$, containing the matrix $[m]$.

ND = Order of the matrices $[k]$ and $[m]$.

The following quantities need to be defined and dimensioned in the program:

U, UI, UTI      = Arrays of size $N \times N$, indicating the matrices $[U]$, $[U]^{-1}$, and $([U]^{-1})^T$ respectively.

BMU, UMU, EV   = Arrays of size $N \times N$ each.

XF              = Array of size $N \times N$. The computed eigenvectors $\vec{X}^{(i)}$ are stored columnwise in XF.

To illustrate the use of the program, the matrices of Example 7.2 are used with

$$[k] = \begin{bmatrix} 2 & -1 & 0 \\ -1 & 2 & -1 \\ 0 & -1 & 1 \end{bmatrix} \quad \text{and} \quad [m] = \begin{bmatrix} 1 & 0 & 0 \\ 0 & 1 & 0 \\ 0 & 0 & 1 \end{bmatrix}$$

The listing of the program and the output are given below.

```
C ================================================================
C
C PROGRAM 16
C DETERMINATION OF EIGENVALUES AND EIGENVECTORS BY FINDING THE MATRIX
C [D] ACCORDING TO THE RELATION [D] = [UTI][M][UI]
C
C ================================================================
C FOLLOWING 5 LINES CONTAIN PROBLEM-DEPENDENT DATA
      DIMENSION BK(3,3),BM(3,3),U(3,3),UI(3,3),UTI(3,3),BMU(3,3),
     2   UMU(3,3),XF(3,3),EV(3,3)
      DATA BK/2.0,-1.0,0.0,-1.0,2.0,-1.0,0.0,-1.0,1.0/
      DATA BM/1.0,0.0,0.0,0.0,1.0,0.0,0.0,0.0,1.0/
      ND=3
C END OF PROBLEM-DEPENDENT DATA
C DECOMPOSING THE MATRIX [BK] INTO TRIANGULAR MATRICES
      CALL DECOMP (BK,U,ND)
      PRINT 100
```

```
100    FORMAT (//,29H UPPER TRIANGULAR MATRIX [U]:,/)
       DO 110 I=1,ND
110    PRINT 120,(U(I,J),J=1,ND)
120    FORMAT (4(1X,E15.6))
       DO 130 I=1,ND
       DO 130 J=1,ND
130    UI(I,J)=0.0
       DO 140 I=1,ND
140    UI(I,I)=1.0/U(I,I)
       DO 160 J=1,ND
       DO 160 II=1,ND
       I=ND-II+1
       IF (I .GE. J) GO TO 160
       IP=I+1
       SUM=0.0
       DO 150 K=IP,J
150    SUM =SUM+U(I,K)*UI(K,J)
       UI(I,J)=-SUM/U(I,I)
160    CONTINUE
       PRINT 170
170    FORMAT (/,46H INVERSE OF THE UPPER TRIANGULAR MATRIX, [UI],/)
       DO 180 I=1,ND
180    PRINT 120,(UI(I,J),J=1,ND)
       DO 190 I=1,ND
       DO 190 J=1,ND
190    UTI(I,J)=UI(J,I)
       CALL MATMUL (BMU,BM,UI,ND,ND,ND)
       CALL MATMUL (UMU,UTI,BMU,ND,ND,ND)
       PRINT 200
200    FORMAT (/,29H MATRIX [UMU] = [UTI][M][UI]:,/)
       DO 210 I=1,ND
210    PRINT 120,(UMU(I,J),J=1,ND)
       CALL JACOBI (UMU,ND,EV,1.0E-05,200)
       PRINT 230
230    FORMAT (/,13H EIGENVALUES:,/)
       PRINT 120,(UMU(I,I),I=1,ND)
       PRINT 240
240    FORMAT (/,27H EIGENVECTORS (COLUMNWISE):,/)
       CALL MATMUL (XF,UI,EV,ND,ND,ND)
       DO 250 I=1,ND
250    PRINT 120,(XF(I,J),J=1,ND)
       STOP
       END
```

UPPER TRIANGULAR MATRIX [U]:

```
    0.141421E+01    -0.707107E+00     0.000000E+00
    0.000000E+00     0.122474E+01    -0.816497E+00
    0.000000E+00     0.000000E+00     0.577350E+00
```

INVERSE OF THE UPPER TRIANGULAR MATRIX, [UI],

```
    0.707107E+00     0.408248E+00     0.577350E+00
    0.000000E+00     0.816497E+00     0.115470E+01
    0.000000E+00     0.000000E+00     0.173205E+01
```

```
MATRIX [UMU] = [UTI][M][UI]:

    0.500000E+00    0.288675E+00    0.408248E+00
    0.288675E+00    0.833333E+00    0.117851E+01
    0.408248E+00    0.117851E+01    0.466667E+01

EIGENVALUES:

    0.504892E+01    0.643104E+00    0.307979E+00

EIGENVECTORS (COLUMNWISE):

    0.736973E+00   -0.590976E+00    0.328051E+00
    0.132799E+01   -0.263064E+00   -0.408952E+00
    0.165597E+01    0.473971E+00    0.181988E+00
```

## REFERENCES

**7.1.** S. Dunkerley, "On the whirling and vibration of shafts," *Philosophical Transactions of the Royal Society of London*, 1894, Series A, Vol. 185, Part I, pp. 279–360.

**7.2.** B. Atzori, "Dunkerley's formula for finding the lowest frequency of vibration of elastic systems," letter to the editor, *Journal of Sound and Vibration*, 1974, Vol. 36, pp. 563–564.

**7.3.** H. H. Jeffcott, "The periods of lateral vibration of loaded shafts—The rational derivation of Dunkerley's empirical rule for determining whirling speeds," *Proceedings of the Royal Society of London*, 1919, Series A, Vol. 95, No. A666, pp. 106–115.

**7.4.** M. Endo and O. Taniguchi, "An extension of the Southwell–Dunkerley methods for synthesizing frequencies," *Journal of Sound and Vibration*, 1976, "Part I: Principles," Vol. 49, pp. 501–516, and "Part II: Applications," Vol. 49, pp. 517–533.

**7.5.** A. Rutenberg, "A lower bound for Dunkerley's formula in continuous elastic systems," *Journal of Sound and Vibration*, 1976, Vol. 45, pp. 249–252.

**7.6.** G. Temple and W. G. Bickley, *Rayleigh's Principle and Its Applications to Engineering*, Dover, New York, 1956.

**7.7.** N. G. Stephen, "Rayleigh's, Dunkerley's and Southwell's methods," *International Journal of Mechanical Engineering Education*, January 1983, Vol. 11, pp. 45–51.

**7.8.** A. Rutenberg, "Dunkerley's formula and alternative approximations," letter to the editor, *Journal of Sound and Vibration*, 1975, Vol. 39, pp. 530–531.

**7.9.** R. Jones, "Approximate expressions for the fundamental frequency of vibration of several dynamic systems," *Journal of Sound and Vibration*, 1976, Vol. 44, pp. 475–478.

**7.10.** R. W. Fitzgerald, *Mechanics of Materials* (2nd Ed.), Addison-Wesley, Reading, Mass., 1982.

**7.11.** H. Holzer, *Die Berechnung der Drehschwin gungen*, Julius Springer, Berlin, 1921.

**7.12.** H. E. Fettis, "A modification of the Holzer method for computing uncoupled torsion and bending modes," *Journal of the Aeronautical Sciences*, October 1949, pp. 625–634; May 1954, pp. 359–360.

**7.13.** S. H. Crandall and W. G. Strang, "An improvement of the Holzer Table based on a suggestion of Rayleigh's," *Journal of Applied Mechanics*, 1957, Vol. 24, p. 228.

**7.14.** S. Mahalingam, "An improvement of the Holzer method," *Journal of Applied Mechanics*, 1958, Vol. 25, p. 618.

**7.15.** S. Mahalingam, "Iterative procedures for torsional vibration analysis and their relationships," *Journal of Sound and Vibration*, 1980, Vol. 68, pp. 465–467.

**7.16.** L. Meirovitch, *Computational Methods in Structural Dynamics*, Sijthoff and Noordhoff, The Netherlands, 1980.

**7.17.** J. H. Wilkinson and G. Reinsch, *Linear Algebra*, Springer Verlag, New York, 1971.

**7.18.** J. W. Wilkinson, *The Algebraic Eigenvalue Problem*, Oxford University Press, London, 1965.

**7.19.** R. S. Martin and J. H. Wilkinson, "Reduction of a symmetric eigenproblem $Ax = \lambda Bx$ and related problems to standard form," *Numerical Mathematics*, 1968, Vol. 11, pp. 99–110.

**7.20.** G. B. Haggerty, *Elementary Numerical Analysis with Programming*, Allyn & Bacon, Boston, 1972.

**7.21.** A. Jennings, "Eigenvalue methods for vibration analysis," *Shock and Vibration Digest*, Part I, February 1980, Vol. 12, pp. 3–16; Part II, January 1984, Vol. 16, pp. 25–33.

**7.22.** K. J. Bathe and E. L. Wilson, "Solution methods for eigenvalue problems in structural mechanics," *International Journal for Numerical Methods in Engineering*, 1973, Vol. 6, pp. 213–226.

**7.23.** E. Cohen and H. McCallion, "Economical methods for finding eigenvalues and eigenvectors," *Journal of Sound and Vibration*, 1967, Vol. 5, pp. 397–406.

**7.24.** A. Jennings and M. Clint, "The evaluation of eigenvalues and eigenvectors of real symmetric matrices by simultaneous iteration," *Computer Journal*, 1970, Vol. 13, pp. 76–80.

**7.25.** A. J. Fricker, "A method for solving high-order real symmetric eigenvalue problems," *International Journal for Numerical Methods in Engineering*, 1983, Vol. 19, pp. 1131–1138.

**7.26.** G. C. Wright and G. A. Miles, "An economical method for determining the smallest eigenvalues of large linear systems," *International Journal for Numerical Methods in Engineering*, 1971, Vol. 3, pp. 25–33.

**7.27.** S. B. Dong, J. A. Wolf, Jr., and F. E. Peterson, "On a direct-iterative eigensolution technique," *International Journal for Numerical Methods in Engineering*, 1972, Vol. 4, pp. 155–161.

**7.28.** G. Longbottom and K. F. Gill, "The estimation of natural frequencies by use of Sturm sequences," *International Journal of Mechanical Engineering Education*, 1976, Vol. 4, pp. 319–329.

**7.29.** K. K. Gupta, "Solution of eigenvalue problems by Sturm sequence method," *International Journal for Numerical Methods in Engineering*, 1972, Vol. 4, pp. 379–404.

**7.30.** W. K. Chen and F. Y. Chen, "Topological analysis of a class of lumped vibrational systems," *Journal of Sound and Vibration*, 1969, Vol. 10, pp. 198–207.

# REVIEW QUESTIONS

**7.1.** Name a few methods for finding the fundamental natural frequency of a multidegree of freedom system.

**7.2.** What is the basic assumption made in deriving Dunkerley's formula?

**7.3.** What is Rayleigh's principle?

**7.4.** State whether we get a lower bound or an upper bound to the fundamental natural frequency if we use (a) Dunkerley's formula and (b) Rayleigh's method.

**7.5.** What is Rayleigh's quotient?

**7.6.** What is the basic principle used in Holzer's method?

**7.7.** What is the matrix iteration method?

**7.8.** Can we use any trial vector $\vec{X}_1$ in the matrix iteration method to find the largest natural frequency?

**7.9.** How do you find the intermediate natural frequencies using the matrix iteration method?

**7.10.** What is the difference between the matrix iteration method and Jacobi's method?

**7.11.** What is a rotation matrix? What is its purpose in Jacobi's method?

**7.12.** What is a standard eigenvalue problem?

**7.13.** What is the role of Choleski decomposition in deriving a standard eigenvalue problem?

**7.14.** How do you find the inverse of an upper triangular matrix?

## PROBLEMS

The problem assignments are organized as follows:

| Problems | Section covered | Topic covered |
|---|---|---|
| 7.1–7.6 | 7.2 | Dunkerley's formula |
| 7.7–7.12 | 7.3 | Rayleigh's method |
| 7.13–7.17 | 7.4 | Holzer's method |
| 7.18–7.23 | 7.5 | Matrix iteration method |
| 7.24–7.25, 7.29 | 7.6 | Jacobi's method |
| 7.26–7.28, 7.30–7.31 | 7.7 | Standard eigenvalue problem |
| 7.32–7.38 | 7.8 | Computer programs |

**7.1.** Estimate the fundamental frequency of the beam shown in Fig. 6.3 using Dunkerley's formula for the following data:
   (a) $m_1 = m_3 = 5m$, $m_2 = m$.    (b) $m_1 = m_3 = m$, $m_2 = 5m$

**7.2.** Find the fundamental frequency of the torsional system shown in Fig. 6.5, using Dunkerley's formula for the following data:
   (a) $J_1 = J_2 = J_3 = J_0$; $k_{t1} = k_{t2} = k_{t3} = k_t$.
   (b) $J_1 = J_0$, $J_2 = 2J_0$, $J_3 = 3J_0$; $k_{t1} = k_t$, $k_{t2} = 2k_t$, $k_{t3} = 3k_t$.

**7.3.** Estimate the fundamental frequency of the beam system shown in Fig. 7.2, using Dunkerley's formula for the following data: $m_1 = m$, $m_2 = 2m$, $l_1 = l_2 = l_3 = l/3$.

**7.4.** The natural frequency of an airplane wing in torsion is found to be 50 Hz. A fuel tank is hung from the center line of the airplane such that its mass moment of inertia about the torsional axis is 250 kg-m². If the torsional stiffness of the wing at this point is $4 \times 10^6$ N-m/rad, find the new torsional frequency of the wing, using Dunkerley's formula.

**7.5.** A mass $M$ is attached at the free end of a cantilever beam of mass $m$, as shown in Fig. 7.9. Estimate the fundamental frequency of the beam when $M = 2m$, using Dunkerley's formula.

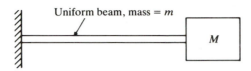

**Figure 7.9**

**7.6.** Determine the fundamental natural frequency of the stretched string system shown in Fig. 5.22 with $m_1 = m_2 = m$ and $l_1 = l_2 = l_3 = l$, using Dunkerley's formula.

**7.7.** Determine the first natural frequency of vibration of the system shown in Fig. 7.1, using Rayleigh's method. Assume $k_1 = k$, $k_2 = 2k$, $k_3 = 3k$, and $m_1 = m$, $m_2 = 2m$, $m_3 = 3m$.

**7.8.** Find the fundamental natural frequency of the torsional system shown in Fig. 6.5 using Rayleigh's method. Assume that $J_1 = J_0$, $J_2 = 2J_0$, $J_3 = 3J_0$, and $k_{t1} = k_{t2} = k_{t3} = k_t$.

**7.9.** Solve Problem 7.6 using Rayleigh's method.

**7.10.** Determine the fundamental natural frequency of the system shown in Fig. 5.22 when $m_1 = m$, $m_2 = 5m$, and $l_1 = l_2 = l_3 = l$, using Rayleigh's method.

**7.11.** Calculate the first natural frequency of the building frame of Problem 5.15 using Rayleigh's method. Assume the first mode configuration to be the same as the static equilibrium shape due to loads proportional to the girder weights.

**7.12.** Prove that Rayleigh's quotient is never higher than the highest eigenvalue.

**7.13.** Find the natural frequencies and mode shapes of the system shown in Fig. 6.8, with $m_1 = 100$ kg, $m_2 = 20$ kg, $m_3 = 200$ kg, $k_1 = 8000$ N/m, and $k_2 = 4000$ N/m, using Holzer's method.

**7.14.** The stiffness and mass matrices of a vibrating system are given by

$$[k] = k \begin{bmatrix} 2 & -1 & 0 \\ -1 & 2 & -1 \\ 0 & -1 & 3 \end{bmatrix}, \qquad [m] = m \begin{bmatrix} 1 & 0 & 0 \\ 0 & 1 & 0 \\ 0 & 0 & 2 \end{bmatrix}$$

Determine all the principal modes and the natural frequencies, using Holzer's method.

**7.15.** For the torsional system shown in Fig. 6.5, determine a principal mode and the corresponding frequency by Holzer's method. Assume $k_{t1} = k_{t2} = k_{t3} = k_t$ and $J_1 = J_2 = J_3 = J_0$.

**7.16.** Find the natural frequencies and mode shapes of the building frame shown in Fig. 5.21, using Holzer's method. Assume that $m_1 = 2m$, $m_2 = m$, that $EI_1 = 2EI$, $EI_2 = EI$, and that the ends of the columns are constrained against rotation.

**7.17.** Find the natural frequencies and mode shapes of the system shown in Fig. 6.22 with $J_1 = 10$ kg-m², $J_2 = 5$ kg-m², $J_3 = 1$ kg-m², and $k_{t1} = k_{t2} = 1 \times 10^6$ N-m/rad, using Holzer's method.

**7.18.** The largest eigenvalue of the matrix

$$[D] = \begin{bmatrix} 2.5 & -1 & 0 \\ -1 & 5 & -\sqrt{2} \\ 0 & -\sqrt{2} & 10 \end{bmatrix}$$

is given by $\lambda_1 = 10.38068$. Find the other eigenvalues and all the eigenvectors of the matrix, using the matrix iteration method. Assume $[m] = [I]$.

**7.19.** The mass and stiffness matrices of a spring-mass system are known to be

$$[m] = m\begin{bmatrix} 1 & 0 & 0 \\ 0 & 1 & 0 \\ 0 & 0 & 2 \end{bmatrix} \quad \text{and} \quad [k] = k\begin{bmatrix} 2 & -1 & 0 \\ -1 & 3 & -2 \\ 0 & -2 & 2 \end{bmatrix}$$

Find the natural frequencies and mode shapes of the system, using the matrix iteration method.

**7.20.** Find the natural frequencies and mode shapes of the system shown in Fig. 6.2 with $k_1 = k$, $k_2 = 2k$, $k_3 = 3k$, and $m_1 = m_2 = m_3 = m$, using the matrix iteration method.

**7.21.** Find the natural frequencies of the system shown in Fig. 6.11, using the matrix iteration method. Assume that $J_{d1} = J_{d2} = J_{d3} = J_0$, $l_i = l$, and $(GJ)_i = GJ$ for $i = 1$ to 4.

**7.22.** Solve Problem 7.6 using the matrix iteration method.

**7.23.** The stiffness and mass matrices of a vibrating system are given by

$$[k] = k\begin{bmatrix} 4 & -2 & 0 & 0 \\ -2 & 3 & -1 & 0 \\ 0 & -1 & 2 & -1 \\ 0 & 0 & -1 & 1 \end{bmatrix} \quad \text{and} \quad [m] = m\begin{bmatrix} 3 & 0 & 0 & 0 \\ 0 & 2 & 0 & 0 \\ 0 & 0 & 1 & 0 \\ 0 & 0 & 0 & 1 \end{bmatrix}$$

Find the fundamental frequency and the mode shape of the system, using the matrix iteration method.

**7.24.** Find the eigenvalues and eigenvectors of the matrix

$$[D] = \begin{bmatrix} 3 & -2 & 0 \\ -2 & 5 & -3 \\ 0 & -3 & 3 \end{bmatrix}$$

using Jacobi's method.

**7.25.** Find the eigenvalues and eigenvectors of the matrix

$$[D] = \begin{bmatrix} 3 & 2 & 1 \\ 2 & 2 & 1 \\ 1 & 1 & 1 \end{bmatrix}$$

using Jacobi's method.

**7.26.** Decompose the matrix

$$[A] = \begin{bmatrix} 4 & -2 & 6 & 4 \\ -2 & 2 & -1 & 3 \\ 6 & -1 & 22 & 13 \\ 4 & 3 & 13 & 46 \end{bmatrix}$$

using the Choleski decomposition technique.

**7.27.** Find the inverse of the following matrix, using the decomposition $[A] = [U]^T[U]$:

$$[A] = \begin{bmatrix} 5 & -1 & 1 \\ -1 & 6 & -4 \\ 1 & -4 & 3 \end{bmatrix}$$

**7.28.** Find the inverse of the following matrix, using Choleski decomposition:

$$[A] = \begin{bmatrix} 2 & 5 & 8 \\ 5 & 16 & 28 \\ 8 & 28 & 54 \end{bmatrix}$$

**7.29.** Find the eigenvalues of the matrix $[A]$ given in Problem 7.26, using Jacobi's method.

**7.30.** Convert Problem 7.23 to a standard eigenvalue problem with a symmetric matrix.

**7.31.** Using the Choleski decomposition technique, express the following matrix as the product of two triangular matrices:

$$[A] = \begin{bmatrix} 16 & -20 & -24 \\ -20 & 89 & -50 \\ -24 & -50 & 280 \end{bmatrix}$$

**7.32.** Find the eigenvalues and eigenvectors of the matrix $[D]$ given in problem 7.18, using subroutine JACOBI.

**7.33.** Solve the general eigenvalue problem stated in Problem 7.19, using the computer program of Sec. 7.8.4.

**7.34.** Solve the general eigenvalue problem stated in Problem 7.23, using the computer program of Sec. 7.8.4.

**7.35.** Find the eigenvalues and eigenvectors of Problem 7.14, using the computer program of Sec. 7.8.4.

**7.36.** Solve Problem 7.25, using subroutine MITER.

**7.37.** Find the eigensolution of the matrix $[A]$ of Problem 7.27, using subroutine JACOBI.

**7.38.** Find the eigensolution of the matrix $[A]$ given in Problem 7.31, using subroutine JACOBI.

# CHAPTER 8

# Continuous Systems

Stephen Prokf'yevich Timoshenko (1878–1972), a Russian-born engineer who emigrated to the U.S.A., was one of the most widely known authors of books in the field of elasticity, strength of materials, and vibrations. He held the chair of mechanics at the University of Michigan and later at Stanford University, and is regarded as the father of engineering mechanics in the U.S.A. The improved theory he presented in 1921 for the vibration of beams has become known as the Timoshenko beam theory.

Courtesy Stanford University School of Engineering

## 8.1 INTRODUCTION

We have so far dealt with discrete systems where mass, damping, and elasticity were assumed to be present only at certain discrete points in the system. There are many cases, known as *distributed* or *continuous systems*, in which it is not possible to identify discrete masses, dampers, or springs. We must then consider the continuous distribution of the mass, damping, and elasticity and assume that each of the infinite number of points of the system can vibrate. This is why a continuous system is also called a *system of infinite degrees of freedom.*

If a system is modeled as a discrete one, the governing equations are ordinary differential equations, which are relatively easy to solve. On the other hand, if the system is modeled as a continuous one, the governing equations are partial differential equations, which are more difficult. However, the information obtained from a discrete model of a system may not be as accurate as that obtained from a continuous model. The choice between the two models must be made carefully, with due consideration of factors such as the purpose of the analysis, the influence of the analysis on design, and the computational time available.

In this chapter, we shall consider the vibration of simple continuous systems—strings, bars, shafts, beams, and membranes. A more specialized treatment of the vibration of continuous structural elements is given in Refs. [8.1–8.3]. In general, the frequency equation of a continuous system is a transcendental equation that yields an infinite number of natural frequencies and normal modes. This is in contrast to the behavior of discrete systems, which yield a finite number of such frequencies and modes. We need to apply boundary conditions to find the natural frequencies of a continuous system. The question of boundary conditions does not arise in the case of discrete systems except in an indirect way, because the influence coefficients depend on the manner in which the system is supported.

## 8.2 TRANSVERSE VIBRATION OF A STRING OR CABLE

### 8.2.1 Equation of Motion

Consider a tightly stretched elastic string or cable of length $l$ subjected to a transverse force $f(x, t)$ per unit length, as shown in Fig. 8.1(a). The transverse displacement of the string, $w(x, t)$, is assumed to be small. Equilibrium of the forces in the $z$ direction gives (see Fig. 8.1(b)):

net force acting on an element = inertia force acting on the element

or

$$(P + dP)\sin(\theta + d\theta) + f\,dx - P\sin\theta = \rho\,dx\,\frac{\partial^2 w}{\partial t^2} \tag{8.1}$$

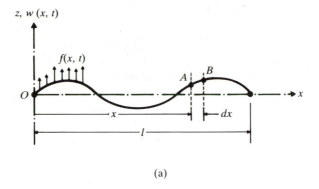

(a)

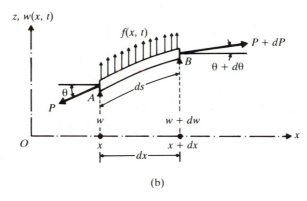

(b)

**Figure 8.1**  A vibrating string.

where $P$ is the tension, $\rho$ is the mass per unit length, and $\theta$ is the angle the deflected string makes with the $x$ axis. For an elemental length $dx$,

$$dP = \frac{\partial P}{\partial x} \cdot dx \tag{8.2}$$

$$\sin\theta \simeq \tan\theta = \frac{\partial w}{\partial x} \tag{8.3}$$

and

$$\sin(\theta + d\theta) \simeq \tan(\theta + d\theta) = \frac{\partial w}{\partial x} + \frac{\partial^2 w}{\partial x^2} \cdot dx \tag{8.4}$$

Hence the forced vibration equation of the nonuniform string, Eq. (8.1), can be simplified to

$$\frac{\partial}{\partial x}\left[P\frac{\partial w(x,t)}{\partial x}\right] + f(x,t) = \rho(x)\frac{\partial^2 w(x,t)}{\partial t^2} \tag{8.5}$$

If the string is uniform and the tension is constant, Eq. (8.5) reduces to

$$P\frac{\partial^2 w(x,t)}{\partial x^2} + f(x,t) = \rho\frac{\partial^2 w(x,t)}{\partial t^2} \tag{8.6}$$

If $f(x,t) = 0$, we obtain the free vibration equation

$$P\frac{\partial^2 w(x,t)}{\partial x^2} = \rho\frac{\partial^2 w(x,t)}{\partial t^2} \tag{8.7}$$

or

$$c^2\frac{\partial^2 w}{\partial x^2} = \frac{\partial^2 w}{\partial t^2} \tag{8.8}$$

where

$$c = \left(\frac{P}{\rho}\right)^{1/2} \tag{8.9}$$

Equation (8.8) is also known as the *wave equation*.

### 8.2.2    Initial and Boundary Conditions

The equation of motion, Eq. (8.5) or its special forms (8.6) and (8.7), is a partial differential equation of the second order. Since the order of the highest derivative of $w$ with respect to $x$ and $t$ in this equation is two, we need to specify two boundary and two initial conditions in finding the solution $w(x,t)$. If the string has a known deflection $w_0(x)$ and velocity $\dot{w}_0(x)$ at time $t = 0$, the initial conditions are specified as

$$w(x, t=0) = w_0(x)$$

$$\frac{\partial w}{\partial t}(x, t=0) = \dot{w}_0(x) \tag{8.10}$$

If the string is fixed at an end, say $x = 0$, the displacement $w$ must always be zero, and so the boundary condition is

$$w(x=0, t) = 0, \qquad t \geq 0 \tag{8.11}$$

Sometimes the string or cable is fastened to a smooth pin or bearing that is free to move in a smooth slot in a direction perpendicular to the string, as shown in Fig. 8.2. This type of end will not support any force at right angles to the cable, so the boundary condition can be stated as

$$P(x)\frac{\partial w(x,t)}{\partial x} = 0 \tag{8.12}$$

If the end $x = 0$ is free and $P$ is a constant, then Eq. (8.12) becomes

$$\frac{\partial w(0,t)}{\partial x} = 0, \qquad t \geq 0 \tag{8.13}$$

If the end $x = l$ is constrained elastically as shown in Fig. 8.3, the boundary

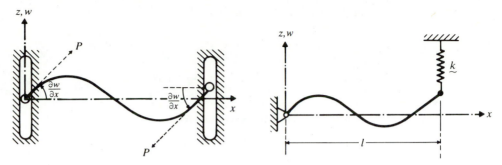

**Figure 8.2**                                    **Figure 8.3**

condition becomes

$$P(x) \frac{\partial w(x,t)}{\partial x}\bigg|_{x=l} = -\underset{\sim}{k}w(x,t)\big|_{x=l}, \qquad t \geq 0 \tag{8.14}$$

where $\underset{\sim}{k}$ is the spring constant.

### 8.2.3  Free Vibration of a Uniform String

The free vibration equation, Eq. (8.8), can be solved by the method of separation of variables. In this method, the solution is written as the product of a function $W(x)$ (which depends only on $x$) and a function $T(t)$ (which depends only on $t$) [8.4]:

$$w(x,t) = W(x)T(t) \tag{8.15}$$

Substitution of Eq. (8.15) into Eq. (8.8) leads to

$$\frac{c^2}{W}\frac{d^2W}{dx^2} = \frac{1}{T}\frac{d^2T}{dt^2} \tag{8.16}$$

Since the left-hand side of this equation depends only on $x$ and the right-hand side depends only on $t$, their common value must be a constant—say, $a$—so that

$$\frac{c^2}{W}\frac{d^2W}{dx^2} = \frac{1}{T}\frac{d^2T}{dt^2} = a \tag{8.17}$$

The equations implied in Eq. (8.17) can be written as

$$\frac{d^2W}{dx^2} - \frac{a}{c^2}W = 0 \tag{8.18}$$

$$\frac{d^2T}{dt^2} - aT = 0 \tag{8.19}$$

Since the constant $a$ is generally negative (see Problem 8.11), we can set $a = -\omega^2$

and write Eqs. (8.18) and (8.19) as

$$\frac{d^2W}{dx^2} + \frac{\omega^2}{c^2} W = 0 \tag{8.20}$$

$$\frac{d^2T}{dt^2} + \omega^2 T = 0 \tag{8.21}$$

The solutions of these equations are given by

$$W(x) = A\cos\frac{\omega x}{c} + B\sin\frac{\omega x}{c} \tag{8.22}$$

$$T(t) = C\cos\omega t + D\sin\omega t \tag{8.23}$$

where $\omega$ is the frequency of vibration and the constants $A$, $B$, $C$, and $D$ can be evaluated from the boundary and initial conditions.

### 8.2.4   Free Vibration of a String with Both Ends Fixed

If the string is fixed at both ends, the boundary conditions are $w(0, t) = w(l, t) = 0$ for all time $t \geq 0$. Hence, from Eq. (8.15), we obtain

$$W(0) = 0 \tag{8.24}$$

$$W(l) = 0 \tag{8.25}$$

In order to satisfy Eq. (8.24), $A$ must be zero in Eq. (8.22). Equation (8.25) requires that

$$B\sin\frac{\omega l}{c} = 0 \tag{8.26}$$

Since $B$ cannot be zero for a nontrivial solution, we have

$$\sin\frac{\omega l}{c} = 0 \tag{8.27}$$

Equation (8.27) is called the *frequency or characteristic equation* and is satisfied by several values of $\omega$. The values of $\omega$ are called the *eigenvalues* (or *natural frequencies* or *characteristic values*) of the problem. The $n$th natural frequency is given by

$$\frac{\omega_n l}{c} = n\pi, \qquad n = 1, 2, \ldots$$

or

$$\omega_n = \frac{nc\pi}{l}, \qquad n = 1, 2, \ldots \tag{8.28}$$

The solution $w_n(x, t)$ corresponding to $\omega_n$ can be expressed as

$$w_n(x, t) = W_n(x)T_n(t) = \sin\frac{n\pi x}{l}\left[C_n\cos\frac{nc\pi t}{l} + D_n\sin\frac{nc\pi t}{l}\right] \tag{8.29}$$

where $C_n$ and $D_n$ are arbitrary constants. The solution $w_n(x, t)$ is called the *nth mode of vibration* or *nth harmonic* or *nth normal mode* of the string. In this mode, each point of the string vibrates with an amplitude proportional to the value of $W_n$ at that point, with the circular frequency $\omega_n = (nc\pi)/l$. The first three modes of vibration are shown in Fig. 8.4. The mode corresponding to $n=1$ is called the *fundamental mode*, and $\omega_1$ is called the *fundamental frequency*. The fundamental period is

$$\tau_1 = \frac{2\pi}{\omega_1} = \frac{2l}{c}$$

The points at which $w_n = 0$ for all times are called *nodes*. Thus the fundamental mode has two nodes, at $x=0$ and $x=l$; the second mode has three nodes, at $x=0$, $x=l/2$, and $x=l$; etc.

The general solution of Eq. (8.8), which satisfies the boundary conditions of Eqs. (8.24) and (8.25), is given by the superposition of all $w_n(x,t)$:

$$w(x,t) = \sum_{n=1}^{\infty} w_n(x,t)$$

$$= \sum_{n=1}^{\infty} \sin\frac{n\pi x}{l}\left[C_n\cos\frac{nc\pi t}{l} + D_n\sin\frac{nc\pi t}{l}\right] \tag{8.30}$$

This equation gives all possible vibrations of the string; the particular vibration that occurs is uniquely determined by the specified initial conditions. The initial conditions give unique values of the constants $C_n$ and $D_n$. If the initial conditions are

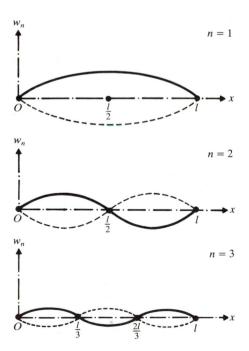

**Figure 8.4** Mode shapes of a string.

specified as in Eq. (8.10), we obtain

$$\sum_{n=1}^{\infty} C_n \sin \frac{n\pi x}{l} = w_0(x) \tag{8.31}$$

$$\sum_{n=1}^{\infty} \frac{nc\pi}{l} D_n \sin \frac{n\pi x}{l} = \dot{w}_0(x) \tag{8.32}$$

which can be seen to be Fourier sine series expansions of $w_0(x)$ and $\dot{w}_0(x)$ in the interval $0 \le x \le l$. The values of $C_n$ and $D_n$ can be determined by multiplying Eqs. (8.31) and (8.32) by $\sin(n\pi x/l)$ and integrating with respect to $x$ from 0 to $l$:

$$C_n = \frac{2}{l} \int_0^l w_0(x) \sin \frac{n\pi x}{l} \, dx \tag{8.33}$$

$$D_n = \frac{2}{nc\pi} \int_0^l \dot{w}_0(x) \sin \frac{n\pi x}{l} \, dx \tag{8.34}$$

### EXAMPLE 8.1

If a string of length $l$, fixed at both ends, is plucked at its midpoint as shown in Fig. 8.5 and then released, determine its subsequent motion.

**Solution.** The solution is given by Eq. (8.30) with $C_n$ and $D_n$ given by Eqs. (8.33) and (8.34), respectively. Since there is no initial velocity, $\dot{w}_0(x) = 0$, and so $D_n = 0$. Thus the solution of Eq. (8.30) reduces to

$$w(x, t) = \sum_{n=1}^{\infty} C_n \sin \frac{n\pi x}{l} \cos \frac{nc\pi t}{l} \tag{E.1}$$

where

$$C_n = \frac{2}{l} \int_0^l w_0(x) \sin \frac{n\pi x}{l} \, dx \tag{E.2}$$

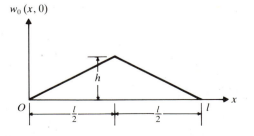

**Figure 8.5** Initial deflection of the string.

The initial deflection $w_0(x)$ is given by

$$w_0(x) = \begin{cases} \dfrac{2hx}{l} & \text{for } 0 \le x \le \dfrac{l}{2} \\[2mm] \dfrac{2h(l-x)}{l} & \text{for } \dfrac{l}{2} \le x \le l \end{cases} \tag{E.3}$$

By substituting Eq. (E.3) into Eq. (E.2), $C_n$ can be evaluated:

$$C_n = \frac{2}{l}\left\{ \int_0^{l/2} \frac{2hx}{l} \sin\frac{n\pi x}{l}\, dx + \int_{l/2}^{l} \frac{2h}{l}(l-x)\sin\frac{n\pi x}{l}\, dx \right\}$$

$$= \begin{cases} \dfrac{8h}{\pi^2 n^2} \sin\dfrac{n\pi}{2} & \text{for } n = 1,3,5,\dots \\[2mm] 0 & \text{for } n = 2,4,6,\dots \end{cases} \tag{E.4}$$

By using the relation

$$\sin\frac{n\pi}{2} = (-1)^{(n-1)/2}; \quad n = 1,3,5,\dots \tag{E.5}$$

the desired solution can be expressed as

$$w(x,t) = \frac{8h}{\pi^2}\left\{ \sin\frac{\pi x}{l}\cos\frac{\pi ct}{l} - \frac{1}{9}\sin\frac{3\pi x}{l}\cos\frac{3\pi ct}{l} + \cdots \right\} \tag{E.6}$$

In this case, no even harmonics are excited.

### 8.2.5 Traveling-Wave Solution

The solution of the wave equation, Eq. (8.8), for a string of infinite length can be expressed as [8.5]

$$w(x,t) = w_1(x - ct) + w_2(x + ct) \tag{8.35}$$

where $w_1$ and $w_2$ are arbitrary functions of $(x - ct)$ and $(x + ct)$, respectively. To show that Eq. (8.35) is a proper solution of Eq. (8.8), we first differentiate Eq. (8.35):

$$\frac{\partial^2 w(x,t)}{\partial x^2} = w_1''(x - ct) + w_2''(x + ct) \tag{8.36}$$

$$\frac{\partial^2 w(x,t)}{\partial t^2} = c^2 w_1''(x - ct) + c^2 w_2''(x + ct) \tag{8.37}$$

Substitution of these equations into Eq. (8.8) reveals that the wave equation is satisfied. In Eq. (8.35), $w_1(x - ct)$ and $w_2(x + ct)$ represent waves that propagate in the positive and negative directions of the $x$ axis, respectively, with a velocity $c$.

For a given problem, the arbitrary functions $w_1$ and $w_2$ are determined from the initial conditions, Eq. (8.10). Substitution of Eq. (8.35) into Eq. (8.10) gives, at $t = 0$,

$$w_1(x) + w_2(x) = w_0(x) \tag{8.38}$$

$$-cw_1'(x) + cw_2'(x) = \dot{w}_0(x) \tag{8.39}$$

where the prime indicates differentiation with respect to the respective argument at $t = 0$ (that is, with respect to $x$). Integration of Eq. (8.39) yields

$$-w_1(x) + w_2(x) = \frac{1}{c} \int_{x_0}^{x} \dot{w}_0(x') \, dx' \tag{8.40}$$

where $x_0$ is a constant. Solution of Eqs. (8.38) and (8.40) gives $w_1$ and $w_2$:

$$w_1(x) = \frac{1}{2}\left[ w_0(x) - \frac{1}{c} \int_{x_0}^{x} \dot{w}_0(x') \, dx' \right] \tag{8.41}$$

$$w_2(x) = \frac{1}{2}\left[ w_0(x) + \frac{1}{c} \int_{x_0}^{x} \dot{w}_0(x') \, dx' \right] \tag{8.42}$$

By replacing $x$ by $(x - ct)$ and $(x + ct)$, respectively, in Eqs. (8.41) and (8.42), we obtain the total solution:

$$
\begin{aligned}
w(x, t) &= w_1(x - ct) + w_2(x + ct) \\
&= \frac{1}{2}\left[ w_0(x - ct) + w_0(x + ct) \right] + \frac{1}{2c} \int_{x - ct}^{x + ct} \dot{w}_0(x') \, dx'
\end{aligned}
\tag{8.43}
$$

Note:

1. As can be seen from Eq. (8.43), there is no need to apply boundary conditions to the problem.

2. The solution given by Eq. (8.43) can be expressed as

$$w(x, t) = w_D(x, t) + w_V(x, t) \tag{8.44}$$

where $w_D(x, t)$ denotes the waves propagating due to the known initial displacement $w_0(x)$ with zero initial velocity, and $w_V(x, t)$ represents waves traveling due only to the known initial velocity $\dot{w}_0(x)$ with zero initial displacement.

The transverse vibration of a string fixed at both ends excited by the transverse impact of an elastic load at an intermediate point was considered in [8.6]. A review of the literature on the dynamics of cables and chains was given by Triantafyllou [8.7].

## 8.3   LONGITUDINAL VIBRATION OF A BAR OR ROD

Consider an elastic bar of length $l$ with varying cross-sectional area $A(x)$, as shown in Fig. 8.6. The forces acting on the cross sections of a small element of the bar are given by $P$ and $P + dP$ with

$$P = \sigma A = EA \frac{\partial u}{\partial x} \tag{8.45}$$

where $\sigma$ is the axial stress, $E$ is Young's modulus, $u$ is the axial displacement, and $\partial u / \partial x$ is the axial strain. If $f(x, t)$ denotes the external force per unit length, the

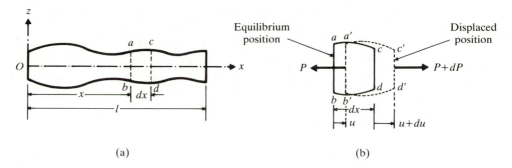

**Figure 8.6**  Longitudinal vibration of a bar.

summation of the forces in the $x$ direction gives the equation of motion:

$$(P + dP) + f\,dx - P = \rho A\,dx\,\frac{\partial^2 u}{\partial t^2} \tag{8.46}$$

where $\rho$ is the mass density of the bar. By using the relation $dP = (\partial P/\partial x)\,dx$ and Eq. (8.45), the equation of motion for the forced longitudinal vibration of a nonuniform bar, Eq. (8.46), can be expressed as

$$\frac{\partial}{\partial x}\left[EA(x)\frac{\partial u(x,t)}{\partial x}\right] + f(x,t) = \rho(x)A(x)\frac{\partial^2 u}{\partial t^2}(x,t) \tag{8.47}$$

For a uniform bar, Eq. (8.47) reduces to

$$EA\frac{\partial^2 u}{\partial x^2}(x,t) + f(x,t) = \rho A\frac{\partial^2 u}{\partial t^2}(x,t) \tag{8.48}$$

The free vibration equation can be obtained from Eq. (8.48), by setting $f = 0$, as

$$c^2\frac{\partial^2 u}{\partial x^2}(x,t) = \frac{\partial^2 u}{\partial t^2}(x,t) \tag{8.49}$$

where

$$c = \sqrt{\frac{E}{\rho}} \tag{8.50}$$

Note that Eqs. (8.47) to (8.50) can be seen to be similar to Eqs. (8.5), (8.6), (8.8), and (8.9), respectively. The solution of Eq. (8.49), which can be obtained as in the case of Eq. (8.8), can thus be written as

$$u(x,t) = \left(\underset{\sim}{A}\cos\frac{\omega x}{c} + \underset{\sim}{B}\sin\frac{\omega x}{c}\right)(C\cos\omega t + D\sin\omega t)* \tag{8.51}$$

If the bar has known initial axial displacement $u_0(x)$ and initial velocity $\dot{u}_0(x)$, the

---

*We use $\underset{\sim}{A}$ and $\underset{\sim}{B}$ in this section; $A$ is used to denote the cross-sectional area of the bar.

| Boundary condition | At left end ($x = 0$) | At right end ($x = l$) |
|---|---|---|
| Fixed end | $x = 0$; $u(0, t) = 0$ | $x = l$; $u(l, t) = 0$ |
| Free end | $x = 0$; $\frac{\partial u}{\partial x}(0, t) = 0$ | $x = l$; $\frac{\partial u}{\partial x}(l, t) = 0$ |
| End spring (spring constant $= \underset{\sim}{k}$) | $\underset{\sim}{k}$; $x = 0$; $AE\frac{\partial u}{\partial x}(0, t) = \underset{\sim}{k}u(0, t)$ | $\underset{\sim}{k}$; $x = l$; $AE\frac{\partial u}{\partial x}(l, t) = -\underset{\sim}{k}\,u(l, t)$ |
| End damper (damping constant $= \underset{\sim}{c}$) | $\underset{\sim}{c}$; $x = 0$; $AE\frac{\partial u}{\partial x}(0, t) = \underset{\sim}{c}\frac{\partial u}{\partial t}(0, t)$ | $\underset{\sim}{c}$; $x = l$; $AE\frac{\partial u}{\partial x}(l, t) = -\underset{\sim}{c}\frac{\partial u}{\partial t}(l, t)$ |
| End mass (mass $= \underset{\sim}{m}$) | $\underset{\sim}{m}$; $x = 0$; $AE\frac{\partial u}{\partial x}(0, t) = \underset{\sim}{m}\frac{\partial^2 u}{\partial t^2}(0, t)$ | $\underset{\sim}{m}$; $x = l$; $AE\frac{\partial u}{\partial x}(l, t) = -\underset{\sim}{m}\frac{\partial^2 u}{\partial t^2}(l, t)$ |

**Figure 8.7** Boundary conditions for a bar in longitudinal vibration.

initial conditions can be stated as

$$u(x, t = 0) = u_0(x)$$

$$\frac{\partial u}{\partial t}(x, t = 0) = \dot{u}_0(x) \tag{8.52}$$

The possible boundary conditions of the bar are shown in Fig. 8.7.

## EXAMPLE 8.2

Find the natural frequencies and normal modes of a bar fixed at one end and free at the other.

***Solution.*** Let the bar be fixed at $x = 0$ and free at $x = l$, so that the boundary conditions can be expressed as

$$u(0, t) = 0, \qquad t \geq 0 \tag{E.1}$$

$$\frac{\partial u}{\partial x}(l, t) = 0, \qquad t \geq 0 \tag{E.2}$$

The use of Eq. (E.1) in Eq. (8.51) gives $\underset{\sim}{A} = 0$, while the use of Eq. (E.2) gives the frequency equation

$$B\frac{\omega}{c}\cos\frac{\omega l}{c}=0 \qquad \text{or} \qquad \cos\frac{\omega l}{c}=0 \tag{E.3}$$

The eigenvalues or natural frequencies are given by

$$\frac{\omega_n l}{c}=(2n+1)\frac{\pi}{2}, \qquad n=0,1,2,\dots$$

or

$$\omega_n=\frac{(2n+1)\pi c}{2l}, \qquad n=0,1,2,\dots \tag{E.4}$$

Thus the total solution of Eq. (8.49) can be written as

$$u(x,t)=\sum_{n=0}^{\infty}u_n(x,t)=\sum_{n=0}^{\infty}\sin\frac{(2n+1)\pi x}{2l}\left[C_n\cos\frac{(2n+1)\pi ct}{2l}\right.$$
$$\left.+D_n\sin\frac{(2n+1)\pi ct}{2l}\right] \tag{E.5}$$

where the values of the constants $C_n$ and $D_n$ can be determined from the initial conditions, as in Eqs. (8.33) and (8.34):

$$C_n=\frac{2}{l}\int_0^l u_0(x)\sin\frac{(2n+1)\pi x}{2l}\,dx \tag{E.6}$$

$$D_n=\frac{4}{(2n+1)\pi c}\int_0^l \dot{u}_0(x)\sin\frac{(2n+1)\pi x}{2l}\,dx \tag{E.7}$$

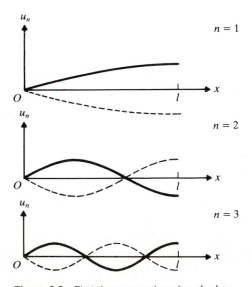

**Figure 8.8**   First three normal modes of a bar.

The first three normal mode shapes of the bar, defined by the sine term before the square brackets in Eq. (E.5), are shown in Fig. 8.8. It is to be noted that the physical displacement occurs along the length of the bar.

**EXAMPLE 8.3**

Find the natural frequencies of a bar with one end fixed and a mass attached at the other end, as in Fig. 8.9.

*Solution.* The equation governing the axial vibration of the bar is given by Eq. (8.49) and the solution by Eq. (8.51). The boundary condition at the fixed end ($x = 0$)

$$u(0, t) = 0 \tag{E.1}$$

leads to $\underset{\sim}{A} = 0$ in Eq. (8.51). At the end $x = l$, the tensile force in the bar must be equal to the inertia force of the vibrating mass $M$, and so

$$AE\frac{\partial u}{\partial x}(l, t) = -M\frac{\partial^2 u}{\partial t^2}(l, t) \tag{E.2}$$

With the help of Eq. (8.51), this equation can be expressed as

$$AE\frac{\omega}{c}\cos\frac{\omega l}{c}(C\cos\omega t + D\sin\omega t) = M\omega^2\sin\frac{\omega l}{c}(C\cos\omega t + D\sin\omega t)$$

That is,

$$\frac{AE\omega}{c}\cos\frac{\omega l}{c} = M\omega^2\sin\frac{\omega l}{c}$$

or

$$\alpha\tan\alpha = \beta \tag{E.3}$$

where

$$\alpha = \frac{\omega l}{c} \tag{E.4}$$

and

$$\beta = \frac{AEl}{c^2 M} = \frac{A\rho l}{M} = \frac{m}{M} \tag{E.5}$$

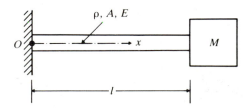

**Figure 8.9**

**TABLE 8.1**

| | Value of the mass ratio $\beta$ | | | | |
|---|---|---|---|---|---|
| | **0.01** | **0.1** | **1.0** | **10.0** | **100.0** |
| Value of $\alpha_1$ $\left(\omega_1 = \dfrac{\alpha_1 c}{l}\right)$ | 0.1000 | 0.3113 | 0.8602 | 1.4291 | 1.5549 |
| Value of $\alpha_2$ $\left(\omega_2 = \dfrac{\alpha_2 c}{l}\right)$ | 3.1448 | 3.1736 | 3.4267 | 4.3063 | 4.6658 |

where $m$ is the mass of the bar. Equation (E.3) is the frequency equation (in the form of a transcendental equation) whose solution gives the natural frequencies of the system. The first two natural frequencies are given in Table 8.1 for different values of the parameter $\beta$.

Note: If the mass of the bar is negligible compared to the mass attached, $m \simeq 0$,

$$c = \left(\frac{E}{\rho}\right)^{1/2} = \left(\frac{EAl}{m}\right)^{1/2} \to \infty \quad \text{and} \quad \alpha = \frac{\omega l}{c} \to 0$$

In this case

$$\tan \frac{\omega l}{c} \simeq \frac{\omega l}{c}$$

and the frequency equation (E.3) can be taken as

$$\left(\frac{\omega l}{c}\right)^2 = \beta$$

This gives the approximate value of the fundamental frequency:

$$\omega_1 = \frac{c}{l} \beta^{1/2} = \frac{c}{l} \left(\frac{\rho Al}{M}\right)^{1/2} = \left(\frac{EA}{lM}\right)^{1/2} = \left(\frac{g}{\delta_s}\right)^{1/2}$$

where

$$\delta_s = \frac{Mgl}{EA}$$

represents the static elongation of the bar under the action of the load $Mg$.

## EXAMPLE 8.4

A bar of uniform cross-sectional area $A$, density $\rho$, modulus of elasticity $E$, and length $l$ is fixed at one end and free at the other end. It is subjected to an axial force $F_0$ at its free end, as shown in Fig. 8.10(a). Study the resulting vibrations if the force $F_0$ is suddenly removed.

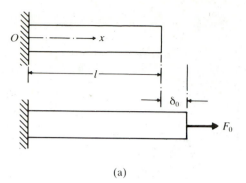

(a)

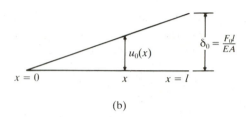

(b)

**Figure 8.10**

**Solution.** The tensile strain induced in the bar due to $F_0$ is

$$\varepsilon = \frac{F_0}{EA}$$

Thus the displacement of the bar just before the force $F_0$ is removed (initial displacement) is given by (see Fig. 8.10(b))

$$u_0 = u(x,0) = \varepsilon x = \frac{F_0 x}{EA}, \qquad 0 \le x \le l \tag{E.1}$$

Since the initial velocity is zero, we have

$$\dot{u}_0 = \frac{\partial u}{\partial t}(x,0) = 0, \qquad 0 \le x \le l \tag{E.2}$$

The general solution of a bar fixed at one end and free at the other end is given by Eq. (E.5) of Example 8.2:

$$u(x,t) = \sum_{n=0}^{\infty} u_n(x,t) = \sum_{n=0}^{\infty} \sin \frac{(2n+1)\pi x}{2l} \left[ C_n \cos \frac{(2n+1)\pi ct}{2l} \right. $$
$$\left. + D_n \sin \frac{(2n+1)\pi ct}{2l} \right] \tag{E.3}$$

where $C_n$ and $D_n$ are given by Eqs. (E.6) and (E.7) of Example 8.2. Since $\dot{u}_0 = 0$, we obtain $D_n = 0$. By using the initial displacement of Eq. (E.1) in Eq. (E.6) of Example

8.2, we obtain

$$C_n = \frac{2}{l} \int_0^l \frac{F_0 x}{EA} \cdot \sin \frac{(2n+1)\pi x}{2l} dx = \frac{8F_0 l}{EA\pi^2} \frac{(-1)^n}{(2n+1)^2} \tag{E.4}$$

Thus the solution becomes

$$u(x,t) = \frac{8F_0 l}{EA\pi^2} \sum_{n=0}^{\infty} \frac{(-1)^n}{(2n+1)^2} \sin \frac{(2n+1)\pi x}{2l} \cos \frac{(2n+1)\pi ct}{2l} \tag{E.5}$$

Equations (E.3) and (E.5) indicate that a typical point at $x = x_0$ on the bar vibrates with an amplitude of

$$C_n \sin \frac{(2n+1)\pi x_0}{2l}$$

and a circular frequency of

$$\frac{(2n+1)\pi c}{2l}$$

## 8.4 TORSIONAL VIBRATION OF A SHAFT OR ROD

Figure 8.11 represents a nonuniform shaft subjected to an external torque $f(x,t)$ per unit length. If $\theta(x,t)$ denotes the angle of twist of the cross section, the relation between the torsional deflection and the twisting moment $M_t(x,t)$ is given by [8.8]

$$M_t(x,t) = GJ(x)\frac{\partial\theta}{\partial x}(x,t) \tag{8.53}$$

where $G$ is the shear modulus and $GJ(x)$ is the torsional stiffness, with $J(x)$ denoting the polar moment of inertia of the cross section in the case of a circular section. If the mass polar moment of inertia of the shaft per unit length is $J_0$, the inertia torque acting on an element of length $dx$ becomes

$$J_0 \, dx \, \frac{\partial^2\theta}{\partial t^2}$$

If an external torque $f(x,t)$ acts on the shaft per unit length, the application of Newton's second law yields the equation of motion:

$$(M_t + dM_t) + f \, dx - M_t = J_0 \, dx \frac{\partial^2\theta}{\partial t^2} \tag{8.54}$$

By expressing $dM_t$ as

$$\frac{\partial M_t}{\partial x} dx$$

and using Eq. (8.53), the forced torsional vibration equation for a nonuniform shaft

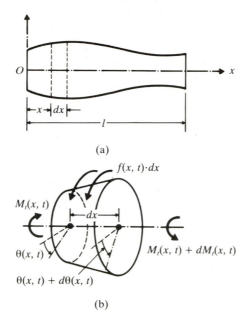

$$O \xrightarrow{\hspace{3cm}} x$$

$\leftarrow x \rightarrow | dx | \leftarrow$

$\longleftarrow l \longrightarrow$

(a)

$f(x, t) \cdot dx$

$M_t(x, t)$

$\leftarrow dx \rightarrow$

$M_t(x, t) + dM_t(x, t)$

$\theta(x, t)$

$\theta(x, t) + d\theta(x, t)$

(b)

**Figure 8.11**   Torsional vibration of a shaft.

can be obtained:

$$\frac{\partial}{\partial x}\left[ GJ(x)\frac{\partial\theta}{\partial x}(x,t)\right] + f(x,t) = J_0(x)\frac{\partial^2\theta}{\partial t^2}(x,t) \tag{8.55}$$

For a uniform shaft, Eq. (8.55) takes the form

$$GJ\frac{\partial^2\theta}{\partial x^2}(x,t) + f(x,t) = J_0\frac{\partial^2\theta}{\partial t^2}(x,t) \tag{8.56}$$

which, in the case of free vibration, reduces to

$$c^2\frac{\partial^2\theta}{\partial x^2}(x,t) = \frac{\partial^2\theta}{\partial t^2}(x,t) \tag{8.57}$$

where

$$c = \sqrt{\frac{GJ}{J_0}} \tag{8.58}$$

Notice that Eqs. (8.55) to (8.58) are similar to the equations derived in the cases of transverse vibration of a string and longitudinal vibration of a bar. If the shaft has a uniform cross section, $J_0 = \rho J$. Hence Eq. (8.58) becomes

$$c = \sqrt{\frac{G}{\rho}} \tag{8.59}$$

| Boundary condition | At left end ($x = 0$) | At right end ($x = l$) |
|---|---|---|
| Fixed end | $\theta(0, t) = 0$ <br> $x = 0$ | $\theta(l, t) = 0$ <br> $x = l$ |
| Free end | $\frac{\partial \theta}{\partial x}(0, t) = 0$ <br> $x = 0$ | $\frac{\partial \theta}{\partial x}(l, t) = 0$ <br> $x = l$ |
| End torsional spring (spring constant = $k_t$) | $k_t$ <br> $GJ\frac{\partial \theta}{\partial x}(0, t) = k_t\theta(0, t)$ <br> $x = 0$ | $k_t$ <br> $GJ\frac{\partial \theta}{\partial x}(l, t) = -k_t\theta(l, t)$ <br> $x = l$ |
| End torsional damper (damping constant = $c_t$) | $c_t$ <br> $GJ\frac{\partial \theta}{\partial x}(0, t) = c_t\frac{\partial \theta}{\partial t}(0, t)$ <br> $x = 0$ | $c_t$ <br> $GJ\frac{\partial \theta}{\partial x}(l, t) = -c_t\frac{\partial \theta}{\partial t}(l, t)$ <br> $x = l$ |
| End inertia (inertia = $J_0$) | $J_0$ <br> $GJ\frac{\partial \theta}{\partial x}(0, t) = J_0\frac{\partial^2 \theta}{\partial t^2}(0, t)$ <br> $x = 0$ | $J_0$ <br> $GJ\frac{\partial \theta}{\partial x}(l, t) = -J_0\frac{\partial^2 \theta}{\partial t^2}(l, t)$ <br> $x = l$ |

**Figure 8.12** Boundary conditions for a shaft (rod) subjected to torsional vibration.

If the shaft is given an angular displacement $\theta_0(x)$ and an angular velocity $\dot{\theta}_0(x)$ at $t = 0$, the initial conditions can be stated as

$$\theta(x, t = 0) = \theta_0(x)$$

$$\frac{\partial \theta}{\partial t}(x, t = 0) = \dot{\theta}_0(x) \tag{8.60}$$

The possible boundary conditions are indicated in Fig. 8.12. The general solution of Eq. (8.57) can be expressed as

$$\theta(x, t) = \left(A \cos \frac{\omega x}{c} + B \sin \frac{\omega x}{c}\right)(C \cos \omega t + D \sin \omega t) \tag{8.61}$$

## EXAMPLE 8.5

Find the natural frequencies and normal modes of torsional vibration for a free-free shaft.

***Solution.*** Since both ends of the shaft are free, the boundary conditions are

$$\frac{\partial \theta}{\partial x}(0, t) = 0, \qquad t \geq 0 \tag{E.1}$$

$$\frac{\partial \theta}{\partial x}(l, t) = 0, \qquad t \geq 0 \tag{E.2}$$

Equation (8.61) gives

$$\frac{\partial \theta}{\partial x} = \frac{\omega}{c}\left( -A\sin\frac{\omega x}{c} + B\cos\frac{\omega x}{c} \right)(C\cos\omega t + D\sin\omega t) \qquad (\text{E.3})$$

Equation (E.1) leads to $B = 0$, while Eq. (E.2) requires

$$\frac{\omega}{c}A\sin\frac{\omega l}{c} = 0 \qquad \text{or} \qquad \sin\frac{\omega l}{c} = 0 \qquad (\text{E.4})$$

This frequency equation gives the natural frequencies as

$$\frac{\omega_n l}{c} = n\pi, \qquad n = 1, 2, \ldots$$

or

$$\omega_n = \frac{n\pi c}{l}, \qquad n = 1, 2, \ldots \qquad (\text{E.5})$$

The normal modes of torsional vibration can be expressed as

$$\theta_n(x, t) = \cos\frac{n\pi x}{l}\left[ C_n\cos\frac{n\pi ct}{l} + D_n\sin\frac{n\pi ct}{l} \right] \qquad (\text{E.6})$$

The first three normal modes, given by the cosine term before the square brackets in Eq. (E.6), are shown in Fig. 8.13.

## EXAMPLE 8.6

The drill pipe of an oil well is connected to a cutter, which is in the form of a rigid circular disc, as shown in Fig. 8.14. Find the natural frequencies of the system by assuming the drill pipe to be uniform and fixed at its upper end.

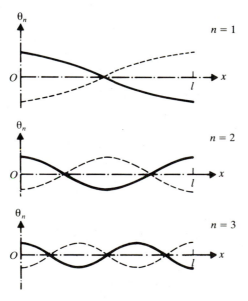

**Figure 8.13**  First three normal modes of a shaft.

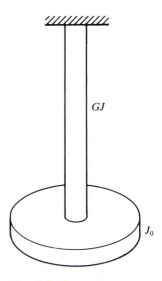

**Figure 8.14**

***Solution.*** The general solution is given by Eq. (8.61). By using the fixed boundary condition $\theta(0, t) = 0$, we obtain from Eq. (8.61), $A = 0$. The boundary condition at $x = l$ can be stated as

$$GJ\frac{\partial \theta}{\partial x}(l, t) = -J_0\frac{\partial^2\theta}{\partial t^2}(l, t) \tag{E.1}$$

That is,

$$BGJ\frac{\omega}{c}\cos\frac{\omega l}{c} = BJ_0\omega^2\sin\frac{\omega l}{c}$$

or

$$\frac{\omega l}{c}\tan\frac{\omega l}{c} = \frac{J\rho l}{J_0} = \frac{J_{\text{rod}}}{J_0} \tag{E.2}$$

Equation (E.2) can be expressed as

$$\alpha\tan\alpha = \beta \qquad \text{where } \alpha = \frac{\omega l}{c} \quad \text{and} \quad \beta = \frac{J_{\text{rod}}}{J_0} \tag{E.3}$$

The solution of Eq. (E.3), and thus the natural frequencies of the system, can be obtained as in the case of Example 8.3.

## 8.5   LATERAL VIBRATION OF BEAMS

### 8.5.1   Equation of Motion

To derive the differential equation of motion for the lateral vibration of beams, consider the forces and moments acting on an element of the beam shown in Fig. 8.15. $M(x, t)$ is the bending moment, $V(x, t)$ is the shear force, and $f(x, t)$ is the external force per unit length of the beam. Since the inertia force acting on the

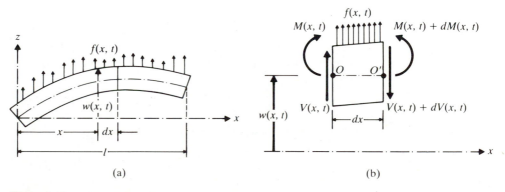

**Figure 8.15**   A beam in bending.

element of the beam is

$$\rho \cdot A(x) \cdot dx \cdot \frac{\partial^2 w}{\partial t^2}(x,t)$$

the force equation of motion in the $z$ direction gives

$$-(V+dV)+f(x,t)\,dx+V=\rho A(x)\,dx\frac{\partial^2 w}{\partial t^2}(x,t) \qquad (8.62)$$

where $\rho$ is the mass density and $A(x)$ is the cross-sectional area of the beam. The moment equation of motion about the $y$ axis passing through the point 0 in Fig. 8.15 leads to

$$(M+dM)-(V+dV)\,dx+f(x,t)\,dx\frac{dx}{2}-M=0 \qquad (8.63)$$

By writing

$$dV=\frac{\partial V}{\partial x}\,dx \qquad \text{and} \qquad dM=\frac{\partial M}{\partial x}\,dx$$

and disregarding terms involving second powers in $dx$, Eqs. (8.62) and (8.63) can be written as

$$-\frac{\partial V}{\partial x}(x,t)+f(x,t)=\rho A(x)\frac{\partial^2 w}{\partial t^2}(x,t) \qquad (8.64)$$

$$\frac{\partial M}{\partial x}(x,t)-V(x,t)=0 \qquad (8.65)$$

By using the relation $V=\partial M/\partial x$ from Eq. (8.65), Eq. (8.64) becomes

$$-\frac{\partial^2 M}{\partial x^2}(x,t)+f(x,t)=\rho A(x)\frac{\partial^2 w}{\partial t^2}(x,t) \qquad (8.66)$$

From the elementary theory of bending of beams (also known as the *Euler-Bernoulli* or *thin beam theory*), the relationship between bending moment and deflection can be expressed as [8.8]

$$M(x,t)=EI(x)\frac{\partial^2 w}{\partial x^2}(x,t) \qquad (8.67)$$

where $E$ is Young's modulus and $I(x)$ is the moment of inertia of the beam cross section about the $y$ axis. Inserting Eq. (8.67) into Eq. (8.66), we obtain the equation of motion for the forced lateral vibration of a nonuniform beam:

$$\frac{\partial^2}{\partial x^2}\left[EI(x)\frac{\partial^2 w}{\partial x^2}(x,t)\right]+\rho A(x)\frac{\partial^2 w}{\partial t^2}(x,t)=f(x,t) \qquad (8.68)$$

For a uniform beam, Eq. (8.68) reduces to

$$EI\frac{\partial^4 w}{\partial x^4}(x,t)+\rho A\frac{\partial^2 w}{\partial t^2}(x,t)=f(x,t) \qquad (8.69)$$

For free vibration, $f(x, t) = 0$, and so the equation of motion becomes

$$c^2 \frac{\partial^4 w}{\partial x^4}(x, t) + \frac{\partial^2 w}{\partial t^2}(x, t) = 0 \qquad (8.70)$$

where

$$c = \sqrt{\frac{EI}{\rho A}} \qquad (8.71)$$

### 8.5.2 Initial and Boundary Conditions

Since the equation of motion involves a second order derivative with respect to time and a fourth order derivative with respect to $x$, two initial conditions and four boundary conditions are needed for finding a unique solution for $w(x, t)$. Usually, the values of lateral displacement and velocity are specified as $w_0(x)$ and $\dot{w}_0(x)$ at $t = 0$, so that the initial conditions become

$$w(x, t = 0) = w_0(x)$$

$$\frac{\partial w}{\partial t}(x, t = 0) = \dot{w}_0(x) \qquad (8.72)$$

If an end of the beam is fixed, the displacement and slope are zero. For a pinned or simply supported end, the displacement and bending moment are zero. The bending moment and shear force are zero for a free end. These and some other boundary conditions are shown in Fig. 8.16.

### 8.5.3 Free Vibration

The free vibration solution can be found using the method of separation of variables as

$$w(x, t) = W(x)T(t) \qquad (8.73)$$

Substituting Eq. (8.73) into Eq. (8.70) and rearranging leads to

$$\frac{c^2}{W(x)} \frac{d^4 W(x)}{dx^4} = -\frac{1}{T(t)} \frac{d^2 T(t)}{dt^2} = a = \omega^2 \qquad (8.74)$$

where $a = \omega^2$ is a positive constant (see Problem 8.43). Equation (8.74) can be written as two equations:

$$\frac{d^4 W(x)}{dx^4} - \beta^4 W(x) = 0 \qquad (8.75)$$

$$\frac{d^2 T(t)}{dt^2} + \omega^2 T(t) = 0 \qquad (8.76)$$

where

$$\beta^4 = \frac{\omega^2}{c^2} = \frac{\rho A \omega^2}{EI} \qquad (8.77)$$

The solution of Eq. (8.76) can be expressed as

$$T(t) = A \cos \omega t + B \sin \omega t \qquad (8.78)$$

| Boundary condition | At left end ($x = 0$) | At right end ($x = l$) |
|---|---|---|
| Free end (bending moment = 0, shear force = 0) |  $EI\dfrac{\partial^2 w}{\partial x^2}(0,t)=0$ <br> $\dfrac{\partial}{\partial x}\left(EI\dfrac{\partial^2 w}{\partial x^2}\right)\Big|_{(0,t)}=0$ <br> $x=0$ | $EI\dfrac{\partial^2 w}{\partial x^2}(l,t)=0$ <br> $\dfrac{\partial}{\partial x}\left(EI\dfrac{\partial^2 w}{\partial x^2}\right)\Big|_{(l,t)}=0$ <br> $x=l$ |
|  Fixed end (deflection = 0, slope = 0) | 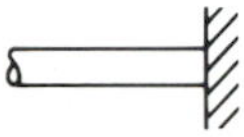 $w(0,t)=0$ <br> $\dfrac{\partial w}{\partial x}(0,t)=0$ <br> $x=0$ | $w(l,t)=0$ <br> $\dfrac{\partial w}{\partial x}(l,t)=0$ <br> $x=l$ |
| 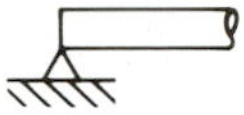 Simply supported end (deflection = 0, bending moment = 0) | 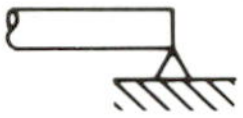 $w(0,t)=0$ <br> $EI\dfrac{\partial^2 w}{\partial x^2}(0,t)=0$ <br> $x=0$ | $w(l,t)=0$ <br> $EI\dfrac{\partial^2 w}{\partial x^2}(l,t)=0$ <br> $x=l$ |
| 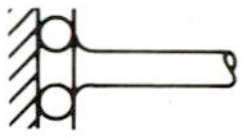 Sliding end (slope = 0, shear force = 0) | $\dfrac{\partial w}{\partial x}(0,t)=0$ <br> $\dfrac{\partial}{\partial x}\left(EI\dfrac{\partial^2 w}{\partial x^2}\right)\Big|_{(0,t)}=0$ <br> $x=0$ | $\dfrac{\partial w}{\partial x}(l,t)=0$ <br> $\dfrac{\partial}{\partial x}\left(EI\dfrac{\partial^2 w}{\partial x^2}\right)\Big|_{(l,t)}=0$ <br> $x=l$ |
| End spring (spring constant = $\underset{\sim}{k}$) | 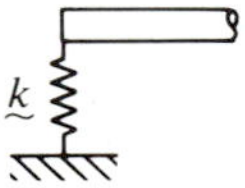 $\dfrac{\partial}{\partial x}\left(EI\dfrac{\partial^2 w}{\partial x^2}\right)\Big|_{(0,t)}=$ <br> $-\underset{\sim}{k}\,w(0,t)$ <br> $EI\dfrac{\partial^2 w}{\partial x^2}(0,t)=0$ <br> $x=0$ | 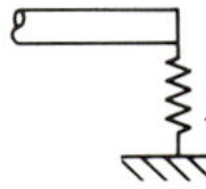 $\dfrac{\partial}{\partial x}\left(EI\dfrac{\partial^2 w}{\partial x^2}\right)\Big|_{(l,t)}=$ <br> $+\underset{\sim}{k}\,w(l,t)$ <br> $EI\dfrac{\partial^2 w}{\partial x^2}(l,t)=0$ <br> $x=l$ |
| End damper (damping constant = $\underset{\sim}{c}$) | 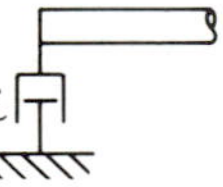 $\dfrac{\partial}{\partial x}\left(EI\dfrac{\partial^2 w}{\partial x^2}\right)\Big|_{(0,t)}=$ <br> $-\underset{\sim}{c}\dfrac{\partial w}{\partial t}(0,t)$ <br> $EI\dfrac{\partial^2 w}{\partial x^2}(0,t)=0$ <br> $x=0$ | 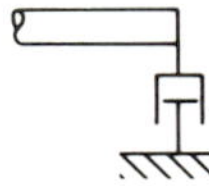 $\dfrac{\partial}{\partial x}\left(EI\dfrac{\partial^2 w}{\partial x^2}\right)\Big|_{(l,t)}=$ <br> $+\underset{\sim}{c}\dfrac{\partial w}{\partial t}(l,t)$ <br> $EI\dfrac{\partial^2 w}{\partial x^2}(l,t)=0$ <br> $x=l$ |
| End mass (mass = $\underset{\sim}{m}$ with negligible moment of inertia) |  $\dfrac{\partial}{\partial x}\left(EI\dfrac{\partial^2 w}{\partial x^2}\right)\Big|_{(0,t)}=$ <br> $-\underset{\sim}{m}\dfrac{\partial^2 w}{\partial t^2}(0,t)$ <br> $EI\dfrac{\partial^2 w}{\partial x^2}(0,t)=0$ <br> $x=0$ | 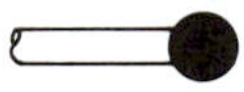 $\dfrac{\partial}{\partial x}\left(EI\dfrac{\partial^2 w}{\partial x^2}\right)\Big|_{(l,t)}=$ <br> $+\underset{\sim}{m}\dfrac{\partial^2 w}{\partial t^2}(l,t)$ <br> $EI\dfrac{\partial^2 w}{\partial x^2}(l,t)=0$ <br> $x=l$ |
| 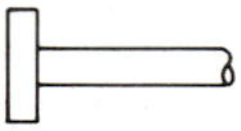 End mass with moment of inertia (mass = $\underset{\sim}{m}$, moment of inertia = $\underset{\sim}{J_0}$) | $EI\dfrac{\partial^2 w}{\partial x^2}(0,t)=$ <br> $\underset{\sim}{J_0}\dfrac{\partial^3 w}{\partial x\partial t^2}(0,t)$ <br> $\dfrac{\partial}{\partial x}\left(EI\dfrac{\partial^2 w}{\partial x^2}\right)\Big|_{(0,t)}=$ <br> $\underset{\sim}{m}\dfrac{\partial^2 w}{\partial t^2}(0,t)$ <br> $x=0$ |  $EI\dfrac{\partial^2 w}{\partial x^2}(l,t)=$ <br> $-\underset{\sim}{J_0}\dfrac{\partial^3 w}{\partial x\partial t^2}(l,t)$ <br> $\dfrac{\partial}{\partial x}\left(EI\dfrac{\partial^2 w}{\partial x^2}\right)\Big|_{(l,t)}=$ <br> $-\underset{\sim}{m}\dfrac{\partial^2 w}{\partial t^2}(l,t)$ <br> $x=l$ |

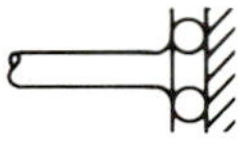

**Figure 8.16** Boundary conditions for the transverse vibration of a beam.

where $A$ and $B$ are constants that can be found from the initial conditions. For the solution of Eq. (8.75), we assume

$$W(x) = Ce^{st}$$ (8.79)

where $C$ and $s$ are constants, and derive the auxiliary equation as

$$s^4 - \beta^4 = 0$$ (8.80)

The roots of this equation are

$$s_{1,2} = \pm \beta, \qquad s_{3,4} = \pm i\beta$$ (8.81)

Hence the solution of Eq. (8.75) becomes

$$W(x) = C_1 e^{\beta x} + C_2 e^{-\beta x} + C_3 e^{i\beta x} + C_4 e^{-i\beta x}$$ (8.82)

where $C_1$, $C_2$, $C_3$, and $C_4$ are constants. Equation (8.82) can also be expressed as

$$W(x) = C_1 \cos \beta x + C_2 \sin \beta x + C_3 \cosh \beta x + C_4 \sinh \beta x$$ (8.83)

or

$$W(x) = C_1(\cos \beta x + \cosh \beta x) + C_2(\cos \beta x - \cosh \beta x)$$
$$+ C_3(\sin \beta x + \sinh \beta x) + C_4(\sin \beta x - \sinh \beta x)$$ (8.84)

where $C_1$, $C_2$, $C_3$, and $C_4$, in each case, are different constants. The constants $C_1$, $C_2$, $C_3$, and $C_4$ can be found from the boundary conditions. The natural frequencies of the beam are computed from Eq. (8.77) as

$$\omega = \beta^2 \sqrt{\frac{EI}{\rho A}} = (\beta l)^2 \sqrt{\frac{EI}{\rho A l^4}}$$ (8.85)

**EXAMPLE 8.7**

Determine the natural frequencies of vibration of a uniform beam fixed at $x = 0$ and simply supported at $x = l$.

*Solution.* The boundary conditions can be stated as

$$W(0) = 0$$ (E.1)

$$\frac{dW}{dx}(0) = 0$$ (E.2)

$$W(l) = 0$$ (E.3)

$$EI \frac{d^2W}{dx^2}(l) = 0 \quad \text{or} \quad \frac{d^2W}{dx^2}(l) = 0$$ (E.4)

Condition (E.1) leads to

$$C_1 + C_3 = 0$$ (E.5)

in Eq. (8.83) while Eqs. (E.2) and (8.83) give

$$\frac{dW}{dx}\bigg|_{x=0} = \beta\left[-C_1\sin\beta x + C_2\cos\beta x + C_3\sinh\beta x + C_4\cosh\beta x\right]_{x=0} = 0$$

or

$$\beta[C_2 + C_4] = 0 \tag{E.6}$$

Thus the solution, Eq. (8.83), becomes

$$W(x) = C_1(\cos\beta x - \cosh\beta x) + C_2(\sin\beta x - \sinh\beta x) \tag{E.7}$$

Applying conditions (E.3) and (E.4) to Eq. (E.7) yields

$$C_1(\cos\beta l - \cosh\beta l) + C_2(\sin\beta l - \sinh\beta l) = 0 \tag{E.8}$$

$$-C_1(\cos\beta l + \cosh\beta l) - C_2(\sin\beta l + \sinh\beta l) = 0 \tag{E.9}$$

For a nontrivial solution of $C_1$ and $C_2$, the determinant of their coefficients must be zero—that is,

$$\begin{vmatrix} (\cos\beta l - \cosh\beta l) & (\sin\beta l - \sinh\beta l) \\ -(\cos\beta l + \cosh\beta l) & -(\sin\beta l + \sinh\beta l) \end{vmatrix} = 0 \tag{E.10}$$

Expanding the determinant gives the frequency equation

$$\cos\beta l \sinh\beta l - \sin\beta l \cosh\beta l = 0$$

or

$$\tan\beta l = \tanh\beta l \tag{E.11}$$

The roots of this equation, $\beta_n l$, give the natural frequencies of vibration:

$$\omega_n = (\beta_n l)^2 \left(\frac{EI}{\rho A l^4}\right)^{1/2}, \qquad n = 1, 2, \ldots \tag{E.12}$$

where the values of $\beta_n l$, $n = 1, 2, \ldots$ satisfying Eq. (E.11) are given in Table 8.2. If the value of $C_2$ corresponding to $\beta_n$ is denoted as $C_{2n}$, it can be expressed in terms of

**TABLE 8.2**

| End conditions | Frequency equation | $\beta_1 l$ | $\beta_2 l$ | $\beta_3 l$ | $\beta_4 l$ |
|---|---|---|---|---|---|
| Pinned-pinned | $\sin\beta_n l = 0$ | $\pi$ | $2\pi$ | $3\pi$ | $4\pi$ |
| Free-free | $\cos\beta_n l \cosh\beta_n l = 1$ | 4.730041 | 7.853205 | 10.995608 | 14.137165 |
| | ($\beta l = 0$ for rigid body mode) | | | | |
| Fixed-fixed | $\cos\beta_n l \cosh\beta_n l = 1$ | 4.730041 | 7.853205 | 10.995608 | 14.137165 |
| Fixed-free | $\cos\beta_n l \cosh\beta_n l = -1$ | 1.875104 | 4.694091 | 7.854757 | 10.995541 |
| Fixed-pinned | $\tan\beta_n l - \tanh\beta_n l = 0$ | 3.926602 | 7.068583 | 10.210176 | 13.351768 |
| Pinned-free | $\tan\beta_n l - \tanh\beta_n l = 0$ | 3.926602 | 7.068583 | 10.210176 | 13.351768 |
| | ($\beta l = 0$ for rigid body mode) | | | | |

$C_{1n}$ from Eq. (E.8) as

$$C_{2n} = -C_{1n}\left(\frac{\cos\beta_n l - \cosh\beta_n l}{\sin\beta_n l - \sinh\beta_n l}\right) \tag{E.13}$$

Hence Eq. (E.7) can be written as

$$W_n(x) = C_{1n}\left[(\cos\beta_n x - \cosh\beta_n x) - \left(\frac{\cos\beta_n l - \cosh\beta_n l}{\sin\beta_n l - \sinh\beta_n l}\right)(\sin\beta_n x - \sinh\beta_n x)\right] \tag{E.14}$$

The normal modes of vibration can be obtained by the use of Eq. (8.73):

$$w_n(x,t) = W_n(x)(A_n\cos\omega_n t + B_n\sin\omega_n t) \tag{E.15}$$

with $W_n(x)$ given by Eq. (E.14). The general or total solution of the fixed-simply supported beam can be expressed by the sum of the normal modes:

$$w(x,t) = \sum_{n=1}^{\infty} w_n(x,t) \tag{E.16}$$

### 8.5.4 Effect of Axial Force

The problem of vibrations of a beam under the action of axial force finds application in the study of vibrations of cables and guy wires. For example, although the vibrations of a cable can be found by treating it as an equivalent string, many cables have failed due to fatigue caused by alternating flexure. The alternating flexure is produced by the regular shedding of vortices from the cable in a light wind. We must therefore consider the effects of axial force and bending stiffness on lateral vibrations in the study of fatigue failure of cables.

To find the effect of an axial force $P(x,t)$ on the bending vibrations of a beam, consider the equation of motion of an element of the beam, as shown in Fig. 8.17. For the vertical motion, we have

$$-(V+dV) + f\,dx + V + (P+dP)\sin(\theta+d\theta) - P\sin\theta = \rho A\,dx\frac{\partial^2 w}{\partial t^2} \tag{8.86}$$

and for the rotational motion about 0,

$$(M+dM) - (V+dV)\,dx + f\,dx\frac{dx}{2} - M = 0 \tag{8.87}$$

For small deflections,

$$\sin(\theta+d\theta) \simeq \theta + d\theta = \theta + \frac{\partial\theta}{\partial x}\,dx = \frac{\partial w}{\partial x} + \frac{\partial^2 w}{\partial x^2}\,dx$$

With this, Eqs. (8.86), (8.87), and (8.67) can be combined to obtain a single differential equation of motion:

$$\frac{\partial^2}{\partial x^2}\left[EI\frac{\partial^2 w}{\partial x^2}\right] + \rho A\frac{\partial^2 w}{\partial t^2} - P\frac{\partial^2 w}{\partial x^2} = f \tag{8.88}$$

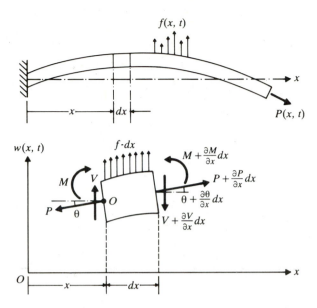

**Figure 8.17**   An element of a beam under axial load.

For the free vibration of a uniform beam, Eq. (8.88) reduces to

$$EI\frac{\partial^4 w}{\partial x^4} + \rho A\frac{\partial^2 w}{\partial t^2} - P\frac{\partial^2 w}{\partial x^2} = 0 \tag{8.89}$$

The solution of Eq. (8.89) can be obtained using the method of separation of variables as

$$w(x,t) = W(x)(A\cos\omega t + B\sin\omega t) \tag{8.90}$$

Substitution of Eq. (8.90) into Eq. (8.89) gives

$$EI\frac{d^4 W}{dx^4} - P\frac{d^2 W}{dx^2} - \rho A\omega^2 W = 0 \tag{8.91}$$

By assuming the solution $W(x)$ to be

$$W(x) = Ce^{sx} \tag{8.92}$$

in Eq. (8.91), the auxiliary equation can be obtained:

$$s^4 - \frac{P}{EI}s^2 - \frac{\rho A\omega^2}{EI} = 0 \tag{8.93}$$

The roots of Eq. (8.93) are

$$s_1^2, s_2^2 = \frac{P}{2EI} \pm \left(\frac{P^2}{4E^2I^2} + \frac{\rho A\omega^2}{EI}\right)^{1/2} \tag{8.94}$$

and so the solution can be expressed as (with absolute value of $s_2$)

$$W(x) = C_1\cosh s_1 x + C_2\sinh s_1 x + C_3\cos s_2 x + C_4\sin s_2 x \tag{8.95}$$

where the constants $C_1$ to $C_4$ are to be determined from the boundary conditions.

---

### EXAMPLE 8.8

Find the natural frequencies of a simply supported beam subjected to an axial compressive force.

*Solution.* The boundary conditions are

$$W(0) = 0 \tag{E.1}$$

$$\frac{d^2 W}{dx^2}(0) = 0 \tag{E.2}$$

$$W(l) = 0 \tag{E.3}$$

$$\frac{d^2 W}{dx^2}(l) = 0 \tag{E.4}$$

Equations (E.1) and (E.2) require that $C_1 = C_3 = 0$ in Eq. (8.95), and so

$$W(x) = C_2\sinh s_1 x + C_4\sin s_2 x \tag{E.5}$$

The application of Eqs. (E.3) and (E.4) to Eq. (E.5) leads to

$$\sinh s_1 l \cdot \sin s_2 l = 0 \tag{E.6}$$

Since $\sinh s_1 l > 0$ for all values of $s_1 l \neq 0$, the only roots to this equation are

$$s_2 l = n\pi, \qquad n = 0,1,2,\dots \tag{E.7}$$

Thus Eqs. (E.7) and (8.94) give the natural frequencies of vibration:

$$\omega_n = \frac{\pi^2}{l^2}\sqrt{\frac{EI}{\rho A}}\left(n^4 + \frac{n^2 P l^2}{\pi^2 EI}\right)^{1/2} \tag{E.8}$$

Since the axial force $P$ is compressive, $P$ is negative. Further, from strength of materials, the smallest Euler buckling load for a simply supported beam is given by [8.9]

$$P_{\text{cri}} = \frac{\pi^2 EI}{l^2} \tag{E.9}$$

Thus Eq. (E.8) can be written as

$$\omega_n = \frac{\pi^2}{l^2}\left(\frac{EI}{\rho A}\right)^{1/2}\left(n^4 - n^2 \frac{P}{P_{\text{cri}}}\right)^{1/2} \tag{E.10}$$

The following observations can be made from the present example:

1. If $P = 0$, the natural frequency will be same as that of a simply supported beam given in Table 8.2.

**2.** If $EI = 0$, the natural frequency (see Eq. (E.8)) reduces to that of a taut string.

**3.** If $P > 0$, the natural frequency increases as the tensile force stiffens the beam.

**4.** As $P \to P_{cri}$, the natural frequency approaches zero.

### 8.5.5 Effects of Rotary Inertia and Shear Deformation

If the cross-sectional dimensions are not small compared to the length of the beam, we need to consider the effects of rotary inertia and shear deformation. The procedure, presented by Timoshenko [8.10], is known as the *thick beam theory* or *Timoshenko beam theory*. Consider the element of the beam shown in Fig. 8.18. If the effect of shear deformation is disregarded, the tangent to the deflected center line $O'T$ coincides with the normal to the face $Q'R'$ (since cross sections normal to the center line remain normal even after deformation). Due to shear deformation, the tangent to the deformed center line $O'T$ will not be perpendicular to the face $Q'R'$. The angle $\gamma$ between the tangent to the deformed center line $(O'T)$ and the normal to the face $(O'N)$ denotes the shear deformation of the element. Since positive shear on the right face $Q'R'$ acts downwards, we have, from Fig. 8.18,

$$\gamma = \phi - \frac{\partial w}{\partial x} \tag{8.96}$$

where $\phi$ denotes the slope of the deflection curve due to bending deformation alone. Note that because of shear alone, the element undergoes distortion but no rotation.

The bending moment $M$ and the shear force $V$ are related to $\phi$ and $w$ by the formulas*

$$M = EI \frac{\partial \phi}{\partial x} \tag{8.97}$$

and

$$V = kAG\gamma = kAG\left(\phi - \frac{\partial w}{\partial x}\right) \tag{8.98}$$

where $G$ denotes the modulus of rigidity of the material of the beam and $k$ is a constant, also known as *Timoshenko's shear coefficient*, which depends on the shape of the cross section. For a rectangular section the value of $k$ is 5/6; for a circular section it is 9/10 [8.11].

---

*Equation (8.97) is similar to Eq. (8.67). Eq. (8.98) can be obtained as follows:

$$\text{shear force} = \text{shear stress} \times \text{area} = \text{shear strain} \times \text{shear modulus} \times \text{area}$$

or

$$V = \gamma GA$$

This equation is modified as $V = kAG\gamma$ by introducing a factor $k$ on the right-hand side to take care of the shape of the cross section.

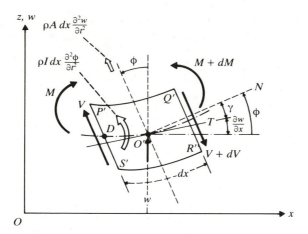

**Figure 8.18**

The equations of motion for the element shown in Fig. 8.18 can be derived as follows:

For translation in the $z$ direction:

$$-[V(x,t)+dV(x,t)]+f(x,t)\,dx+V(x,t)=\rho A(x)\,dx\frac{\partial^2 w}{\partial t^2}(x,t)$$

$$\equiv \text{translational inertia of the element}$$

$$(8.99)$$

For rotation about a line passing through point $D$ and parallel to the $y$ axis:

$$[M(x,t)+dM(x,t)]-[V(x,t)+dV(x,t)]\,dx+f(x,t)\,dx\frac{dx}{2}-M(x,t)$$

$$=\rho I(x)\,dx\frac{\partial^2\phi}{\partial t^2}\equiv \text{rotary inertia of the element} \qquad (8.100)$$

Using the relations

$$dV=\frac{\partial V}{\partial x}\,dx \qquad \text{and} \qquad dM=\frac{\partial M}{\partial x}\,dx$$

along with Eqs. (8.97) and (8.98) and disregarding terms involving second powers in $dx$, Eqs. (8.99) and (8.100) can be expressed as

$$-kAG\left(\frac{\partial\phi}{\partial x}-\frac{\partial^2 w}{\partial x^2}\right)+f(x,t)=\rho A\frac{\partial^2 w}{\partial t^2} \qquad (8.101)$$

$$EI\frac{\partial^2\phi}{\partial x^2}-kAG\left(\phi-\frac{\partial w}{\partial x}\right)=\rho I\frac{\partial^2\phi}{\partial t^2} \qquad (8.102)$$

By solving Eq. (8.101) for $\partial\phi/\partial x$ and substituting the result in Eq. (8.102), we obtain the desired equation of motion for the forced vibration of a uniform beam:

$$EI\frac{\partial^4 w}{\partial x^4} + \rho A\frac{\partial^2 w}{\partial t^2} - \rho I\left(1 + \frac{E}{kG}\right)\frac{\partial^4 w}{\partial x^2 \partial t^2} + \frac{\rho^2 I}{kG}\frac{\partial^4 w}{\partial t^4}$$

$$+ \frac{EI}{kAG}\frac{\partial^2 f}{\partial x^2} - \frac{\rho I}{kAG}\frac{\partial^2 f}{\partial t^2} - f = 0 \tag{8.103}$$

For free vibration, $f = 0$, and Eq. (8.103) reduces to

$$EI\frac{\partial^4 w}{\partial x^4} + \rho A\frac{\partial^2 w}{\partial t^2} - \rho I\left(1 + \frac{E}{kG}\right)\frac{\partial^4 w}{\partial x^2 \partial t^2} + \frac{\rho^2 I}{kG}\frac{\partial^4 w}{\partial t^4} = 0 \tag{8.104}$$

The following boundary conditions are to be applied in the solution of Eq. (8.103) or (8.104).

**1.** Fixed end:

$$\phi = w = 0$$

**2.** Simply supported end:

$$EI\frac{\partial\phi}{\partial x} = w = 0$$

**3.** Free end:

$$kAG\left(\frac{\partial w}{\partial x} - \phi\right) = EI\frac{\partial\phi}{\partial x} = 0$$

---

### EXAMPLE 8.9

Determine the effects of rotary inertia and shear deformation on the natural frequencies of a simply supported uniform beam.

*Solution.* By defining

$$\alpha^2 = \frac{EI}{\rho A} \qquad \text{and} \qquad r^2 = \frac{I}{A} \tag{E.1}$$

Eq. (8.104) can be written as

$$\alpha^2\frac{\partial^4 w}{\partial x^4} + \frac{\partial^2 w}{\partial t^2} - r^2\left(1 + \frac{E}{kG}\right)\frac{\partial^4 w}{\partial x^2 \partial t^2} + \frac{\rho r^2}{kG}\frac{\partial^4 w}{\partial t^4} = 0 \tag{E.2}$$

We can express the solution of Eq. (E.2) as

$$w(x, t) = C\sin\frac{n\pi x}{l}\cos\omega_n t \tag{E.3}$$

which satisfies the necessary boundary conditions at $x = 0$ and $x = l$. Here, $C$ is a constant and $\omega_n$ is the $n$th natural frequency. By substituting Eq. (E.3) into Eq.

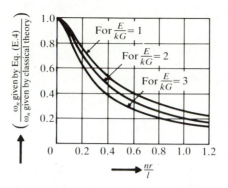

**Figure 8.19**

(E.2), we obtain the frequency equation:

$$\omega_n^4\left(\frac{\rho r^2}{kG}\right) - \omega_n^2\left(1 + \frac{n^2\pi^2 r^2}{l^2} + \frac{n^2\pi^2 r^2}{l^2}\cdot\frac{E}{kG}\right) + \left(\frac{\alpha^2 n^4\pi^4}{l^4}\right) = 0 \qquad \text{(E.4)}$$

It can be seen that Eq. (E.4) is a quadratic equation in $\omega_n^2$ for any given $n$; there are two values of $\omega_n$ that satisfy Eq. (E.4). The smaller value corresponds to the bending deformation mode, while the larger one corresponds to the shear deformation mode.

The values of the ratio of $\omega_n$ given by Eq. (E.4) to the natural frequency given by the classical theory (in Table 8.2)* are plotted for three values of $E/kG$ in Fig. 8.19 [8.22].

Note the following aspects of rotary inertia and shear deformation:

1. If the effect of rotary inertia alone is considered, the resulting equation of motion does not contain any term involving the shear coefficient $k$. Hence we obtain (from Eq. (8.104)):

$$EI\frac{\partial^4 w}{\partial x^4} + \rho A\frac{\partial^2 w}{\partial t^2} - \rho I\frac{\partial^4 w}{\partial x^2\,\partial t^2} = 0 \qquad \text{(E.5)}$$

In this case the frequency equation (E.4) reduces to

$$\omega_n^2 = \frac{\alpha^2 n^4\pi^4}{l^4\left(1 + \dfrac{n^2\pi^2 r^2}{l^2}\right)} \qquad \text{(E.6)}$$

2. If the effect of shear deformation alone is considered, the resulting equation of motion does not contain the terms originating from $\rho I(\partial^2\phi/\partial t^2)$ in Eq. (8.102).

---

*The theory used for the derivation of the equation of motion (8.68), which disregards the effects of rotary inertia and shear deformation, is called the *classical* or *Euler-Bernoulli theory*.

Thus we obtain the equation of motion:

$$EI \frac{\partial^4 w}{\partial x^4} + \rho A \frac{\partial^2 w}{\partial t^2} - \frac{EI\rho}{kG} \frac{\partial^4 w}{\partial x^2 \partial t^2} = 0 \qquad \text{(E.7)}$$

and the corresponding frequency equation:

$$\omega_n^2 = \frac{\alpha^2 n^4 \pi^4}{l^4 \left( 1 + \frac{n^2 \pi^2 r^2}{l^2} \cdot \frac{E}{kG} \right)} \qquad \text{(E.8)}$$

**3.** If both the effects of rotary inertia and shear deformation are disregarded, Eq. (8.104) reduces to the classical equation of motion, Eq. (8.70):

$$EI \frac{\partial^4 w}{\partial x^4} + \rho A \frac{\partial^2 w}{\partial t^2} = 0 \qquad \text{(E.9)}$$

and Eq. (E.4) to

$$\omega_n^2 = \frac{\alpha^2 n^4 \pi^4}{l^4} \qquad \text{(E.10)}$$

---

### 8.5.6 Other Effects

The transverse vibration of tapered beams is presented in Refs. [8.12–8.14]. The natural frequencies of continuous beams are discussed by Wang [8.15]. The dynamic response of beams resting on elastic foundation is considered in Refs. [8.16, 8.17]. The effect of support flexibility on the natural frequencies of beams is presented in [8.18, 8.19]. A treatment of the problem of natural vibrations of a system of elastically connected Timoshenko beams is given in Ref. [8.20]. A comparison of the exact and approximate solutions of vibrating beams is made by Hutchinson [8.30]. The steady-state vibration of damped beams is considered in Ref. [8.21].

## 8.6 VIBRATION OF MEMBRANES

A membrane is a plate that is subjected to tension and has negligible bending resistance. Thus a membrane bears the same relationship to a plate as a string bears to a beam. A drumhead is an example of a membrane.

### 8.6.1 Equation of Motion

To derive the equation of motion of a membrane, consider the membrane to be bounded by a plane curve $S$ in the $xy$ plane, as shown in Fig. 8.20. Let $f(x, y, t)$ denote the pressure loading acting in the $z$ direction and $P$ the intensity of tension at a point that is equal to the product of the tensile stress and the thickness of the membrane. The magnitude of $P$ is usually constant throughout the membrane, as in a drumhead. If we consider an elemental area $dx\,dy$, forces of magnitude $P\,dx$ and

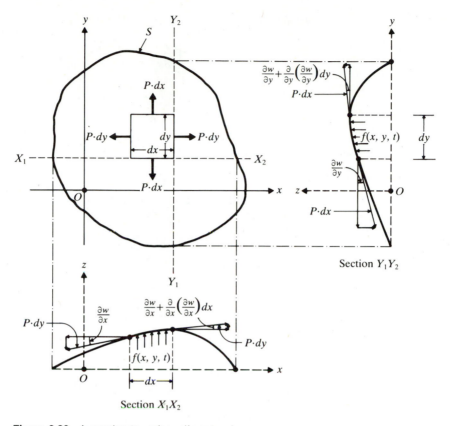

**Figure 8.20**   A membrane under uniform tension.

$P\,dy$ act on the sides parallel to the $y$ and $x$ axes, respectively, as shown in Fig. 8.20. The net forces acting along the $z$ direction due to these forces are

$$\left( P\frac{\partial^2 w}{\partial y^2}\,dx\,dy \right) \qquad \text{and} \qquad \left( P\frac{\partial^2 w}{\partial x^2}\,dx\,dy \right)$$

The pressure force along the $z$ direction is $f(x, y, t)\,dx\,dy$, and the inertia force is

$$\rho(x, y)\frac{\partial^2 w}{\partial t^2}\,dx\,dy$$

where $\rho(x, y)$ is the mass per unit area. The equation of motion for the forced transverse vibration of the membrane can be obtained:

$$P\left( \frac{\partial^2 w}{\partial x^2} + \frac{\partial^2 w}{\partial y^2} \right) + f = \rho\frac{\partial^2 w}{\partial t^2} \qquad (8.105)$$

If the external force $f(x, y, t) = 0$, Eq. (8.105) gives the free vibration equation

$$c^2 \left( \frac{\partial^2 w}{\partial x^2} + \frac{\partial^2 w}{\partial y^2} \right) = \frac{\partial^2 w}{\partial t^2} \tag{8.106}$$

where

$$c = \left( \frac{P}{\rho} \right)^{1/2} \tag{8.107}$$

Equations (8.105) and (8.106) can be expressed as

$$P \cdot \nabla^2 w + f = \rho \frac{\partial^2 w}{\partial t^2} \tag{8.108}$$

and

$$c^2 \cdot \nabla^2 w = \frac{\partial^2 w}{\partial t^2} \tag{8.109}$$

where

$$\nabla^2 = \frac{\partial^2}{\partial x^2} + \frac{\partial^2}{\partial y^2} \tag{8.110}$$

is the Laplacian operator.

### 8.6.2 Initial and Boundary Conditions

Since the equation of motion, Eq. (8.105) or (8.106), involves second order partial derivatives with respect to each of $t$, $x$, and $y$, we need to specify two initial conditions and four boundary conditions to find a unique solution of the problem. Usually, the displacement and velocity of the membrane at $t = 0$ are specified as $w_0(x, y)$ and $\dot{w}_0(x, y)$. Hence the initial conditions are given by

$$w(x, y, 0) = w_0(x, y)$$

$$\frac{\partial w}{\partial t}(x, y, 0) = \dot{w}_0(x, y) \tag{8.111}$$

The boundary conditions are of the following types:

1. If the membrane is fixed at any point $(x_1, y_1)$ on a segment of the boundary, we have

$$w(x_1, y_1, t) = 0, \qquad t \geq 0 \tag{8.112}$$

2. If the membrane is free to deflect transversely (in the $z$ direction) at a different point $(x_2, y_2)$ of the boundary, then the force component in the $z$ direction must be zero. Thus

$$P \frac{\partial w}{\partial n}(x_2, y_2, t) = 0, \qquad t \geq 0 \tag{8.113}$$

where $\partial w / \partial n$ represents the derivative of $w$ with respect to a direction $n$ normal to the boundary at the point $(x_2, y_2)$.

The solution of the equation of motion of the vibrating membrane was presented in Refs. [8.23–8.25].

---

**EXAMPLE 8.10**

Find the free vibration solution of a rectangular membrane of sides $a$ and $b$ along the $x$ and $y$ axes, respectively.

***Solution.*** By using the method of separation of variables, $w(x, y, t)$ can be assumed to be

$$w(x, y, t) = W(x, y)T(t) = X(x)Y(y)T(t) \tag{E.1}$$

By using Eqs. (E.1) and (8.106), we obtain

$$\frac{d^2X(x)}{dx^2} + \alpha^2 X(x) = 0 \tag{E.2}$$

$$\frac{d^2Y(y)}{dy^2} + \beta^2 Y(y) = 0 \tag{E.3}$$

$$\frac{d^2T(t)}{dt^2} + \omega^2 T(t) = 0 \tag{E.4}$$

where $\alpha^2$ and $\beta^2$ are constants related to $\omega^2$ as follows:

$$\beta^2 = \frac{\omega^2}{c^2} - \alpha^2 \tag{E.5}$$

The solutions of Eqs. (E.2) to (E.4) are given by

$$X(x) = C_1 \cos \alpha x + C_2 \sin \alpha x \tag{E.6}$$

$$Y(y) = C_3 \cos \beta y + C_4 \sin \beta y \tag{E.7}$$

$$T(t) = A \cos \omega t + B \sin \omega t \tag{E.8}$$

where the constants $C_1$ to $C_4$, $A$, and $B$ can be determined from the boundary and initial conditions.

---

## 8.7 RAYLEIGH'S METHOD

Rayleigh's method can be applied to find the fundamental natural frequency of continuous systems. This method is much simpler than exact analysis for systems with varying distributions of mass and stiffness. Although the method is applicable to all continuous systems, we shall apply it only to beams in this section.* Consider

---

*An integral equation approach for the determination of the fundamental frequency of vibrating beams is presented by Penny and Reed [8.26].

the beam shown in Fig. 8.15. In order to apply Rayleigh's method, we need to derive expressions for the maximum kinetic and potential energies and Rayleigh's quotient. The kinetic energy of the beam can be expressed as

$$T = \frac{1}{2}\int_0^l \dot{w}^2 \, dm = \frac{1}{2}\int_0^l \dot{w}^2 \rho A(x) \, dx \tag{8.114}$$

The maximum kinetic energy can be found by assuming a harmonic variation $w(x,t) = W(x)\cos\omega t$:

$$T_{max} = \frac{\omega^2}{2}\int_0^l \rho A(x) W^2(x) \, dx \tag{8.115}$$

The potential energy of the beam can be found from the work done by the elastic forces to the neutral configuration. By disregarding the work done by the shear forces, we have

$$V = \frac{1}{2}\int_0^l M\,d\theta = \frac{1}{2}\int_0^l \left(EI\frac{\partial^2 w}{\partial x^2}\right)\frac{\partial^2 w}{\partial x^2}\cdot dx = \frac{1}{2}\int_0^l EI\left(\frac{\partial^2 w}{\partial x^2}\right)^2 \cdot dx \tag{8.116}$$

Since the maximum value of $w(x,t)$ is $W(x)$, the maximum value of $V$ is given by

$$V_{max} = \frac{1}{2}\int_0^l EI(x)\left(\frac{d^2 W(x)}{dx^2}\right)^2 dx \tag{8.117}$$

By equating $T_{max}$ to $V_{max}$, we obtain Rayleigh's quotient:

$$R(\omega) = \omega^2 = \frac{\int_0^l EI\left(\dfrac{d^2 W(x)}{dx^2}\right)^2 \cdot dx}{\int_0^l \rho A(W(x))^2 \cdot dx} \tag{8.118}$$

Thus the natural frequency of the beam can be found once the deflection $W(x)$ is known. In general, $W(x)$ is not known and must therefore be assumed. Generally, the static equilibrium shape is assumed for $W(x)$ to obtain the fundamental frequency. It is to be noted that the assumed shape $W(x)$ unintentionally introduces a constraint on the system (which amounts to adding additional stiffness to the system), and so the frequency given by Eq. (8.118) is higher than the exact value [8.27].

For a stepped beam, Eq. (8.118) can be more conveniently written as

$$R(\omega) = \omega^2 = \frac{E_1 I_1 \int_0^{l_1}\left(\dfrac{d^2 W}{dx^2}\right)^2 dx + E_2 I_2 \int_{l_1}^{l_2}\left(\dfrac{d^2 W}{dx^2}\right)^2 dx + \cdots}{\rho A_1 \int_0^{l_1} W^2\,dx + \rho A_2 \int_{l_1}^{l_2} W^2\,dx + \cdots} \tag{8.119}$$

## EXAMPLE 8.11

Find the fundamental frequency of transverse vibration of the nonuniform cantilever beam shown in Fig. 8.21, using the deflection shape $W(x) = (1 - x/l)^2$.

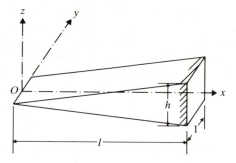

**Figure 8.21**

***Solution.*** The given deflection shape can be verified to satisfy the boundary conditions of the beam. The cross-sectional area $A$ and the moment of inertia $I$ of the beam can be expressed as

$$A(x) = \frac{hx}{l}, \qquad I(x) = \frac{1}{12}\left(\frac{hx}{l}\right)^3 \tag{E.1}$$

Thus Rayleigh's quotient gives

$$\omega^2 = \frac{\int_0^l E\left(\frac{h^3 x^3}{12 l^3}\right)\left(\frac{2}{l^2}\right)^2 dx}{\int_0^l \rho\left(\frac{hx}{l}\right)\left(1 - \frac{x}{l}\right)^4 dx} = 2.5\frac{Eh^2}{\rho l^4}$$

or

$$\omega = 1.5811\left(\frac{Eh^2}{\rho l^4}\right)^{1/2} \tag{E.2}$$

The exact value of the frequency, for this case [8.2], is known to be

$$\omega_1 = 1.5343\left(\frac{Eh^2}{\rho l^4}\right)^{1/2} \tag{E.3}$$

Thus the value of $\omega_1$ given by Rayleigh's method can be seen to be 3.0503 percent higher than the exact value.

## 8.8  THE RAYLEIGH-RITZ METHOD

The Rayleigh-Ritz method can be considered an extension of Rayleigh's method. It is based on the premise that a closer approximation to the exact natural modes can be obtained by superposing a number of assumed functions than by using a single assumed function, as in Rayleigh's method. If the assumed functions are suitably chosen, this method provides not only the approximate value of the fundamental frequency but also the approximate values of the higher natural frequencies and the mode shapes. An arbitrary number of functions can be used, and the number of frequencies that can be obtained is equal to the number of functions used. A large number of functions, although it involves more computational work, leads to more accurate results.

In the case of transverse vibration of beams, if $n$ functions are chosen for approximating the deflection $W(x)$, we can write

$$W(x) = c_1 w_1(x) + c_2 w_2(x) + \cdots + c_n w_n(x) \tag{8.120}$$

where $w_1(x), w_2(x), \ldots, w_n(x)$ are known linearly independent functions of the spatial coordinate $x$, which satisfy all the boundary conditions of the problem, and $c_1, c_2, \ldots, c_n$ are coefficients to be found. The coefficients $c_i$ are to be determined so that the assumed functions $w_i(x)$ provide the best possible approximation to the natural modes. To obtain such approximations, the coefficients $c_i$ are adjusted and the natural frequency is made stationary at the natural modes. For this we substitute Eq. (8.120) in Rayleigh's quotient, Eq. (8.118), and the resulting expression is partially differentiated with respect to each of the coefficients $c_i$. To make the natural frequency stationary, we set each of the partial derivatives equal to zero and obtain

$$\frac{\partial(\omega^2)}{\partial c_i} = 0, \qquad i = 1, 2, \ldots, n \tag{8.121}$$

Equation (8.121) denotes a set of $n$ linear algebraic equations in the coefficients $c_1, c_2, \ldots, c_n$ and also contains the undetermined quantity $\omega^2$. This defines an algebraic eigenvalue problem similar to the ones that arise in multidegree of freedom systems. The solution of this eigenvalue problem generally gives $n$ natural frequencies $\omega_i^2$, $i = 1, 2, \ldots, n$ and $n$ eigenvectors, each containing a set of numbers for $c_1, c_2, \ldots, c_n$. For example, the $i$th eigenvector corresponding to $\omega_i$ may be expressed as

$$\vec{C}^{(i)} = \begin{Bmatrix} c_1^{(i)} \\ c_2^{(i)} \\ \vdots \\ c_n^{(i)} \end{Bmatrix} \tag{8.122}$$

When this eigenvector—the values of $c_1^{(i)}, c_2^{(i)}, \ldots, c_n^{(i)}$—is substituted into Eq. (8.120), we obtain the best possible approximation to the $i$th mode of the beam. A method of reducing the size of the eigenproblem in the Rayleigh-Ritz method is presented in Ref. [8.28]. A new approach, which combines the advantages of the Rayleigh-Ritz analysis and the finite element method is given in Ref. [8.29]. The basic Rayleigh-Ritz procedure is illustrated with the help of the following example.

| **EXAMPLE 8.12** |
|---|

Find the natural frequencies of the tapered cantilever beam of Example 8.11 by using the Rayleigh-Ritz method.

***Solution.*** We assume the deflection functions $w_i(x)$ to be

$$w_1(x) = \left(1 - \frac{x}{l}\right)^2 \tag{E.1}$$

$$w_2(x) = \frac{x}{l}\left(1 - \frac{x}{l}\right)^2 \tag{E.2}$$

$$w_3(x) = \frac{x^2}{l^2}\left(1 - \frac{x}{l}\right)^2 \tag{E.3}$$

$$\vdots$$

If we use a one-term approximation,

$$W(x) = c_1\left(1 - \frac{x}{l}\right)^2 \tag{E.4}$$

the fundamental frequency will be the same as the one found in Example 8.11. Now we use a two-term approximation,

$$W(x) = c_1\left(1 - \frac{x}{l}\right)^2 + c_2\frac{x}{l}\left(1 - \frac{x}{l}\right)^2 \tag{E.5}$$

Rayleigh's quotient is given by

$$R[W(x)] = \omega^2 = \frac{X}{Y} \tag{E.6}$$

where

$$X = \int_0^l EI(x)\left(\frac{d^2W(x)}{dx^2}\right)^2 dx \tag{E.7}$$

and

$$Y = \int_0^l \rho A(x)[W(x)]^2 dx \tag{E.8}$$

If Eq. (E.5) is substituted, Eq. (E.6) becomes a function of $c_1$ and $c_2$. The conditions that make $\omega^2$ or $R[W(x)]$ stationary are

$$\frac{\partial(\omega^2)}{\partial c_1} = \frac{Y\dfrac{\partial X}{\partial c_1} - X\dfrac{\partial Y}{\partial c_1}}{Y^2} = 0 \tag{E.9}$$

$$\frac{\partial(\omega^2)}{\partial c_2} = \frac{Y\dfrac{\partial X}{\partial c_2} - X\dfrac{\partial Y}{\partial c_2}}{Y^2} = 0 \tag{E.10}$$

These equations can be rewritten as

$$\frac{\partial X}{\partial c_1} - \frac{X}{Y}\frac{\partial Y}{\partial c_1} = \frac{\partial X}{\partial c_1} - \omega^2 \frac{\partial Y}{\partial c_1} = 0 \tag{E.11}$$

$$\frac{\partial X}{\partial c_2} - \frac{X}{Y}\frac{\partial Y}{\partial c_2} = \frac{\partial X}{\partial c_2} - \omega^2 \frac{\partial Y}{\partial c_2} = 0 \tag{E.12}$$

By substituting Eq. (E.5) into Eqs. (E.7) and (E.8), we obtain

$$X = \frac{Eh^3}{3l^3}\left(\frac{c_1^2}{4} + \frac{c_2^2}{10} + \frac{c_1 c_2}{5}\right) \tag{E.13}$$

$$Y = \rho h l\left(\frac{c_1^2}{30} + \frac{c_2^2}{280} + \frac{2c_1 c_2}{105}\right) \tag{E.14}$$

With the help of Eqs. (E.13) and (E.14), Eqs. (E.11) and (E.12) can be expressed as

$$\begin{bmatrix} \left(\dfrac{1}{2} - \underset{\sim}{\omega}^2 \cdot \dfrac{1}{15}\right) & \left(\dfrac{1}{5} - \underset{\sim}{\omega}^2 \cdot \dfrac{2}{105}\right) \\ \left(\dfrac{1}{5} - \underset{\sim}{\omega}^2 \cdot \dfrac{2}{105}\right) & \left(\dfrac{1}{5} - \underset{\sim}{\omega}^2 \cdot \dfrac{1}{140}\right) \end{bmatrix}\begin{Bmatrix} c_1 \\ c_2 \end{Bmatrix} = \begin{Bmatrix} 0 \\ 0 \end{Bmatrix} \tag{E.15}$$

where

$$\underset{\sim}{\omega}^2 = \frac{3\omega^2 \rho l^4}{Eh^2} \tag{E.16}$$

By setting the determinant of the matrix in Eq. (E.15) equal to zero, we obtain the frequency equation:

$$\frac{1}{8820}\underset{\sim}{\omega}^4 - \frac{13}{1400}\underset{\sim}{\omega}^2 + \frac{3}{50} = 0 \tag{E.17}$$

The roots of Eq. (E.17) are given by $\underset{\sim}{\omega}_1 = 2.6599$ and $\underset{\sim}{\omega}_2 = 8.6492$. Thus the natural frequencies of the tapered beam are

$$\omega_1 \simeq 1.5367\left(\frac{Eh^2}{\rho l^4}\right)^{1/2} \tag{E.18}$$

and

$$\omega_2 \simeq 4.9936\left(\frac{Eh^2}{\rho l^4}\right)^{1/2} \tag{E.19}$$

## 8.9 COMPUTER PROGRAM

A Fortran computer program, in the form of subroutine NONEQN, is given for finding the roots of a nonlinear equation of the form $F(Y) = 0$. The following arguments are used in this subroutine:

N       =   Number of roots to be determined. Input data.

X       =   Array of dimension $N$. Contains the computed values of the roots. Output.

XS      =   Initial guess for the first root. Input data.

XINC    =   Initial increment to be used in searching for the root. Input data.

NINT    =   Maximum number of subintervals to be used. A value on the order of 50 is to be used. Input data.

ITER    =   Maximum number of iterations permitted in finding a root. A value on the order of 100 is to be used. Input data.

EPS     =   Convergence requirement. A small value on the order of $10^{-6}$ is to be used. Input data.

An external function routine, "Function $F(Y)$" is to be provided by the user. In it, the nonlinear function $F$ must be defined in terms of the independent parameter $Y$.

For illustration, the roots of the frequency equation of a fixed-pinned beam are determined using subroutine NONEQN. Since the frequency equation [see Table 8.2] is given by

$$\tan \beta_n l - \tanh \beta_n l = 0$$

the external function is defined as

$$F = TAN(Y) - TANH(Y)$$

Five roots are found (N = 5) by using an initial guess value of XS = 2.0 with an increment of XINC = 0.1. The main program, subroutine NONEQN, and the output of the program are given below.

```
C===============================================================
C
C PROGRAM 17
C MAIN PROGRAM FOR CALLING THE SUBROUTINE NONEQN
C
C===============================================================
C FOLLOWING 3 LINES CONTAIN PROBLEM-DEPENDENT DATA
      DIMENSION X(5)
      DATA N,XS,XINC,NINT,ITER,EPS
     2   /5,2.0,0.1,50,100,0.000001/
C END OF PROBLEM-DEPENDENT DATA
      PRINT 10,N,XS,XINC,NINT,ITER,EPS
```

```
10    FORMAT (//,28H ROOTS OF NONLINEAR EQUATION,//,6H DATA:,/,
     2    2H N,4X,2H =,I4,/,3H XS,3X,2H =,E15.6,/,8H XINC  =,E15.6,/,
     3    8H NINT  =,I4,/,8H ITER  =,I4,/,8H EPS    =,E15.6)
      CALL NONEQN (N,X,XS,XINC,NINT,ITER,EPS)
      PRINT 20,(X(I),I=1,N)
20    FORMAT (//,7H ROOTS:,/,(E15.6))
      STOP
      END
C================================================================
C
C
C SUBROUTINE NONEQN
C
C================================================================
      SUBROUTINE NONEQN (N,X,XS,XINC,NINT,ITER,EPS)
      DIMENSION X(N)
      CC=0.0
      DELX=0.0
      II=0
      NP=1
      A=F(XS)
      IF (A .NE. 0.0 .OR. XS .NE. 0.0) GO TO 10
      II=II+1
      X(II)=0.0
      XS=XS+XINC
      A=F(XS)
10    AA=A
      A=F(XS)
      CC=AMAX1(XS,CC)
      IF (NP .GT. NINT) GO TO 50
      NP=NP+1
      AAA=A*AA
      XS=XS+XINC
      IF (AAA) 20,40,10
20    XS=XS-XINC
      XINCN=XINC/10.0
      I=1
30    B=F(XS)
      CON=(F(XS+XINCN) - F(XS))/XINCN
      DELX=B/CON
      XS=XS-DELX
      DELX=ABS(DELX)
      ITER1=I
      IF (DELX .LT. EPS) GO TO 40
      CC=AMAX1(XS,CC)
      I=I+1
      IF (I .GT. ITER) GO TO 40
      GO TO 30
40    II=II+1
      X(II)=XS
      XS=CC+XINC
      IF (II .GE. N) GO TO 50
      A=F(XS)
      IF (A .EQ. 0.0) A=F(XS+XINC)
      XS=XS+XINC
```

```
       GO TO 10
50     RETURN
       END
C=====================================================================
C
C FUNCTION F(Y)
C THIS FUNCTION IS PROBLEM-DEPENDENT
C
C=====================================================================
       FUNCTION F(Y)
       F = TAN(Y)-TANH(Y)
       RETURN
       END
```

```
ROOTS OF NONLINEAR EQUATION

DATA:
N      =    5
XS     =    0.200000E+01
XINC   =    0.100000E+00
NINT   =   50
ITER   =  100
EPS    =    0.100000E-05


ROOTS:
   0.392660E+01
   0.706858E+01
   0.102102E+02
   0.133518E+02
   0.164934E+02
```

## REFERENCES

**8.1.** S. K. Clark, *Dynamics of Continuous Elements*, Prentice-Hall, Englewood Cliffs, N.J., 1972.

**8.2.** S. Timoshenko, D. H. Young, and W. Weaver, Jr., *Vibration Problems in Engineering* (4th Ed.), John Wiley, New York, 1974.

**8.3.** A. Leissa, *Vibration of Plates*, NASA SP-160, Washington, D.C., 1969.

**8.4.** I. S. Habib, *Engineering Analysis Methods*, Lexington Books, Lexington, Mass., 1975.

**8.5.** J. D. Achenbach, *Wave Propagation in Elastic Solids*, North-Holland Publishing Company, Amsterdam, 1973.

**8.6.** K. K. Deb, "Dynamics of a string and an elastic hammer," *Journal of Sound and Vibration*, Vol. 40, 1975, pp. 243–248.

**8.7.** M. S. Triantafyllou, "Linear dynamics of cables and chains," *Shock and Vibration Digest*, Vol. 16, March 1984, pp. 9–17.

**8.8.** R. W. Fitzgerald, *Mechanics of Materials* (2nd Ed.), Addison-Wesley, Reading, Mass., 1982.

**8.9.** S. P. Timoshenko and J. Gere, *Theory of Elastic Stability* (2nd Ed.), McGraw-Hill, New York, 1961.

**8.10.** S. P. Timoshenko, "On the correction for shear of the differential equation for transverse vibration of prismatic bars," *Philosophical Magazine*, Series 6, Vol. 41, 1921, pp. 744–746.

**8.11.** G. R. Cowper, "The shear coefficient in Timoshenko's beam theory," *Journal of Applied Mechanics*, Vol. 33, 1966, pp. 335–340.

**8.12.** G. W. Housner and W. O. Keightley, "Vibrations of linearly tapered beams," Part I, *Transactions of ASCE*, Vol. 128, 1963, pp. 1020–1048.

**8.13.** E. T. Granch and A. A. Adler, "Bending vibrations of variable section beams," *Journal of Applied Mechanics*, Vol. 23, 1956, pp. 103–108.

**8.14.** J. H. Gaines and E. Volterra, "Transverse vibrations of cantilever bars of variable cross section," *Journal of the Acoustical Society of America*, Vol. 39, 1966, pp. 674–679.

**8.15.** T. M. Wang, "'Natural frequencies of continuous Timoshenko beams," *Journal of Sound and Vibration*, Vol. 13, 1970, pp. 409–414.

**8.16.** S. L. Grassie, R. W. Gregory, D. Harrison, and K. L. Johnson, "The dynamic response of railway track to high frequency vertical excitation," *Journal of Mechanical Engineering Science*, Vol. 24, June 1982, pp. 77–90.

**8.17.** M. N. Pavlovic and G. B. Wylie, "Vibration of beams on non-homogeneous elastic foundations," *Earthquake Engineering and Structural Dynamics*, Vol. 11, 1983, pp. 797–808.

**8.18.** T. Justine and A. Krishnan, "Effect of support flexibility on fundamental frequency of beams," *Journal of Sound and Vibration*, Vol. 68, 1980, pp. 310–312.

**8.19.** K. A. R. Perkins, "The effect of support flexibility on the natural frequencies of a uniform cantilever," *Journal of Sound and Vibration*, Vol. 4, 1966, pp. 1–8.

**8.20.** S. S. Rao, "Natural frequencies of systems of elastically connected Timoshenko beams," *Journal of the Acoustical Society of America*, Vol. 55, 1974, pp. 1232–1237.

**8.21.** A. M. Ebner and D. P. Billington, "Steady state vibration of damped Timoshenko beams," *Journal of Structural Division* (ASCE), Vol. 3, 1968, p. 737.

**8.22.** M. Levinson and D. W. Cooke, "On the frequency spectra of Timoshenko beams," *Journal of Sound and Vibration*, Vol. 84, 1982, pp. 319–326.

**8.23.** N. Y. Olcer, "General solution to the equation of the vibrating membrane," *Journal of Sound and Vibration*, Vol. 6, 1967, pp. 365–374.

**8.24.** G. R. Sharp, "Finite transform solution of the vibrating annular membrane," *Journal of Sound and Vibration*, Vol. 6, 1967, pp. 117–128.

**8.25.** J. Mazumdar, "A review of approximate methods for determining the vibrational modes of membranes," *Shock and Vibration Digest*, Vol. 14, February 1982, pp. 11–17.

**8.26.** J. E. Penny and J. R. Reed, "An integral equation approach to the fundamental frequency of vibrating beams," *Journal of Sound and Vibration*, Vol. 19, 1971, pp. 393–400.

**8.27.** G. Temple and W. G. Bickley, *Rayleigh's Principle and Its Application to Engineering*, Dover, New York, 1956.

**8.28.** W. L. Craver, Jr. and D. M. Egle, "A method for selection of significant terms in the assumed solution in a Rayleigh-Ritz analysis," *Journal of Sound and Vibration*, Vol. 22, 1972, pp. 133–142.

**8.29.** L. Klein, "Transverse vibrations of non-uniform beams," *Journal of Sound and Vibration*, Vol. 37, 1974, pp. 491–505.

**8.30.** J. R. Hutchinson, "Transverse vibrations of beams: Exact versus approximate solutions," *Journal of Applied Mechanics*, Vol. 48, 1981, pp. 923–928.

## REVIEW QUESTIONS

**8.1.** How does a continuous system differ from a discrete system in the nature of its equation of motion?

**8.2.** How many natural frequencies does a continuous system have?

**8.3.** Are the boundary conditions important in a discrete system? Why?

**8.4.** What is a wave equation? What is a traveling-wave solution?

**8.5.** What is the significance of wave velocity?

**8.6.** State the boundary conditions to be specified at a simply supported end of a beam if (a) thin beam theory is used and (b) Timoshenko beam theory is used.

**8.7.** State the possible boundary conditions at the ends of a string.

**8.8.** What is the main difference in the nature of the frequency equations of a discrete system and a continuous system?

**8.9.** What is the effect of a tensile force on the natural frequencies of a beam?

**8.10.** Under what circumstances does the frequency of vibration of a beam subjected to an axial load become zero?

**8.11.** Why does the natural frequency of a beam become lower if the effects of shear deformation and rotary inertia are considered?

**8.12.** Give two practical examples of the vibration of membranes.

**8.13.** What is the basic principle used in Rayleigh's method?

**8.14.** Why is the natural frequency given by Rayleigh's method always larger than the true value of $\omega_1$?

**8.15.** What is the difference between Rayleigh's method and the Rayleigh-Ritz method?

**8.16.** What is Rayleigh's quotient?

## PROBLEMS

The problem assignments are organized as follows:

| Problems | Section covered | Topic covered |
|----------|-----------------|---------------|
| 8.1–8.11 | 8.2 | Transverse vibration of strings |
| 8.12–8.17 | 8.3 | Longitudinal vibration of bars |
| 8.18–8.26 | 8.4 | Torsional vibration of shafts |
| 8.27–8.43 | 8.5 | Beams |
| 8.44–8.48 | 8.6 | Membranes |
| 8.49–8.57 | 8.7 | Rayleigh's method |
| 8.58–8.62 | 8.8 | The Rayleigh-Ritz method |
| 8.63–8.65 | 8.9 | Computer program |

**8.1.** Determine the velocity of wave propagation in an underwater cable of mass $\rho = 5$ kg/m when stretched by a tension $P = 4000$ N.

**8.2.** A steel wire of 2 mm diameter is fixed between two points located 2 m apart. The tensile force in the wire is 250 N. Determine (a) the fundamental frequency of vibration and (b) the velocity of wave propagation in the wire.

**8.3.** The stretched string of a musical instrument is 0.5 m long and has a fundamental frequency of 3000 Hz. Determine (a) the frequency of the fourth mode and (b) the fundamental frequency if the tension is increased by 20%.

**8.4.** Find the time it takes for a transverse wave to travel along a transmission line from one tower to another one 300 m away. Assume the horizontal component of the cable tension as 30000 N and the mass of the cable as 2 kg/m of length.

**8.5.** A cord of length $l$ and mass $\rho$ per unit length is stretched under a tension $P$. One end of the cord is fixed while the other end is fastened to a mass $m$, which can move in a frictionless slot, as shown in Fig. 8.22. The mass $m$ is also connected to a spring of stiffness $k$. Derive an expression for the natural frequencies of transverse vibration of the cord.

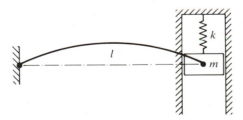

**Figure 8.22**

**8.6.** Determine the tension needed in order that a cord of a musical instrument of diameter $d = 0.5$ mm and density 7800 kg/m³ will have a fundamental frequency of transverse vibration 200 Hz. The cord is fixed at its ends and has a length of 1 m.

**8.7.** A cord of length $l$ and mass $\rho$ per unit length is stretched under a tension $P$. One end of the cord is fixed and the other end is connected to a pin, which can move in a frictionless slot. Find the natural frequencies of vibration of the cord.

**8.8.** Two masses $m/2$ each are connected at the ends of a cord of length $l$ and mass density $\rho$, as shown in Fig. 8.23. The assembly is made to rotate at an angular velocity $\omega_0$. Disregarding the variation in tension in the cord, show that the differential equation for the lateral vibration of the cord is given by

$$\frac{\partial^2 w}{\partial x^2} = \frac{4\rho}{ml\omega_0^2}\left(\frac{\partial^2 w}{\partial t^2} - \omega_0^2 w\right)$$

If the cord assumes a full sine wave form in the first principal mode, determine the corresponding frequency of vibration.

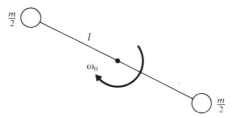

**Figure 8.23**

**8.9.** Find the free vibration solution of a cord fixed at both ends when its initial conditions are given by

$$w(x,0) = 0, \qquad \frac{\partial w}{\partial t}(x,0) = \frac{2ax}{l} \qquad \text{for} \quad 0 \le x \le \frac{l}{2}$$

and

$$\frac{\partial w}{\partial t}(x,0) = 2a\left(1 - \frac{x}{l}\right) \qquad \text{for} \quad \frac{l}{2} \le x \le l$$

**8.10.** A flexible cable of mass $\rho$ per unit length is fixed at its upper end and is free to oscillate under the influence of gravity as shown in Fig. 8.24. Derive the equation of motion for transverse vibration.

**8.11.** Prove that the constant $a$ in Eqs. (8.18) and (8.19) is negative for common boundary conditions. Hint: Multiply Eq. (8.18) by $W(x)$ and integrate with respect to $x$ from 0 to $l$.

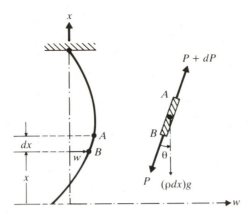

**Figure 8.24**

**8.12.** Derive an equation for the principal modes of longitudinal vibration of a uniform bar having both ends free.

**8.13.** Find the fundamental frequency for the longitudinal vibration of a uniform steel bar having a cross section 20 mm square and length 1 m when one end is fixed and the other end is free.

**8.14.** A spring of stiffness $k = 10^5$ N/m and mass 10 kg is placed between two walls. If the distance between the walls is 200 mm, find the natural frequencies of longitudinal vibration of the spring.

**8.15.** Derive the frequency equation for the longitudinal vibration of the systems shown in Fig. 8.25.

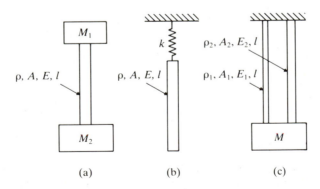

**Figure 8.25**

**8.16.** A thin bar of length $l$ and mass $m$ is clamped at one end and free at the other. What mass $M$ must be attached to the free end in order to decrease the fundamental frequency of longitudinal vibration by 50% from its fixed-free value?

**8.17.** Derive the frequency equation for the longitudinal vibration of a stepped bar having two different cross-sectional areas $A_1$ and $A_2$ over lengths $l_1$ and $l_2$, respectively. Assume fixed-free end conditions.

**8.18.** Show that the natural frequencies of torsional vibration of a uniform shaft of length $l$, fixed at the middle point and free at the two ends, are given by

$$\omega_n = (2n+1)\pi\sqrt{\frac{G}{\rho l^2}}, \qquad n = 0,1,2,\ldots$$

where $G$ is the shear modulus and $\rho$ is the mass density of the shaft.

**8.19.** A uniform shaft of length $l$, mass density $\rho$, and torsional stiffness $GJ$ is fixed at $x = 0$ and is connected to a torsional spring of stiffness $k_t$ at $x = l$. Determine the frequency equation for torsional vibration. Verify the correctness of the equation by considering the special cases of $k_t = 0$ and $k_t = \infty$.

**8.20.** A torsional system consists of a shaft with a disc of mass moment of inertia $J_0$ mounted at its center. If both ends of the shaft are fixed, find the response of the system in free torsional vibration of the shaft. Assume that the disc is given an initial angular displacement of $\theta_0$ and an initial velocity of zero.

**8.21.** Find the natural frequencies for torsional vibration of a fixed-fixed shaft.

**8.22.** Solve Problem 8.21 if one end of the shaft is fixed and the other free.

**8.23.** Derive the frequency equation for the torsional vibration of a uniform shaft carrying rotors of mass moment of inertia $J_1$ and $J_2$, one at each end.

**8.24.** An external torque $M_t(t) = M_{t0}\cos \omega t$ is applied at the free end of a fixed-free uniform shaft. Find the steady-state vibration of the shaft.

**8.25.** Find the fundamental frequency for torsional vibration of a shaft of length 2 m and diameter 50 mm when both the ends are fixed. The density of the material is 7800 kg/m$^3$ and the modulus of rigidity is $8.0 \times 10^{11}$ N/m$^2$.

**8.26.** A uniform shaft, supported at $x = 0$ and rotating at an angular velocity $\omega$, is suddenly stopped at the end $x = 0$. If the end $x = l$ is free, determine the subsequent angular displacement response of the shaft.

**8.27.** Compute the first three natural frequencies of the transverse vibrations of a uniform beam of rectangular cross-section (100 mm $\times$ 200 mm) with $l = 2$ m, $E = 20.5 \times 10^{11}$ N/m$^2$, and $\rho = 7.83 \times 10^3$ kg/m$^3$ for the following cases:
a. When both ends are simply supported
b. When both ends are built-in (clamped)
c. When one end is fixed and the other end is free
d. When both ends are free.

**8.28.** Derive an expression for the natural frequencies for the lateral vibration of a uniform fixed-fixed beam.

**8.29.** Derive an expression for the natural frequencies for the transverse vibration of a uniform beam with both ends simply supported.

**8.30.** Derive the expression for the natural frequencies for the lateral vibration of a uniform beam suspended as a pendulum.

**8.31.** A simply supported beam has a cross-sectional area 4000 mm², moment of inertia $20 \times 10^6$ mm⁴, mass 28 kg/m, and length 2 m. If $E = 2.05 \times 10^{11}$ N/m², find the first three natural frequencies for the lateral vibration of the beam.

**8.32.** A uniform beam, simply supported at both ends, is found to vibrate in its first mode with an amplitude of 10 mm at its center. If $A = 120$ mm², $I = 1000$ mm⁴, $E = 20.5 \times 10^{11}$ N/m², $\rho = 7.83 \times 10^3$ kg/m³, and $l = 1$ m, determine the maximum bending moment in the beam.

**8.33.** Derive the frequency equation for the transverse vibration of a uniform beam resting on springs at both ends, as shown in Fig. 8.26. The springs can deflect vertically only and the beam is horizontal in the equilibrium position.

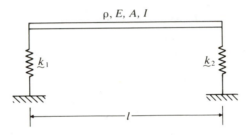

**Figure 8.26**

**8.34.** A simply supported uniform beam of length $l$ carries a mass $M$ at the center of the beam. Assuming the mass $M$ to be a point mass, obtain the frequency equation of the system.

**8.35.** A uniform fixed-fixed beam of length $2l$ is simply supported at the middle point. Derive the frequency equation for the transverse vibration of the beam.

**8.36.** A uniform beam of length $l$ and mass density $\rho$ is fixed at one end and carries a concentrated mass $M$ at the other end. Derive the frequency equation of the beam.

**8.37.** A simply supported beam carries initially a uniformly distributed load of intensity $f_0$. Find the vibration response of the beam if the load is suddenly removed.

**8.38.** Estimate the fundamental frequency of a cantilever beam whose cross-sectional area and moment of inertia vary as

$$A(x) = A_0 \frac{x}{l} \quad \text{and} \quad I(x) = I_0 \frac{x}{l}$$

where $x$ is measured from the free end.

**8.39.** (a) Derive a general expression for the response of a uniform beam subjected to an arbitrary force.

(b)   Use the result of part (a) to find the response of a uniform simply supported beam under the harmonic force $F_0 \sin \omega t$ applied at $x = a$. Assume the initial conditions as $w(x,0) = (\partial w / \partial t)(x,0) = 0$.

**8.40.** Derive Eqs. (E.5) and (E.6) of Example 8.9.

**8.41.** Derive Eqs. (E.7) and (E.8) of Example 8.9.

**8.42.** Prove that the normal modes of a beam are orthogonal, that is,

$$\int_0^l W_i(x) W_j(x) \, dx = 0, \qquad i \neq j$$

**8.43.** Prove that the constant $a$ in Eq. (8.74) is positive for common boundary conditions. Hint: Multiply Eq. (8.75) by $W(x)$ and integrate with respect to $x$ from 0 to $l$.

**8.44.** Starting from fundamentals, show that the equation for the lateral vibration of a circular membrane is given by

$$\frac{\partial^2 w}{\partial r^2} + \frac{1}{r} \frac{\partial w}{\partial r} + \frac{1}{r^2} \frac{\partial^2 w}{\partial \theta^2} = \frac{\rho}{P} \frac{\partial^2 w}{\partial t^2}$$

**8.45.** Using the equation of motion given in Problem 8.44, find the natural frequencies of a circular membrane of radius $R$ clamped around the boundary at $r = R$.

**8.46.** Consider a rectangular membrane of sides $a$ and $b$ supported along all the edges.
(a)   Derive an expression for the deflection $w(x, y, t)$ under an arbitrary pressure $f(x, y, t)$.
(b)   Find the response when a uniformly distributed pressure $f_0$ is applied to a membrane that is initially at rest.

**8.47.** If all the edges of a rectangular membrane of sides $a$ and $b$ parallel to the $x$ and $y$ axes, respectively, are fixed, show that its free vibration solution can be expressed as

$$w(x, y, t) = \sum_{m=1}^{\infty} \sum_{n=1}^{\infty} \sin \frac{m \pi x}{a} \sin \frac{n \pi y}{b} (A_{mn} \sin \omega_{mn} t + B_{mn} \cos \omega_{mn} t)$$

and its natural frequencies $\omega_{mn}$ as

$$\omega_{mn}^2 = c^2 \pi^2 \left( \frac{m^2}{a^2} + \frac{n^2}{b^2} \right), \qquad m, n = 1, 2, \dots$$

**8.48.** Find the free vibration response of a rectangular membrane of sides $a$ and $b$ subjected to the following initial conditions:

$$\left. \begin{array}{l} w(x, y, 0) = 0 \\[2mm] \dfrac{\partial w}{\partial t}(x, y, 0) = \dot{w}_0 \sin \dfrac{\pi x}{a} \sin \dfrac{2 \pi y}{b} \end{array} \right\}, \qquad \begin{array}{l} 0 \leq x \leq a \\ 0 \leq y \leq b \end{array}$$

Assume that the edges of the membrane are fixed.

**8.49.** Find the fundamental natural frequency of a uniform cantilever beam using the static deflection curve

$$W(x) = \frac{c_0}{24EI}(x^4 - 4lx^3 + 6l^2x^2)$$

where $c_0$ is a constant.

**8.50.** Solve Problem 8.49 using the deflection equation

$$W(x) = c_0\left(1 - \cos\frac{\pi x}{2l}\right)$$

**8.51.** Solve Problem 4.49 using the deflection shape $W(x) = c_0 x^2/l^2$.

**8.52.** Determine the fundamental frequency of a uniform fixed-fixed beam carrying a mass $M$ at the middle by applying Rayleigh's method. Use the static deflection curve for $W(x)$.

**8.53.** Applying Rayleigh's method, determine the fundamental frequency of a cantilever beam (fixed at $x = l$) whose cross-sectional area $A(x)$ and moment of inertia $I(x)$ vary as $A(x) = A_0 x/l$ and $I(x) = I_0 x/l$.

**8.54.** Using Rayleigh's method, find the fundamental frequency for the lateral vibration of the beam shown in Fig. 8.27. The restoring force in the spring $k$ is proportional to the deflection, and the restoring moment in the spring $k_t$ is proportional to the angular deflection.

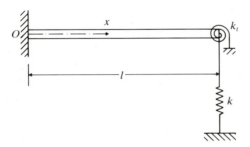

**Figure 8.27**

**8.55.** Using Rayleigh's method, estimate the fundamental frequency for the lateral vibration of a uniform beam fixed at both the ends. Assume the deflection curve to be

$$W(x) = c_1\left(1 - \cos\frac{2\pi x}{l}\right)$$

**8.56.** Find the fundamental frequency of longitudinal vibration of the tapered bar shown in Fig. 8.28, using Rayleigh's method with the mode shape

$$U(x) = c_1\sin\frac{\pi x}{2l}$$

The mass per unit length is given by

$$m(x) = 2m_0\left(1 - \frac{x}{l}\right)$$

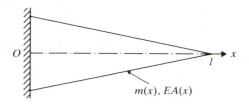

**Figure 8.28**

and the stiffness by

$$EA(x) = 2EA_0\left(1 - \frac{x}{l}\right)$$

**8.57.** Approximate the fundamental frequency of a rectangular membrane supported along all the edges by using Rayleigh's method with

$$W(x, y) = c_1 xy(x - a)(y - b).$$

Hint:

$$V = \frac{P}{2} \iint \left[\left(\frac{\partial w}{\partial x}\right)^2 + \left(\frac{\partial w}{\partial y}\right)^2\right] dx\, dy \qquad \text{and} \qquad T = \frac{\rho}{2} \iint \left(\frac{\partial w}{\partial t}\right)^2 dx\, dy$$

**8.58.** Estimate the fundamental frequency of a fixed-fixed string, assuming the mode shape
(a)  $W(x) = c_1 x(l - x)$
(b)  $W(x) = c_1 x(l - x) + c_2 x^2(l - x)^2$

**8.59.** Estimate the fundamental frequency for the longitudinal vibration of a uniform bar fixed at $x = 0$ and free at $x = l$ by assuming the mode shapes as (a) $U(x) = c_1(x/l)$ and (b) $U(x) = c_1(x/l) + c_2(x/l)^2$.

**8.60.** A stepped bar, fixed at $x = 0$ and free at $x = l$, has a cross-sectional area of $2A$ for $0 \le x \le l/3$ and $A$ for $l/3 \le x \le l$. Assuming the mode shape

$$U(x) = c_1 \sin\frac{\pi x}{2l} + c_2 \sin\frac{3\pi x}{2l}$$

estimate the first two natural frequencies of longitudinal vibration.

**8.61.** Solve Problem 8.56 using the Rayleigh-Ritz method with the mode shape

$$U(x) = c_1 \sin\frac{\pi x}{2l} + c_2 \sin\frac{3\pi x}{2l}$$

**8.62.** Find the first two natural frequencies of a fixed-fixed uniform string of mass density $\rho$ per unit length stretched between $x = 0$ and $x = l$ with an initial tension $P$. Assume the deflection functions

$$w_1(x) = x(l - x)$$
$$w_2(x) = x^2(l - x)^2$$

**8.63.** Find the values of $\alpha_1$ and $\alpha_2$ for $\beta = 0.01$ and 1.0 in Example 8.3 using the subroutine NONEQN.

**8.64.** Find the first five natural frequencies of a thin fixed-fixed beam using subroutine NONEQN.

**8.65.** Write a computer program for finding numerically the mode shapes of thin fixed-simply supported beams by using the known values of the natural frequencies.

# CHAPTER 9

# Vibration Control

Leonhard Euler (1707–1783) was a Swiss mathematician who became a court mathematician and later a professor of mathematics in St. Petersburg, Russia. He produced many works in algebra and geometry and was interested in the geometrical form of deflection curves in strength of materials. Euler's column buckling load is quite familiar to mechanical and civil engineers, and Euler's constant and Euler's coordinate system are well known to mathematicians. He derived the equation of motion for the bending vibrations of a rod (Euler-Bernoulli theory) and presented a series form of solution, as well as studying the dynamics of a vibrating ring. (By permission of Brown Brothers, Sterling, PA 18463.)

Courtesy Brown Brothers

## 9.1  INTRODUCTION

There are numerous sources of vibration in an industrial environment: impact processes such as pile driving and blasting; rotating or reciprocating machinery such as engines, compressors, and motors; transportation vehicles such as trucks, trains, and aircraft; the flow of fluids; and many others. The presence of vibration often leads to undesirable effects such as structural or mechanical failure, frequent and costly maintenance of machines, and human pain and discomfort [9.1]. Vibration can sometimes be eliminated on the basis of theoretical analysis. However, the manufacturing costs involved in eliminating the vibration may be too high; a designer must compromise between an acceptable amount of vibration and a reasonable manufacturing cost. In some cases the excitation or shaking force is inherent in the machine. As seen earlier, even a relatively small excitation force can cause an undesirably large response near resonance, especially in lightly damped systems. In these cases, the magnitude of the response can be significantly reduced by the use of isolators and auxiliary mass absorbers [9.2–9.4]. In this chapter, we shall consider various techniques of vibration control—that is, methods involving the elimination or reduction of vibration.

## 9.2  REDUCTION OF VIBRATION AT THE SOURCE

Methods of controlling vibration at the source involve minimization of the excitation forces. A common source of vibration in rotating and reciprocating machines is unbalance. We shall consider the analysis of these machines in the presence of unbalance as well as the means of controlling the vibrations that result from unbalanced forces.

## 9.3  BALANCING OF ROTATING MACHINES

The presence of an eccentric or unbalanced mass in a rotating disc causes vibration, which may be acceptable up to a certain level. Figure 9.1 is a vibration-severity chart that can be used to determine acceptable vibration levels at various frequencies or operating speeds [9.5]. If the vibration caused by an unbalanced mass is not acceptable, it can be eliminated either by removing the eccentric mass or by adding an equal mass in such a position that it cancels the effect of the unbalance. Although this solution appears to be simple, the problem becomes complicated because the amount and location of the eccentric mass are not easily determined in real machinery. Practical machines are unbalanced due to such irregularities as machining errors, nonuniform material density, and variation in the sizes of bolts and rivets. Furthermore, the machine may have a long rotor, which cannot be idealized as a thin disc. In all these cases, the unbalance can be determined by measuring the resulting vibration. In this section, we shall consider two types of balancing: *single-plane* or *static balancing* and *two-plane or dynamic balancing* [9.6–9.9].

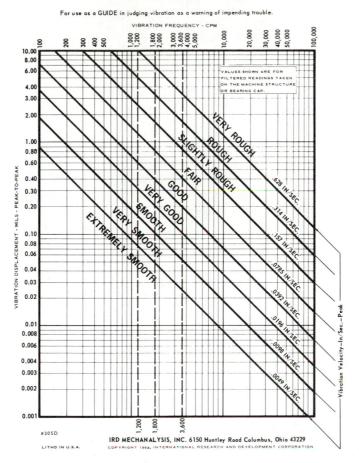

**Figure 9.1**  (Reprinted with permission of IRD Mechanalysis, Inc. Copyright 1964.)

### 9.3.1  Single-Plane Balancing

Single-plane balancing can be used for any rotating component in the form of a thin disc. For example, grinding wheels and impellers that are mounted on an overhung shaft can be balanced using the single-plane balancing procedure. The procedure is illustrated with reference to Fig. 9.2. The grinding wheel is attached to a rotating shaft that has bearing at $A$ and is driven by an electric motor. The shaft and the wheel rotate at an angular velocity $\omega$ rad/sec or $f = \omega/2\pi$ Hz.

Before starting the procedure, *reference marks*, also known as *phase marks*, are placed both on the rotor (wheel) and the stator, as shown in Fig. 9.3(a). A vibration pickup is placed in contact with the bearing, as shown in Fig. 9.2, and the vibration analyzer is set to a frequency corresponding to the angular velocity of the grinding wheel. The vibration signal (for example, the displacement amplitude) produced by

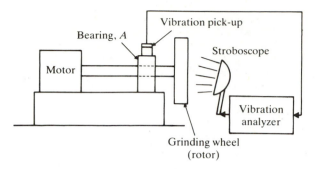

**Figure 9.2**

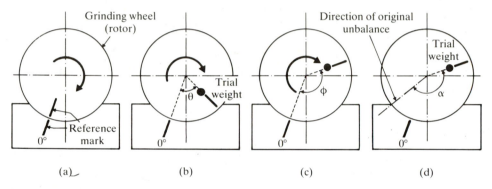

**Figure 9.3**

the unbalance can be read from the indicating meter of the vibration analyzer. A stroboscopic light is fired by the vibration analyzer at the frequency of the rotating wheel. When the rotor rotates at speed $f$, the phase mark on the rotor appears stationary under the stroboscopic light but is positioned at an angle $\theta$ from the mark on the stator, as shown in Fig. 9.3(b), due to phase lag in the response. Both the angle $\theta$ and the amplitude $A_u$ (read from the vibration analyzer) caused by the original unbalance are noted. The rotor is then stopped, and a known trial weight $W$ is attached to the rotor, as shown in Fig. 9.3(b). When the rotor runs at speed $f$, the new angular position of the rotor phase mark $\phi$ and the vibration amplitude $A_{u+w}$, caused by the combined unbalance of rotor and trial weight, are noted (see Fig. 9.3(c)).*

Now we construct a vector diagram to find the magnitude and location of the correction mass for balancing the wheel. The original unbalance vector $\vec{A}_u$ is drawn in an arbitrary direction, with its length equal to $A_u$, as shown in Fig. 9.4. Then the

---

*Note that if the trial weight is placed in a position that shifts the net unbalance in a clockwise direction, the stationary position of the phase mark will be shifted by exactly the same amount in the counterclockwise direction, and vice versa.

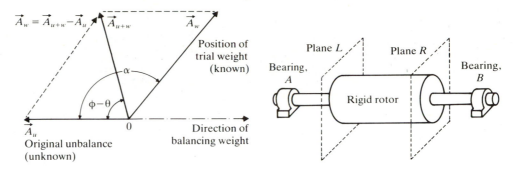

**Figure 9.4**

**Figure 9.5**

combined unbalance vector is drawn as $\vec{A}_{u+w}$ at an angle $\phi - \theta$ from the direction of $\vec{A}_u$ with a length of $A_{u+w}$. The difference vector $\vec{A}_w = \vec{A}_{u+w} - \vec{A}_u$ in Fig. 9.4 then represents the unbalance vector due to the trial weight $W$. The magnitude of $\vec{A}_w$ can be computed using the law of cosines:

$$A_w = \left[ A_u^2 + A_{u+w}^2 - 2 A_u A_{u+w} \cos(\phi - \theta) \right]^{1/2} \tag{9.1}$$

Since the magnitude of the trial weight $W$ and its direction relative to the original unbalance ($\alpha$ in Fig. 9.4) are known, the original unbalance itself must be at an angle $\alpha$ away from the position of the trial weight, as shown in Fig. 9.3(d). The angle $\alpha$ can be obtained from the law of cosines:

$$\alpha = \cos^{-1} \left[ \frac{A_u^2 + A_w^2 - A_{u+w}^2}{2 A_u A_w} \right] \tag{9.2}$$

The magnitude of the original unbalance is $W_0 = (A_u / A_w) \cdot W$, located at the same radial distance from the rotation axis of the rotor as the weight $W$. Once the location and magnitude of the original unbalance are known, correction weight can be added to balance the wheel properly.

### 9.3.2 Two-Plane Balancing

The single-plane balancing procedure can be used for balancing in one plane—that is, for rotors of the rigid disc type. If the rotor is an elongated rigid body as shown in Fig. 9.5, the unbalance shows up in the form of a relatively large vibration displacement at a frequency corresponding to the rotational speed of the rotor. The effective position of the unbalance may be anywhere along the length of the rotor. In this case, balancing can be achieved by adding balancing weights in any two planes [9.10]. For convenience, the two planes are usually chosen as the end planes of the rotor (shown by dotted lines in Fig. 9.5).

For illustration, let the rotor have an unbalanced mass $m$ located on the outer surface at a distance $l/3$ from the right end, as shown in Fig. 9.6(a). The force exerted by this unbalance on the rotor is $F = m\omega^2 R$, where $\omega$ is the angular velocity

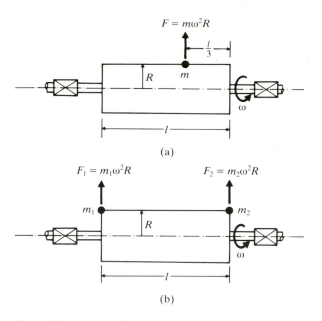

Figure 9.6

and $R$ is the radius of the rotor. The unbalanced mass $m$ can be replaced by two masses $m_1$ and $m_2$, located at the ends of the rotor, as shown in Fig. 9.6(b). The forces exerted on the rotor by these masses are $F_1 = m_1\omega^2 R$ and $F_2 = m_2\omega^2 R$. For the equivalence of the two force systems, the resultant force and the resultant moment about any point must be the same for each system. For the equivalence of force, we have

$$m\omega^2 R = m_1\omega^2 R + m_2\omega^2 R \qquad \text{or} \qquad m = m_1 + m_2 \tag{9.3}$$

and for the equivalence of moments about, say, the right end, we have

$$m\omega^2 R \frac{l}{3} = m_1\omega^2 R l \qquad \text{or} \qquad m = 3m_1 \tag{9.4}$$

Equations (9.3) and (9.4) give $m_1 = m/3$ and $m_2 = 2m/3$. This example shows that any unbalanced mass can be offset by two equivalent unbalanced masses on the end planes of the rotor.

Figure 9.7 shows a general condition of unbalance in a long rotor. Here, $U_L$ and $U_R$ represent the weights of the original unbalanced masses in the left and the right planes, respectively. At the rotor's operating speed $\omega$, the vibration amplitude and phase due to the original unbalance are measured at the two bearings $A$ and $B$, and the results are recorded as vectors $\vec{V}_A$ and $\vec{V}_B$. The magnitude of the vibration vector is taken as the vibration amplitude, while the direction of the vector is taken as the negative of the phase angle observed under stroboscopic light with reference to the

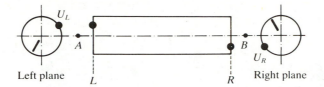

**Figure 9.7**

stator reference line. The measured vectors $\vec{V}_A$ and $\vec{V}_B$ can be expressed as

$$\vec{V}_A = \vec{A}_{AL}\vec{U}_L + \vec{A}_{AR}\vec{U}_R \tag{9.5}$$

$$\vec{V}_B = \vec{A}_{BL}\vec{U}_L + \vec{A}_{BR}\vec{U}_R \tag{9.6}$$

where $\vec{A}_{ij}$ is the vector operator, which reflects the effect of the unbalance in plane $j$ ($j = L, R$) on the vibration at bearing $i$ ($i = A, B$). Note that $\vec{U}_L, \vec{U}_R$, and all the vector operators $\vec{A}_{ij}$ are unknown in Eqs. (9.5) and (9.6).

As in the case of single-plane balancing, we add known trial weights and take additional measurements to obtain information about the unbalanced masses. First we add a known trial weight $\vec{W}_L$ in the left plane at a known angular position. We then measure the displacement and phase of vibration at the two bearings while the rotor is rotating at speed $\omega$. We denote these measured vibrations as vectors as

$$\vec{V}'_A = \vec{A}_{AL}(\vec{U}_L + \vec{W}_L) + \vec{A}_{AR}\vec{U}_R \tag{9.7}$$

$$\vec{V}'_B = \vec{A}_{BL}(\vec{U}_L + \vec{W}_L) + \vec{A}_{BR}\vec{U}_R \tag{9.8}$$

By subtracting Eqs. (9.5) and (9.6) from Eqs. (9.7) and (9.8), respectively, and solving, we obtain*

$$\vec{A}_{AL} = \frac{\vec{V}'_A - \vec{V}_A}{\vec{W}_L} \tag{9.9}$$

$$\vec{A}_{BL} = \frac{\vec{V}'_B - \vec{V}_B}{\vec{W}_L} \tag{9.10}$$

Next we remove $\vec{W}_L$ and add a known trial weight $\vec{W}_R$ in the right plane at a known

---

*It can be seen that complex subtraction, division, and multiplication are often used in the computation of the balancing weights. If

$$\vec{A} = a\underline{/\theta_A} \qquad \text{and} \qquad \vec{B} = b\underline{/\theta_B}$$

we can rewrite $\vec{A}$ and $\vec{B}$ as $\vec{A} = a_1 + ia_2$ and $\vec{B} = b_1 + ib_2$, where $a_1 = a\cos\theta_A$, $a_2 = a\sin\theta_A$, $b_1 = b\cos\theta_B$, and $b_2 = b\sin\theta_B$. Then the formulas for complex subtraction, division, and multiplication are [9.11]:

$$\vec{A} - \vec{B} = (a_1 - b_1) + i(a_2 - b_2)$$

$$\frac{\vec{A}}{\vec{B}} = \frac{(a_1 b_1 + a_2 b_2) + i(a_2 b_1 - a_1 b_2)}{(b_1^2 + b_2^2)}$$

$$\vec{A} \cdot \vec{B} = (a_1 b_1 - a_2 b_2) + i(a_2 b_1 + a_1 b_2)$$

angular position and measure the resulting vibrations while the rotor is running at speed $\omega$. The measured vibrations can be denoted as vectors:

$$\vec{V}_A'' = \vec{A}_{AR}(\vec{U}_R + \vec{W}_R) + \vec{A}_{AL}\vec{U}_L \tag{9.11}$$

$$\vec{V}_B'' = \vec{A}_{BR}(\vec{U}_R + \vec{W}_R) + \vec{A}_{BL}\vec{U}_L \tag{9.12}$$

As before, we subtract Eqs. (9.5) and (9.6) from Eqs. (9.11) and (9.12), respectively, to find

$$\vec{A}_{AR} = \frac{\vec{V}_A'' - \vec{V}_A}{\vec{W}_R} \tag{9.13}$$

$$\vec{A}_{BR} = \frac{\vec{V}_B'' - \vec{V}_B}{\vec{W}_R} \tag{9.14}$$

Once the vector operators $\vec{A}_{ij}$ are known, Eqs. (9.5) and (9.6) can be solved to find the unbalance vectors $\vec{U}_L$ and $\vec{U}_R$:

$$\vec{U}_L = \frac{\vec{A}_{BR}\vec{V}_A - \vec{A}_{AR}\vec{V}_B}{\vec{A}_{BR}\vec{A}_{AL} - \vec{A}_{AR}\vec{A}_{BL}} \tag{9.15}$$

$$\vec{U}_R = \frac{\vec{A}_{BL}\vec{V}_A - \vec{A}_{AL}\vec{V}_B}{\vec{A}_{BL}\vec{A}_{AR} - \vec{A}_{AL}\vec{A}_{BR}} \tag{9.16}$$

**Figure 9.8**    (Courtesy of Bruel & Kjaer Instruments, Inc., Marlborough, Mass.)

The rotor can now be balanced by adding equal and opposite balancing weights in each plane. The balancing weights in the left and right planes can be denoted vectorially as $\vec{B}_L = -\vec{U}_L$ and $\vec{B}_R = -\vec{U}_R$. It can be seen that the two plane balancing procedure is a straightforward extension of the single-plane balancing procedure. In practice, however, care must be exercised with regard to the angles associated with the various vectors. Figure 9.8 shows a practical example of two-plane balancing.

---

**EXAMPLE 9.1**

In the two-plane balancing of a turbine rotor, the data obtained from measurement of the original unbalance, the right-plane trial weight, and the left-plane trial weight is shown below. The displacement amplitudes are in mils (1/1000 inch). Determine the size and location of the balance weights required.

| Condition | Vibration (displacement) amplitude | | Phase angle | |
| --- | --- | --- | --- | --- |
| | At bearing $A$ | At bearing $B$ | At bearing $A$ | At bearing $B$ |
| Original unbalance | 8.5 | 6.5 | 60° | 205° |
| $W_L = 10.0$ oz added at 170° from reference mark | 6.0 | 4.5 | 125° | 230° |
| $W_R = 12.0$ oz added at 180° from reference mark | 6.0 | 10.5 | 35° | 160° |

*Solution.* The given data can be expressed in vector notation as

$$\vec{V}_A = 8.5 \bigg/ 60° = 4.2500 + i7.3612$$

$$\vec{V}_B = 6.5 \bigg/ 205° = -5.8910 - i2.7470$$

$$\vec{V}'_A = 6.0 \bigg/ 125° = -3.4415 + i4.9149$$

$$\vec{V}'_B = 4.5 \bigg/ 230° = -2.8926 - i3.4472$$

$$\vec{V}''_A = 6.0 \bigg/ 35° = 4.9149 + i3.4415$$

$$\vec{V}''_B = 10.5 \bigg/ 160° = -9.8668 + i3.5912$$

$$\vec{W}_L = 10.0 \bigg/ 270° = 0.0000 - i10.0000$$

$$\vec{W}_R = 12.0 \bigg/ 180° = -12.0000 + i0.0000$$

Equations (9.9) and (9.10) give

$$\vec{A}_{AL} = \frac{\vec{V}_A' - \vec{V}_A}{\vec{W}_L} = \frac{-7.6915 - i2.4463}{0.0000 - i10.0000} = 0.2446 - i0.7691$$

$$\vec{A}_{BL} = \frac{\vec{V}_B' - \vec{V}_B}{\vec{W}_L} = \frac{2.9985 - i0.7002}{0.0000 - i10.0000} = 0.0700 + i0.2998$$

The use of Eqs. (9.13) and (9.14) leads to

$$\vec{A}_{AR} = \frac{\vec{V}_A'' - \vec{V}_A}{\vec{W}_R} = \frac{0.6649 - i3.9198}{-12.0000 + i0.0000} = -0.0554 + i0.3266$$

$$\vec{A}_{BR} = \frac{\vec{V}_B'' - \vec{V}_B}{\vec{W}_R} = \frac{-3.9758 + i6.3382}{-12.0000 + i0.0000} = 0.3313 - i0.5282$$

The unbalance weights can be determined from Eqs. (9.15) and (9.16):

$$\vec{U}_L = \frac{(5.2962 + i0.1941) - (1.2237 - i1.7721)}{(-0.3252 - i0.3840) - (-0.1018 + i0.0063)} = \frac{(4.0725 + i1.9661)}{(-0.2234 - i0.3903)}$$

$$= -8.2930 + i5.6879$$

$$\vec{U}_R = \frac{(-1.9096 + i1.7898) - (-3.5540 + i3.8590)}{(-0.1018 + i0.0063) - (-0.3252 - i0.3840)} = \frac{(1.6443 - i2.0693)}{(0.2234 + i0.3903)}$$

$$= -2.1773 - i5.4592$$

Thus the required balance weights are given by

$$\vec{B}_L = -\vec{U}_L = (8.2930 - i5.6879) = 10.0561 \big/ 145.5548°$$

$$\vec{B}_R = -\vec{U}_R = (2.1773 + i5.4592) = 5.8774 \big/ 248.2559°$$

Physically, the turbine rotor will be balanced if a 10.0561 oz weight is added to the left plane at 145.5548° and a 5.8774 oz weight is added to the right plane at 248.2559° from the reference position. Note that the balance weights must be added at the same radial position as the trial weights. If a balance weight needs to be placed at a different radial position, the required balance weight must be changed in inverse proportion to the radial distance from the axis of rotation.

## 9.4 WHIRLING OF ROTATING SHAFTS

In the previous section, the rotor system—the shaft as well as the rotating body—was assumed to be rigid. In practice, however, all rotating shafts are flexible and therefore tend to bow out at certain speeds and whirl in a complicated manner. *Whirling* can be defined as the rotation of the plane made by the bent shaft and the line joining the centers of the bearings. Whirling is caused by such factors as mass

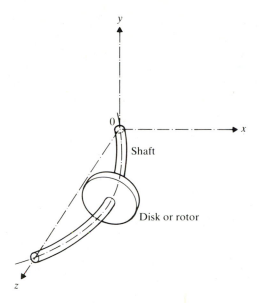

**Figure 9.9**

unbalance, fluid friction in bearings, and hysteresis damping in the shaft. We shall now consider a simple analysis for the whirling of shafts [9.12, 9.13].

### 9.4.1 Equations of Motion

Consider a shaft supported by two bearings and carrying a rotor or disc of mass $m$ at the middle, as shown in Fig. 9.9. We shall assume that the rotor is subjected to a steady-state excitation due to mass unbalance. The forces acting on the rotor are the inertia force due to the acceleration of the mass center, the spring force due to the elasticity of the shaft, and the external and internal damping forces.*

Let $O$ denote the equilibrium position of the shaft when balanced perfectly (Fig. 9.10). Let the shaft rotate with an angular velocity $\omega$. We use a fixed coordinate system ($x$ and $y$ fixed to the earth) with $O$ as the origin for deriving the equations of motion. The points $P$ and $Q$ denote the geometric center and mass center of the disc, respectively. The equation of motion of the system (mass $m$) can be written as

$$\text{inertia force } (\vec{F_i}) = \text{elastic force } (\vec{F_e}) + \text{internal damping force } (\vec{F_{di}})$$

$$+ \text{external damping force } (\vec{F_{de}}) \tag{9.17}$$

---

*Any rotating system responds in two different ways to damping or friction forces, depending upon whether the forces rotate with the shaft or not. When the positions at which the forces act remain fixed in space, as in the case of damping forces (which cause energy losses) in the bearing support structure, the damping is called *stationary or external damping*. On the other hand, if the positions at which they act rotate with the shaft in space, as in the case of internal friction of the shaft material, the damping is called *rotary or internal damping*.

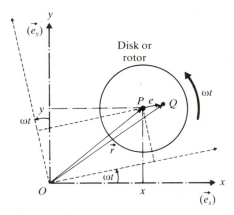

**Figure 9.10**

The various forces in Eq. (9.17) can be expressed as follows:

$$\text{Inertia force:} \qquad \vec{F}_i = m\ddot{\vec{r}} \tag{9.18}$$

where $\vec{r}$ denotes the radius vector of the mass center $Q$ given by

$$\vec{r} = (x + e\cos\omega t)\vec{e}_x + (y + e\sin\omega t)\vec{e}_y \tag{9.19}$$

with $x$ and $y$ representing the coordinates of the geometric center and $\vec{e}_x$ and $\vec{e}_y$ denoting the unit vectors along the $x$ and $y$ coordinates, respectively. Equations (9.18) and (9.19) lead to

$$\vec{F}_i = m\left[(\ddot{x} - e\omega^2\cos\omega t)\vec{e}_x + (\ddot{y} - e\omega^2\sin\omega t)\vec{e}_y\right] \tag{9.20}$$

$$\text{Elastic force:} \qquad \vec{F}_e = -k(x\vec{e}_x + y\vec{e}_y) \tag{9.21}$$

where $k$ is the stiffness of the shaft.

$$\text{Internal damping force:} \qquad \vec{F}_{di} = -c_i\left[(\dot{x} + \omega y)\vec{e}_x + (\dot{y} - \omega x)\vec{e}_y\right] \tag{9.22}$$

where $c_i$ is the internal (rotary) damping coefficient.

$$\text{External damping force:} \qquad \vec{F}_{de} = -c_e(\dot{x}\vec{e}_x + \dot{y}\vec{e}_y) \tag{9.23}$$

where $c_e$ is the external damping coefficient. By substituting Eqs. (9.20) to (9.23) into Eq. (9.17), we obtain the equations of motion in scalar form:

$$m\ddot{x} + (c_i + c_e)\dot{x} + kx + c_i\omega y = m\omega^2 e\cos\omega t \tag{9.24}$$

$$m\ddot{y} + (c_i + c_e)\dot{y} + ky - c_i\omega x = m\omega^2 e\sin\omega t \tag{9.25}$$

These equations indicate that the equations of motion for the lateral vibration of the rotor are coupled and are dependent on the speed of the steady-state rotation. By defining a complex quantity $w$ as

$$w = x + iy \tag{9.26}$$

where $i = (-1)^{1/2}$ and by adding Eq. (9.24) to Eq. (9.25) multiplied by $i$, we obtain a single equation of motion:

$$m\ddot{w} + (c_i + c_e)\dot{w} + kw - i\omega c_i w = m\omega^2 ee^{i\omega t} \tag{9.27}$$

### 9.4.2  Critical Speeds

Critical speeds correspond to resonant frequencies of a system. If a periodic component of the forcing function is equal to the natural frequency of the system, resonance will occur. The speeds at which resonance occurs are called the *critical speeds*. They correspond to the eigenvalues of the equations that define the dynamic state of the system [9.14]. The undamped natural frequency of the rotor system can be obtained by solving Eqs. (9.24), (9.25), or (9.27), retaining only the homogeneous part with $c_i = c_e = 0$. This gives the natural frequency of the system (or critical speed of the undamped system):

$$\omega_n = \left(\frac{k}{m}\right)^{1/2} \tag{9.28}$$

If the running speed is equal to this critical speed, the rotor undergoes large deflections, and the force transmitted to the bearings can cause bearing failures. A rapid transition of the running speed through a critical speed is expected to limit the whirl amplitudes, while a slow transition of the critical speed aids the development of large amplitudes. Reference [9.16] gives a Fortran computer program for calculating the critical speeds of rotating shafts.

### 9.4.3  Response of the System

We shall now consider the response of the rotor shown in Fig. 9.10 to harmonic excitation at the rotor speed frequency (due to mass unbalance) by assuming the internal damping $c_i$ to be zero. For this, we solve Eqs. (9.24) and (9.25) and find the rotor's dynamic whirl amplitudes resulting from the mass unbalance. With $c_i = 0$, Eqs. (9.24) and (9.25) reduce to

$$m\ddot{x} + c_e\dot{x} + kx = m\omega^2 e \cos \omega t \tag{9.29}$$

$$m\ddot{y} + c_e\dot{y} + ky = m\omega^2 e \sin \omega t \tag{9.30}$$

The solution of Eqs. (9.29) and (9.30) can be expressed as

$$x(t) = C_1 e^{-\alpha t}\cos(\beta t + \phi_1) + Dm\omega^2 e \cos(\omega t + \phi_2) \tag{9.31}$$

$$y(t) = C_2 e^{-\alpha t}\cos(\beta t + \phi_3) + Dm\omega^2 e \cos(\omega t + \phi_4) \tag{9.32}$$

where $C_1$, $C_2$, $D$, $\alpha$, $\beta$, $\phi_1$, $\phi_2$, $\phi_3$, and $\phi_4$ are constants. The first term in Eqs. (9.31) and (9.32) contains a decaying exponential term representing a transient solution, and the second term denotes a steady-state circular whirl. By substituting the steady-state part of Eq. (9.31) into Eq. (9.29), we can find the amplitude of the circular whirl:

$$m\omega^2 eD = \frac{m\omega^2 e}{\left[(k - m\omega^2)^2 + \omega^2 c_e^2\right]^{1/2}} \tag{9.33}$$

with

$$D = \frac{1}{\left[(k - m\omega^2)^2 + \omega^2 c_e^2\right]^{1/2}} \tag{9.34}$$

The phase angle is given by

$$\phi_2 = \phi_4 = -\tan^{-1}\left(\frac{c_e\omega}{k - m\omega^2}\right) \tag{9.35}$$

By differentiating Eq. (9.34) with respect to $\omega$ and setting the result equal to zero, we can find the rotational speed $\omega$ at which the whirl amplitude becomes a maximum:

$$\omega = \frac{\omega_n}{\left\{1 - \frac{1}{2}\left(\frac{c_e}{\omega_n}\right)^2\right\}^{1/2}} \tag{9.36}$$

where

$$\omega_n = \left(\frac{k}{m}\right)^{1/2} \tag{9.37}$$

It can be seen that the critical speed corresponds exactly to the natural frequency $\omega_n$ only when the damping $(c_e)$ is zero. Furthermore, Eq. (9.36) shows that the presence of damping, in general, increases the value of the critical speed compared to $\omega_n$. A plot of Eqs. (9.33) and (9.35) is shown in Fig. 9.11. Since the forcing function is proportional to $\omega^2$, we normally expect the vibration amplitude to increase with the speed $\omega$. However, the actual amplitude appears as shown in Fig. 9.11. From Eq. (9.33), we note that the response at low speeds is determined by the spring constant $k$, since all the other terms are small. Also, the value of the phase angle $\phi_2$ is 0° for small values of $\omega$. As $\omega$ increases, the amplitude of the response reaches a peak, since resonance occurs at $k - m\omega^2 = 0$. Around resonance, the response is essentially limited by the damping term. The phase lag is 90° at resonance. As the speed

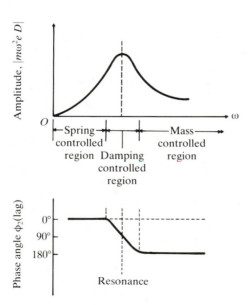

**Figure 9.11**

$\omega$ increases beyond $\omega_n$, the response is primarily determined by the mass-acceleration term, which is 180° out of phase with the unbalanced force. Thus the shaft moves in a direction opposite to that of the unbalanced force, so the response is limited.

The vibration and balancing of unbalanced flexible rotors are presented in Refs. [9.17 and 9.18].

### 9.4.4  Stability Analysis

Instability in a flexible rotor system can occur due to such factors as internal friction, asymmetry in the rotating system, action of an oil-lubricated journal in its bearing (such as oil whip), and nonlinearities in the rotational system. As seen earlier, the stability of the system can be investigated by considering the equation governing the dynamics of the system. The stability of a system governed by an $n$th order linear differential equation with constant coefficients was examined in Sec. 6.14. The procedure can be extended to equations with complex coefficients [9.19]. For this, the characteristic equation is first expressed as

$$(a_0 + ib_0)\lambda^n + (a_1 + ib_1)\lambda^{n-1} + \cdots + (a_{n-1} + ib_{n-1})\lambda + (a_n + ib_n) = 0 \qquad (9.38)$$

A necessary and sufficient condition for the system to be stable is that the set of determinants generated below are positive.

$$-\begin{vmatrix} a_0 & a_1 \\ b_0 & b_1 \end{vmatrix} > 0 \qquad (9.39)$$

$$\begin{vmatrix} a_0 & a_1 & a_2 & 0 \\ b_0 & b_1 & b_2 & 0 \\ 0 & a_0 & a_1 & a_2 \\ 0 & b_0 & b_1 & b_2 \end{vmatrix} > 0 \qquad (9.40)$$

$$\vdots$$

$$(-1)^n \begin{vmatrix} a_0 & a_1 & a_2 & \cdots & a_n & 0 & 0 & \cdots & 0 \\ b_0 & b_1 & b_2 & \cdots & b_n & 0 & 0 & \cdots & 0 \\ 0 & a_0 & a_1 & \cdots & a_{n-1} & a_n & 0 & \cdots & 0 \\ 0 & b_0 & b_1 & \cdots & b_{n-1} & b_n & 0 & \cdots & 0 \\ 0 & 0 & a_0 & \cdots & a_{n-2} & a_{n-1} & a_n & \cdots & 0 \\ 0 & 0 & b_0 & \cdots & b_{n-2} & b_{n-1} & b_n & \cdots & 0 \\ \vdots & & & & & & & & \\ 0 & 0 & 0 & \cdots & 0 & a_0 & a_1 & \cdots & a_n \\ 0 & 0 & 0 & \cdots & 0 & b_0 & b_1 & \cdots & b_n \end{vmatrix} > 0 \qquad (9.41)$$

To apply this stability criterion to Eq. (9.27), we write its characteristic equation as

$$s^2 + \frac{c_i + c_e}{m}s + \frac{k - i\omega c_i}{m} = 0 \qquad (9.42)$$

By substituting $s = i\lambda$ in Eq. (9.42) and applying Eqs. (9.39) to (9.41), we obtain

$$\left(\frac{c_i + c_e}{m}\right) > 0 \tag{9.43}$$

and

$$-\left(\frac{c_i + c_e}{m}\right)^2 \frac{k}{m} + \frac{\omega^2 c_i^2}{m^2} < 0 \tag{9.44}$$

Equation (9.43) is automatically satisfied, while Eq. (9.44) yields the condition

$$\left(1 + \frac{c_e}{c_i}\right)\sqrt{\frac{k}{m}} - \omega > 0 \tag{9.45}$$

This equation also shows that internal and external friction can cause instability at rotating speeds above the first critical speed.

## 9.5 BALANCING OF RECIPROCATING ENGINES

The essential moving elements of a reciprocating engine are the piston, the crank, and the connecting rod. Vibrations in reciprocating engines arise due to (1) periodic variations of the gas pressure in the cylinder and (2) inertia forces associated with the moving parts [9.20]. We shall now analyze a reciprocating engine and find the unbalanced forces caused by these factors.

### 9.5.1 Unbalanced Forces Due to Fluctuations in Gas Pressure

Figure 9.12(a) is a schematic diagram of a cylinder of a reciprocating engine. The engine is driven by the expanding gas confined between the cylinder head and the piston. The expanding gas exerts on the piston a pressure force $F$, which is transmitted to the crankshaft through the connecting rod. The reaction to the force $F$ can be resolved into two components: one of magnitude $F/\cos\phi$, acting along the connecting rod, and the other of magnitude $F\tan\phi$, acting in a horizontal direction. The force $F/\cos\phi$ induces a torque $M_t$, which tends to rotate the crankshaft (In Fig. 9.12(b), $M_t$ acts about an axis perpendicular to the plane of the paper and passes through point $Q$.)

$$M_t = \left(\frac{F}{\cos\phi}\right) \cdot r\cos\theta \tag{9.46}$$

The force $(F/\cos\phi)$ induced at the bearing of the crankshaft can be resolved into two components: one of magnitude $F$ in the vertical direction and the other of magnitude $F\tan\phi$ in the horizontal direction, as shown in Fig. 9.12(b). Finally, the forces transmitted to the stationary parts of the engine are:

1. force $F$ acting upwards at the cylinder head
2. force $F\tan\phi$ acting toward the right at the cylinder head

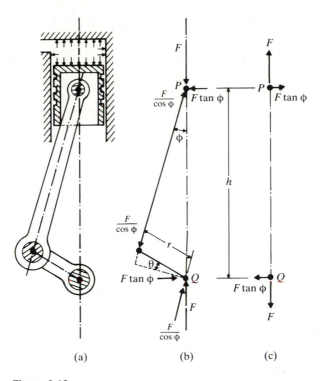

**Figure 9.12**

**3.** force $F$ acting downwards at the crankshaft bearing $Q$

**4.** force $F \tan \phi$ acting toward the left at the crankshaft bearing

These forces are shown in Fig. 9.12(c). Although the total resultant force is zero, there is a resultant torque $M_Q = Fh \tan \phi$ on the chassis of the engine, where $h$ can be found from the geometry of the system:

$$h = \frac{r \cos \theta}{\sin \phi} \tag{9.47}$$

Thus the resultant torque is given by

$$M_Q = \frac{Fr \cos \theta}{\cos \phi} \tag{9.48}$$

It can be seen that $M_t$ and $M_Q$ given by Eqs. (9.46) and (9.48) are identical. This indicates that the torque induced on the crankshaft due to the gas pressure in the cylinder is reflected at the support of the engine. The magnitude of torque $M_Q$ varies with time, along with the variation of force $F$. The magnitude of force $F$ changes from a maximum to a minimum at a frequency governed by the number of cylinders in the engine, the type of the operating cycle, and the rotating speed of the engine.

### 9.5.2  Unbalanced Forces Due to Inertia of the Moving Parts

**Acceleration of the Piston.** Figure 9.13 shows the crank (of length $r$), the connecting rod (of length $l$), and the piston of a reciprocating engine. The crank is assumed to rotate in an anticlockwise direction at a constant angular speed of $\omega$, as shown in Fig. 9.13. If we consider the origin of the $x$ axis ($O$) as the uppermost position of the piston, the displacement of the piston $P$ corresponding to an angular displacement of the crank of $\theta = \omega t$ can be expressed as

$$x_p = r + l - r\cos\theta - l\cos\phi = r + l - r\cos\omega t - l\sqrt{1 - \sin^2\phi} \tag{9.49}$$

But

$$l\sin\phi = r\sin\theta = r\sin\omega t$$

and hence

$$\cos\phi = \left(1 - \frac{r^2}{l^2}\sin^2\omega t\right)^{1/2} \tag{9.50}$$

By substituting Eq. (9.50) into Eq. (9.49), we obtain

$$x_p = r + l - r\cos\omega t - l\sqrt{1 - \frac{r^2}{l^2}\sin^2\omega t} \tag{9.51}$$

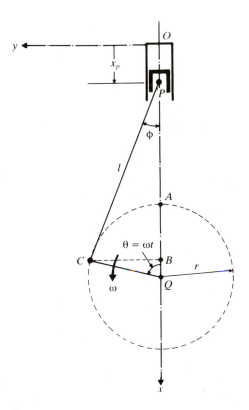

**Figure 9.13**

Due to the presence of the term involving the square root, Eq. (9.51) is not very convenient in further calculation. Equation (9.51) can be simplified by noting that, in general, $r/l < 1/4$ and by using the expansion relation

$$\sqrt{1-\varepsilon} \simeq 1 - \frac{\varepsilon}{2} \tag{9.52}$$

Hence Eq. (9.51) can be approximated as

$$x_p \simeq r(1 - \cos\omega t) + \frac{r^2}{2l}\sin^2\omega t \tag{9.53}$$

or, equivalently,

$$x_p \simeq r\left(1 + \frac{r}{2l}\right) - r\left(\cos\omega t + \frac{r}{4l}\cos 2\omega t\right) \tag{9.54}$$

Equation (9.54) can be differentiated with respect to time to obtain expressions for the velocity and the acceleration of the piston:

$$\dot{x}_p = r\omega\left(\sin\omega t + \frac{r}{2l}\sin 2\omega t\right) \tag{9.55}$$

$$\ddot{x}_p = r\omega^2\left(\cos\omega t + \frac{r}{l}\cos 2\omega t\right) \tag{9.56}$$

**Acceleration of the Crankpin.** The crankpin $C$ moves in a circular path with the crankshaft axis $Q$ as the center. With respect to the $xy$ coordinate axes shown in Fig. 9.13, the vertical and horizontal displacement components of the crankpin $C$ are given by

$$x_c = OA + AB = l + r(1 - \cos\omega t) \tag{9.57}$$

$$\dot{x}_c = r\omega\sin\omega t \tag{9.58}$$

$$\ddot{x}_c = r\omega^2\cos\omega t \tag{9.59}$$

$$y_c = CB = r\sin\omega t \tag{9.60}$$

$$\dot{y}_c = r\omega\cos\omega t \tag{9.61}$$

$$\ddot{y}_c = -r\omega^2\sin\omega t \tag{9.62}$$

**Inertia Forces.** Although the mass of the connecting rod is distributed throughout its length, it is generally idealized as a massless link with two masses concentrated at its ends—the piston end and the crankpin end. If $m_p$ and $m_c$ denote the total mass of the piston and of the crankpin (including the concentrated mass of the connecting rod) respectively, the vertical component of the inertia force ($F_x$) for one cylinder is given by

$$F_x = m_p\ddot{x}_p + m_c\ddot{x}_c \tag{9.63}$$

By substituting Eqs. (9.56) and (9.59) for the accelerations of $P$ and $C$, Eq. (9.63) becomes

$$F_x = (m_p + m_c)r\omega^2\cos\omega t + m_p\frac{r^2\omega^2}{l}\cos 2\omega t \tag{9.64}$$

It can be observed that the vertical component of the inertia force consists of two parts. One part, known as the *primary part*, has a frequency equal to the rotational frequency of the crank $\omega$. The other part, known as the *secondary part*, has a frequency equal to twice the rotational frequency of the crank.

Similarly, the horizontal component of inertia force for a cylinder can be obtained:

$$F_y = m_p \ddot{y}_p + m_c \ddot{y}_c \qquad (9.65)$$

where $\ddot{y}_p = 0$ and $\ddot{y}_c$ is given by Eq. (9.62). Thus

$$F_y = - m_c r \omega^2 \sin \omega t \qquad (9.66)$$

The horizontal component of the inertia force can be observed to have only a primary part.

### 9.5.3  Balancing of Reciprocating Engines

The unbalanced or inertia forces on a single cylinder are given by Eqs. (9.64) and (9.66). In these equations, $m_p$ and $m_c$ represent the equivalent reciprocating and rotating masses, respectively. The mass $m_p$ is always positive, but $m_c$ can be made zero by counterbalancing the crank. It is therefore possible to reduce the horizontal inertia force $F_y$ to zero, but the vertical unbalanced force always exists. Thus a single cylinder engine is inherently unbalanced.

In a multicylinder engine, it is possible to balance some or all of the inertia forces and torques by proper arrangement of the cranks. The conditions necessary for such balancing can be derived with reference to the six-cylinder engine shown in Fig. 9.14. The angular and axial positions of crank $i$ are indicated by $\alpha_i$ and $l_i$, measured from the position of the first crank, as shown in Fig. 9.14. For force balance, the total inertia force in the vertical and horizontal directions must be zero.

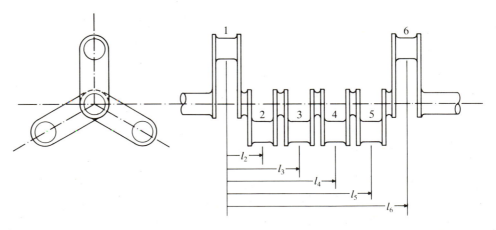

**Figure 9.14**

Thus

$$(F_x)_{\text{total}} = \sum_{i=1}^{n} (F_x)_i = 0 \tag{9.67}$$

$$(F_y)_{\text{total}} = \sum_{i=1}^{n} (F_y)_i = 0 \tag{9.68}$$

where $(F_x)_i$ and $(F_y)_i$ are the vertical and horizontal components of inertia force of cylinder $i$ given by (see Eqs. (9.64) and (9.66)):

$$(F_x)_i = (m_p + m_c)_i r\omega^2 \cos(\omega t + \alpha_i)$$

$$+ (m_p)_i \frac{r^2 \omega^2}{l} \cos(2\omega t + 2\alpha_i) \tag{9.69}$$

$$(F_y)_i = -(m_c)_i r\omega^2 \sin(\omega t + \alpha_i) \tag{9.70}$$

and $n$ is the number of cylinders in the engine. If the reciprocating and rotating masses for each cylinder are assumed to be respectively equal (that is, $(m_p)_i = m_p$ and $(m_c)_i = m_c$ for $i = 1, 2, \ldots, n$) and the instant $t$ at which Eqs. (9.67) and (9.68) are applied is taken to be zero, we obtain the conditions necessary for the total force balance:

$$\sum_{i=1}^{n} \cos \alpha_i = 0 \quad \text{and} \quad \sum_{i=1}^{n} \cos 2\alpha_i = 0 \tag{9.71}$$

$$\sum_{i=1}^{n} \sin \alpha_i = 0 \tag{9.72}$$

The inertia forces $(F_x)_i$ and $(F_y)_i$ of the $i$th cylinder induce moments about the horizontal and vertical transverse axes, respectively. If we consider the $x$ and $y$ axes passing through crank 1 as reference axes and assume that $(m_p)_i = m_p$ and $(m_c)_i = m_c$ for $i = 1$ to $n$, the inertia moments can be balanced if the following equations are satisfied. Moment about $z$ axis:

$$M_z = \sum_{i=2}^{n} (F_x)_i l_i = 0 \tag{9.73}$$

or

$$\sum_{i=2}^{n} l_i \cos \alpha_i = 0 \quad \text{and} \quad \sum_{i=2}^{n} l_i \cos 2\alpha_i = 0 \tag{9.74}$$

Moment about $x$ axis:

$$M_x = \sum_{i=2}^{n} (F_y)_i l_i = 0 \tag{9.75}$$

or

$$\sum_{i=2}^{n} l_i \sin \alpha_i = 0 \tag{9.76}$$

Thus we can arrange the cylinders of a multicylinder reciprocating engine so as to satisfy Eqs. (9.71), (9.72), (9.74), and (9.76); it will be completely balanced against the inertia forces and moments.

## 9.6  CONTROL OF VIBRATION

In many practical situations, it is possible to reduce but not eliminate the dynamic forces that cause vibrations. Several methods can be used to control vibrations. Among them, the following are important.

1. By controlling the natural frequencies of the system and avoiding resonance under external excitations.
2. By preventing excessive response of the system, even at resonance, by introducing a damping or energy-dissipating mechanism.
3. By reducing the transmission of the excitation forces from one part of the machine to another, by the use of vibration isolators.
4. By reducing the response of the system, by the addition of an auxiliary mass neutralizer or vibration absorber.

We shall now consider the details of these methods.

## 9.7  CONTROL OF NATURAL FREQUENCIES

It is well known that whenever the frequency of excitation coincides with one of the natural frequencies of the system, resonance occurs. The most prominent feature of resonance is a large displacement. In most mechanical and structural systems, large displacements indicate undesirably large strains and stresses, which can lead to the failure of the system. Hence resonance conditions must be avoided in any system. In most cases, the excitation frequency cannot be controlled, because it is imposed by the functional requirements of the system or machine. We must concentrate on controlling the natural frequencies of the system to avoid resonance.

As indicated by Eq. (2.11), the natural frequency of a system can be changed either by varying the mass $m$ or the stiffness $k$.* In many practical cases, however, the mass cannot be changed easily, since its value is determined by the functional requirements of the system. For example, the mass of a flywheel on a shaft is determined by the amount of energy it must store in one cycle. Therefore, the stiffness of the system is the factor that is most often changed to alter its natural frequencies. For example, the stiffness of a rotating shaft can be altered by varying one or more of its parameters, such as materials or number and location of support points (bearings).

---

*Although this statement is made with reference to a single degree of freedom system, it is generally true even for multidegree of freedom and continuous systems.

## 9.8   INTRODUCTION OF DAMPING

Although damping is disregarded so as to simplify the analysis, especially in finding the natural frequencies, most systems possess damping to some extent. The presence of damping is helpful in many cases. In systems such as automobile shock absorbers and many vibration-measuring instruments, damping must be introduced to fulfill the functional requirements [9.21–9.23].

If the system undergoes forced vibration, its response or amplitude of vibration tends to become large near resonance if there is no damping. The presence of damping always limits the amplitude of vibration. If the forcing frequency is known, it may be possible to avoid resonance by changing the natural frequency of the system. However, the system or the machine may be required to operate over a range of speeds, as in the case of a variable-speed electric motor or an internal combustion engine. It may not be possible to avoid resonance under all operating conditions. In such cases, we can introduce damping into the system to control its response, by the use of structural materials having high internal damping, such as cast iron or laminated or sandwich materials.

## 9.9   USE OF VIBRATION ISOLATORS

Vibration isolation methods are used to reduce the undesirable effects of vibration [9.24–9.30]. In its most elementary form, a vibration isolator can be considered as a resilient member placed between the vibrating system or equipment and its supporting structure or foundation. The resilient member is made up of an isolating material such as rubber, cork, felt, or metallic springs. The function of an isolator is to reduce the magnitude of the force transmitted from the equipment to its foundation or to reduce the motion transmitted from a vibrating foundation to the equipment. Figure 9.15 shows typical spring and pneumatic mounts which can be used as isolators, and Fig. 9.16 illustrates the use of isolators in the mounting of a high-speed punch press [9.33]. Active control mechanisms are also proposed for vibration isolation [9.31, 9.32]. The optimal synthesis of vibration isolators is presented in Refs. [9.34–9.38].

We shall first analyze the vibration isolation systems by assuming the foundation to be rigid.

### 9.9.1   Vibration Isolation System with Rigid Foundation

When a machine or equipment is placed on a resilient member resting on a rigid foundation or support, the system can be idealized as a single degree of freedom system, as shown in Fig. 9.17(a). The resilient member is assumed to have both elasticity and damping and is modeled as a spring $k$ and a dashpot $c$, as shown in Fig. 9.17(b). It is assumed that the operation of the machine gives rise to a harmonically varying force $F(t) = F_0 \cos \omega t$. The equation of motion of the machine

**Figure 9.15** (a) Undamped spring mount; (b) damped spring mount; (c) pneumatic rubber mount. (Courtesy of *Sound and Vibration*.)

(of mass $m$) is given by

$$m\ddot{x} + c\dot{x} + kx = F_0\cos\omega t \tag{9.77}$$

Since the transient solution dies out after some time, only the steady-state solution will be left. The steady-state solution of Eq. (9.77) is given by (see Eq. (3.24))

$$x(t) = X\cos(\omega t - \phi) \tag{9.78}$$

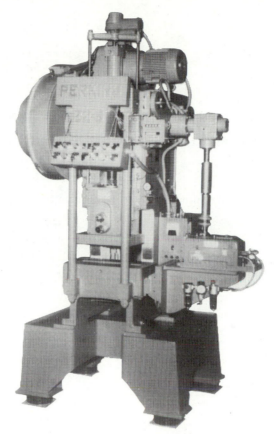

**Figure 9.16**   High-speed punch press mounted on pneumatic rubber mounts.   (Courtesy of *Sound and Vibration*.)

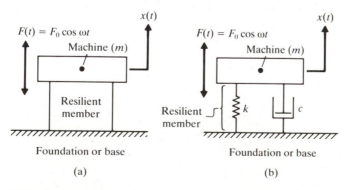

**Figure 9.17**   Machine and resilient member on rigid foundation.

where

$$X = \frac{F_0}{\left[(k - m\omega^2)^2 + \omega^2 c^2\right]^{1/2}} \tag{9.79}$$

and

$$\phi = \tan^{-1}\left(\frac{\omega c}{k - m\omega^2}\right) \tag{9.80}$$

The force transmitted through the spring and the dashpot are, respectively,

$$kx(t) = kX\cos(\omega t - \phi) \tag{9.81}$$

$$c\dot{x}(t) = -c\omega X\sin(\omega t - \phi) \tag{9.82}$$

The magnitude of the total transmitted force ($F_t$) is given by

$$F_t = \left[(kx)^2 + (c\dot{x})^2\right]^{1/2} = X\sqrt{k^2 + \omega^2 c^2}$$

$$= \frac{F_0(k^2 + \omega^2 c^2)^{1/2}}{\left[(k - m\omega^2)^2 + \omega^2 c^2\right]^{1/2}} \tag{9.83}$$

The transmissibility or transmission ratio of the isolator ($T_r$) is defined as the ratio of the magnitude of the force transmitted to that of the exciting force:

$$T_r = \frac{F_t}{F_0} = \left\{\frac{k^2 + \omega^2 c^2}{(k - m\omega^2)^2 + \omega^2 c^2}\right\}^{1/2}$$

$$= \left\{\frac{1 + \left(2\zeta\frac{\omega}{\omega_n}\right)^2}{\left[1 - \left(\frac{\omega}{\omega_n}\right)^2\right]^2 + \left(2\zeta\frac{\omega}{\omega_n}\right)^2}\right\}^{1/2} \tag{9.84}$$

The variation of $T_r$ with the frequency ratio $\omega/\omega_n$ is shown in Fig. 9.18.

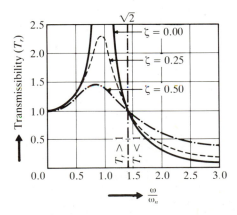

**Figure 9.18**   Variation of transmission ratio ($T_r$) with $\omega$.

### 9.9.2    Vibration Isolation System with Flexible Foundation

In many practical situations, the structure or foundation to which the isolator is connected moves when the machine mounted on the isolator operates. For example, in the case of a turbine supported on the hull of a ship or an aircraft engine mounted on the wing of an airplane, the area surrounding the point of support also moves with the isolator. In such cases, the system can be represented as having two degrees of freedom. In Fig. 9.19, $m_1$ and $m_2$ denote the masses of the machine and the supporting structure that moves with the isolator, respectively. The isolator is represented by a spring $k$, and the damping is disregarded for the sake of simplicity. The equations of motion of the masses $m_1$ and $m_2$ are

$$m_1\ddot{x}_1 + k(x_1 - x_2) = F_0\cos \omega t \tag{9.85}$$

$$m_2\ddot{x}_2 + k(x_2 - x_1) = 0 \tag{9.86}$$

By assuming a harmonic solution of the form

$$x_j = X_j\cos \omega t, \qquad j = 1, 2 \tag{9.87}$$

Eqs. (9.85) and (9.86) give

$$\left. \begin{array}{l} X_1(k - m_1\omega^2) - X_2 k = F_0 \\ -X_1 k + X_2(k - m_2\omega^2) = 0 \end{array} \right\} \tag{9.88}$$

The natural frequencies of the system are given by the roots of the equation

$$\begin{vmatrix} (k - m_1\omega^2) & -k \\ -k & (k - m_2\omega^2) \end{vmatrix} = 0 \tag{9.89}$$

The roots of Eq. (9.89) are given by

$$\omega_1^2 = 0, \qquad \omega_2^2 = \frac{(m_1 + m_2)k}{m_1 m_2} \tag{9.90}$$

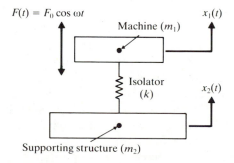

$F(t) = F_0 \cos \omega t$

Machine ($m_1$)

$x_1(t)$

Isolator ($k$)

$x_2(t)$

Supporting structure ($m_2$)

**Figure 9.19**   Machine with isolator on a flexible foundation.

The value $\omega_1 = 0$ corresponds to rigid-body motion since the system is unconstrained. In the steady state, the amplitudes of $m_1$ and $m_2$ are governed by Eq. (9.88), whose solution yields

$$X_1 = \frac{(k - m_2\omega^2)F_0}{\left[(k - m_1\omega^2)(k - m_2\omega^2) - k^2\right]} \tag{9.91}$$

$$X_2 = \frac{kF_0}{\left[(k - m_1\omega^2)(k - m_2\omega^2) - k^2\right]} \tag{9.92}$$

The force transmitted to the supporting structure $(F_t)$ is given by the amplitude of $m_2\ddot{x}_2$:

$$F_t = -m_2\omega^2 X_2 = \frac{-m_2 k\omega^2 F_0}{\left[(k - m_1\omega^2)(k - m_2\omega^2) - k^2\right]} \tag{9.93}$$

The transmissibility of the isolator $(T_r)$ is given by

$$T_r = \frac{F_t}{F_0} = \frac{-m_2 k\omega^2}{\left[(k - m_1\omega^2)(k - m_2\omega^2) - k^2\right]}$$

$$= \frac{1}{\left(\dfrac{m_1 + m_2}{m_2} - \dfrac{m_1\omega^2}{k}\right)} = \frac{m_2}{(m_1 + m_2)}\left(\frac{1}{1 - \dfrac{\omega^2}{\omega_2^2}}\right) \tag{9.94}$$

where $\omega_2$ is the natural frequency of the system given by Eq. (9.90). Equation (9.94) shows, as in the case of an isolator on a rigid base, that the force transmitted to the foundation becomes less as the natural frequency of the system $\omega_2$ is reduced.

The value of $T_r$ of an isolator with rigid foundation is plotted in Fig. 9.18. It can be seen that the force transmitted to the foundation cannot become infinite at resonance unless damping is zero. If the force transmitted to the foundation is to be small, $\omega/\omega_n$ should be large—that is, the isolator should have a natural frequency much smaller than the operating speed of the machine it supports. Although damping reduces the amplitude of motion $(X)$ for all frequencies, it reduces the maximum force transmitted $(F_t)$ only if $\omega/\omega_n < \sqrt{2}$. Above that value, the addition of damping increases the force transmitted. Furthermore, the force transmitted to the foundation is larger than the force applied by the machine $(T_r > 1)$ when $\omega/\omega_n < \sqrt{2}$. These observations suggest that the damping of the isolator should be made as small as possible, so as to reduce the force transmitted. If the speed of the machine (forcing frequency) varies, we must compromise in choosing the amount of damping to minimize the force transmitted. The amount of damping should be sufficient to limit the amplitude $X$ and the force transmitted $F_t$ while passing through the resonance, but not so much as to increase unnecessarily the force transmitted at the operating speeds.

## EXAMPLE 9.2

A refrigeration unit operating at 1000 rpm is to be supported by four springs, each having a stiffness of $K$. If only 10 percent of the shaking force of the unit is to be transmitted to the base, what should be the value of $K$? Assume the mass of the refrigeration unit to be 40 kg.

**Solution.** Since the transmissibility has to be 0.1, we have, from Eq. (9.84),

$$0.1 = \left[ \frac{1 + \left(2\zeta \frac{\omega}{\omega_n}\right)^2}{\left\{1 - \left(\frac{\omega}{\omega_n}\right)^2\right\}^2 + \left(2\zeta \frac{\omega}{\omega_n}\right)^2} \right]^{1/2} \tag{E.1}$$

where the forcing frequency is given by

$$\omega = \frac{1000 \times 2\pi}{60} = 104.72 \text{ rad/sec} \tag{E.2}$$

and the natural frequency of the system by

$$\omega_n = \left(\frac{k}{m}\right)^{1/2} = \left(\frac{4K}{40}\right)^{1/2} = \frac{\sqrt{K}}{3.1623} \tag{E.3}$$

By assuming the damping ratio to be $\zeta = 0$, we obtain from Eq. (E.1),

$$0.1 = \frac{\pm 1}{\left\{1 - \left(\frac{104.72 \times 3.1623}{\sqrt{K}}\right)^2\right\}} \tag{E.4}$$

To avoid imaginary values, we need to consider the negative sign on the right-hand side of Eq. (E.4). This leads to

$$\frac{331.1561}{\sqrt{K}} = 3.3166$$

or

$$K = 9969.6365 \text{ N/m}$$

## EXAMPLE 9.3

A vibrating system is to be isolated from its supporting base. Find the required damping ratio that must be achieved by the isolator to limit the transmissibility at resonance to $T_r = 4$. Assume the system to have a single degree of freedom.

**Solution.** By setting $\omega = \omega_n$, Eq. (9.84) gives

$$T_r = \frac{\sqrt{1 + (2\zeta)^2}}{2\zeta}$$

or

$$\zeta = \frac{1}{2\sqrt{T_r^2 - 1}} = \frac{1}{2\sqrt{15}} = 0.1291$$

### 9.9.3    Vibration Isolation System
### with Partially Flexible Foundation

We shall now consider a more realistic situation. Figure 9.20 shows an isolator whose base, instead of being completely rigid or completely flexible, is partially flexible [9.39]. We can define the mechanical impedance of the base structure, $Z(\omega)$, as the force at frequency $\omega$ required to produce a unit displacement of the base (as in Sec. 3.5):

$$Z(\omega) = \frac{\text{applied force of frequency } \omega}{\text{displacement}}$$

The equations of motion are given by*

$$m_1 \ddot{x}_1 + k(x_1 - x_2) = F_0 \cos \omega t \tag{9.95}$$

$$k(x_2 - x_1) = -x_2 Z(\omega) \tag{9.96}$$

By substituting the harmonic solution

$$x_j(t) = X_j \cos \omega t, \qquad j = 1, 2 \tag{9.97}$$

into Eqs. (9.95) and (9.96), $X_1$ and $X_2$ can be obtained as in the previous case:

$$X_1 = \frac{[k + Z(\omega)] X_2}{k} = \frac{[k + Z(\omega)] F_0}{[Z(\omega)(k - m_1\omega^2) - km_1\omega^2]}$$

$$X_2 = \frac{kF_0}{[Z(\omega)(k - m_1\omega^2) - km_1\omega^2]} \tag{9.98}$$

The amplitude of the force transmitted is given by

$$F_t = X_2 Z(\omega) = \frac{kZ(\omega) F_0}{[Z(\omega)(k - m_1\omega^2) - km_1\omega^2]} \tag{9.99}$$

and the transmissibility of the isolator by

$$T_r = \frac{F_t}{F_0} = \frac{kZ(\omega)}{[Z(\omega)(k - m_1\omega^2) - km_1\omega^2]} \tag{9.100}$$

---

*If the base is completely flexible with an unconstrained mass of $m_2$, $Z(\omega) = -\omega^2 m_2$, and Eqs. (9.96) and (9.97) lead to Eq. (9.86).

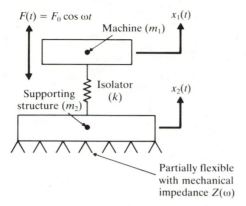

**Figure 9.20**   Machine with isolator on a partially flexible foundation.

In practice, the mechanical impedance $Z(\omega)$ depends on the nature of the base structure. It can be found experimentally by measuring the displacement produced by a vibrator that applies a harmonic force on the base structure. In some cases, such as in the case of an isolator resting on a concrete raft on soil, the mechanical impedance at any frequency $\omega$ can be found in terms of the spring-mass-dashpot model of the soil.

## 9.10   USE OF VIBRATION ABSORBERS

A machine or system may experience excessive vibration if it is acted upon by a force whose excitation frequency nearly coincides with a natural frequency of the machine or system. In such cases, the vibration of the machine or system can be reduced by using a *vibration neutralizer* or *dynamic vibration absorber*. This is simply another spring-mass system. We shall consider the analysis of a dynamic vibration absorber by idealizing the machine as a single degree of freedom system.

### 9.10.1   Dynamic Vibration Absorber

When we couple an auxiliary mass $m_2$ to a machine of mass $m_1$ through a spring of stiffness $k_2$, the resulting two degree of freedom system will look as shown in Fig. 9.21. The equations of motion of the masses $m_1$ and $m_2$ are

$$m_1\ddot{x}_1 + k_1 x_1 + k_2(x_1 - x_2) = F_0 \sin \omega t$$

$$m_2\ddot{x}_2 + k_2(x_2 - x_1) = 0 \qquad (9.101)$$

By assuming harmonic solution,

$$x_j(t) = X_j \sin \omega t, \qquad j = 1, 2 \qquad (9.102)$$

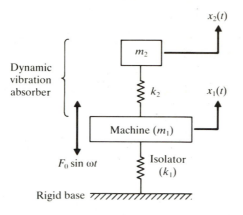

**Figure 9.21** Dynamic vibration absorber.

we can obtain

$$X_1 = \frac{\left(k_2 - m_2\omega^2\right)F_0}{\left(k_1 + k_2 - m_1\omega^2\right)\left(k_2 - m_2\omega^2\right) - k_2^2} \tag{9.103}$$

$$X_2 = \frac{k_2 F_0}{\left(k_1 + k_2 - m_1\omega^2\right)\left(k_2 - m_2\omega^2\right) - k_2^2} \tag{9.104}$$

We are primarily interested in reducing the amplitude of the machine ($X_1$). In order to make the amplitude of $m_1$ zero, the numerator of Eq. (9.103) should be set equal to zero. This gives

$$\omega^2 = \frac{k_2}{m_2} \tag{9.105}$$

If the machine, before the addition of the dynamic vibration absorber, operates near its resonance, $\omega^2 \simeq \omega_1^2 = k_1/m_1$. Thus if the absorber is designed such that

$$\omega^2 = \frac{k_2}{m_2} = \frac{k_1}{m_1} \tag{9.106}$$

the amplitude of vibration of the machine, while operating at its original resonant frequency, will be zero. By defining

$$\delta_{st} = \frac{F_0}{k_1}, \qquad \omega_1 = \left(\frac{k_1}{m_1}\right)^{1/2}$$

as the natural frequency of the machine or main system, and

$$\omega_2 = \left(\frac{k_2}{m_2}\right)^{1/2} \tag{9.107}$$

as the natural frequency of the absorber or auxiliary system, Eqs. (9.103) and (9.104) can be rewritten as

$$\frac{X_1}{\delta_{st}} = \frac{1 - \left(\dfrac{\omega}{\omega_2}\right)^2}{\left[1 + \dfrac{k_2}{k_1} - \left(\dfrac{\omega}{\omega_1}\right)^2\right]\left[1 - \left(\dfrac{\omega}{\omega_2}\right)^2\right] - \dfrac{k_2}{k_1}} \tag{9.108}$$

$$\frac{X_2}{\delta_{st}} = \frac{1}{\left[1 + \dfrac{k_2}{k_1} - \left(\dfrac{\omega}{\omega_1}\right)^2\right]\left[1 - \left(\dfrac{\omega}{\omega_2}\right)^2\right] - \dfrac{k_2}{k_1}} \tag{9.109}$$

Figure 9.22 shows the variation of the amplitude of vibration of the machine $(X_1/\delta_{st})$ with the machine speed $(\omega/\omega_1)$. The two peaks correspond to the two natural frequencies of the composite system. As seen before, $X_1 = 0$ at $\omega = \omega_1$. At this frequency, Eq. (9.109) gives

$$X_2 = -\frac{k_1}{k_2}\delta_{st} = -\frac{F_0}{k_2} \tag{9.110}$$

This shows that the force exerted by the auxiliary spring is opposite to the impressed force $(k_2 X_2 = -F_0)$ and neutralizes it, thus reducing $X_1$ to zero. The size of the dynamic vibration absorber can be found from Eqs. (9.110) and (9.106):

$$k_2 X_2 = m_2 \omega^2 X_2 = -F_0 \tag{9.111}$$

Thus the values of $k_2$ and $m_2$ depend on the allowable value of $X_2$.

It can be seen from Fig. 9.22 that the dynamic vibration absorber, while eliminating vibration at the known impressed frequency $\omega$, introduces two resonant frequencies $\Omega_1$ and $\Omega_2$, at which the amplitude of the machine is infinite. In practice, the operating frequency $\omega$ must therefore be kept away from the frequencies $\Omega_1$ and

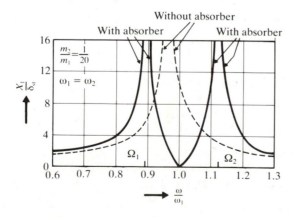

**Figure 9.22**   Effect of undamped vibration absorber on the response of machine.

$\Omega_2$. The values of $\Omega_1$ and $\Omega_2$ can be found by equating the denominator of Eq. (9.108) to zero. Noting that

$$\frac{k_2}{k_1} = \frac{k_2}{m_2} \frac{m_2}{m_1} \frac{m_1}{k_1} = \frac{m_2}{m_1} \left( \frac{\omega_2}{\omega_1} \right)^2 \tag{9.112}$$

and setting the denominator of Eq. (9.108) to zero leads to

$$\left( \frac{\omega}{\omega_2} \right)^4 \left( \frac{\omega_2}{\omega_1} \right)^2 - \left( \frac{\omega}{\omega_2} \right)^2 \left[ 1 + \left( 1 + \frac{m_2}{m_1} \right) \left( \frac{\omega_2}{\omega_1} \right)^2 \right] + 1 = 0 \tag{9.113}$$

The two roots of this equation are given by

$$\left. \begin{array}{c} \left( \frac{\Omega_1}{\omega_2} \right)^2 \\[2mm] \left( \frac{\Omega_2}{\omega_2} \right)^2 \end{array} \right\} = \frac{\left[ 1 + \left( 1 + \frac{m_2}{m_1} \right) \left( \frac{\omega_2}{\omega_1} \right)^2 \right] \mp \left\{ \left[ 1 + \left( 1 + \frac{m_2}{m_1} \right) \left( \frac{\omega_2}{\omega_1} \right)^2 \right]^2 - 4 \left( \frac{\omega_2}{\omega_1} \right)^2 \right\}^{1/2}}{2 \left( \frac{\omega_2}{\omega_1} \right)^2}$$

$$\tag{9.114}$$

which can be seen to be functions of $(m_2/m_1)$ and $(\omega_2/\omega_1)$.

## EXAMPLE 9.4

A large four-pole synchronous motor weighing 3000 N operates at a synchronous speed of $n = 3000$ rpm. The vibration frequency of the machine is given by $f = pn/120$ Hz, where $p$ is the number of poles. It has been observed that this unit induces vibration into the surrounding area through its pedestal mount. Determine the parameters of the vibration absorber that will reduce the vibration when mounted on the pedestal. The magnitude of the exciting force is 250 N, and the amplitude of motion of the auxiliary mass is to be limited to 2 mm.

*Solution.* The frequency of vibration of the machine is

$$f = \frac{4 \times 3000}{120} = 100 \text{ Hz} \qquad \text{or} \qquad \omega = 628.32 \text{ rad/sec}$$

Since the motion of the pedestal is to be made equal to zero, the amplitude of motion of the auxiliary mass should be equal and opposite to that of the exciting force. Thus from Eq. (9.111), we obtain

$$|F_0| = m_2 \omega^2 X_2 \tag{E.1}$$

Substitution of the given data yields

$$250 = m_2 (628.32)^2 (0.002)$$

Therefore $m_2 = 0.31665$ kg. The spring stiffness $k_2$ can be determined from Eq. (9.106):

$$\omega^2 = \frac{k_2}{m_2}$$

Therefore, $k_2 = (628.32)^2 (0.31665) = 125009$ N/m.

---

**EXAMPLE 9.5**

In a certain heat exchanger, a section of pipe that carries coolant is found to vibrate violently at a compressor speed of 240 rpm. To eliminate this difficulty, we propose to clamp a cantilever spring-mass system to the pipe, to act as a vibration absorber. In a trial test, an absorber of 1 kg mass, tuned to 240 cpm, resulted in two natural frequencies of 200 and 288 cpm for the system. If the absorber is to be designed so that the natural frequencies of the system lie outside the region 160 to 320 cpm, find the necessary mass and spring stiffness of the absorber.

**Solution.** The natural frequencies $\omega_1$ of the pipe and $\omega_2$ of the absorber are given by

$$\omega_1 = \sqrt{\frac{k_1}{m_1}}, \qquad \omega_2 = \sqrt{\frac{k_2}{m_2}} \qquad \text{(E.1)}$$

The resonant frequencies $\Omega_1$ and $\Omega_2$ of the combined system are given by Eq. (9.114). Since the absorber ($m = 1$ kg) is tuned, $\omega_1 = \omega_2 = 25.1328$ rad/sec (corresponding to 240 cpm). Using the notation

$$\mu = \frac{m_2}{m_1}, \qquad r_1 = \frac{\Omega_1}{\omega_2}, \qquad \text{and} \qquad r_2 = \frac{\Omega_2}{\omega_2}$$

Eq. (9.114) becomes

$$r_1^2, r_2^2 = \left(1 + \frac{\mu}{2}\right) \mp \sqrt{\left(1 + \frac{\mu}{2}\right)^2 - 1} \qquad \text{(E.2)}$$

Since $\Omega_1$ and $\Omega_2$ are known to be 20.9440 rad/sec (or 200 cpm) and 30.1594 rad/sec (or 288 cpm), respectively, we find that

$$r_1 = \frac{\Omega_1}{\omega_2} = \frac{20.944}{25.1328} = 0.8333$$

$$r_2 = \frac{\Omega_2}{\omega_2} = \frac{30.1594}{25.1328} = 1.20$$

Hence

$$r_1^2 = \left(1 + \frac{\mu}{2}\right) - \sqrt{\left(1 + \frac{\mu}{2}\right)^2 - 1}$$

or

$$\mu = \left(\frac{r_1^4 + 1}{r_1^2}\right) - 2 \qquad (E.3)$$

Since $r_1 = 0.8333$, Eq. (E.3) gives $\mu = m_2/m_1 = 0.1345$ and $m_1 = m_2/0.1345 = 7.4349$ kg. The specified lower limit of $\Omega_1$ is 160 cpm or 16.7552 rad/sec, and so

$$r_1 = \frac{\Omega_1}{\omega_2} = \frac{16.7552}{25.1328} = 0.6667$$

Equation (E.3) gives $\mu = m_2/m_1 = 0.6942$ and $m_2 = m_1(0.6942) = 5.1613$ kg. With these values, the second resonant frequency can be found from

$$r_2^2 = \left(1 + \frac{\mu}{2}\right) + \sqrt{\left(1 + \frac{\mu}{2}\right)^2 - 1} = 2.2497$$

which gives $\Omega_2 \approx 360$ cpm larger than the specified upper limit of 320 cpm. The spring stiffness of the absorber is given by

$$k_2 = \omega_2^2 m_2 = (25.1328)^2(5.1613) = 3260.1746 \text{ N/m}$$

### 9.10.2    Damped Dynamic Vibration Absorber

The dynamic vibration absorber described in the previous section removes the original resonance peak in the response curve of the machine but introduces two new peaks. If it is necessary to reduce the amplitude of vibration of the machine over a

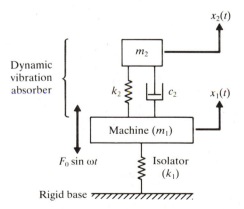

**Figure 9.23**   Damped dynamic vibration absorber.

range of frequencies, we can use a damped dynamic vibration absorber, as shown in Fig. 9.23. The equations of motion of the two masses are given by

$$m_1\ddot{x}_1 + k_1x_1 + k_2(x_1 - x_2) + c_2(\dot{x}_1 - \dot{x}_2) = F_0\sin \omega t \tag{9.115}$$

$$m_2\ddot{x}_2 + k_2(x_2 - x_1) + c_2(\dot{x}_2 - \dot{x}_1) = 0 \tag{9.116}$$

By assuming the solution to be

$$x_j(t) = X_je^{i\omega t}, \qquad j = 1, 2 \tag{9.117}$$

the solution of Eqs. (9.115) and (9.116) can be obtained:

$$X_1 = \frac{F_0(k_2 - m_2\omega^2 + ic_2\omega)}{[(k_1 - m_1\omega^2)(k_2 - m_2\omega^2) - m_2k_2\omega^2] + i\omega c_2(k_1 - m_1\omega^2 - m_2\omega^2)} \tag{9.118}$$

$$X_2 = \frac{X_1(k_2 + i\omega c_2)}{(k_2 - m_2\omega^2 + i\omega c_2)} \tag{9.119}$$

By defining

$$\mu = m_2/m_1 = \text{mass ratio} = \text{absorber mass/main mass} \tag{9.120}$$

$$\delta_{st} = F_0/k_1 = \text{static deflection of the system} \tag{9.121}$$

$$\omega_a^2 = k_2/m_2 = \text{square of natural frequency of the absorber} \tag{9.122}$$

$$\omega_n^2 = k_1/m_1 = \text{square of natural frequency of main mass} \tag{9.123}$$

$$f = \omega_a/\omega_n = \text{ratio of natural frequencies} \tag{9.124}$$

$$g = \omega/\omega_n = \text{forced frequency ratio} \tag{9.125}$$

$$c_c = 2m_2\omega_n = \text{critical damping constant} \tag{9.126}$$

$$\zeta = c_2/c_c = \text{damping ratio} \tag{9.127}$$

the magnitude of $X_1$ can be expressed as

$$\frac{X_1}{\delta_{st}} = \left[\frac{(2\zeta g)^2 + (g^2 - f^2)^2}{(2\zeta g)^2(g^2 - 1 + \mu g^2)^2 + \{\mu f^2g^2 - (g^2 - 1)(g^2 - f^2)\}^2}\right]^{1/2} \tag{9.128}$$

This equation shows that the amplitude of vibration of the main mass is a function

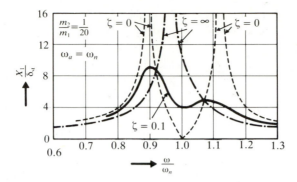

**Figure 9.24**   Effect of damped vibration absorber on the response of the machine.

of $\mu$, $f$, $g$, and $\zeta$. The graph of

$$\left| \frac{X_1}{\delta_{st}} \right|$$

against the forced frequency ratio $g = \omega/\omega_n$ is shown in Fig. 9.24 for $f = 1$ and $\mu = 1/20$ for a few different values of $\zeta$. For $c_2 = 0$ ($\zeta = 0$), we have the same case as Fig. 9.22, a known result. When the damping becomes infinite ($\zeta = \infty$), the two masses $m_1$ and $m_2$ are virtually clamped together, and the system behaves essentially as a single degree of freedom system with a mass of $(21/20)m_1$. Thus the peak of $X_1$ is infinite for $c_2 = 0$ as well as for $c_2 = \infty$. Somewhere in between these limits, the peak of $X_1$ is a minimum. It is possible to find this optimum value of $c_2$ [9.40]. Additional work relating to the optimum design of vibration absorbers can be found in Refs. [9.41 and 9.42].

## 9.11   COMPUTER PROGRAM

A Fortran subroutine BALAN is given for two-plane balancing. The arrays VA, VB, VAP, VBP, VAPP, VBPP, WL, WR, BL, and BR, each of dimension 2, are used to represent the vectors $\vec{V}_A$, $\vec{V}_B$, $\vec{V}_A'$, $\vec{V}_B'$, $\vec{V}_A''$, $\vec{V}_B''$, $\vec{W}_L$, $\vec{W}_R$, $\vec{B}_L$, and $\vec{B}_R$, respectively. The arrays BL and BR give the magnitude and position of the balancing weights in the left and the right planes.

For illustration, Example 9.1 is solved using this subroutine. The main program for this example, subroutine BALAN, and the output of the program are given on the following page.

```
C ================================================================
C
C PROGRAM 18
C TWO-PLANE BALANCING
C
C ================================================================
      DIMENSION VA(2),VB(2),VAP(2),VBP(2),VAPP(2),VBPP(2),WL(2),WR(2),
     2   BL(2),BR(2)
C FOLLOWING 8 LINES CONTAIN PROBLEM-DEPENDENT DATA
      DATA VA/8.5,60.0/
      DATA VAP/6.0,125.0/
      DATA WL/10.0,270.0/
      DATA VB/6.5,205.0/
      DATA VBP/4.5,230.0/
      DATA VAPP/6.0,35.0/
      DATA VBPP/10.5,160.0/
      DATA WR/12.0,180.0/
C END OF PROBLEM-DEPENDENT DATA
      CALL BALAN (VA,VB,VAP,VBP,VAPP,VBPP,WL,WR,BL,BR)
      PRINT 10
10    FORMAT (//,31H RESULTS OF TWO-PLANE BALANCING)
      PRINT 20,BL(1),BL(2)
20    FORMAT (//,28H LEFT-PLANE BALANCING WEIGHT,//,11H MAGNITUDE=,
     2   E15.8,/,7H ANGLE=,4X,E15.8)
      PRINT 30,BR(1),BR(2)
30    FORMAT (//,29H RIGHT-PLANE BALANCING WEIGHT,//,
     2   11H MAGNITUDE=,E15.8,/,7H ANGLE=,4X,E15.8)
      STOP
      END
C ================================================================
C
C SUBROUTINE BALAN
C
C ================================================================
      SUBROUTINE BALAN (VA,VB,VAP,VBP,VAPP,VBPP,WL,WR,BL,BR)
      DIMENSION VA(2),VB(2),VAP(2),VBP(2),VAPP(2),VBPP(2),WL(2),WR(2),
     2   BL(2),BR(2),P(2),Q(2),R(2),S(2),AAL(2),ABL(2),AAR(2),ABR(2),
     3   UL(2),UR(2)
      PI=180.0/3.1415926
      VA(2)=VA(2)/PI
      P(1)=VA(1)
      P(2)=VA(2)
      VA(1)=P(1)*COS(P(2))
      VA(2)=P(1)*SIN(P(2))
      VB(2)=VB(2)/PI
      P(1)=VB(1)
      P(2)=VB(2)
      VB(1)=P(1)*COS(P(2))
      VB(2)=P(1)*SIN(P(2))
      VAP(2)=VAP(2)/PI
      P(1)=VAP(1)
      P(2)=VAP(2)
      VAP(1)=P(1)*COS(P(2))
      VAP(2)=P(1)*SIN(P(2))
```

```
VBP(2)=VBP(2)/PI
P(1)=VBP(1)
P(2)=VBP(2)
VBP(1)=P(1)*COS(P(2))
VBP(2)=P(1)*SIN(P(2))
VAPP(2)=VAPP(2)/PI
P(1)=VAPP(1)
P(2)=VAPP(2)
VAPP(1)=P(1)*COS(P(2))
VAPP(2)=P(1)*SIN(P(2))
VBPP(2)=VBPP(2)/PI
P(1)=VBPP(1)
P(2)=VBPP(2)
VBPP(1)=P(1)*COS(P(2))
VBPP(2)=P(1)*SIN(P(2))
WL(2)=WL(2)/PI
P(1)=WL(1)
P(2)=WL(2)
WL(1)=P(1)*COS(P(2))
WL(2)=P(1)*SIN(P(2))
WR(2)=WR(2)/PI
P(1)=WR(1)
P(2)=WR(2)
WR(1)=P(1)*COS(P(2))
WR(2)=P(1)*SIN(P(2))
CALL VSUB (VAP,VA,R)
CALL VDIV (R,WL,AAL)
CALL VSUB (VBP,VB,S)
CALL VDIV (S,WL,ABL)
CALL VSUB (VAPP,VA,P)
CALL VDIV (P,WR,AAR)
CALL VSUB (VBPP,VB,Q)
CALL VDIV (Q,WR,ABR)
AR1=SQRT(AAR(1)**2+AAR(2)**2)
AR2=ATAN(AAR(2)/AAR(1))*PI
AL1=SQRT(AAL(1)**2+AAL(2)**2)
AL2=ATAN(AAL(2)/AAL(1))*PI
CALL VMULT (ABL,VA,R)
CALL VMULT (AAL,VB,S)
CALL VSUB (R,S,VAP)
CALL VMULT (AAR,ABL,R)
CALL VMULT (AAL,ABR,S)
CALL VSUB (R,S,VBP)
CALL VDIV (VAP,VBP,UR)
CALL VMULT (ABR,VA,R)
CALL VMULT (AAR,VB,S)
CALL VSUB (R,S,VAP)
CALL VMULT (ABR,AAL,R)
CALL VMULT (AAR,ABL,S)
CALL VSUB (R,S,VBP)
CALL VDIV (VAP,VBP,UL)
BL(1)=SQRT(UL(1)**2+UL(2)**2)
A1=UL(2)/UL(1)
BL(2)=ATAN(UL(2)/UL(1))
BR(1)=SQRT(UR(1)**2+UR(2)**2)
```

```
        A2=UR(2)/UR(1)
        BR(2)=ATAN(UR(2)/UR(1))
        BL(2)=BL(2)*PI
        BR(2)=BR(2)*PI
        BL(2)=BL(2)+180.0
        BR(2)=BR(2)+180.0
        RETURN
        END
C ============================================================
C
C SUBROUTINE VSUB (A,B,C)
C
C ============================================================
        SUBROUTINE VSUB (A,B,C)
        DIMENSION A(2),B(2),C(2)
        C(1)=A(1)-B(1)
        C(2)=A(2)-B(2)
        RETURN
        END
C ============================================================
C
C SUBROUTINE VDIV
C
C ============================================================
        SUBROUTINE VDIV (A,B,C)
        DIMENSION A(2),B(2),C(2)
        C(1)=(A(1)*B(1)+A(2)*B(2))/(B(1)**2+B(2)**2)
        C(2)=(A(2)*B(1)-A(1)*B(2))/(B(1)**2+B(2)**2)
        RETURN
        END
C ============================================================
C
C SUBROUTINE VMULT
C
C ============================================================
        SUBROUTINE VMULT (A,B,C)
        DIMENSION A(2),B(2),C(2)
        C(1)=A(1)*B(1)-A(2)*B(2)
        C(2)=A(2)*B(1)+A(1)*B(2)
        RETURN
        END
```

```
RESULTS OF TWO-PLANE BALANCING

LEFT-PLANE BALANCING WEIGHT

MAGNITUDE= 0.10056140E+02
ANGLE=     0.14555478E+03

RIGHT-PLANE BALANCING WEIGHT

MAGNITUDE= 0.58773613E+01
ANGLE=     0.24825594E+03
```

## REFERENCES

**9.1.** M. J. Griffin and E. M. Whitham, "The discomfort produced by impulsive whole-body vibration," *Journal of the Acoustical Society of America*, Vol. 68, 1980, pp. 1277–1284.

**9.2.** R. L. Eshleman, "Identification and correction of machinery vibration problems," *Sound and Vibration*, Vol. 15, April 1981, pp. 12–18.

**9.3.** J. E. Ruzicka, "Fundamental concepts of vibration control," *Sound and Vibration*, Vol. 5, July 1971, pp. 16–22.

**9.4.** J. Schnerre and J. G. Bollinger, "How to solve a typical machinery vibration problem," *Sound and Vibration*, Vol. 5, July 1971, pp. 24–28.

**9.5.** R. L. Fox, "Machinery vibration monitoring and analysis techniques," *Sound and Vibration*, Vol. 5, November 1971, pp. 35–40.

**9.6.** J. N. Macduff, "A procedure for field balancing rotating machinery," *Sound and Vibration*, Vol. 1, 1967, pp. 16–20.

**9.7.** J. F. G. Wort, "Industrial balancing machines," *Sound and Vibration*, Vol. 13, October 1979, pp. 14–19.

**9.8.** D. G. Stadelbauer, "Dynamic balancing with microprocessors," *Shock and Vibration Digest*, Vol. 14, December 1982, pp. 3–7.

**9.9.** A. S. Babkin and J. J. Anderson, "Mechanical signature analysis," *Sound and Vibration*, Vol. 7, April 1973, pp. 35–42.

**9.10.** R. L. Baxter, "Dynamic balancing," *Sound and Vibration*, Vol. 6, April 1972, pp. 30–33.

**9.11.** J. H. Harter and W. D. Beitzel, *Mathematics Applied to Electronics*, Reston Publishing Co., Reston, Virginia, 1980.

**9.12.** R. G. Loewy and V. J. Piarulli, "Dynamics of rotating shafts," *Shock and Vibration Monograph SVM-4*, Shock and Vibration Information Center, Naval Research Laboratory, Washington, D.C., 1969.

**9.13.** M. L. Adams and J. Padovan, "Insight into linearized rotor dynamics," *Journal of Sound and Vibration*, Vol. 76, 1981, pp. 129–142.

**9.14.** H. N. Ozguven, "On the critical speed of continuous shaft-disk systems," *Journal of Vibration, Acoustics, Stress, and Reliability in Design*, Vol. 106, 1984, pp. 59–61.

**9.15.** R. L. Eshleman and R. A. Eubanks, "On the critical speeds of a continuous shaft-disk system," *Journal of Engineering for Industry*, Vol. 89, 1967, pp. 645–652.

**9.16.** R. J. Trivisonno, "Fortran IV computer program for calculating critical speeds of rotating shafts," NASA TN D-7385, 1973.

**9.17.** R. E. D. Bishop and G. M. L. Gladwell, "The vibration and balancing of an unbalanced flexible rotor," *Journal of Mechanical Engineering Science*, Vol. 1, 1959, pp. 66–77.

**9.18.** A. G. Parkinson, "The vibration and balancing of shafts rotating in asymmetric bearings," *Journal of Sound and Vibration*, Vol. 2, 1965, pp. 477–501.

**9.19.** J. D. Irwin and E. R. Graf, *Industrial Noise and Vibration Control*, Prentice-Hall, Englewood Cliffs, N.J., 1979.

**9.20.** C. E. Crede, *Vibration and Shock Isolation*, Wiley, New York, 1951.

**9.21.** W. E. Purcell, "Materials for noise and vibration control," *Sound and Vibration*, Vol. 16, July 1982, pp. 6–31.

**9.22.** J. C. Snowdon, "Rubberlike materials, their internal damping and role in vibration isolation," *Journal of Sound and Vibration*, Vol. 2, 1965, pp. 175–193.

**9.23.** D. R. McAuliffe, T. D. Agne, and J. I. Hammond, "Materials for noise and vibration control," *Sound and Vibration*, Vol. 6, July 1972, pp. 20–24.

**9.24.** D. E. Baxa and R. A. Dykstra, "Pneumatic isolation systems control forging hammer vibration," *Sound and Vibration*, Vol. 14, May 1980, pp. 22–25.

**9.25.** E. I. Rivin, "Vibration isolation of industrial machinery—Basic considerations," *Sound and Vibration*, Vol. 12, November 1978, pp. 14–19.

**9.26.** R. Adair, "The design and application of pneumatic vibration isolators," *Sound and Vibration*, Vol. 8, August 1974, pp. 24–27.

**9.27.** E. Rivin, "Vibration isolation of precision machinery," *Sound and Vibration*, Vol. 13, August 1979, pp. 18–23.

**9.28.** E. I. Rivin, "Design of vibration isolation systems for forging hammers," *Sound and Vibration*, Vol. 12, April 1978, pp. 12–15.

**9.29.** R. A. Waller, "Building on springs," *Sound and Vibration*, Vol. 9, March 1975, pp. 22–25.

**9.30.** R. M. Hochheiser, "How to select vibration isolators for OEM machinery and equipment," *Sound and Vibration*, Vol. 8, August 1974, pp. 14–23.

**9.31.** P. C. Calcaterra, "Active vibration isolation for aircraft seating," *Sound and Vibration*, Vol. 6, March 1972, pp. 18–23.

**9.32.** O. Vilnay, "Active control of machinery foundation," *Journal of Engineering Mechanics*, ASCE, Vol. 110, 1984, pp. 273–281.

**9.33.** C. M. Salerno and R. M. Hochheiser, "How to select vibration isolators for use as machinery mounts," *Sound and Vibration*, Vol. 7, August 1973, pp. 22–28.

**9.34.** C. A. Mercer and P. L. Rees, "An optimum shock isolator," *Journal of Sound and Vibration*, Vol. 18, 1971, pp. 511–520.

**9.35.** M. L. Munjal, "A rational synthesis of vibration isolators," *Journal of Sound and Vibration*, Vol. 39, 1975, pp. 247–263.

**9.36.** C. Ng and P. F. Cunniff, "Optimization of mechanical vibration isolation systems with multi-degrees of freedom," *Journal of Sound and Vibration*, Vol. 36, 1974, pp. 105–117.

**9.37.** S. K. Hati and S. S. Rao, "Cooperative solution in the synthesis of multidegree of freedom shock isolation systems," *Journal of Vibration, Acoustics, Stress, and Reliability in Design*, Vol. 105, 1983, pp. 101–103.

**9.38.** S. S. Rao and S. K. Hati, "Optimum design of shock and vibration isolation systems using game theory," *Journal of Engineering Optimization*, Vol. 4, 1980, pp. 1–8.

**9.39.** J. I. Soliman and M. G. Hallam, "Vibration isolation between non-rigid machines and non-rigid foundations," *Journal of Sound and Vibration*, Vol. 8, 1968, pp. 329–351.

**9.40.** J. Ormondroyd and J. P. Den Hartog, "The theory of the dynamic vibration absorber," *Transactions of ASME*, Vol. 50, 1928, p. APM-241.

**9.41.** H. Puksand, "Optimum conditions for dynamic vibration absorbers for variable speed systems with rotating and reciprocating unbalance," *International Journal of Mechanical Engineering Education*, Vol. 3, April 1975, pp. 145–152.

**9.42.** A. Soom and M-S. Lee, "Optimal design of linear and nonlinear absorbers for damped systems," *Journal of Vibration, Acoustics, Stress, and Reliability in Design*, Vol. 105, 1983, pp. 112–119.

## REVIEW QUESTIONS

**9.1.** Name some sources of industrial vibrations.

**9.2.** What are the various methods available for vibration control?

**9.3.** What is single-plane balancing?

**9.4.** Describe the two-plane balancing procedure.

**9.5.** What is whirling?

**9.6.** What is the difference between stationary damping and rotary damping?

**9.7.** How is the critical speed of a shaft determined?

**9.8.** What causes instability in a rotor system?

**9.9.** What considerations are to be taken into account for the balancing of a reciprocating engine?

**9.10.** What is the function of a vibration isolator?

**9.11.** What is a vibration absorber?

**9.12.** What is the difference between a vibration isolator and a vibration absorber?

**9.13.** Does spring mounting always reduce the vibration of the foundation of a machine?

**9.14.** Is it better to use a soft spring in the flexible mounting of a machine? Why?

**9.15.** Is the shaking force proportional to the square of the speed of a machine? Does the vibratory force transmitted to the foundation increase with the speed of the machine?

**9.16.** Why does dynamic balancing imply static balancing?

**9.17.** Explain why dynamic balancing can never be achieved by a static test alone.

**9.18.** Why does a rotating shaft always vibrate? What is the source of the shaking force?

**9.19.** Is it always advantageous to include a damper in the secondary system of a dynamic vibration absorber?

## PROBLEMS

The problem assignments are organized as follows:

| Problems | Section covered | Topic covered |
|---|---|---|
| 9.1–9.8 | 9.3 | Balancing of rotating machines |
| 9.9–9.10 | 9.4 | Whirling of shafts |

**9.1.** The amplitude and phase angle due to original unbalance in a grinding wheel operating at 1200 rpm are found to be 10 mils and 40° counterclockwise from the phase mark. When a trial weight $W = 6$ oz is added at 65° clockwise from the phase mark and at a radial distance 2.5 in. from the center of rotation, the amplitude and phase angle are observed to be 19 mils and 150° counterclockwise. Find the magnitude and angular position of the balancing weight if it is to be located 2.5 in. radially from the center of rotation.

**9.2.** An unbalanced flywheel shows a displacement of 6.5 mils and a phase angle of 15° clockwise from the phase mark. With the addition of a trial weight of magnitude 2 oz at an angular position 45° counterclockwise from the phase mark, the displacement becomes 8.8 mils and the phase angle 35° counterclockwise. Find the magnitude and angular position of the balancing weight required.

**9.3.** The data obtained in a two-plane balancing procedure is given below. Determine the magnitude and angular position of the balancing weights, assuming that all angles are measured from an arbitrary phase mark and all weights are added at the same radius.

| Condition | Amplitude | | Phase angle | |
| --- | --- | --- | --- | --- |
| | Bearing A | Bearing B | Bearing A | Bearing B |
| Original unbalance | 5 | 4 | 100° | 200° |
| $W_L = 2$ oz added at 15° in the left plane | 6 | 4.5 | 120° | 130° |
| $W_R = 2$ oz added at 0° in the right plane | 5.5 | 5 | 90° | 40° |

**9.4.** Repeat Problem 9.3 for the following data:

| Condition | Amplitude | | Phase angle | |
| --- | --- | --- | --- | --- |
| | Bearing A | Bearing B | Bearing A | Bearing B |
| Original unbalance | 0.22 | 0.13 | 65° | 160° |
| $W_L = 0.3$ kg at 90° in the left plane | 0.16 | 0.11 | 125° | 135° |
| $W_R = 0.35$ kg at 180° in the right plane | 0.20 | 0.25 | 40° | 165° |

**9.5.** A grinding wheel rotates clockwise at 2400 rpm. Using a vibration analyzer, we find an original unbalance of amplitude 4 mils and phase angle 45°. When a trial weight $W = 4$ oz is added at 20° clockwise from the phase mark, the amplitude becomes 8 mils and the phase angle 145°. If the phase angles are measured counterclockwise from the right-hand horizontal, calculate the magnitude and location of the necessary balancing weight.

**9.6.** Figure 9.25 shows a rotating system in which the shaft is supported in bearings at $A$ and $B$. The three masses $m_1$, $m_2$, and $m_3$ are connected to the shaft as indicated in the figure. a. Find the bearing reactions at $A$ and $B$ if the speed of the shaft is 1000 rpm. b. Determine the locations and magnitudes of the balancing masses to be placed at a radius of 0.25 m in the planes $L$ and $R$, which can be assumed to pass through the bearings $A$ and $B$.

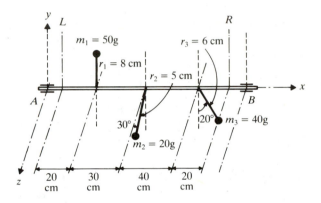

**Figure 9.25**

**9.7.** A circular disc rotating about its geometric axis has three holes $A$, $B$, and $C$ drilled through it. The diameters ($d$), radial positions ($r$) and angular locations ($\theta$) of these holes are given by $d_A = 0.5$ cm, $r_A = 10$ cm, $\theta_A = 30°$, $d_B = 1.5$ cm, $r_B = 20$ cm, $\theta_B = 120°$, $d_C = 1.0$ cm, $r_C = 15$ cm, and $\theta_C = 150°$. Find the diameter and angular location of a fourth hole at 20 cm radius that will balance the disc.

**9.8.** A turbine rotor is run at the natural frequency of the system. A stroboscope indicates that the maximum displacement of the rotor occurs at an angle 229°, as indicated on a protractor attached to the rotor and reading positively in the direction of rotation. At what angular position must mass be removed from the rotor in order to improve its balancing?

**9.9.** A solid disc of mass 15 kg is mounted at the center of a steel shaft of diameter 20 mm, which is supported in two bearings separated by a distance of 1 m. Assuming the shaft to be simply supported at the bearings, determine the lowest critical speed.

**9.10.** Derive the expression for the stress induced in a shaft with an unbalanced concentrated mass located midway between two bearings.

**9.11.** A single-cylinder engine has a total mass of 150 kg. Its reciprocating mass is 5 kg, and the rotating mass is 2.5 kg. The stroke $(2r)$ is 15 cm, and the speed is 600 rpm. a. If the engine is mounted floating on very weak springs, what is the amplitude of vertical vibration of the engine? b. If the engine is mounted solidly on a rigid foundation, what is the alternating force amplitude transmitted? Assume the connecting rod to be of infinite length.

**9.12.** For the two-cylinder in-line engine shown in Fig. 9.26, the unbalance in each cylinder is equivalent to an eccentric mass $m$ at a radius $r$ from the axis of rotation. Determine the counterweights necessary in planes $L$ and $R$ if they are also to be located at a radial distance $r$ from the axis of rotation.

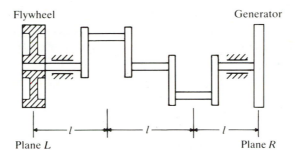

Flywheel

Generator

Plane $L$

Plane $R$

**Figure 9.26**

**9.13.** A small machine mounted on a table can be modeled as a single degree of freedom system. If the forcing frequency and the damping ratio of the system are 20 Hz and 0.15, respectively, what is the required natural frequency of the system in order to obtain 90% isolation—that is, a transmissibility of 0.1?

**9.14.** A reciprocating engine is subjected to a fluctuating exciting force whose frequency is the same as the running speed of the engine. The engine has a mass of 200 kg and is supported on a vibration isolation mounting, which has an equivalent spring constant of 10 kN/m, and a dashpot having a damping ratio of 0.15. a. Find the speed range over which the amplitude of the fluctuating force transmitted to the foundation will be larger than the exciting force. b. Find the speed range over which the transmitted force amplitude will be less than 10% of the exciting force amplitude.

**9.15.** A machine of mass 200 kg is supported on four springs of total stiffness 500 kN/m and has an unbalanced rotating element that results in a disturbing force of 300 N at a speed of 2400 rpm. Assuming a damping factor of $\zeta = 0.1$, determine a. its amplitude of motion due to the unbalance, b. the transmissibility, and c. the transmitted force.

**9.16.** A helicopter transmitter weighing 85 N is to be isolated from engine vibrations ranging in frequency from 25 Hz to 35 Hz. Find the necessary static deflection of the isolator for 80% isolation.

**9.17.** A sensitive electronic instrument of mass 90 kg is to be installed on a platform that is subjected to an acceleration of 200 mm/s$^2$ at a frequency of 25 Hz. If the instrument is mounted on a rubber pad having an equivalent stiffness of 300 N/mm and a damping ratio of 0.1, determine the acceleration transmitted to the foundation.

**9.18.** If the instrument of Problem 9.17 can only tolerate an acceleration of 30 mm/s$^2$, suggest a solution, assuming that the same rubber pad is the only isolator available.

**9.19.** A machine of mass 2500 kg operates at a frequency of 1800 rpm. It is supported by twelve metallic springs, each of which has a stiffness of $2 \times 10^6$ N/m. Determine a. the transmissibility, b. the percentage of isolation, and c. the spring constant that would give 80% isolation.

**9.20.** A reciprocating machine of mass 30 kg runs at a constant speed of 4000 rpm. After installation, the forcing frequency is found to be too close to the natural frequency of the system. Design a dynamic vibration absorber if the nearest natural frequency of the combined system is at least 25% from the excitation frequency.

**9.21.** In a certain refrigeration plant, a section of pipe carrying the refrigerant has been found to vibrate violently at a compressor speed of 300 rpm. We propose to eliminate this difficulty by clamping a cantilever spring-mass system to the pipe to act as an absorber. In a trial test, a 1 kg absorber tuned to 300 rpm results in two natural frequencies of 260 and 350 cpm. If the absorber is to be designed so that the natural frequencies lie outside the region 240 to 370 cpm, what must be the mass and spring stiffness of the absorber?

**9.22.** A jig used to size gravel in a concrete mixing plant contains a sieve that reciprocates with a frequency of 1200 cycles per minute. The jig has a mass of 200 kg and a natural frequency of 800 cycles per minute. If an absorber of mass 50 kg is to be added to eliminate the vibrations of the jig frame, determine the spring stiffness of the absorber. Also find the resulting two natural frequencies of the system.

**9.23.** An air compressor of mass 200 kg, with an unbalance of 0.01 kg-m, is found to have a large amplitude of vibration while running at 1200 rpm. Determine the mass and spring constant of the absorber to be added if the natural frequencies of the system are to be at least 20% from the impressed frequency.

**9.24.** Plot the graphs of $(\Omega_1/\omega_2)$ against $(m_2/m_1)$ and $(\Omega_2/\omega_2)$ against $(m_2/m_1)$ as $(m_2/m_1)$ varies from 0 to 1.0 when $\omega_2/\omega_1 = 1$.

**9.25.** Determine the operating range of the frequency ratio $\omega/\omega_2$ for an undamped vibration absorber to limit the value of $|X_1/\delta_{st}|$ to 0.5. Assume that $\omega_1 = \omega_2$ and $m_2 = 0.1m_1$.

**9.26.** In a chemical plant, a reciprocating pump discharges a fluid into a pipeline. It is observed that the pulsations of the fluid coincide with the natural frequency of a section of the pipeline when the pump speed is 232 rpm. We propose to clamp a cantilever carrying an end mass of 0.9 kg to absorb the vibrations. With this arrangement, the resulting natural frequencies of the system are found to be 200 and 320 cpm. Because of plant conditions, these frequencies should be kept below 160 and above 360 cpm. Determine the mass to be used and the stiffness of the cantilever needed to achieve this.

**9.27.** The natural frequency of vertical vibration of a fan and its supporting structure coincides with the constant speed of operation of the fan, which is 1500 rpm. The fan has an unbalance of 2 kg-cm. We propose to reduce the resonant condition by attaching a dynamic vibration absorber to the system, as shown in Fig. 9.27. It is specified that

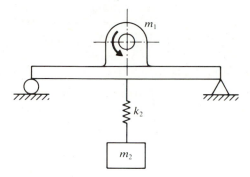

**Figure 9.27**

the fundamental natural frequency of the resulting system shall be 0.75 times the forced frequency. Determine the required ratio of the absorber mass to the main mass. If the mass of the main system (the fan, its housing, the motor, and part of the supporting structure) is 300 kg, determine the mass and spring stiffness of the absorber. Also, find the amplitude of vibration of the absorber mass.

**9.28.** In a chemical processing plant, an installation is found to vibrate violently at 500 cycles/minute. As a trial remedy, two vibration absorbers, each having a mass of 4 kg, are attached to the installation; the resulting two resonant frequencies of the combined system are found to be 450 and 560 cycles/minute. How many vibration absorbers are required to ensure that no resonance occurs between 390 and 650 cycles/minute?

**9.29.** The arrangement of a torsional dynamic absorber is shown in Fig. 9.28. The dynamic absorber, having a mass moment of inertia $J_2$, is free to rotate on the shaft and is

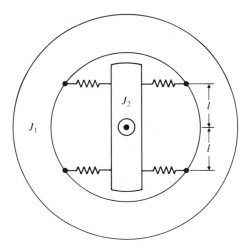

**Figure 9.28**

connected to a flywheel of mass moment of inertia $J_1$ by four springs of stiffness $k$ each. Derive the equations of motion for the system. Find the response of the flywheel if a harmonically varying torque of frequency $\omega$ is applied to the flywheel.

**9.30.** Solve Problem 9.3 using subroutine BALAN.

**9.31.** Solve Problem 9.4 using subroutine BALAN.

**9.32.** Write a computer program to find the displacements of the main mass and the auxiliary mass of a damped dynamic vibration absorber. Use this program to generate the results of Fig. 9.24.

# CHAPTER 10

# Numerical Integration Methods in Vibration Analysis

Nathan Newmark (1910–1981) was an American engineer and a professor of civil engineering at the University of Illinois at Champaign-Urbana. His research in earthquake resistant structures and structural dynamics is widely known. The numerical method he presented in 1959 for the dynamic response computation of linear and nonlinear systems is quite famous as the "Newmark $\beta$-method."

## 10.1  INTRODUCTION

When the differential equation of motion of a vibrating system cannot be integrated in closed form, a numerical approach must be used. Several numerical methods are available for the solution of vibration problems [10.1–10.3].* Numerical integration methods have two fundamental characteristics. First, they are not intended to satisfy the governing differential equation(s) at all time $t$, but only at discrete time intervals $\Delta t$ apart. Second, a suitable type of variation of the displacement $x$, velocity $\dot{x}$, and acceleration $\ddot{x}$ is assumed within each time interval $\Delta t$. Different numerical integration methods can be obtained, depending on the type of variation assumed for the displacement, velocity, and acceleration within each time interval $\Delta t$. We shall assume that the values of $x$ and $\dot{x}$ are known to be $x_0$ and $\dot{x}_0$, respectively, at time $t = 0$ and that the solution of the problem is required from $t = 0$ to $t = T$. In the following, we subdivide the time duration $T$ into $n$ equal steps $\Delta t$ so that $\Delta t = T/n$ and seek the solution at $t_0 = 0, t_1 = \Delta t, t_2 = 2\Delta t, \ldots, t_n = n\Delta t = T$. We shall derive formulas for finding the solution at $t_i = i\Delta t$ from the known solution at $t_{i-1} = (i-1)\Delta t$ according to five different numerical integration schemes: the finite difference method, the Runge-Kutta method, the Houbolt method, the Wilson method, and the Newmark method. References [10.4–10.6] deal with other numerical methods applicable to vibration problems.

## 10.2  FINITE DIFFERENCE METHOD

The main idea in the finite difference method is to use approximations to derivatives. Thus the governing differential equation of motion and the associated boundary conditions, if applicable, are replaced by the corresponding finite difference equations. Three types of formulas—forward, backward, and central difference formulas —can be used to derive the finite difference equations [10.7–10.10]. We shall consider only the central difference formulas in this chapter, since they are most accurate.

In the finite difference method, we replace the solution domain (over which the solution of the given differential equation is required) with a finite number of points, referred to as *mesh* or *grid points*, and seek to determine the values of the desired solution at these points. The grid points are usually considered to be equally spaced along each of the independent coordinates (see Fig. 10.1). By using Taylor's series expansion, $x_{i+1}$ and $x_{i-1}$ can be expressed about the grid point $i$ as

$$x_{i+1} = x_i + h\dot{x}_i + \frac{h^2}{2}\ddot{x}_i + \frac{h^3}{6}\dddot{x}_i + \cdots \tag{10.1}$$

$$x_{i-1} = x_i - h\dot{x}_i + \frac{h^2}{2}\ddot{x}_i - \frac{h^3}{6}\dddot{x} + \cdots \tag{10.2}$$

where $x_i = x(t = t_i)$ and $h = t_{i+1} - t_i = \Delta t$. By taking two terms only and subtracting Eq. (10.2) from Eq. (10.1), we obtain the central difference approximation to the first

---

*A numerical procedure using different types of interpolation functions for approximating the forcing function $F(t)$ was presented in Sec. 4.8.

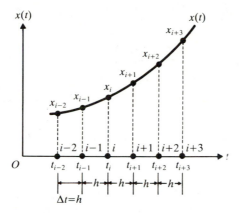

**Figure 10.1**

derivative of $x$ at $t = t_i$:

$$\dot{x}_i = \left.\frac{dx}{dt}\right|_{t_i} = \frac{1}{2h}(x_{i+1} - x_{i-1}) \tag{10.3}$$

By taking terms up to the second derivative and adding Eqs. (10.1) and (10.2), we obtain the central difference formula for the second derivative:

$$\ddot{x}_i = \left.\frac{d^2x}{dt^2}\right|_{t_i} = \frac{1}{h^2}(x_{i+1} - 2x_i + x_{i-1}) \tag{10.4}$$

## 10.3 CENTRAL DIFFERENCE METHOD FOR SINGLE DEGREE OF FREEDOM SYSTEMS

The governing equation of a viscously damped single degree of freedom system is

$$m\frac{d^2x}{dt^2} + c\frac{dx}{dt} + kx = F(t) \tag{10.5}$$

Let the duration over which the solution of Eq. (10.5) is required be divided into $n$ equal parts of interval $h = \Delta t$ each. To obtain a satisfactory solution, we must select a time step $\Delta t$ that is smaller than a critical time step $\Delta t_{\text{cri}}$.* Let the initial conditions be given by $x(t = 0) = x_0$ and $\dot{x}(t = 0) = \dot{x}_0$.

---

*Numerical methods that require the use of a time step ($\Delta t$) smaller than a critical time step ($\Delta t_{\text{cri}}$) are said to be *conditionally stable* [10.11, 10.12]. If $\Delta t$ is taken to be larger than $\Delta t_{\text{cri}}$, the method becomes unstable. This means that the truncation of higher-order terms in the derivation of Eqs. (10.3) and (10.4) (or rounding-off in the computer) causes errors that grow and make the response computations worthless in most cases. The critical time step is given by $\Delta t_{\text{cri}} = \tau_n/\pi$, where $\tau_n$ is the natural period of the system or the smallest such period in the case of a multidegree of freedom system [10.13]. Naturally, the accuracy of the solution always depends on the size of the time step. By using an unconditionally stable method, we can choose the time step with regard to accuracy only, not with regard to stability. This usually allows a much larger time step to be used for any given accuracy.

Replacing the derivatives by the central differences and writing Eq. (10.5) at grid point $i$ gives

$$m\left\{\frac{x_{i+1}-2x_i+x_{i-1}}{(\Delta t)^2}\right\}+c\left\{\frac{x_{i+1}-x_{i-1}}{2\Delta t}\right\}+kx_i=F_i \tag{10.6}$$

where $x_i=x(t_i)$ and $F_i=F(t_i)$. Solution of Eq. (10.6) for $x_{i+1}$ yields

$$x_{i+1}=\left\{\frac{1}{\dfrac{m}{(\Delta t)^2}+\dfrac{c}{2\Delta t}}\right\}\left[\left\{\frac{2m}{(\Delta t)^2}-k\right\}x_i+\left\{\frac{c}{2\Delta t}-\frac{m}{(\Delta t)^2}\right\}x_{i-1}+F_i\right] \tag{10.7}$$

This is called the *recurrence formula*. It permits us to calculate the displacement of the mass ($x_{i+1}$) if we know the previous history of displacements at $t_i$ and $t_{i-1}$, as well as the present external force $F_i$. Repeated application of Eq. (10.7) yields the complete time history of the behavior of the system. Note that the solution of $x_{i+1}$ is based on the use of the equilibrium equation at time $t_i$—that is, Eq. (10.6). For this reason, this integration procedure is called an *explicit integration method*. Certain care has to be exercised in applying Eq. (10.7) for $i=0$. Since both $x_0$ and $x_{-1}$ are needed in finding $x_1$, and the initial conditions provide only the values of $x_0$ and $\dot{x}_0$, we need to find the value of $x_{-1}$. Thus the method is not self-starting. However, we can generate the value of $x_{-1}$ by using Eqs. (10.3) and (10.4) as follows. By substituting the known values of $x_0$ and $\dot{x}_0$ into Eq. (10.5), $\ddot{x}_0$ can be found:

$$\ddot{x}_0=\frac{1}{m}\left[F(t=0)-c\dot{x}_0-kx_0\right] \tag{10.8}$$

Application of Eqs. (10.3) and (10.4) at $i=0$ yields the value of $x_{-1}$:

$$x_{-1}=x_0-\Delta t\dot{x}_0+\frac{(\Delta t)^2}{2}\ddot{x}_0 \tag{10.9}$$

---

### EXAMPLE 10.1

Find the response of a viscously damped single degree of freedom system subjected to a force

$$F(t)=F_0\left(1-\sin\frac{\pi t}{2t_0}\right)$$

with the following data: $F_0=1$, $t_0=\pi$, $m=1$, $c=0.2$, and $k=1$. Assume the values of the displacement and velocity of the mass at $t=0$ to be zero.

*Solution.* The governing differential equation is

$$m\ddot{x}+c\dot{x}+kx=F(t)=F_0\left(1-\sin\frac{\pi t}{2t_0}\right) \tag{E.1}$$

The finite difference solution of Eq. (E.1) is given by Eq. (10.7). Since the initial conditions are $x_0 = \dot{x}_0 = 0$, Eq. (10.8) yields $\ddot{x}_0 = 1$; hence Eq. (10.9) gives $x_{-1} = (\Delta t)^2/2$. Thus the solution of Eq. (E.1) can be found from the recurrence relation

$$x_{i+1} = \frac{1}{\left[ \dfrac{m}{(\Delta t)^2} + \dfrac{c}{2\Delta t} \right]} \left[ \left\{ \frac{2m}{(\Delta t)^2} - k \right\} x_i + \left\{ \frac{c}{2\Delta t} - \frac{m}{(\Delta t)^2} \right\} x_{i-1} + F_i \right]; \quad i = 0,1,2,\ldots$$

(E.2)

with $x_0 = 0$, $x_{-1} = (\Delta t)^2/2$, $x_i = x(t_i) = x(i\Delta t)$, and

$$F_i = F(t_i) = F_0 \left( 1 - \sin\frac{i\pi\Delta t}{2t_0} \right)$$

The undamped natural frequency and the natural period of the system are given by

$$\omega_n = \left( \frac{k}{m} \right)^{1/2} = 1 \tag{E.3}$$

and

$$\tau_n = \frac{2\pi}{\omega_n} = 2\pi \tag{E.4}$$

Thus the time step $\Delta t$ must be less than $\tau_n/\pi = 2.0$. We shall find the solution of Eq. (E.1) by using the time steps $\Delta t = \tau_n/20$ and $\tau_n/40$. The values of the response $x_i$ obtained at different instants of time $t_i$ are shown in Table 10.1.

   This example can be seen to be identical to Example 4.11. The results obtained by idealization 4 (piecewise linear type interpolation) of Example 4.11 are shown in

**TABLE 10.1   Comparison of Solutions of Example 10.1**

| Time ($t_i$) | Value of $x_i = x(t_i)$ obtained with | | Value of $x_i$ given by idealization 4 of Example 4.11 |
| --- | --- | --- | --- |
| | $\Delta t = \dfrac{\tau_n}{20}$ | $\Delta t = \dfrac{\tau_n}{40}$ | |
| 0 | 0.00000 | 0.00000 | 0.00000 |
| $\pi/10$ | 0.04935 | 0.04638 | 0.04541 |
| $2\pi/10$ | 0.17169 | 0.16569 | 0.16377 |
| $3\pi/10$ | 0.33627 | 0.32767 | 0.32499 |
| $4\pi/10$ | 0.51089 | 0.50056 | 0.49746 |
| $5\pi/10$ | 0.66544 | 0.65456 | 0.65151 |
| $6\pi/10$ | 0.77492 | 0.76485 | 0.76238 |
| $7\pi/10$ | 0.82185 | 0.81395 | 0.81255 |
| $8\pi/10$ | 0.79771 | 0.79314 | 0.79323 |
| $9\pi/10$ | 0.70339 | 0.70296 | 0.70482 |
| $\pi$ | 0.54869 | 0.55275 | 0.55647 |

the last column of Table 10.1. It can be observed that the finite difference method gives reasonably accurate results.

## 10.4 RUNGE-KUTTA METHOD FOR SINGLE DEGREE OF FREEDOM SYSTEMS

The Runge-Kutta method is self-starting and gives quite accurate results. In this method, we first reduce the second order differential equation to two first order equations. For example, Eq. (10.5) can be rewritten as

$$\ddot{x} = \frac{1}{m}\left[F(t) - c\dot{x} - kx\right] = f(x, \dot{x}, t) \tag{10.10}$$

By defining $x_1 = x$ and $x_2 = \dot{x}$, Eq. (10.10) can be written as two first order equations:

$$\dot{x}_1 = x_2$$
$$\dot{x}_2 = f(x_1, x_2, t) \tag{10.11}$$

By defining

$$\vec{X}(t) = \begin{Bmatrix} x_1(t) \\ x_2(t) \end{Bmatrix} \quad \text{and} \quad \vec{F}(t) = \begin{Bmatrix} x_2 \\ f(x_1, x_2, t) \end{Bmatrix}$$

the following recurrence formula is used to find the values of $\vec{X}(t)$ at different grid points $t_i$ according to the fourth order Runge-Kutta method [10.14]:

$$\vec{X}_{i+1} = \vec{X}_i + \tfrac{1}{6}\left[\vec{K}_1 + 2\vec{K}_2 + 2\vec{K}_3 + \vec{K}_4\right] \tag{10.12}$$

where

$$\vec{K}_1 = h\vec{F}(\vec{X}_i, t_i) \tag{10.13}$$

$$\vec{K}_2 = h\vec{F}(\vec{X}_i + \tfrac{1}{2}\vec{K}_1, t_i + \tfrac{1}{2}h) \tag{10.14}$$

$$\vec{K}_3 = h\vec{F}(\vec{X}_i + \tfrac{1}{2}\vec{K}_2, t_i + \tfrac{1}{2}h) \tag{10.15}$$

$$\vec{K}_4 = h\vec{F}(\vec{X}_i + \vec{K}_3, t_{i+1}) \tag{10.16}$$

### EXAMPLE 10.2

Find the solution of Example 10.1 using the Runge-Kutta method.

**Solution.** In this problem,

$$\vec{X}(t) = \begin{Bmatrix} x_1(t) \\ x_2(t) \end{Bmatrix} = \begin{Bmatrix} x(t) \\ \dot{x}(t) \end{Bmatrix}$$

**TABLE 10.2**

| Step $i$ | Time $t_i$ | $x_1 = x$ | $x_2 = \dot{x}$ |
|---|---|---|---|
| 1 | 0.3142 | 0.045406 | 0.275591 |
| 2 | 0.6283 | 0.163726 | 0.461502 |
| 3 | 0.9425 | 0.324850 | 0.547296 |
| ⋮ | | | |
| 19 | 5.9690 | −0.086558 | 0.765737 |
| 20 | 6.2832 | 0.189886 | 0.985565 |

and

$$\vec{F}(t) = \left\{ \begin{array}{c} x_2 \\ f(x_1, x_2, t) \end{array} \right\} = \left\{ \begin{array}{c} \dot{x}(t) \\ \dfrac{1}{m}\left[ F_0\left( 1 - \sin\dfrac{\pi t}{2t_0}\right) - c\dot{x}(t) - kx(t)\right] \end{array} \right\}$$

From the known initial conditions, we have

$$\vec{X}_0 = \left\{ \begin{array}{c} 0 \\ 0 \end{array} \right\}$$

The values of $\vec{X}_{i+1}$, $i = 0, 1, 2, \ldots$ obtained according to Eq. (10.12) are shown in Table 10.2.

## 10.5 CENTRAL DIFFERENCE METHOD FOR MULTIDEGREE OF FREEDOM SYSTEMS

The equation of motion of a viscously damped multidegree of freedom system (see Eq. (6.108)) can be expressed as

$$[m]\ddot{\vec{x}} + [c]\dot{\vec{x}} + [k]\vec{x} = \vec{F} \tag{10.17}$$

where $[m]$, $[c]$ and $[k]$ are the mass, damping and stiffness matrices, $\vec{x}$ is the displacement vector and $\vec{F}$ is the force vector. The procedure indicated for the case of a single degree of freedom system can be directly extended to this case [10.15, 10.16], and the central difference method can be described by the following steps.

**1.** From the known initial conditions $\vec{x}(t = 0) = \vec{x}_0$ and $\dot{\vec{x}}(t = 0) = \dot{\vec{x}}_0$, compute $\ddot{\vec{x}}(t = 0) = \ddot{\vec{x}}_0$:

$$\ddot{\vec{x}}_0 = [m]^{-1}\left[ \vec{F}(t = 0) - [c]\dot{\vec{x}}_0 - [k]\vec{x}_0\right] \tag{10.18}$$

**2.** Select a time step $\Delta t$ such that $\Delta t < \Delta t_{\text{cri}}$.

**3.** Compute $\vec{x}_{-1}$:

$$\vec{x}_{-1} = \vec{x}_0 - \Delta t\, \dot{\vec{x}}_0 + \frac{(\Delta t)^2}{2}\, \ddot{\vec{x}}_0 \tag{10.19}$$

**4.** Find $\vec{x}_{i+1} = \vec{x}(t = t_{i+1})$, starting with $i = 0$:

$$\vec{x}_{i+1} = \left[ \frac{1}{(\Delta t)^2}[m] + \frac{1}{2\Delta t}[c] \right]^{-1} \left[ \vec{F}_i - \left( [k] - \frac{2}{(\Delta t)^2}[m] \right)\vec{x}_i \right.$$
$$\left. - \left( \frac{1}{(\Delta t)^2}[m] - \frac{1}{2\Delta t}[c] \right)\vec{x}_{i-1} \right] \tag{10.20}$$

where

$$\vec{F}_i = \vec{F}(t = t_i) \tag{10.21}$$

If required, evaluate accelerations and velocities at $t_i$:

$$\ddot{\vec{x}}_i = \frac{1}{(\Delta t)^2}[\vec{x}_{i+1} - 2\vec{x}_i + \vec{x}_{i-1}] \tag{10.22}$$

and

$$\dot{\vec{x}}_i = \frac{1}{2\Delta t}[\vec{x}_{i+1} - \vec{x}_{i-1}] \tag{10.23}$$

The stability of the finite difference scheme for solving matrix equations is discussed in Ref. [10.17].

---

**EXAMPLE 10.3**

Find the response of the two degree of freedom system shown in Fig. 10.2 when the forcing functions are given by $F_1(t) = 0$ and $F_2(t) = 10$. Assume the value of $c$ as zero and the initial conditions as $\vec{x}(t = 0) = \dot{\vec{x}}(t = 0) = \vec{0}$.

**Solution.** The equations of motion are given by

$$[m]\ddot{\vec{x}}(t) + [c]\dot{\vec{x}}(t) + [k]\vec{x}(t) = \vec{F}(t) \tag{E.1}$$

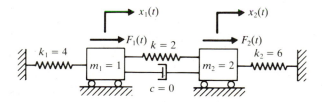

**Figure 10.2**

where

$$[m] = \begin{bmatrix} m_1 & 0 \\ 0 & m_2 \end{bmatrix} = \begin{bmatrix} 1 & 0 \\ 0 & 2 \end{bmatrix} \qquad \text{(E.2)}$$

$$[c] = \begin{bmatrix} c & -c \\ -c & c \end{bmatrix} = \begin{bmatrix} 0 & 0 \\ 0 & 0 \end{bmatrix} \qquad \text{(E.3)}$$

$$[k] = \begin{bmatrix} k_1 + k & -k \\ -k & k + k_2 \end{bmatrix} = \begin{bmatrix} 6 & -2 \\ -2 & 8 \end{bmatrix} \qquad \text{(E.4)}$$

$$\vec{F}(t) = \begin{Bmatrix} F_1(t) \\ F_2(t) \end{Bmatrix} = \begin{Bmatrix} 0 \\ 10 \end{Bmatrix} \qquad \text{(E.5)}$$

and

$$\vec{x}(t) = \begin{Bmatrix} x_1(t) \\ x_2(t) \end{Bmatrix} \qquad \text{(E.6)}$$

**TABLE 10.3**

| Time ($t_i = i\Delta t$) | $\vec{x}_i = \vec{x}(t = t_i)$ |
|---|---|
| $t_1$ | $\begin{Bmatrix} 0 \\ 0.1466 \end{Bmatrix}$ |
| $t_2$ | $\begin{Bmatrix} 0.0172 \\ 0.5520 \end{Bmatrix}$ |
| $t_3$ | $\begin{Bmatrix} 0.0931 \\ 1.1222 \end{Bmatrix}$ |
| $t_4$ | $\begin{Bmatrix} 0.2678 \\ 1.7278 \end{Bmatrix}$ |
| $t_5$ | $\begin{Bmatrix} 0.5510 \\ 2.2370 \end{Bmatrix}$ |
| $t_6$ | $\begin{Bmatrix} 0.9027 \\ 2.5470 \end{Bmatrix}$ |
| $t_7$ | $\begin{Bmatrix} 1.2354 \\ 2.6057 \end{Bmatrix}$ |
| $t_8$ | $\begin{Bmatrix} 1.4391 \\ 2.4189 \end{Bmatrix}$ |
| $t_9$ | $\begin{Bmatrix} 1.4202 \\ 2.0422 \end{Bmatrix}$ |
| $t_{10}$ | $\begin{Bmatrix} 1.1410 \\ 1.5630 \end{Bmatrix}$ |
| $t_{11}$ | $\begin{Bmatrix} 0.6437 \\ 1.0773 \end{Bmatrix}$ |
| $t_{12}$ | $\begin{Bmatrix} 0.0463 \\ 0.6698 \end{Bmatrix}$ |

The undamped natural frequencies and mode shapes of the system can be found by solving the eigenvalue problem

$$\left[ -\omega^2 \begin{bmatrix} 1 & 0 \\ 0 & 2 \end{bmatrix} + \begin{bmatrix} 6 & -2 \\ -2 & 8 \end{bmatrix} \right] \begin{Bmatrix} X_1 \\ X_2 \end{Bmatrix} = \begin{Bmatrix} 0 \\ 0 \end{Bmatrix} \tag{E.7}$$

The solution of Eq. (E.7) is given by

$$\omega_1 = 1.807747, \qquad \vec{X}^{(1)} = \begin{Bmatrix} 1.0000 \\ 1.3661 \end{Bmatrix} \tag{E.8}$$

$$\omega_2 = 2.594620, \qquad \vec{X}^{(2)} = \begin{Bmatrix} 1.0000 \\ -0.3661 \end{Bmatrix} \tag{E.9}$$

Thus the natural periods of the system are

$$\tau_1 = \frac{2\pi}{\omega_1} = 3.4757 \quad \text{and} \quad \tau_2 = \frac{2\pi}{\omega_2} = 2.4216$$

We shall select the time step $(\Delta t)$ as $\tau_2/10 = 0.24216$. The initial value of $\ddot{\vec{x}}$ can be found as follows:

$$\ddot{\vec{x}}_0 = [m]^{-1}\{\vec{F} - [k]\vec{x}_0\} = \begin{bmatrix} 1 & 0 \\ 0 & 2 \end{bmatrix}^{-1} \begin{Bmatrix} 0 \\ 10 \end{Bmatrix} = \frac{1}{2}\begin{bmatrix} 2 & 0 \\ 0 & 1 \end{bmatrix}\begin{Bmatrix} 0 \\ 10 \end{Bmatrix} = \begin{Bmatrix} 0 \\ 5 \end{Bmatrix} \tag{E.10}$$

and the value of $\vec{x}_{-1}$ as follows:

$$\vec{x}_{-1} = \vec{x}_0 - \Delta t \dot{\vec{x}}_0 + \frac{(\Delta t)^2}{2}\ddot{\vec{x}}_0 = \begin{Bmatrix} 0 \\ 0.1466 \end{Bmatrix} \tag{E.11}$$

Now Eq. (10.20) can be applied recursively to obtain $\vec{x}_1, \vec{x}_2, \ldots$. The results are shown in Table 10.3.

## 10.6  FINITE DIFFERENCE METHOD FOR CONTINUOUS SYSTEMS

### 10.6.1  Longitudinal Vibration of Bars

**Equation of Motion.** The equation of motion governing the free longitudinal vibration of a uniform bar (see Eqs. (8.49) and (8.20)) can be expressed as

$$\frac{d^2U}{dx^2} + \alpha^2 U = 0 \tag{10.24}$$

where

$$\alpha^2 = \frac{\omega^2}{c^2} = \frac{\rho\omega^2}{E} \tag{10.25}$$

To obtain the finite difference approximation of Eq. (10.24), we first divide the bar of length $l$ into $n-1$ equal parts each of length $h = l/(n-1)$ and denote the mesh

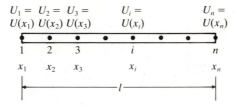

**Figure 10.3**   Division of a bar for finite difference approximation.

points as $1, 2, 3, \ldots, i, \ldots, n$, as shown in Fig. 10.3. Then, by denoting the value of $U$ at mesh point $i$ as $U_i$ and using a formula for the second derivative similar to Eq. (10.4), Eq. (10.24) for mesh point $i$ can be written as

$$\frac{1}{h^2}(U_{i+1} - 2U_i + U_{i-1}) + \alpha^2 U_i = 0$$

or

$$U_{i+1} - (2 - \lambda)U_i + U_{i-1} = 0 \tag{10.26}$$

where $\lambda = h^2 \alpha^2$. The application of Eq. (10.26) at mesh points $i = 2, 3, \ldots, n-1$ leads to the equations

$$U_3 - (2 - \lambda)U_2 + U_1 = 0$$

$$U_4 - (2 - \lambda)U_3 + U_2 = 0$$

$$\vdots$$

$$U_n - (2 - \lambda)U_{n-1} + U_{n-2} = 0 \tag{10.27}$$

which can be stated in matrix form as

$$
\begin{bmatrix}
-1 & (2-\lambda) & -1 & 0 & 0 & \cdots & 0 & 0 & 0 \\
0 & -1 & (2-\lambda) & -1 & 0 & \cdots & 0 & 0 & 0 \\
0 & 0 & -1 & (2-\lambda) & -1 & \cdots & 0 & 0 & 0 \\
\cdot & \cdot & \cdot & \cdot & \cdot & \cdots & \cdot & \cdot & \cdot \\
\cdot & \cdot & \cdot & \cdot & \cdot & \cdots & \cdot & \cdot & \cdot \\
0 & 0 & 0 & 0 & 0 & \cdots & -1 & (2-\lambda) & -1
\end{bmatrix}
\begin{Bmatrix}
U_1 \\ U_2 \\ U_3 \\ \cdot \\ \cdot \\ U_n
\end{Bmatrix}
=
\begin{Bmatrix}
0 \\ 0 \\ 0 \\ \cdot \\ \cdot \\ 0
\end{Bmatrix}
$$

$$\tag{10.28}$$

**Boundary Conditions**

*Fixed end*. The deflection is zero at a fixed end. Assuming that the bar is fixed at $x = 0$ and $x = l$, we set $U_1 = U_n = 0$ in Eq. (10.28) and obtain the equation

$$[[A] - \lambda[I]]\vec{U} = \vec{0} \tag{10.29}$$

where

$$[A] = \begin{bmatrix} 2 & -1 & 0 & 0 & \cdots & 0 & 0 & 0 \\ -1 & 2 & -1 & 0 & \cdots & 0 & 0 & 0 \\ 0 & -1 & 2 & -1 & \cdots & 0 & 0 & 0 \\ \vdots & \vdots & \vdots & \vdots & \cdots & \vdots & \vdots & \vdots \\ & & & & \cdots & & & \\ 0 & 0 & 0 & 0 & \cdots & 0 & -1 & 2 \end{bmatrix} \tag{10.30}$$

$$\vec{U} = \begin{Bmatrix} U_2 \\ U_3 \\ \vdots \\ U_{n-1} \end{Bmatrix} \tag{10.31}$$

and $[I] =$ identity matrix of order $n - 2$.

Note that the eigenvalue problem of Eq. (10.29) can be solved easily, since the matrix $[A]$ is a tridiagonal matrix [10.18, 10.20].

*Free end.* The stress is zero at a free end, so $(dU)/(dx) = 0$. We can use a formula for the first derivative similar to Eq. (10.3). To illustrate the procedure, let the bar be free at $x = 0$ and fixed at $x = l$. The boundary conditions can then be stated as

$$\frac{dU}{dx}\Big|_1 \simeq \frac{U_2 - U_{-1}}{2h} = 0 \quad \text{or} \quad U_{-1} = U_2 \tag{10.32}$$

$$U_n = 0 \tag{10.33}$$

In order to apply Eq. (10.32), we need to imagine the function $U(x)$ to be continuous beyond the length of the bar and create a fictitious mesh point $-1$ so that $U_{-1}$ becomes the fictitious displacement of the point $x_{-1}$. The application of Eq. (10.26) at mesh point $i = 1$ yields

$$U_2 - (2 - \lambda)U_1 + U_{-1} = 0 \tag{10.34}$$

By incorporating the condition $U_{-1} = U_2$ (Eq. (10.32)), Eq. (10.34) can be written as

$$(2 - \lambda)U_1 - 2U_2 = 0 \tag{10.35}$$

By adding Eqs. (10.35) and (10.28), we obtain the final equations:

$$[[A] - \lambda[I]]\vec{U} = \vec{0} \tag{10.36}$$

where

$$[A] = \begin{bmatrix} 2 & -2 & 0 & 0 & \cdots & 0 & 0 & 0 \\ -1 & 2 & -1 & 0 & \cdots & 0 & 0 & 0 \\ 0 & -1 & 2 & -1 & \cdots & 0 & 0 & 0 \\ \vdots & & & & & & & \\ 0 & 0 & 0 & 0 & \cdots & -1 & 2 & -1 \\ 0 & 0 & 0 & 0 & \cdots & 0 & -1 & 2 \end{bmatrix} \tag{10.37}$$

and

$$\vec{U} = \begin{Bmatrix} U_1 \\ U_2 \\ \vdots \\ U_{n-1} \end{Bmatrix}$$

(10.38)

### 10.6.2    Transverse Vibration of Beams

**Equation of Motion.** The governing differential equation for the transverse vibration of a uniform beam is given by Eq. (8.75):

$$\frac{d^4 W}{dx^4} - \beta^4 W = 0$$

(10.39)

where

$$\beta^4 = \frac{\rho A \omega^2}{EI}$$

(10.40)

By using the central difference formula for the fourth derivative,* Eq. (10.39) can be written at any mesh point $i$ as

$$W_{i+2} - 4W_{i+1} + (6 - \lambda) W_i - 4W_{i-1} + W_{i-2} = 0$$

(10.41)

where

$$\lambda = h^4 \beta^4$$

(10.42)

Let the beam be divided into $n - 1$ equal parts with $n$ mesh points and $h = l/(n - 1)$. The application of Eq. (10.41) at the mesh points $i = 3, 4, \ldots, n - 2$ leads to the equations

$$\begin{bmatrix} 1 & -4 & (6-\lambda) & -4 & 1 & 0 & 0 & \cdots & 0 & 0 & 0 & 0 & 0 \\ 0 & 1 & -4 & (6-\lambda) & -4 & 1 & 0 & \cdots & 0 & 0 & 0 & 0 & 0 \\ 0 & 0 & 1 & -4 & (6-\lambda) & -4 & 1 & \cdots & 0 & 0 & 0 & 0 & 0 \\ \vdots & & & & & & & & & & & \\ 0 & 0 & 0 & 0 & 0 & 0 & 0 & \cdots & 1 & -4 & (6-\lambda) & -4 & 1 \end{bmatrix} \begin{Bmatrix} W_1 \\ W_2 \\ W_3 \\ \vdots \\ W_n \end{Bmatrix} = \begin{Bmatrix} 0 \\ 0 \\ 0 \\ \vdots \\ 0 \end{Bmatrix}$$

(10.43)

### Boundary Conditions

*Fixed end.* The deflection $W$ and the slope $(dW)/(dx)$ are zero at a fixed end. If the end $x = 0$ is fixed, we introduce a fictitious node $-1$ on the left-hand side of the

---

*The central difference formula for the fourth derivative (see Problem 10.3) is given by

$$\left. \frac{d^4 f}{dx^4} \right|_i \simeq \frac{1}{h^4} (f_{i+2} - 4f_{i+1} + 6f_i - 4f_{i-1} + f_{i-2})$$

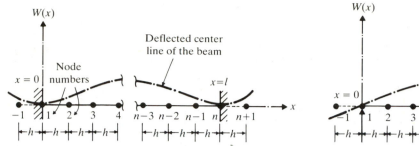

**Figure 10.4**   Beam with fixed ends.

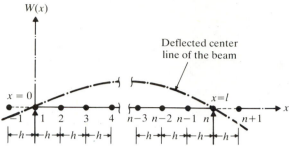

**Figure 10.5**   Beam with simply supported ends.

beam, as shown in Fig. 10.4, and state the boundary conditions, using the central difference formula for $(dW)/(dx)$, as

$$W_1 = 0$$

$$\frac{dW}{dx}\bigg|_1 = \frac{1}{2h}(W_2 - W_{-1}) = 0 \qquad \text{or} \qquad W_{-1} = W_2 \qquad (10.44)$$

where $W_i$ denotes the value of $W$ at node $i$. If the end $x = l$ is fixed, we introduce the fictitious node $n+1$ on the right side of the beam, as shown in Fig. 10.4, and state the boundary conditions as

$$W_n = 0$$

$$\frac{dW}{dx}\bigg|_n = \frac{1}{2h}(W_{n+1} - W_{n-1}) = 0, \qquad \text{or} \qquad W_{n+1} = W_{n-1} \qquad (10.45)$$

*Simply supported end.* If the end $x = 0$ is simply supported (see Fig. 10.5), we have

$$W_1 = 0$$

$$\frac{d^2W}{dx^2}\bigg|_1 = \frac{1}{h^2}(W_2 - 2W_1 + W_{-1}) = 0, \qquad \text{or} \qquad W_{-1} = -W_2 \qquad (10.46)$$

Similar equations can be written if the end $x = l$ is simply supported.

*Free end.* Since bending moment and shear force are zero at a free end, we introduce two fictitious nodes outside the beam, as shown in Fig. 10.6, and use

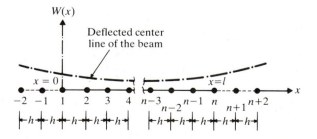

**Figure 10.6**   Beam with free ends.

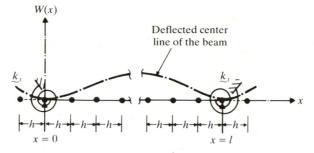

**Figure 10.7**  Beam with elastically restrained ends.

central difference formulas for approximating the second and the third derivatives of the deflection $W$. For example, if the end $x = 0$ is free, we have

$$\left.\frac{d^2W}{dx^2}\right|_1 = \frac{1}{h^2}(W_2 - 2W_1 + W_{-1}) = 0$$

$$\left.\frac{d^3W}{dx^3}\right|_1 = \frac{1}{2h^3}(W_3 - 2W_2 + 2W_{-1} - W_{-2}) = 0 \tag{10.47}$$

*Elastically restrained end*. Let the rotation of the beam be restrained at the end by a torsional spring of stiffness $\underset{\sim}{k}_t$, as shown in Fig. 10.7. If the end $x = l$ is restrained, the boundary conditions to be satisfied are

$$W_n = 0$$

$$M(x = l) = - \underset{\sim}{k}_t \frac{dW}{dx}(x = l) \tag{10.48}$$

Since

$$M(x = l) = EI\frac{d^2W}{dx^2}(x = l) = EI \left.\frac{d^2W}{dx^2}\right|_n = \frac{EI}{h^2}(W_{n+1} - 2W_n + W_{n-1})$$

and

$$\frac{dW}{dx}(x = l) = \left.\frac{dW}{dx}\right|_n = \frac{1}{2h}(W_{n+1} - W_{n-1}) \tag{10.49}$$

the boundary conditions become

$$W_n = 0$$

$$\frac{EI}{h^2}(W_{n+1} + W_{n-1}) = - \frac{\underset{\sim}{k}_t}{2h}(W_{n+1} - W_{n-1}) \tag{10.50}$$

---

**EXAMPLE 10.4**

Find the natural frequencies of the simply supported-fixed beam shown in Fig. 10.8. Assume that the cross section of the beam is constant along its length.

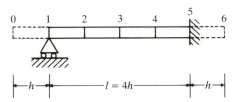

**Figure 10.8**

**Solution.** We shall divide the beam into four segments and express the governing equation

$$\frac{d^4W}{dx^4} - \beta^4 W = 0 \tag{E.1}$$

in finite difference form at each of the interior mesh points. This yields the following equations:

$$W_0 - 4W_1 + (6 - \lambda)W_2 - 4W_3 + W_4 = 0 \tag{E.2}$$

$$W_1 - 4W_2 + (6 - \lambda)W_3 - 4W_4 + W_5 = 0 \tag{E.3}$$

$$W_2 - 4W_3 + (6 - \lambda)W_4 - 4W_5 + W_6 = 0 \tag{E.4}$$

where $W_0$ and $W_6$ denote the values of $W$ at the fictitious nodes 0 and 6, respectively, and

$$\lambda = h^4\beta^4 = \frac{h^4\rho A\omega^2}{EI} \tag{E.5}$$

The boundary conditions at the simply supported end (mesh point 1) are

$$W_1 = 0$$

$$W_0 = -W_2 \tag{E.6}$$

At the fixed end (mesh point 5) the boundary conditions are

$$W_5 = 0$$

$$W_6 = W_4 \tag{E.7}$$

With the help of Eqs. (E.6) and (E.7), Eqs. (E.2) to (E.4) can be reduced to

$$(5 - \lambda)W_2 - 4W_3 + W_4 = 0 \tag{E.8}$$

$$-4W_2 + (6 - \lambda)W_3 - 4W_4 = 0 \tag{E.9}$$

$$W_2 - 4W_3 + (7 - \lambda)W_4 = 0 \tag{E.10}$$

Equations (E.8) to (E.10) can be written in matrix form as

$$\begin{bmatrix} (5-\lambda) & -4 & 1 \\ -4 & (6-\lambda) & -4 \\ 1 & -4 & (7-\lambda) \end{bmatrix} \begin{Bmatrix} W_2 \\ W_3 \\ W_4 \end{Bmatrix} = \begin{Bmatrix} 0 \\ 0 \\ 0 \end{Bmatrix} \qquad \text{(E.11)}$$

The solution of the eigenvalue problem (Eq. (E.11)) gives the following results:

$$\lambda_1 = 0.7135; \qquad \omega_1 = \frac{0.8447}{h^2}\sqrt{\frac{EI}{\rho A}}; \qquad \begin{Bmatrix} W_2 \\ W_3 \\ W_4 \end{Bmatrix}^{(1)} = \begin{Bmatrix} 0.5880 \\ 0.7215 \\ 0.3656 \end{Bmatrix} \qquad \text{(E.12)}$$

$$\lambda_2 = 5.0322; \qquad \omega_2 = \frac{2.2433}{h^2}\sqrt{\frac{EI}{\rho A}}; \qquad \begin{Bmatrix} W_2 \\ W_3 \\ W_4 \end{Bmatrix}^{(2)} = \begin{Bmatrix} 0.6723 \\ -0.1846 \\ -0.7169 \end{Bmatrix} \qquad \text{(E.13)}$$

$$\lambda_3 = 12.2543; \qquad \omega_3 = \frac{3.5006}{h^2}\sqrt{\frac{EI}{\rho A}}; \qquad \begin{Bmatrix} W_2 \\ W_3 \\ W_4 \end{Bmatrix}^{(3)} = \begin{Bmatrix} 0.4498 \\ -0.6673 \\ 0.5936 \end{Bmatrix} \qquad \text{(E.14)}$$

## 10.7 HOUBOLT METHOD

We shall consider the Houbolt method with reference to a multidegree of freedom system. In this method, the following finite difference expansions are employed*

$$\dot{\vec{x}}_{i+1} = \frac{1}{6\Delta t}\left(11\vec{x}_{i+1} - 18\vec{x}_i + 9\vec{x}_{i-1} - 2\vec{x}_{i-2}\right) \qquad (10.51)$$

$$\ddot{\vec{x}}_{i+1} = \frac{1}{(\Delta t)^2}\left(2\vec{x}_{i+1} - 5\vec{x}_i + 4\vec{x}_{i-1} - \vec{x}_{i-2}\right) \qquad (10.52)$$

---

*For a function of one variable $x(t)$, Eqs. (10.51) and (10.52) can be derived by considering the function values $x_{i+1}$, $x_i$, $x_{i-1}$, and $x_{i-2}$ at the equally spaced grid points $t_{i+1}$, $t_i$, $t_{i-1}$, and $t_{i-2}$, respectively, as shown in Fig. 10.9 [10.21]. By considering a cubic curve that passes through these four successive ordinates, the following finite difference equations can be obtained:

$$\dot{x}_{i+1} = \frac{1}{6(\Delta t)}\left(11x_{i+1} - 18x_i + 9x_{i-1} - 2x_{i-2}\right)$$

$$\ddot{x}_{i+1} = \frac{1}{(\Delta t)^2}\left(2x_{i+1} - 5x_i + 4x_{i-1} - x_{i-2}\right)$$

Equations (10.51) and (10.52) represent the vector form of these equations.

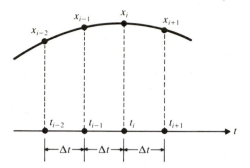

**Figure 10.9**

To find the solution at step $i+1$ ($\vec{x}_{i+1}$), we consider Eq. (10.17) at $t_{i+1}$, so that

$$[m]\ddot{\vec{x}}_{i+1}+[c]\dot{\vec{x}}_{i+1}+[k]\vec{x}_{i+1}=\vec{F}_{i+1}\equiv\vec{F}(t=t_{i+1}) \qquad (10.53)$$

By substituting Eqs. (10.51) and (10.52) into Eq. (10.53), we obtain

$$\left(\frac{2}{(\Delta t)^2}[m]+\frac{11}{6\Delta t}[c]+[k]\right)\vec{x}_{i+1}=\vec{F}_{i+1}+\left(\frac{5}{(\Delta t)^2}[m]+\frac{3}{\Delta t}[c]\right)\vec{x}_i$$

$$-\left(\frac{4}{(\Delta t)^2}[m]+\frac{3[c]}{2\Delta t}\right)\vec{x}_{i-1}$$

$$+\left(\frac{1}{(\Delta t)^2}[m]+\frac{[c]}{3\Delta t}\right)\vec{x}_{i-2} \qquad (10.54)$$

Note that the equilibrium equation at time $t_{i+1}$, Eq. (10.53), is used in finding the solution $\vec{X}_{i+1}$ through Eq. (10.54). This is also true of the Wilson and Newmark methods. For this reason, these methods are called *implicit integration methods*.

It can be seen from Eq. (10.54) that a knowledge of $\vec{x}_i$, $\vec{x}_{i-1}$, and $\vec{x}_{i-2}$ is required to find the solution $\vec{x}_{i+1}$. Thus the values of $\vec{x}_{-1}$ and $\vec{x}_{-2}$ are to be found before attempting to find the vector $\vec{x}_1$ using Eq. (10.54). Since there is no direct method to find $\vec{x}_{-1}$ and $\vec{x}_{-2}$, we do not use Eq. (10.54) to find $\vec{x}_1$ and $\vec{x}_2$. Instead, we use the central difference method described in Sec. 10.5 to find $\vec{x}_1$ and $\vec{x}_2$. Once $\vec{x}_0$ is known from the given initial conditions of the problem and $\vec{x}_1$ and $\vec{x}_2$ are known from the central difference method, the subsequent solutions $\vec{x}_3, \vec{x}_4, \ldots$ can be found by using Eq. (10.54).

The step-by-step procedure to be used in the Houbolt method is given below:

1. From the known initial conditions $\vec{x}(t=0)=\vec{x}_0$ and $\dot{\vec{x}}(t=0)=\dot{\vec{x}}_0$, find $\ddot{\vec{x}}_0=\ddot{\vec{x}}(t=0)$ using Eq. (10.18).

2. Select a suitable time step $\Delta t$.

3. Determine $\vec{x}_{-1}$ using Eq. (10.19).

4. Find $\vec{x}_1$ and $\vec{x}_2$ using the central difference equation (10.20).

5.  Compute $\vec{x}_{i+1}$, starting with $i = 2$ and using Eq. (10.54):

$$\vec{x}_{i+1} = \left[ \frac{2}{(\Delta t)^2} [m] + \frac{11}{6\Delta t} [c] + [k] \right]^{-1} \left\{ \vec{F}_{i+1} + \left( \frac{5}{(\Delta t)^2} [m] + \frac{3}{\Delta t} [c] \right) \vec{x}_i \right.$$

$$\left. - \left( \frac{4}{(\Delta t)^2} [m] + \frac{3}{2\Delta t} [c] \right) \vec{x}_{i-1} + \left( \frac{1}{(\Delta t)^2} [m] + \frac{1}{3\Delta t} [c] \right) \vec{x}_{i-2} \right\}$$

$$(10.55)$$

If required, evaluate the velocity and acceleration vectors $\dot{\vec{x}}_{i+1}$ and $\ddot{\vec{x}}_{i+1}$ using Eqs. (10.51) and (10.52).

## EXAMPLE 10.5

Find the response of the two degree of freedom system considered in Example 10.3 using the Houbolt method.

### TABLE 10.4

| Time $t_i = i\Delta t$ | $\vec{x}_i = \vec{x}(t = t_i)$ |
| --- | --- |
| $t_1$ | $\left\{ \begin{array}{c} 0.0000 \\ 0.1466 \end{array} \right\}$ |
| $t_2$ | $\left\{ \begin{array}{c} 0.0172 \\ 0.5520 \end{array} \right\}$ |
| $t_3$ | $\left\{ \begin{array}{c} 0.0917 \\ 1.1064 \end{array} \right\}$ |
| $t_4$ | $\left\{ \begin{array}{c} 0.2501 \\ 1.6909 \end{array} \right\}$ |
| $t_5$ | $\left\{ \begin{array}{c} 0.4924 \\ 2.1941 \end{array} \right\}$ |
| $t_6$ | $\left\{ \begin{array}{c} 0.7867 \\ 2.5297 \end{array} \right\}$ |
| $t_7$ | $\left\{ \begin{array}{c} 1.0734 \\ 2.6489 \end{array} \right\}$ |
| $t_8$ | $\left\{ \begin{array}{c} 1.2803 \\ 2.5454 \end{array} \right\}$ |
| $t_9$ | $\left\{ \begin{array}{c} 1.3432 \\ 2.2525 \end{array} \right\}$ |
| $t_{10}$ | $\left\{ \begin{array}{c} 1.2258 \\ 1.8325 \end{array} \right\}$ |
| $t_{11}$ | $\left\{ \begin{array}{c} 0.9340 \\ 1.3630 \end{array} \right\}$ |
| $t_{12}$ | $\left\{ \begin{array}{c} 0.5178 \\ 0.9224 \end{array} \right\}$ |

*Solution.* The value of $\ddot{\vec{x}}_0$ can be found using Eq. (10.18):

$$\ddot{\vec{x}} = \begin{Bmatrix} 0 \\ 5 \end{Bmatrix}$$

By using a value of $\Delta t = 0.24216$, Eq. (10.20) can be used to find $\vec{x}_1$ and $\vec{x}_2$, and then Eq. (10.55) can be used recursively to obtain $\vec{x}_3, \vec{x}_4, \ldots$, as shown in Table 10.4.

## **10.8**  WILSON METHOD

The Wilson method assumes that the acceleration of the system varies linearly between two instants of time. In particular, the two instants of time are taken as indicated in Fig. 10.10. Thus the acceleration is assumed to be linear from time $t_i = i\Delta t$ to time $t_{i+\theta} = t_i + \theta \Delta t$, where $\theta \geq 1.0$ [10.22]. For this reason, this method is also called the *Wilson $\theta$ method*. If $\theta = 1.0$, this method reduces to the linear acceleration scheme [10.23].

A stability analysis of the Wilson method shows that it is unconditionally stable provided that $\theta \geq 1.37$. In this section, we shall consider the Wilson method for a multidegree of freedom system.

Since $\ddot{\vec{x}}(t)$ is assumed to vary linearly between $t_i$ and $t_{i+\theta}$, we can predict the value of $\ddot{\vec{x}}$ at any time $t_i + \tau$, $0 \leq \tau \leq \theta \Delta t$:

$$\ddot{\vec{x}}(t_i + \tau) = \ddot{\vec{x}}_i + \frac{\tau}{\theta \Delta t}\left(\ddot{\vec{x}}_{i+\theta} - \ddot{\vec{x}}_i\right) \tag{10.56}$$

By integrating Eq. (10.56), we obtain*

$$\dot{\vec{x}}(t_i + \tau) = \dot{\vec{x}}_i + \ddot{\vec{x}}_i \tau + \frac{\tau^2}{2\theta \Delta t}\left(\ddot{\vec{x}}_{i+\theta} - \ddot{\vec{x}}_i\right) \tag{10.57}$$

and

$$\vec{x}(t_i + \tau) = \vec{x}_i + \dot{\vec{x}}_i \tau + \frac{1}{2}\ddot{\vec{x}}_i \tau^2 + \frac{\tau^3}{6\theta \Delta t}\left(\ddot{\vec{x}}_{i+\theta} - \ddot{\vec{x}}_i\right) \tag{10.58}$$

By substituting $\tau = \theta \Delta t$ into Eqs. (10.57) and (10.58), we obtain

$$\dot{\vec{x}}_{i+\theta} = \dot{\vec{x}}(t_i + \theta \Delta t) = \dot{\vec{x}}_i + \frac{\theta \Delta t}{2}\left(\ddot{\vec{x}}_{i+\theta} + \ddot{\vec{x}}_i\right) \tag{10.59}$$

$$\vec{x}_{i+\theta} = \vec{x}(t_i + \theta \Delta t) = \vec{x}_i + \theta \Delta t \dot{\vec{x}}_i + \frac{\theta^2 (\Delta t)^2}{6}\left(\ddot{\vec{x}}_{i+\theta} + 2\ddot{\vec{x}}_i\right) \tag{10.60}$$

Equation (10.60) can be solved to obtain

$$\ddot{\vec{x}}_{i+\theta} = \frac{6}{\theta^2 (\Delta t)^2}\left(\vec{x}_{i+\theta} - \vec{x}_i\right) - \frac{6}{\theta \Delta t}\dot{\vec{x}}_i - 2\ddot{\vec{x}}_i \tag{10.61}$$

---

*$\dot{\vec{x}}_i$ and $\vec{x}_i$ have been substituted in place of the integration constants in Eqs. (10.57) and (10.58), respectively.

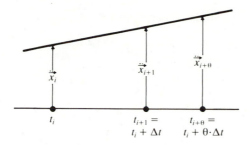

**Figure 10.10**   Linear acceleration assumption of the Wilson method.

By substituting Eq. (10.61) into Eq. (10.59), we obtain

$$\dot{\vec{x}}_{i+\theta} = \frac{3}{\theta \Delta t}\left(\vec{x}_{i+\theta} - \vec{x}_i\right) - 2\dot{\vec{x}}_i - \frac{\theta \Delta t}{2}\ddot{\vec{x}}_i \tag{10.62}$$

To obtain the value of $x_{i+\theta}$, we consider the equilibrium equation (10.17) at time $t_{i+\theta} = t_i + \theta \Delta t$ and write

$$[m]\ddot{\vec{x}}_{i+\theta} + [c]\dot{\vec{x}}_{i+\theta} + [k]\vec{x}_{i+\theta} = \vec{F}_{i+\theta} \tag{10.63}$$

where the force vector $\vec{F}_{i+\theta}$ is also obtained by using the linear assumption:

$$\vec{F}_{i+\theta} = \vec{F}_i + \theta\left(\vec{F}_{i+1} - \vec{F}_i\right) \tag{10.64}$$

The Wilson method can be described by the following steps:

1. From the known initial conditions $\vec{x}_0$ and $\dot{\vec{x}}_0$, find $\ddot{\vec{x}}_0$ using Eq. (10.18).

2. Select a suitable time step $\Delta t$ and a suitable value of $\theta$ ($\theta$ is usually taken as 1.4).

3. Compute the effective load vector $\tilde{\vec{F}}_{i+\theta}$, starting with $i = 0$:

$$\tilde{\vec{F}}_{i+\theta} = \vec{F}_i + \theta\left(\vec{F}_{i+1} - \vec{F}_i\right) + [m]\left(\frac{6}{\theta^2(\Delta t)^2}\vec{x}_i + \frac{6}{\theta \Delta t}\dot{\vec{x}}_i + 2\ddot{\vec{x}}_i\right)$$

$$+ [c]\left(\frac{3}{\theta \Delta t}\vec{x}_i + 2\dot{\vec{x}}_i + \frac{\theta \Delta t}{2}\ddot{\vec{x}}_i\right) \tag{10.65}$$

4. Find the displacement vector at time $t_{i+\theta}$:

$$\vec{x}_{i+\theta} = \left[\frac{6}{\theta^2(\Delta t)^2}[m] + \frac{3}{\theta \Delta t}[c] + [k]\right]^{-1}\tilde{\vec{F}}_{i+\theta} \tag{10.66}$$

**5.** Calculate the acceleration, velocity, and displacement vectors at time $t_{i+1}$:

$$\ddot{\vec{x}}_{i+1} = \frac{6}{\theta^3(\Delta t)^2}\left(\vec{x}_{i+\theta} - \vec{x}_i\right) - \frac{6}{\theta^2 \Delta t}\dot{\vec{x}}_i + \left(1 - \frac{3}{\theta}\right)\ddot{\vec{x}}_i \qquad (10.67)$$

$$\dot{\vec{x}}_{i+1} = \dot{\vec{x}}_i + \frac{\Delta t}{2}\left(\ddot{\vec{x}}_{i+1} + \ddot{\vec{x}}_i\right) \qquad (10.68)$$

$$\vec{x}_{i+1} = \vec{x}_i + \Delta t\dot{\vec{x}}_i + \frac{(\Delta t)^2}{6}\left(\ddot{\vec{x}}_{i+1} + 2\ddot{\vec{x}}_i\right) \qquad (10.69)$$

### EXAMPLE 10.6

Find the response of the system considered in Example 10.3, using the Wilson $\theta$ method with $\theta = 1.4$.

**Solution.** The value of $\ddot{\vec{x}}_0$ can be obtained as in the case of Example 10.3:

$$\ddot{\vec{x}}_0 = \begin{Bmatrix} 0 \\ 5 \end{Bmatrix}$$

### TABLE 10.5

| Time $t_i = i\Delta t$ | $\vec{x}_i = \vec{x}(t = t_i)$ |
|:---:|:---:|
| $t_1$ | $\begin{Bmatrix} 0.0033 \\ 0.1392 \end{Bmatrix}$ |
| $t_2$ | $\begin{Bmatrix} 0.0289 \\ 0.5201 \end{Bmatrix}$ |
| $t_3$ | $\begin{Bmatrix} 0.1072 \\ 1.0579 \end{Bmatrix}$ |
| $t_4$ | $\begin{Bmatrix} 0.2649 \\ 1.6408 \end{Bmatrix}$ |
| $t_5$ | $\begin{Bmatrix} 0.5076 \\ 2.1529 \end{Bmatrix}$ |
| $t_6$ | $\begin{Bmatrix} 0.8074 \\ 2.4981 \end{Bmatrix}$ |
| $t_7$ | $\begin{Bmatrix} 1.1035 \\ 2.6191 \end{Bmatrix}$ |
| $t_8$ | $\begin{Bmatrix} 1.3158 \\ 2.5056 \end{Bmatrix}$ |
| $t_9$ | $\begin{Bmatrix} 1.3688 \\ 2.1929 \end{Bmatrix}$ |
| $t_{10}$ | $\begin{Bmatrix} 1.2183 \\ 1.7503 \end{Bmatrix}$ |
| $t_{11}$ | $\begin{Bmatrix} 0.8710 \\ 1.2642 \end{Bmatrix}$ |
| $t_{12}$ | $\begin{Bmatrix} 0.3897 \\ 0.8208 \end{Bmatrix}$ |

Then, by using Eqs. (10.67) to (10.69) with a time step of $\Delta t = 0.24216$, we obtain the results indicated in Table 10.5.

## 10.9 NEWMARK METHOD

The Newmark integration method is also based on the assumption that the acceleration varies linearly between two instants of time. The resulting expressions for the velocity and displacement vectors $\dot{\vec{x}}_{i+1}$ and $\vec{x}_{i+1}$, for a multidegree of freedom system [10.24], are written as in Eqs. (10.57) and (10.58):

$$\dot{\vec{x}}_{i+1} = \dot{\vec{x}}_i + \left[(1-\beta)\ddot{\vec{x}}_i + \beta\ddot{\vec{x}}_{i+1}\right]\Delta t \tag{10.70}$$

$$\vec{x}_{i+1} = \vec{x}_i + \Delta t\dot{\vec{x}}_i + \left[(\tfrac{1}{2}-\alpha)\ddot{\vec{x}}_i + \alpha\ddot{\vec{x}}_{i+1}\right](\Delta t)^2 \tag{10.71}$$

where the parameters $\alpha$ and $\beta$ indicate how much the acceleration at the end of the interval enters into the velocity and displacement equations at the end of the interval $\Delta t$. In fact, $\alpha$ and $\beta$ can be chosen to obtain the desired accuracy and stability characteristics [10.25]. When $\beta = \tfrac{1}{2}$ and $\alpha = \tfrac{1}{6}$, Eqs. (10.70) and (10.71) correspond to the linear acceleration method (which can also be obtained using $\theta = 1$ in the Wilson method). When $\beta = \tfrac{1}{2}$ and $\alpha = \tfrac{1}{4}$, Eqs. (10.70) and (10.71) correspond to the assumption of constant acceleration between $t_i$ and $t_{i+1}$. To find the value of $\ddot{\vec{x}}_{i+1}$, the equilibrium equation (10.17) is considered at $t = t_{i+1}$ so that

$$[m]\ddot{\vec{x}}_{i+1} + [c]\dot{\vec{x}}_{i+1} + [k]\vec{x}_{i+1} = \vec{F}_{i+1} \tag{10.72}$$

Equation (10.71) can be used to express $\ddot{\vec{x}}_{i+1}$ in terms of $\vec{x}_{i+1}$, and the resulting expression can be substituted into Eq. (10.70) to express $\dot{\vec{x}}_{i+1}$ in terms of $\vec{x}_{i+1}$. By substituting these expressions for $\dot{\vec{x}}_{i+1}$ and $\ddot{\vec{x}}_{i+1}$ into Eq. (10.72), we can obtain a relation for finding $\vec{x}_{i+1}$:

$$\vec{x}_{i+1} = \left[\frac{1}{\alpha(\Delta t)^2}[m] + \frac{\beta}{\alpha\Delta t}[c] + [k]\right]^{-1}$$

$$\times \left\{ \vec{F}_{i+1} + [m]\left(\frac{1}{\alpha(\Delta t)^2}\vec{x}_i + \frac{1}{\alpha\Delta t}\dot{\vec{x}}_i + \left(\frac{1}{2\alpha}-1\right)\ddot{\vec{x}}_i\right) \right.$$

$$\left. + [c]\left(\frac{\beta}{\alpha\Delta t}\vec{x}_i + \left(\frac{\beta}{\alpha}-1\right)\dot{\vec{x}}_i + \left(\frac{\beta}{\alpha}-2\right)\frac{\Delta t}{2}\ddot{\vec{x}}_i\right) \right\} \tag{10.73}$$

The Newmark method can be summarized in the following steps:

**1.** From the known values of $\vec{x}_0$ and $\dot{\vec{x}}_0$, find $\ddot{\vec{x}}_0$ using Eq. (10.18).

**2.** Select suitable values of $\Delta t$, $\alpha$, and $\beta$.

3. Calculate the displacement vector $\vec{x}_{i+1}$, starting with $i = 0$ and using Eq. (10.73).

4. Find the acceleration and velocity vectors at time $t_{i+1}$:

$$\ddot{\vec{x}}_{i+1} = \frac{1}{\alpha(\Delta t)^2}(\vec{x}_{i+1} - \vec{x}_i) - \frac{1}{\alpha\Delta t}\dot{\vec{x}}_i - \left(\frac{1}{2\alpha} - 1\right)\ddot{\vec{x}}_i \qquad (10.74)$$

$$\dot{\vec{x}}_{i+1} = \dot{\vec{x}}_i + (1-\beta)\Delta t\ddot{\vec{x}}_i + \beta\Delta t\ddot{\vec{x}}_{i+1} \qquad (10.75)$$

It is important to note that unless $\beta$ is taken as $\frac{1}{2}$, there is a spurious damping introduced, proportional to $(\beta - \frac{1}{2})$. If $\beta$ is taken as zero, a negative damping results; this involves a self-excited vibration arising solely from the numerical procedure. Similarly, if $\beta$ is greater than $\frac{1}{2}$, a positive damping is introduced. This reduces the magnitude of response even without real damping in the problem [10.24].

### EXAMPLE 10.7

Find the response of the system considered in Example 10.3, using the Newmark method with $\alpha = \frac{1}{6}$ and $\beta = \frac{1}{2}$.

**TABLE 10.6**

| Time $t_i = i\Delta t$ | $\vec{x}_i = \vec{x}(t = t_i)$ |
|---|---|
| $t_1$ | $\left\{ \begin{array}{c} 0.0026 \\ 0.1411 \end{array} \right\}$ |
| $t_2$ | $\left\{ \begin{array}{c} 0.0246 \\ 0.5329 \end{array} \right\}$ |
| $t_3$ | $\left\{ \begin{array}{c} 0.1005 \\ 1.0884 \end{array} \right\}$ |
| $t_4$ | $\left\{ \begin{array}{c} 0.2644 \\ 1.6870 \end{array} \right\}$ |
| $t_5$ | $\left\{ \begin{array}{c} 0.5257 \\ 2.2027 \end{array} \right\}$ |
| $t_6$ | $\left\{ \begin{array}{c} 0.8530 \\ 2.5336 \end{array} \right\}$ |
| $t_7$ | $\left\{ \begin{array}{c} 1.1730 \\ 2.6229 \end{array} \right\}$ |
| $t_8$ | $\left\{ \begin{array}{c} 1.3892 \\ 2.4674 \end{array} \right\}$ |
| $t_9$ | $\left\{ \begin{array}{c} 1.4134 \\ 2.1137 \end{array} \right\}$ |
| $t_{10}$ | $\left\{ \begin{array}{c} 1.1998 \\ 1.6426 \end{array} \right\}$ |
| $t_{11}$ | $\left\{ \begin{array}{c} 0.7690 \\ 1.1485 \end{array} \right\}$ |
| $t_{12}$ | $\left\{ \begin{array}{c} 0.2111 \\ 0.7195 \end{array} \right\}$ |

***Solution.*** The value of $\ddot{\vec{x}}_0$ can be found using Eq. (10.18):

$$\ddot{\vec{x}}_0 = \begin{Bmatrix} 0 \\ 5 \end{Bmatrix}$$

With the values of $\alpha = \frac{1}{6}$, $\beta = 0.5$, and $\Delta t = 0.24216$, Eq. (10.73) gives the values of $\vec{x}_i = \vec{x}(t = t_i)$ as shown in Table 10.6.

## 10.10 COMPUTER PROGRAMS

### 10.10.1 Fourth Order Runge-Kutta Method

A Fortran program in the form of subroutine RK4 is given for solving a system of $N$ first order differential equations of the form

$$\frac{d\vec{X}}{dt} = \vec{F}(\vec{X}, t)$$

Subroutine RK4 is to be called *NSTEP* times to generate the solution one time step at a time. The following arguments are used for the subroutine:

T = Current value of time. Input data.

DT = Time increment. Input data.

N = Number of differential equations. Input data.

XX = Array of size $N$, containing the initial conditions $x_i(0)$. Input data.

F, XI, XJ, XK, XL, UU = Arrays of size $N$ each.

The program requires a user-supplied subroutine FUN(X, F, N, T). This subroutine must evaluate the forcing functions $F(1), F(2), \ldots, F(N)$ at any specified vectors $\vec{X}$ (argument X) and time $t$ (argument T).

Example 10.2 is solved by the use of subroutine RK4. The main program, subroutine RK4, and the results given by the program are shown below.

```
C ==========================================================================
C
C PROGRAM 19
C MAIN PROGRAM FOR CALLING THE SUBROUTINE RK4
C
C ==========================================================================
C FOLLOWING 7 LINES CONTAIN PROBLEM-DEPENDENT DATA
      DIMENSION TIME(40),X(40,2),XX(2),F(2),YI(2),YJ(2),YK(2),YL(2),
     2   UU(2)
      XX(1)=0.0
      XX(2)=0.0
      NEQ=2
      NSTEP=40
```

```
          DT=0.31416/2
          T=0.0
          PRINT 10
    10    FORMAT (//,3X,5H  I  ,10H   TIME(I),7X,5H X(1),12X,5H X(2),/)
          DO 40 I=1,NSTEP
          CALL RK4 (T,DT,NEQ,XX,F,YI,YJ,YK,YL,UU)
          TIME(I)=T
          DO 20 J=1,NEQ
    20    X(I,J)=XX(J)
          PRINT 30, I, TIME(I),(X(I,J),J=1,NEQ)
    30    FORMAT (2X,I5,F10.4,2X,E15.8,2X,E15.8)
    40    CONTINUE
          STOP
          END
C ==================================================================
C
C SUBROUTINE RK4
C
C ==================================================================
          SUBROUTINE RK4 (T,DT,N,XX,F,XI,XJ,XK,XL,UU)
          DIMENSION XI(N),XJ(N),XK(N),XL(N),UU(N),XX(N),F(N)
          DO 10 I=1,N
    10    UU(I)=XX(I)
          CALL FUN (XX,F,N,T)
          DO 20 I=1,N
          XI(I)=F(I)*DT
    20    XX(I)=UU(I)+XI(I)/2.0
          T=T+DT/2.0
          CALL FUN (XX,F,N,T)
          DO 30 I=1,N
          XJ(I)=F(I)*DT
    30    XX(I)=UU(I)+XJ(I)/2.0
          CALL FUN (XX,F,N,T)
          DO 40 I  =1,N
          XK(I)=F(I)*DT
    40    XX(I)=UU(I)+XK(I)
          T=T+DT/2.0
          CALL FUN (XX,F,N,T)
          DO 50 I=1,N
          XL(I)=F(I)*DT
    50    XX(I)=UU(I)+(XI(I)+2.0*XJ(I)+2.0*XK(I)+XL(I))/6.0
          RETURN
          END
C ==================================================================
C
C SUBROUTINE FUN FOR USE IN THE SUBROUTINE RK4
C THIS SUBROUTINE CHANGES FROM PROBLEM TO PROBLEM
C
C ==================================================================
          SUBROUTINE FUN (X,F,N,T)
          DIMENSION X(N),F(N)
          F(1)=X(2)
          F0=1.0
          T0=3.1416
```

```
XM=1.0
XC=0.2
XK=1.0
FT=F0*(1.0-SIN(3.1416*T/(2.0*T0)))
F(2)=(FT-XC*X(2)-XK*X(1))
RETURN
END
```

| I | TIME(I) | X(1) | X(2) |
|---|---------|------|------|
| 1 | 0.1571 | 0.11863151E-01 | 0.14791380E+00 |
| 2 | 0.3142 | 0.45406420E-01 | 0.27559111E+00 |
| 3 | 0.4712 | 0.97257055E-01 | 0.38067484E+00 |
| 4 | 0.6283 | 0.16372624E+00 | 0.46150222E+00 |
| 5 | 0.7854 | 0.24091978E+00 | 0.51712251E+00 |
| 6 | 0.9425 | 0.32485014E+00 | 0.54729646E+00 |
| 7 | 1.0996 | 0.41154701E+00 | 0.55247802E+00 |
| 8 | 1.2566 | 0.49716362E+00 | 0.53377944E+00 |
| 9 | 1.4137 | 0.57807648E+00 | 0.49292117E+00 |
| 10 | 1.5708 | 0.65097594E+00 | 0.43216801E+00 |
| 11 | 1.7279 | 0.71294606E+00 | 0.35425371E+00 |
| 12 | 1.8850 | 0.76153159E+00 | 0.26229632E+00 |
| 13 | 2.0420 | 0.79479134E+00 | 0.15970632E+00 |
| 14 | 2.1991 | 0.81133646E+00 | 0.50090402E-01 |
| 15 | 2.3562 | 0.81035364E+00 | -0.62847026E-01 |
| 16 | 2.5133 | 0.79161268E+00 | -0.17540169E+00 |
| 17 | 2.6704 | 0.75545907E+00 | -0.28396457E+00 |
| 18 | 2.8274 | 0.70279163E+00 | -0.38511199E+00 |
| 19 | 2.9845 | 0.63502711E+00 | -0.47568849E+00 |
| 20 | 3.1416 | 0.55405176E+00 | -0.55288076E+00 |
| 21 | 3.2987 | 0.46216279E+00 | -0.61428112E+00 |
| 22 | 3.4558 | 0.36200023E+00 | -0.65793908E+00 |
| 23 | 3.6128 | 0.25647193E+00 | -0.68240052E+00 |
| 24 | 3.7699 | 0.14867344E+00 | -0.68673331E+00 |
| 25 | 3.9270 | 0.41804813E-01 | -0.67053956E+00 |
| 26 | 4.0841 | -0.60913362E-01 | -0.63395429E+00 |
| 27 | 4.2412 | -0.15632270E+00 | -0.57763124E+00 |
| 28 | 4.3982 | -0.24140659E+00 | -0.50271636E+00 |
| 29 | 4.5553 | -0.31336465E+00 | -0.41080984E+00 |
| 30 | 4.7124 | -0.36968049E+00 | -0.30391818E+00 |
| 31 | 4.8695 | -0.40818110E+00 | -0.18439761E+00 |
| 32 | 5.0266 | -0.42708698E+00 | -0.54890379E-01 |
| 33 | 5.1836 | -0.42505166E+00 | 0.81744030E-01 |
| 34 | 5.3407 | -0.40119010E+00 | 0.22250143E+00 |
| 35 | 5.4978 | -0.35509574E+00 | 0.36430690E+00 |
| 36 | 5.6549 | -0.28684574E+00 | 0.50408858E+00 |
| 37 | 5.8120 | -0.19699481E+00 | 0.63884968E+00 |
| 38 | 5.9690 | -0.86557955E-01 | 0.76573688E+00 |
| 39 | 6.1261 | 0.43017015E-01 | 0.88210326E+00 |
| 40 | 6.2832 | 0.18988648E+00 | 0.98556501E+00 |

### 10.10.2 Central Difference Method

A Fortran subroutine CDIFF is written to find the dynamic response of a multide-gree of freedom system by the use of the central difference method. The following arguments are used:

M = Array of size $N \times N$, denoting the mass matrix. Input data.

C = Array of size $N \times N$, denoting the damping matrix. Input data.

K = Array of size $N \times N$, denoting the stiffness matrix. Input data.

XI = Array of size $N$, containing the initial values $x_i(0)$. Input data.

XDI = Array of size $N$, containing the initial values $\dot{x}_i(0)$. Input data.

XDDI = Array of size $N$.

N = Number of degrees of freedom. Input data.

NSTEP = Number of time steps at which the solution is to be found. Input data.

DELT = Time increment between steps. Input data.

F = Array of size $N$, containing the values of the forcing function at any specified time.

R, RR, XM1, XM2, XP1, XMK, XMI, ZA, ZB, ZC, LA, S = arrays of size $N$ each.

MC, MK, MCI, MMC, MI = arrays of size $N \times N$ each.

X, XD, XDD = arrays of size $NSTEP1 \times N$ each. $X(I, J) = x_j(t_i)$, $XD(I, J) = \dot{x}_j(t_i)$, $XDD(I, J) = \ddot{x}_j(t_i)$. Output.

LB = Array of size $N \times 2$.

NSTEP1 = NSTEP + 1. Input data.

The program requires a user-supplied subroutine EXTFUN(F, TIME, N). This must evaluate the forcing functions $F(1), \ldots, F(N)$ at a given time, TIME.

For illustration, subroutine CDIFF is used to solve Example 10.3. The main program that calls CDIFF, subroutine CDIFF, and the output of the program are given below.

```
C ================================================================
C
C PROGRAM 20
C MAIN PROGRAM WHICH CALLS CDIFF
C
C ================================================================
C FOLLOWING 10 LINES CONTAIN PROBLEM-DEPENDENT DATA
      REAL M(2,2),K(2,2),MC(2,2),MK(2,2),MCI(2,2),MMC(2,2),MI(2,2)
      DIMENSION C(2,2),XI(2),XDI(2),XDDI(2),XM1(2),F(2),R(2),RR(2),
     2    XMK(2),XMI(2),XM2(2),XP1(2),ZA(2),ZB(2),ZC(2),LA(2),LB(2,2),
     3    S(2),X(25,2),XD(25,2),XDD(25,2)
      DATA N,NSTEP,NSTEP1,DELT/2,24,25,0.24216267/
      DATA XI/0.0,0.0/
      DATA XDI/0.0,0.0/
```

```
           DATA M/1.0,0.0,0.0,2.0/
           DATA C/0.0,0.0,0.0,0.0/
           DATA K/6.0,-2.0,-2.0,8.0/
C END OF PROBLEM-DEPENDENT DATA
           CALL CDIFF (M,C,K,XI,XDI,XDDI,N,NSTEP,DELT,F,R,RR,XM1,XM2,XP1,
          2   MC,MK,MCI,XMK,MMC,XMI,ZA,ZB,ZC,LA,LB,S,X,XD,XDD,NSTEP1,MI)
           PRINT 10
   10      FORMAT (//,38H SOLUTION BY CENTRAL DIFFERENCE METHOD,/)
           PRINT 20, N,NSTEP,DELT
   20      FORMAT (12H GIVEN DATA:,/,3H N=,I5,4X,7H NSTEP=,I5,4X,6H DELT=,
          2   E15.8,/)
           PRINT 30
   30      FORMAT (10H SOLUTION:,//,5H STEP,3X,5H TIME,3X,7H X(I,1),3X,
          2   8H XD(I,1),2X,9H XDD(I,1),4X,7H X(I,2),3X,8H XD(I,2),2X,
          3   9H XDD(I,2),/)
           DO 40  I=1,NSTEP1
           TIME=REAL(I-1)*DELT
   40      PRINT 50, I,TIME,X(I,1),XD(I,1),XDD(I,1),X(I,2),XD(I,2),XDD(I,2)
   50      FORMAT (1X,I4,F8.4,6(1X,E10.4))
           STOP
           END
C =======================================================================
C
C SUBROUTINE CDIFF
C
C =======================================================================
           SUBROUTINE CDIFF (M,C,K,XI,XDI,XDDI,N,NSTEP,DELT,F,R,RR,XM1,
          2   XM2,XP1,MC,MK,MCI,XMK,MMC,XMI,ZA,ZB,ZC,LA,LB,S,X,XD,XDD,NSTEP1
          3   ,MI)
           REAL M(N,N),K(N,N),MC(N,N),MK(N,N),MCI(N,N),MMC(N,N),MI(N,N)
           DIMENSION C(N,N),XI(N),XDI(N),XDDI(N),F(N),R(N),RR(N),XM1(N),
          2   XM2(N),XP1(N),XMK(N),XMI(N),ZA(N),ZB(N),ZC(N),LA(N),LB(N,2),S(N
          3   ,X(NSTEP1,N),XD(NSTEP1,N),XDD(NSTEP1,N)
           DO 5 I=1,N
           DO 5 J=1,N
   5       MI(I,J)=M(I,J)
           CALL SIMUL (MI,ZA,N,0,LA,LB,S)
           CALL EXTFUN (F,0.0,N)
           DO 20 I=1,N
           R(I)=F(I)
           DO 10 J=1,N
   10      R(I)=R(I)-C(I,J)*XDI(J)-K(I,J)*XI(J)
   20      CONTINUE
           CALL XMULT (MI,R,XDDI,N)
           DO 25 J=1,N
           X(1,J)=XI(J)
           XD(1,J)=XDI(J)
   25      XDD(1,J)=XDDI(J)
           DO 30 I=1,N
   30      XM1(I)=XI(I)-DELT*XDI(I)+(DELT**2)*XDDI(I)/2.0
           DO 40 I=1,N
           DO 40 J=1,N
           MC(I,J)=(M(I,J)/(DELT**2))+(C(I,J)/(2.0*DELT))
           MK(I,J)=K(I,J)-2.0*M(I,J)/(DELT**2)
```

```
40    MMC(I,J)=(M(I,J)/(DELT**2))-(C(I,J)/(2.0*DELT))
      DO 45 I=1,N
      DO 45 J=1,N
45    MCI(I,J)=MC(I,J)
      CALL SIMUL (MCI,ZA,N,0,LA,LB,S)
      TIME=-DELT
      DO 90 I=1,NSTEP
      CALL XMULT (MK,XI,XMK,N)
      CALL XMULT (MMC,XM1,XMI,N)
      TIME=TIME+DELT
      CALL EXTFUN (F,TIME,N)
      DO 50 J=1,N
50    RR(J)=F(J)-XMK(J)-XMI(J)
      CALL XMULT (MCI,RR,XP1,N)
      DO 60 J=1,N
      XDI(J)=(XP1(J)-XM1(J))/(2.0*DELT)
60    XDDI(J)=(XP1(J)-2.0*XI(J)+XM1(J))/(DELT**2)
      DO 70 J=1,N
      XM1(J)=XI(J)
70    XI(J)=XP1(J)
      DO 80 J=1,N
      X(I+1,J)=XP1(J)
      XD(I+1,J)=XDI(J)
80    XDD(I+1,J)=XDDI(J)
90    CONTINUE
      RETURN
      END
C ==========================================================
C
C
C SUBROUTINE EXTFUN
C THIS SUBROUTINE IS PROBLEM-DEPENDENT
C
C ==========================================================
      SUBROUTINE EXTFUN (F,TIME,N)
      DIMENSION F(N)
      F(1)=0.0
      F(2)=10.0
      RETURN
      END
C ==========================================================
C
C SUBROUTINE XMULT
C
C ==========================================================
      SUBROUTINE XMULT (A,B,BB,N)
      DIMENSION A(N,N),B(N),BB(N)
      DO 10 I=1,N
      BB(I)=0.0
      DO 10 J=1,N
10    BB(I)=BB(I)+A(I,J)*B(J)
      RETURN
      END
```

```
SOLUTION BY CENTRAL DIFFERENCE METHOD
GIVEN DATA:
N=    2    NSTEP=   24    DELT= 0.24216267E+00
SOLUTION:
```

| STEP | TIME | X(I,1) | XD(I,1) | XDD(I,1) | X(I,2) | XD(I,2) | XDD(I,2) |
|---|---|---|---|---|---|---|---|
| 1 | 0.0000 | 0.0000E+00 | 0.0000E+00 | 0.0000E+00 | 0.0000E+00 | 0.0000E+00 | 0.5000E+01 |
| 2 | 0.2422 | 0.0000E+00 | 0.0000E+00 | 0.0000E+00 | 0.1466E+00 | 0.0000E+00 | 0.5000E+01 |
| 3 | 0.4843 | 0.1719E-01 | 0.3550E-01 | 0.2932E+00 | 0.5520E+00 | 0.1140E+01 | 0.4414E+01 |
| 4 | 0.7265 | 0.9309E-01 | 0.1922E+00 | 0.1001E+01 | 0.1122E+01 | 0.2014E+01 | 0.2809E+01 |
| 5 | 0.9687 | 0.2678E+00 | 0.5175E+00 | 0.1686E+01 | 0.1728E+01 | 0.2428E+01 | 0.6043E+00 |
| 6 | 1.2108 | 0.5510E+00 | 0.9455E+00 | 0.1849E+01 | 0.2237E+01 | 0.2302E+01 | -.1643E+01 |
| 7 | 1.4530 | 0.9027E+00 | 0.1311E+01 | 0.1168E+01 | 0.2547E+01 | 0.1692E+01 | -.3397E+01 |
| 8 | 1.6951 | 0.1235E+01 | 0.1413E+01 | -.3219E+00 | 0.2606E+01 | 0.7613E+00 | -.4285E+01 |
| 9 | 1.9373 | 0.1439E+01 | 0.1108E+01 | -.2201E+01 | 0.2419E+01 | -.2646E+00 | -.4188E+01 |
| 10 | 2.1795 | 0.1420E+01 | 0.3814E+00 | -.3797E+01 | 0.2042E+01 | -.1164E+01 | -.3236E+01 |
| 11 | 2.4216 | 0.1141E+01 | -.6155E+00 | -.4437E+01 | 0.1563E+01 | -.1767E+01 | -.1749E+01 |
| 12 | 2.6638 | 0.6437E+00 | -.1603E+01 | -.3720E+01 | 0.1077E+01 | -.1992E+01 | -.1110E+00 |
| 13 | 2.9060 | 0.4627E-01 | -.2260E+01 | -.1708E+01 | 0.6698E+00 | -.1844E+01 | 0.1335E+01 |
| 14 | 3.1481 | -.4889E+00 | -.2339E+01 | 0.1062E+01 | 0.4012E+00 | -.1396E+01 | 0.2367E+01 |
| 15 | 3.3903 | -.8050E+00 | -.1758E+01 | 0.3736E+01 | 0.3030E+00 | -.7575E+00 | 0.2906E+01 |
| 16 | 3.6324 | -.8023E+00 | -.6471E+00 | 0.5436E+01 | 0.3797E+00 | -.4438E-01 | 0.2983E+01 |
| 17 | 3.8746 | -.4728E+00 | 0.6859E+00 | 0.5573E+01 | 0.6135E+00 | 0.6412E+00 | 0.2679E+01 |
| 18 | 4.1168 | 0.9503E-01 | 0.1853E+01 | 0.4064E+01 | 0.9689E+00 | 0.1217E+01 | 0.2073E+01 |
| 19 | 4.3589 | 0.7431E+00 | 0.2510E+01 | 0.1368E+01 | 0.1396E+01 | 0.1615E+01 | 0.1219E+01 |
| 20 | 4.6011 | 0.1293E+01 | 0.2474E+01 | -.1667E+01 | 0.1832E+01 | 0.1782E+01 | 0.1598E+00 |
| 21 | 4.8433 | 0.1603E+01 | 0.1776E+01 | -.4096E+01 | 0.2208E+01 | 0.1676E+01 | -.1035E+01 |
| 22 | 5.0854 | 0.1608E+01 | 0.6502E+00 | -.5205E+01 | 0.2453E+01 | 0.1281E+01 | -.2227E+01 |
| 23 | 5.3276 | 0.1335E+01 | -.5545E+00 | -.4744E+01 | 0.2510E+01 | 0.6238E+00 | -.3202E+01 |
| 24 | 5.5697 | 0.8862E+00 | -.1491E+01 | -.2990E+01 | 0.2350E+01 | -.2124E+00 | -.3704E+01 |
| 25 | 5.8119 | 0.4013E+00 | -.1928E+01 | -.6176E+00 | 0.1984E+01 | -.1086E+01 | -.3513E+01 |

### 10.10.3  Houbolt Method

A Fortran subroutine HOBOLT implements the Houbolt method. The following arguments are used for this subroutine:

| | | |
|---|---|---|
| M | = | Mass matrix of size $N \times N$. Input data. |
| C | = | Damping matrix of size $N \times N$. Input data. |
| K | = | Stiffness matrix of size $N \times N$. Input data. |
| XI, XDI | = | Vectors of size $N$ each, containing the initial values of $x_i$ and $\dot{x}_i$. Input data. |
| XDDI | = | Vector of size $N$. |
| N | = | Order of the matrix $[m]$. Input data. |
| NSTEP | = | Number of time steps at which the solution is to be found. Input data. |
| DELT | = | Time increment between steps. Input data. |

F   =   Vector of size $N$, containing the values of the forcing function at any specified time.

R, RR, XM1, XM2, XP1, XMK, XMI, ZA, ZB, ZC, LA, S = Vectors of size $N$ each.

MC, MK, MCI, MMC, MI = Matrices of size $N \times N$ each.

X, XD, XDD = Matrices of size $NSTEP1 \times N$ each. $X(I,J) = x_j(t_i)$, $XD(I,J) = \dot{x}_j(t_i)$, $XDD(I,J) = \ddot{x}_j(t_i)$. Output.

NSTEP1 =   NSTEP + 1.

LB   =   Matrix of size $N \times 2$.

The program requires a user-supplied subroutine EXTFUN(F, TIME, N). This subroutine must evaluate the forcing functions $F(1), F(2), \ldots, F(N)$ at a given time, *TIME*.

For illustration, Example 10.5 is solved by the use of subroutine HOBOLT. The main program, subroutine HOBOLT, and the results of the program are given below.

```
C ====================================================================
C
C PROGRAM 21
C MAIN PROGRAM WHICH CALLS HOBOLT
C
C ====================================================================
C FOLLOWING 10 LINES CONTAIN PROBLEM-DEPENDENT DATA
      REAL M(2,2),K(2,2),MC(2,2),MK(2,2),MCI(2,2),MMC(2,2),MI(2,2)
      DIMENSION C(2,2),XI(2),XDI(2),XDDI(2),XM1(2),F(2),R(2),RR(2),
     2    XMK(2),XMI(2),XM2(2),XP1(2),ZA(2),ZB(2),ZC(2),LA(2),LB(2,2),
     3    S(2),X(25,2),XD(25,2),XDD(25,2)
      DATA N,NSTEP,NSTEP1,DELT/2,24,25,0.24216267/
      DATA XI/0.0,0.0/
      DATA XDI/0.0,0.0/
      DATA M/1.0,0.0,0.0,2.0/
      DATA C/0.0,0.0,0.0,0.0/
      DATA K/6.0,-2.0,-2.0,8.0/
C END OF PROBLEM-DEPENDENT DATA
      CALL HOBOLT (M,C,K,XI,XDI,XDDI,N,NSTEP,DELT,F,R,RR,XM1,XM2,XP1,
     2    MC,MK,MCI,XMK,MMC,XMI,ZA,ZB,ZC,LA,LB,S,X,XD,XDD,NSTEP1,MI)
      PRINT 10
10    FORMAT (//,27H SOLUTION BY HOUBOLT METHOD,/)
      PRINT 20, N,NSTEP,DELT
20    FORMAT (12H GIVEN DATA:,/,3H N=,I5,4X,7H NSTEP=,I5,4X,6H DELT=,
     2    E15.8,/)
      PRINT 30
30    FORMAT (10H SOLUTION:,//,5H STEP,3X,5H TIME,3X,7H X(I,1),3X,
     2    8H XD(I,1),2X,9H XDD(I,1),4X,7H X(I,2),3X,8H XD(I,2),2X,
     3    9H XDD(I,2),/)
      DO 40  I=1,NSTEP1
      TIME=REAL(I-1)*DELT
40    PRINT 50, I,TIME,X(I,1),XD(I,1),XDD(I,1),X(I,2),XD(I,2),XDD(I,2)
50    FORMAT (1X,I4,F8.4,6(1X,E10.4))
      STOP
      END
```

```
C   =================================================================
C
C  SUBROUTINE HOBOLT
C
C   =================================================================
      SUBROUTINE HOBOLT (M,C,K,XI,XDI,XDDI,N,NSTEP,DELT,F,R,RR,XM1,
     2   XM2,XP1,MC,MK,MCI,XMK,MMC,XMI,ZA,ZB,ZC,LA,LB,S,X,XD,XDD,NSTEP1
     3   ,MI)
      REAL M(N,N),K(N,N),MC(N,N),MK(N,N),MCI(N,N),MMC(N,N),MI(N,N)
      DIMENSION C(N,N),XI(N),XDI(N),XDDI(N),F(N),R(N),RR(N),XM1(N),
     2   XM2(N),XP1(N),XMK(N),XMI(N),ZA(N),ZB(N),ZC(N),X(NSTEP1,N),
     3   XD(NSTEP1,N),XDD(NSTEP1,N),LA(N),LB(N,2),S(N)
      DO 5 I=1,N
      DO 5 J=1,N
5     MI(I,J)=M(I,J)
      CALL SIMUL (MI,ZA,N,0,LA,LB,S)
      CALL EXTFUN (F,0.0,N)
      DO 20 I=1,N
      R(I)=F(I)
      DO 10 J=1,N
10    R(I)=R(I)-C(I,J)*XDI(J)-K(I,J)*XI(J)
20    CONTINUE
      CALL XMULT (MI,R,XDDI,N)
      DO 25 J=1,N
      X(1,J)=XI(J)
      XD(1,J)=XDI(J)
25    XDD(1,J)=XDDI(J)
      DO 30 I=1,N
30    XM1(I)=XI(I)-DELT*XDI(I)+(DELT**2)*XDDI(I)/2.0
      DO 40 I=1,N
      DO 40 J=1,N
      MC(I,J)=(M(I,J)/(DELT**2))+(C(I,J)/(2.0*DELT))
      MK(I,J)=K(I,J)-2.0*M(I,J)/(DELT**2)
40    MMC(I,J)=(M(I,J)/(DELT**2))-(C(I,J)/(2.0*DELT))
      DO 45 I=1,N
      DO 45 J=1,N
45    MCI(I,J)=MC(I,J)
      CALL SIMUL (MCI,ZA,N,0,LA,LB,S)
      TIME=-DELT
      DO 90 I=1,2
      CALL XMULT (MK,XI,XMK,N)
      CALL XMULT (MMC,XM1,XMI,N)
      TIME=TIME+DELT
      CALL EXTFUN (F,TIME,N)
      DO 50 J=1,N
50    RR(J)=F(J)-XMK(J)-XMI(J)
      CALL XMULT (MCI,RR,XP1,N)
      DO 60 J=1,N
      XDI(J)=(XP1(J)-XM1(J))/(2.0*DELT)
60    XDDI(J)=(XP1(J)-2.0*XI(J)+XM1(J))/(DELT**2)
      DO 70 J=1,N
      XM1(J)=XI(J)
70    XI(J)=XP1(J)
      DO 80 J=1,N
```

```
            X(I+1,J)=XP1(J)
            XD(I+1,J)=XDI(J)
     80     XDD(I+1,J)=XDDI(J)
     90     CONTINUE
            DO 160 II=3,NSTEP
            DO 110 I=1,N
            DO 110 J=1,N
            MC(I,J)=M(I,J)*2.0/(DELT**2)+11.0*C(I,J)/(6.0*DELT)+K(I,J)
            MK(I,J)=5.0*M(I,J)/(DELT**2)+3.0*C(I,J)/DELT
            MI(I,J)=4.0*M(I,J)/(DELT**2)+3.0*C(I,J)/(2.0*DELT)
    110     MMC(I,J)=M(I,J)/(DELT**2)+C(I,J)/(3.0*DELT)
            DO 120 I=1,N
            XI(I)=X(II,I)
            XM1(I)=X(II-1,I)
    120     XM2(I)=X(II-2,I)
            CALL XMULT (MK,XI,ZA,N)
            CALL XMULT (MI,XM1,ZB,N)
            CALL XMULT (MMC,XM2,ZC,N)
            TIME=REAL(II)*DELT
            CALL EXTFUN (F,TIME,N)
            DO 130 I=1,N
    130     R(I)=F(I)+ZA(I)-ZB(I)+ZC(I)
            DO 135 I=1,N
            DO 135 J=1,N
    135     MCI(I,J)=MC(I,J)
            CALL SIMUL (MCI,ZA,N,0,LA,LB,S)
            CALL XMULT (MCI,R,XP1,N)
            DO 140 I=1,N
            XDI(I)=(11.0*XP1(I)-18.0*XI(I)+9.0*XM1(I)-2.0*XM2(I))/(6.0*DELT)
    140     XDDI(I)=(2.0*XP1(I)-5.0*XI(I)+4.0*XM1(I)-XM2(I))/(DELT**2)
            DO 150 I=1,N
            X(II+1,I)=XP1(I)
            XD(II+1,I)=XDI(I)
    150     XDD(II+1,I)=XDDI(I)
    160     CONTINUE
            RETURN
            END
C =====================================================================
C
C SUBROUTINE EXTFUN
C
C =====================================================================
            SUBROUTINE EXTFUN (F,TIME,N)
            DIMENSION F(N)
            F(1)=0.0
            F(2)=10.0
            RETURN
            END
C =====================================================================
C
C SUBROUTINE XMULT
C
C =====================================================================
            SUBROUTINE XMULT (A,B,BB,N)
            DIMENSION A(N,N),B(N),BB(N)
```

```
          DO 10 I=1,N
          BB(I)=0.0
          DO 10 J=1,N
    10    BB(I)=BB(I)+A(I,J)*B(J)
          RETURN
          END
```

SOLUTION BY HOUBOLT METHOD

GIVEN DATA:
N=    2    NSTEP=    24    DELT= 0.24216267E+00

SOLUTION:

| STEP | TIME | X(I,1) | XD(I,1) | XDD(I,1) | X(I,2) | XD(I,2) | XDD(I,2) |
|---|---|---|---|---|---|---|---|
| 1 | 0.0000 | 0.0000E+00 | 0.0000E+00 | 0.0000E+00 | 0.0000E+00 | 0.0000E+00 | 0.5000E+01 |
| 2 | 0.2422 | 0.0000E+00 | 0.0000E+00 | 0.0000E+00 | 0.1466E+00 | 0.0000E+00 | 0.5000E+01 |
| 3 | 0.4843 | 0.1719E-01 | 0.3550E-01 | 0.2932E+00 | 0.5520E+00 | 0.1140E+01 | 0.4414E+01 |
| 4 | 0.7265 | 0.9173E-01 | 0.4815E+00 | 0.1662E+01 | 0.1106E+01 | 0.2446E+01 | 0.6661E+00 |
| 5 | 0.9687 | 0.2501E+00 | 0.8635E+00 | 0.1881E+01 | 0.1691E+01 | 0.2312E+01 | -.1513E+01 |
| 6 | 1.2108 | 0.4924E+00 | 0.1174E+01 | 0.1434E+01 | 0.2194E+01 | 0.1757E+01 | -.3284E+01 |
| 7 | 1.4530 | 0.7867E+00 | 0.1278E+01 | 0.3394E+00 | 0.2530E+01 | 0.9207E+00 | -.4332E+01 |
| 8 | 1.6951 | 0.1073E+01 | 0.1087E+01 | -.1143E+01 | 0.2649E+01 | -.2187E-01 | -.4522E+01 |
| 9 | 1.9373 | 0.1280E+01 | 0.5906E+00 | -.2591E+01 | 0.2545E+01 | -.8952E+00 | -.3901E+01 |
| 10 | 2.1795 | 0.1343E+01 | -.1264E+00 | -.3554E+01 | 0.2252E+01 | -.1555E+01 | -.2667E+01 |
| 11 | 2.4216 | 0.1226E+01 | -.9066E+00 | -.3690E+01 | 0.1832E+01 | -.1911E+01 | -.1104E+01 |
| 12 | 2.6638 | 0.9340E+00 | -.1558E+01 | -.2878E+01 | 0.1363E+01 | -.1934E+01 | 0.4820E+00 |
| 13 | 2.9060 | 0.5178E+00 | -.1906E+01 | -.1262E+01 | 0.9224E+00 | -.1652E+01 | 0.1828E+01 |
| 14 | 3.1481 | 0.6242E-01 | -.1845E+01 | 0.7795E+00 | 0.5770E+00 | -.1138E+01 | 0.2754E+01 |
| 15 | 3.3903 | -.3323E+00 | -.1367E+01 | 0.2739E+01 | 0.3725E+00 | -.4914E+00 | 0.3178E+01 |
| 16 | 3.6324 | -.5761E+00 | -.5704E+00 | 0.4115E+01 | 0.3293E+00 | 0.1839E+00 | 0.3107E+01 |
| 17 | 3.8746 | -.6108E+00 | 0.3677E+00 | 0.4552E+01 | 0.4436E+00 | 0.7914E+00 | 0.2615E+01 |
| 18 | 4.1168 | -.4258E+00 | 0.1233E+01 | 0.3934E+01 | 0.6899E+00 | 0.1254E+01 | 0.1815E+01 |
| 19 | 4.3589 | -.6001E-01 | 0.1829E+01 | 0.2413E+01 | 0.1026E+01 | 0.1520E+01 | 0.8341E+00 |
| 20 | 4.6011 | 0.4068E+00 | 0.2026E+01 | 0.3642E+00 | 0.1402E+01 | 0.1563E+01 | -.2027E+00 |
| 21 | 4.8433 | 0.8737E+00 | 0.1790E+01 | -.1716E+01 | 0.1763E+01 | 0.1385E+01 | -.1179E+01 |
| 22 | 5.0854 | 0.1243E+01 | 0.1187E+01 | -.3340E+01 | 0.2058E+01 | 0.1013E+01 | -.1991E+01 |
| 23 | 5.3276 | 0.1441E+01 | 0.3662E+00 | -.4155E+01 | 0.2246E+01 | 0.4955E+00 | -.2543E+01 |
| 24 | 5.5697 | 0.1436E+01 | -.4846E+00 | -.4020E+01 | 0.2299E+01 | -.9674E-01 | -.2760E+01 |
| 25 | 5.8119 | 0.1241E+01 | -.1182E+01 | -.3029E+01 | 0.2209E+01 | -.6813E+00 | -.2593E+01 |

## REFERENCES

**10.1.** G. L. Goudreau and R. L. Taylor, "Evaluation of numerical integration methods in elastodynamics," *Computational Methods in Applied Mechanics and Engineering*, Vol. 2, 1973, pp. 69–97.

**10.2.** S. W. Key, "Transient response by time integration: Review of implicit and explicit operators," in J. Donéa (Ed.), *Advanced Structural Dynamics*, Applied Science Publishers, London, 1980.

**10.3.** R. E. Cornwell, R. R. Craig, Jr., and C. P. Johnson, "On the application of the mode-acceleration method to structural engineering problems," *Earthquake Engineering and Structural Dynamics*, Vol. 11, 1983, pp. 679–688.

**10.4.** E. D. Denman, "A numerical method for coupled differential equations," *International Journal for Numerical Methods in Engineering*, Vol. 4, 1972, pp. 587–596.

**10.5.** A. Jennings and D. R. L. Orr, "Application of the simultaneous iteration method to undamped vibration problems," *International Journal for Numerical Methods in Engineering*, Vol. 3, 1971, pp. 13–24.

**10.6.** S. J. Farlow, "A computer program to find analytical solutions of second order linear differential equations," *International Journal for Numerical Methods in Engineering*, Vol. 6, 1973, pp. 603–606.

**10.7.** T. Wah and L. R. Colcote, *Structural Analysis by Finite Difference Calculus*, Van Nostrand Reinhold, New York, 1970.

**10.8.** R. Ali, "Finite difference methods in vibration analysis," *Shock and Vibration Digest*, Vol. 15, March 1983, pp. 3–7.

**10.9.** M. Reali, R. Rangogni, and V. Pennati, "Compact analytic expressions of two-dimensional finite difference forms," *International Journal for Numerical Methods in Engineering*, Vol. 20, 1984, pp. 121–130.

**10.10.** P. C. M. Lau, "Finite difference approximation for ordinary derivatives," *International Journal for Numerical Methods in Engineering*, Vol. 17, 1981, pp. 663–678.

**10.11.** R. D. Krieg, "Unconditional stability in numerical time integration methods," *Journal of Applied Mechanics*, Vol. 40, 1973, pp. 417–421.

**10.12.** T. Belytschko and R. Mullen, "Stability of explicit-implicit mesh partitions in time integration," *International Journal for Numerical Methods in Engineering*, Vol. 12, 1978, pp. 1575–1586.

**10.13.** S. Levy and W. D. Kroll, "Errors introduced by finite space and time increments in dynamic response computation," *Proceedings, First U.S. National Congress of Applied Mechanics*, 1951, pp. 1–8.

**10.14.** S. Nakamura, *Computational Methods in Engineering and Science*, John Wiley, New York, 1977.

**10.15.** T. Belytschko, "Explicit time integration of structure-mechanical systems," in J. Donéa (Ed.), *Advanced Structural Dynamics*, Applied Science Publishers, London, 1980, pp. 97–122.

**10.16.** S. Levy and J. P. D. Wilkinson, *The Component Element Method in Dynamics with Application to Earthquake and Vehicle Engineering*, McGraw-Hill, New York, 1976.

**10.17.** J. W. Leech, P. T. Hsu, and E. W. Mack, "Stability of a finite-difference method for solving matrix equations," *AIAA Journal*, Vol. 3, 1965, pp. 2172–2173.

**10.18.** S. D. Conte and C. W. DeBoor, *Elementary Numerical Analysis: An Algorithmic Approach* (2nd Ed.), McGraw-Hill, New York, 1972.

**10.19.** C. F. Gerald and P. O. Wheatley, *Applied Numerical Analysis* (3d Ed.), Addison-Wesley, Reading, Mass., 1984.

**10.20.** L. V. Atkinson and P. J. Harley, *Introduction to Numerical Methods with PASCAL*, Addison-Wesley, Reading, Mass., 1984.

**10.21.** J. C. Houbolt, "A recurrence matrix solution for the dynamic response of elastic aircraft," *Journal of Aeronautical Sciences*, Vol. 17, 1950, pp. 540–550, 594.

**10.22.** E. L. Wilson, I. Farhoomand, and K. J. Bathe, "Nonlinear dynamic analysis of complex structures," *International Journal of Earthquake Engineering and Structural Dynamics*, Vol. 1, 1973, pp. 241–252.

**10.23.** S. P. Timoshenko, D. H. Young, and W. Weaver, Jr., *Vibration Problems in Engineering* (4th Ed.), John Wiley, New York, 1974.

**10.24.** N. M. Newmark, "A method of computation for structural dynamics," *ASCE Journal of Engineering Mechanics Division*, Vol. 85, 1959, pp. 67–94.

**10.25.** T. J. R. Hughes, "A note on the stability of Newmark's algorithm in nonlinear structural dynamics," *International Journal for Numerical Methods in Engineering*, Vol. 11, 1976, pp. 383–386.

## REVIEW QUESTIONS

**10.1.** Describe the procedure of the finite difference method.

**10.2.** Derive the central difference formulas for the first and the second derivatives of a function, using Taylor's series expansion.

**10.3.** What is a conditionally stable method?

**10.4.** What is the main difference between the central difference method and the Runge-Kutta method?

**10.5.** Why is it necessary to introduce fictitious mesh points in the finite difference method of solution?

**10.6.** Define a tridiagonal matrix.

**10.7.** What is the basic assumption of the Wilson method?

**10.8.** What is a linear acceleration method?

**10.9.** What is the difference between explicit and implicit integration methods?

**10.10.** Can we use the numerical integration methods discussed in this chapter to solve nonlinear vibration problems?

## PROBLEMS

The problem assignments are organized as follows:

| Problems | Section covered | Topic covered |
|---|---|---|
| 10.1–10.3 | 10.2 | Finite difference approach |
| 10.4–10.10 | 10.3 | Central difference method for single degree of freedom systems |
| 10.11–10.17, 10.21 | 10.4 | Runge-Kutta method |
| 10.18–10.20 | 10.5 | Central difference method for multidegree of freedom systems |
| 10.22–10.27 | 10.6 | Central difference method for continuous systems |
| 10.28–10.30 | 10.7, 10.10 | Houbolt method |
| 10.31–10.34 | 10.8, 10.10 | Wilson method |
| 10.35–10.38 | 10.9, 10.10 | Newmark method |

**10.1.** The forward difference formulas make use of the values of the function to the right of the base grid point. Thus the first derivative at point $i(t = t_i)$ is defined as

$$\frac{dx}{dt} = \frac{x(t + \Delta t) - x(t)}{\Delta t} = \frac{x_{i+1} - x_i}{\Delta t}$$

Derive the forward difference formulas for $(d^2x)/(dt^2)$, $(d^3x)/(dt^3)$, and $(d^4x)/(dt^4)$ at $t_i$.

**10.2.** The backward difference formulas make use of the values of the function to the left of the base grid point. Accordingly, the first derivative at point $i(t = t_i)$ is defined as

$$\frac{dx}{dt} = \frac{x(t) - x(t - \Delta t)}{\Delta t} = \frac{x_i - x_{i-1}}{\Delta t}$$

Derive the backward difference formulas for $(d^2x)/(dt^2)$, $(d^3x)/(dt^3)$, and $(d^4x)/(dt^4)$ at $t_i$.

**10.3.** Derive the formula for the fourth derivative, $(d^4x)/(dt^4)$, according to the central difference method.

**10.4.** Find the free vibratory response of an undamped single degree of freedom system with $m = 1$ and $k = 1$, using the central difference method. Assume $x_0 = 0$ and $\dot{x}_0 = 1$. Compare the results obtained with $\Delta t = 1$ and $\Delta t = 0.5$ with the exact solution $x(t) = \sin t$.

**10.5.** Integrate the differential equation

$$-\frac{d^2x}{dt^2} + 0.1x = 0 \quad \text{for} \quad 0 \le t \le 10$$

using the backward difference formula with $\Delta t = 1$. Assume the initial conditions as $x_0 = 1$ and $\dot{x}_0 = 0$.

**10.6.** Find the free vibration response of a viscously damped single degree of freedom system with $m = k = c = 1$, using the central difference method. Assume that $x_0 = 0$, $\dot{x}_0 = 1$, and $\Delta t = 0.5$.

**10.7.** Solve Problem 10.6 by changing $c$ to 2.

**10.8.** Solve Problem 10.6 by taking the value of $c$ as 4.

**10.9.** Find the solution of the equation $4\ddot{x} + 2\dot{x} + 3000x = F(t)$, where $F(t)$ is as shown in Fig. 10.11 for the duration $0 \le t \le 1$. Assume that $x_0 = \dot{x}_0 = 0$ and $\Delta t = 0.05$.

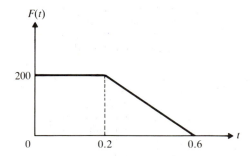

**Figure 10.11**

**10.10.** Find the solution of a spring-mass-damper system governed by the equation $m\ddot{x} + c\dot{x} + kx = F(t) = \delta F$. $t$ with $m = c = k = 1$ and $\delta F = 1$. Assume the initial values of $x$ and $\dot{x}$ to be zero and $\Delta t = 0.5$. Compare the central difference solution with the exact solution given in Example 4.6.

**10.11.** Express the following $n$th order differential equation as a system of $n$ first order differential equations:

$$a_n \frac{d^n x}{dt^n} + a_{n-1} \frac{d^{n-1} x}{dt^{n-1}} + \cdots + a_1 \frac{dx}{dt} = g(x, t)$$

**10.12.** Find the solution of the following equations by using the fourth order Runge-Kutta method with $\Delta t = 0.1$:
(a) $\dot{x} = x - 1.5 e^{-0.5t}$; $x_0 = 1$
(b) $\dot{x} = -tx^2$; $x_0 = 1$.

**10.13.** The second order Runge-Kutta formula is given by

$$\vec{X}_{i+1} = \vec{X}_i + \frac{1}{2}\left(\vec{K}_1 + \vec{K}_2\right)$$

where

$$\vec{K}_1 = h\vec{F}\left(\vec{X}_i, t_i\right) \qquad \text{and} \qquad \vec{K}_2 = h\vec{F}\left(\vec{X}_i + \vec{K}_1, t_i + h\right)$$

Using this formula, solve the problem considered in Example 10.2.

**10.14.** The third order Runge-Kutta formula is given by

$$\vec{X}_{i+1} = \vec{X}_i + \frac{1}{6}\left(\vec{K}_1 + 4\vec{K}_2 + \vec{K}_3\right)$$

where

$$\vec{K}_1 = h\vec{F}\left(\vec{X}_i, t_i\right)$$

$$\vec{K}_2 = h\vec{F}\left(\vec{X}_i + \frac{1}{2}\vec{K}_1, \quad t_i + \frac{1}{2}h\right)$$

and

$$\vec{K}_3 = h\vec{F}\left(\vec{X}_i - \vec{K}_1 + 2\vec{K}_2, \quad t_i + h\right)$$

Using this formula, solve the problem considered in Example 10.2.

**10.15.** Solve the differential equation $\ddot{x} + 1000x = 0$ with the initial conditions $x_0 = 5$ and $\dot{x}_0 = 0$, using the second order Runge-Kutta method. Use $\Delta t = 0.01$.

**10.16.** Solve Problem 10.15 using the third order Runge-Kutta method.

**10.17.** Solve Problem 10.15 using the fourth order Runge-Kutta method.

**10.18.** Find the response of the two degree of freedom system shown in Fig. 10.2 when $c = 2$, $F_1(t) = 0$, $F_2(t) = 10$, using the central difference method.

**10.19.** Find the response of the system shown in Fig. 10.2 when $F_1(t) = 10\sin 5t$ and $F_2(t) = 0$, using the central difference method.

**10.20.** The equations of motion of a two degree of freedom system are given by $2\ddot{x}_1 + 6x_1 - 2x_2 = 5$ and $\ddot{x}_2 - 2x_1 + 4x_2 = 20\sin 5t$. Assuming the initial conditions as $x_1(0) = \dot{x}_1(0) = x_2(0) = \dot{x}_2(0) = 0$, find the response of the system, using the central difference method with $\Delta t = 0.25$.

**10.21.** Solve Problem 10.20 using the fourth order Runge-Kutta method.

**10.22.** Find the natural frequencies of a fixed-fixed bar undergoing longitudinal vibration, using 3 mesh points in the range $0 < x < l$.

**10.23.** Derive the finite difference equations governing the forced longitudinal vibration of a fixed-free uniform bar, using a total of $n$ mesh points. Find the natural frequencies of the bar, using $n = 4$.

**10.24.** Derive the finite difference equations for the forced vibration of a fixed-fixed uniform shaft under torsion, using a total of $n$ mesh points.

**10.25.** Find the first three natural frequencies of a uniform fixed-fixed beam.

**10.26.** Derive the finite difference equations for the forced vibration of a cantilever beam subjected to a transverse force $f(x, t) = f_0 \cos \omega t$ at the free end.

**10.27.** Derive the finite difference equations for the forced vibration analysis of a rectangular membrane, using $m$ and $n$ mesh points in the $x$ and $y$ directions, respectively. Assume the membrane to be fixed along all the edges. Use the central difference formula.

**10.28.** Solve Problem 10.18 using the Houbolt method.

**10.29.** Solve Problem 10.19 using the Houbolt method.

**10.30.** Solve Problem 10.20 using the Houbolt method.

**10.31.** Solve Problem 10.18 using the Wilson method, with $\theta = 1.4$.

**10.32.** Solve Problem 10.19 using the Wilson method with $\theta = 1.4$.

**10.33.** Solve Problem 10.20 using the Wilson method with $\theta = 1.4$.

**10.34.** Write a subroutine WILSON for implementing the Wilson method. Use this program to find the solution of Example 10.6.

**10.35.** Solve Problem 10.18 using the Newmark method, with $\alpha = \frac{1}{6}$ and $\beta = \frac{1}{2}$.

**10.36.** Solve Problem 10.19 using the Newmark method, with $\alpha = \frac{1}{6}$ and $\beta = \frac{1}{2}$.

**10.37.** Solve Problem 10.20 using the Newmark method, with $\alpha = \frac{1}{6}$ and $\beta = \frac{1}{2}$.

**10.38.** Write a subroutine NUMARK for implementing the Newmark method. Use this subroutine to find the solution of Example 10.7.

# CHAPTER 11

# Further Topics in Vibration

Jules Henri Poincaré (1854–1912) was a French mathematician and professor of celestial mechanics at the University of Paris and of mechanics at the École Polytechnique. His contributions to pure and applied mathematics, particularly to celestial mechanics and electrodynamics, are outstanding. His classification of singular points of nonlinear autonomous systems is important in the study of nonlinear vibrations.

Courtesy Culver Pictures

## 11.1  OVERVIEW OF THE TOPICS

In the earlier chapters, small displacements were assumed in considering linear vibration problems. Whenever finite amplitudes of motion are encountered, nonlinear analysis becomes necessary. The superposition principle, which is very useful in linear analysis, does not hold true in the case of nonlinear analysis. Some elementary methods for nonlinear vibration analysis are given in the first part of this chapter (Secs. 11.2–11.8).

The analysis methods considered so far fall into the category of *deterministic vibration*, which implies that the system parameters, forcing function, and response characteristics are completely known. There are many practical vibration problems in which the excitation of the system and hence the response cannot be predicted with precision. An introduction to random vibration analysis is given in the second part of this chapter (Secs. 11.9–11.20).

The finite element method is a numerical method that can be used for the accurate solution of complex mechanical and structural vibration problems. In this method, the actual structure is replaced by several pieces or elements, in each of which a simple form of solution is assumed. The idea is that if the solutions of the various elements are selected properly, they can be made to converge to the exact solution of the total structure as the element size is reduced. An introduction to finite element analysis is given in the final part of the chapter (Secs. 11.21–11.26).

## 11.2  INTRODUCTION TO NONLINEAR VIBRATION

In the preceding chapters, the equation of motion contained displacement or its derivatives only to the first degree, and no square or higher powers of displacement or velocity were involved. For this reason, the governing differential equations of motion and the corresponding systems were called *linear*. For convenience of analysis, most systems are modeled as linear systems, but real systems are actually more often nonlinear than linear [11.1–11.4]. Since mass, damper, and spring are the basic components of a vibratory system, nonlinearity may be introduced into the system by any of these components. In many cases, linear analysis is insufficient to describe the behavior of the physical system adequately. One of the main reasons for modeling a physical system as a nonlinear one is that totally unexpected phenomena sometimes occur in nonlinear systems—phenomena that are not predicted or even hinted at by linear theory.

## 11.3  EXAMPLES OF NONLINEAR VIBRATION PROBLEMS

The following examples are given to illustrate the origin of nonlinearity in some physical systems.

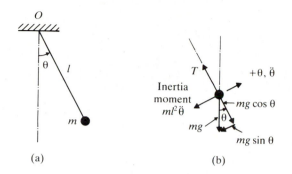

**Figure 11.1**

### 11.3.1 Simple Pendulum

For nonlinearity in the elastic component, consider a simple pendulum of length $l$, having a bob of mass $m$, as shown in Fig. 11.1(a). The differential equation governing the free vibration of the pendulum can be derived from Fig. 11.1(b):

$$ml^2\ddot{\theta} + mgl \sin\theta = 0 \tag{11.1}$$

By replacing $\sin\theta$ by its approximate value $\theta - \theta^3/6$, we can write this equation as

$$ml^2\ddot{\theta} + mgl\left(\theta - \frac{\theta^3}{6}\right) = 0 \tag{11.2}$$

It can be seen that Eq. (11.2) is nonlinear because of the term involving $\theta^3$. If the oscillations are assumed to be small about the position of equilibrium, we can disregard the term involving $\theta^3$, and Eq. (11.2) can be reduced to a linear equation:

$$ml^2\ddot{\theta} + mgl\theta = 0 \tag{11.3}$$

### 11.3.2 Mechanical Chatter, Belt Friction System

Nonlinearity in the damping member is illustrated by Fig. 11.2(a). The system behaves nonlinearly because of the dry friction between the mass $m$ and the moving belt. For this system, there are two friction coefficients: the static coefficient of friction ($\mu_s$), corresponding to the force required to initiate the motion of the body held by dry friction; and the kinetic coefficient of friction ($\mu_k$), corresponding to the force required to maintain the body in motion. In either case, the component of the applied force tangent to the friction surface ($F$) is the product of the appropriate friction coefficient and the force normal to the surface.

The sequence of motion of the system shown in Fig. 11.2(a) is as follows. The mass is initially at rest on the belt. Due to the displacement of the mass $m$ along with the belt, the spring elongates. As the spring extends, the spring force on the mass increases until the static friction force is overcome and the mass begins to slide. It slides rapidly towards the right, thereby relieving the spring force until the kinetic friction force halts it. The spring then begins to build up the spring force again. The variation of the damping force with the velocity of the mass is shown in Fig. 11.2(b).

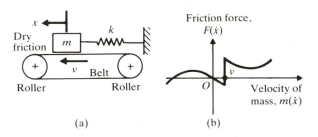

**Figure 11.2**

The equation of motion can be expressed as

$$m\ddot{x} + F(\dot{x}) + kx = 0 \tag{11.4}$$

where the friction force $F$ is a nonlinear function of $\dot{x}$, as shown in Fig. 11.2(b).

For large values of $\dot{x}$, the damping force is positive (the curve has a positive slope) and energy is removed from the system. On the other hand, for small values of $\dot{x}$, the damping force is negative (the curve has a negative slope) and energy is put into the system. Although there is no external stimulus, the system can have an oscillatory motion; it corresponds to a nonlinear self-excited system. This phenomenon of self-excited vibration is called *mechanical chatter*.

### 11.3.3 Variable Mass System

Nonlinearity in the mass element is illustrated by Fig. 11.3. The mass of the system depends on the displacement $x$, and so the equation of motion becomes

$$m(x)\ddot{x} + kx = 0 \tag{11.5}$$

Note that this is a nonlinear differential equation due to the nonlinearity of the first term.

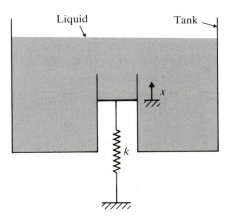

**Figure 11.3**

## 11.4 EXACT METHODS FOR NONLINEAR SYSTEMS

An exact solution is possible only for a relatively few nonlinear systems whose motion is governed by specific types of second order nonlinear differential equations. The solutions are exact in the sense that they are given either in closed form or in the form of an expression that can be numerically evaluated to any degree of accuracy. In this section, we shall consider a simple nonlinear system for which the exact solution is available. For a single degree of freedom system with a general restoring (spring) force $F(x)$, the free vibration equation can be expressed as

$$\ddot{x} + a^2 F(x) = 0 \tag{11.6}$$

where $a^2$ is a constant. Equation (11.6) can be rewritten as

$$\frac{d}{dx}(\dot{x}^2) + 2a^2 F(x) = 0 \tag{11.7}$$

Integration of Eq. (11.7) yields

$$\dot{x}^2 = 2a^2 \int_x^{x_0} F(\eta)\, d\eta \quad \text{or} \quad |\dot{x}| = \sqrt{2}\, a \left\{ \int_x^{x_0} F(\eta)\, d\eta \right\}^{1/2} \tag{11.8}$$

where $\eta$ is the integration variable and $x_0$ is the value of $x$ when $\dot{x} = (dx)/(dt) = 0$. Equation (11.8), when integrated again, gives

$$t - t_0 = \frac{1}{\sqrt{2}\, a} \int_0^x \frac{d\xi}{\left\{ \int_\xi^{x_0} F(\eta)\, d\eta \right\}^{1/2}} \tag{11.9}$$

where $\xi$ is the new integration variable and $t_0$ corresponds to the time when $x = 0$. Equation (11.9) thus gives the exact solution of Eq. (11.6) in all those situations where the integrals of Eq. (11.9) can be evaluated in closed form. After evaluating the integrals of Eq. (11.9), one can invert the result and obtain the displacement-time relation. If $F(x)$ is an odd function,

$$F(-x) = -F(x) \tag{11.10}$$

By considering Eq. (11.9) from zero displacement to maximum displacement, the period of vibration $\tau$ can be obtained:

$$\tau = \frac{4}{\sqrt{2}\, a} \int_0^{x_0} \frac{d\xi}{\left\{ \int_\xi^{x_0} F(\eta)\, d\eta \right\}^{1/2}} \tag{11.11}$$

For illustration, let $F(x) = x^n$. In this case Eqs. (11.9) and (11.11) become

$$t - t_0 = \frac{1}{a}\sqrt{\frac{n+1}{2}} \int_0^{x_0} \frac{d\xi}{\left(x_0^{n+1} - \xi^{n+1}\right)^{1/2}} \tag{11.12}$$

and

$$\tau = \frac{4}{a}\sqrt{\frac{n+1}{2}} \int_0^{x_0} \frac{d\xi}{\left(x_0^{n+1} - \xi^{n+1}\right)^{1/2}} \tag{11.13}$$

By setting $y = \xi/x_0$, Eq. (11.13) can be written as

$$\tau = \frac{4}{a} \frac{1}{\left(x_0^{n-1}\right)^{1/2}} \sqrt{\frac{n+1}{2}} \int_0^1 \frac{dy}{\left(1 - y^{n+1}\right)^{1/2}} \tag{11.14}$$

This expression can be evaluated numerically to any desired level of accuracy.

## 11.5  APPROXIMATE ANALYTICAL METHODS FOR NONLINEAR SYSTEMS

In the absence of an exact analytical solution to a nonlinear vibration problem, we wish to find at least an approximate solution. Both analytical and numerical methods are available for approximate solution of nonlinear vibration problems [11.5, 11.6]. We shall now consider two analytical techniques: the iterative and the Ritz averaging methods.

### 11.5.1   Iterative Method

We shall illustrate the iterative method of finding the periodic solutions of Duffing's equation, which represents the equation of motion of a damped, harmonically excited, single degree of freedom system with a nonlinear spring. First we consider the solution of the undamped equation.

**Solution of the Undamped Equation.** If damping is disregarded, Duffing's equation becomes

$$\ddot{x} + \alpha x \pm \beta x^3 = F\cos \omega t$$

or

$$\ddot{x} = -\alpha x \mp \beta x^3 + F\cos \omega t \tag{11.15}$$

As a first approximation, we assume the solution to be

$$x_1(t) = A\cos \omega t \tag{11.16}$$

where $A$ is an unknown. By substituting Eq. (11.16) into Eq. (11.15), we obtain the differential equation for the second approximation:

$$\ddot{x}_2 = -A\alpha \cos \omega t \mp A^3 \beta \cos^3 \omega t + F\cos \omega t \tag{11.17}$$

By using the identity

$$\cos^3 \omega t = \tfrac{3}{4}\cos \omega t + \tfrac{1}{4}\cos 3\omega t \tag{11.18}$$

Eq. (11.17) can be expressed as

$$\ddot{x}_2 = -\left(A\alpha \pm \tfrac{3}{4}A^3\beta - F\right)\cos \omega t \mp \tfrac{1}{4}A^3\beta \cos 3\omega t \tag{11.19}$$

By integrating this equation and setting the constants of integration to zero (so as to make the solution harmonic with period $\tau = 2\pi/\omega$) we obtain the second approximation:

$$x_2(t) = \frac{1}{\omega^2}\left(A\alpha \pm \tfrac{3}{4}A^3\beta - F\right)\cos \omega t \pm \frac{A^3\beta}{36\omega^2}\cos 3\omega t \tag{11.20}$$

Duffing [11.1] reasoned at this point that if $x_1(t)$ and $x_2(t)$ are good approximations to the solution $x(t)$, the coefficients of $\cos \omega t$ in the two equations (11.16) and (11.20) should not be very different. Thus by equating these coefficients, we obtain

$$A = \frac{1}{\omega^2}\left( A\alpha \pm \frac{3}{4}A^3\beta - F \right)$$

or

$$\omega^2 = \alpha \pm \frac{3}{4}A^2\beta - \frac{F}{A} \tag{11.21}$$

For present purposes, we will stop the procedure with the second approximation. It can be verified that this procedure yields the exact solution for the case of a linear spring (with $\beta = 0$):

$$A = \frac{F}{\alpha - \omega^2} \tag{11.22}$$

where $A$ denotes the amplitude of the harmonic response of the linear system.

For a nonlinear system (with $\beta \neq 0$), Eq. (11.21) shows that the frequency $\omega$ is a function of $\beta$, $A$, and $F$. Note that the quantity $A$, in the case of a nonlinear system, is not the amplitude of the harmonic response but only the coefficient of the first term of its solution. However, it is commonly taken as the amplitude of the harmonic response of the system. For the free vibration of the nonlinear system, $F = 0$ and Eq. (11.21) reduces to

$$\omega^2 = \alpha \pm \frac{3}{4}A^2\beta \tag{11.23}$$

This equation shows that the frequency of the response increases with the amplitude $A$ for the hardening spring and decreases for the softening spring.

For both linear and nonlinear systems, when $F \neq 0$ (forced vibration), there are two values of the frequency $\omega$ for any given amplitude $|A|$. One of these values of $\omega$ is smaller and the other larger than the corresponding frequency of free vibration at that amplitude. For the smaller value of $\omega$, $A > 0$ and the harmonic response of the system is in phase with the external force. For the larger value of $\omega$, $A < 0$ and the response is 180° out of phase with the external force. Note that only the harmonic solutions of Duffing's equation—that is, solutions for which the frequency is the same as that of the external force $F\cos \omega t$—have been considered in the present analysis. It has been observed [11.2] that oscillations whose frequency is a fraction, such as $\frac{1}{2}, \frac{1}{3}, \ldots, \frac{1}{n}$, of that of the applied force, are also possible for Duffing's equation. Such oscillations are termed *subharmonic response* [11.7].

**Solution of the Damped Equation.** If we consider viscous damping, we obtain Duffing's equation:

$$\ddot{x} + c\dot{x} + \alpha x \pm \beta x^3 = F\cos \omega t \tag{11.24}$$

For a damped system, it was observed in earlier chapters that there is a phase difference between the applied force and the response or solution. The usual procedure is to prescribe the applied force first and then determine the phase of the solution. In the present case, however, it is more convenient to fix the phase of the solution and keep the phase of the applied force as a quantity to be determined. We take the differential equation, Eq. (11.24), in the form

$$\ddot{x} + c\dot{x} + \alpha x \pm \beta x^3 = F\cos(\omega t + \phi)$$

$$= A_1\cos \omega t - A_2\sin \omega t \tag{11.25}$$

in which the amplitude $F = (A_1^2 + A_2^2)^{1/2}$ of the applied force is considered fixed, but the ratio $A_1/A_2 = \tan^{-1}\phi$ is left to be determined. We assume that $c$, $A_1$, and $A_2$ are all small, of order $\beta$. As with Eq. (11.15), we assume the first approximation to the solution to be

$$x_1 = A\cos \omega t \tag{11.26}$$

where $A$ is assumed fixed and $\omega$ to be determined. By substituting Eq. (11.26) into Eq. (11.25) and making use of the relation (11.18), we obtain

$$\left[(\alpha - \omega^2)A \pm \frac{3}{4}\beta A^3\right]\cos \omega t - c\omega A \sin \omega t \pm \frac{\beta A^3}{4}\cos 3\omega t = A_1\cos \omega t - A_2\sin \omega t$$

$$\tag{11.27}$$

By disregarding the term involving $\cos 3\omega t$ and equating the coefficients of $\cos \omega t$ and $\sin \omega t$ on both sides of Eq. (11.27), we obtain the following relations:

$$(\alpha - \omega^2)A \pm \frac{3}{4}\beta A^3 = A_1$$

$$c\omega A = A_2 \tag{11.28}$$

The relation between the amplitude of the applied force and the quantities $A$ and $\omega$ can be obtained by squaring and adding the equations (11.28):

$$\left[(\alpha - \omega^2)A \pm \frac{3}{4}\beta A^3\right]^2 + (c\omega A)^2 = A_1^2 + A_2^2 = F^2 \tag{11.29}$$

Equation (11.29) can be rewritten as

$$S^2(\omega, A) + c^2\omega^2 A^2 = F^2 \tag{11.30}$$

where

$$S(\omega, A) = (\alpha - \omega^2)A \pm \frac{3}{4}\beta A^3 \tag{11.31}$$

It can be seen that for $c = 0$, Eq. (11.30) reduces to $S(\omega, A) = F$, which is the same as Eq. (11.21). The response curves given by Eq. (11.30) are shown in Fig. 11.4.

**Jump Phenomenon.** As mentioned earlier, nonlinear systems exhibit phenomena that cannot occur in linear systems. For example, the amplitude of vibration of the system described by Eq. (11.25) has been found to increase or decrease suddenly as the excitation frequency $\omega$ is increased or decreased, as shown in Fig. 11.5. For a constant magnitude of $F$, the amplitude of vibration will increase along the points

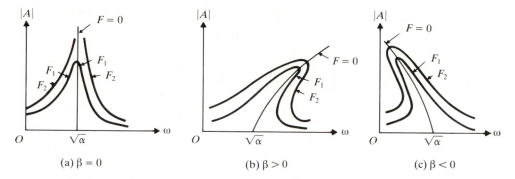

(a) $\beta = 0$                    (b) $\beta > 0$                    (c) $\beta < 0$

**Figure 11.4**   Response curves of Duffing's equation.

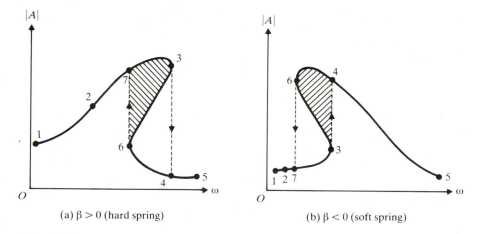

(a) $\beta > 0$ (hard spring)                    (b) $\beta < 0$ (soft spring)

**Figure 11.5**

$1, 2, 3, 4, 5$ on the curve when the excitation frequency $\omega$ is slowly increased. The amplitude of vibration jumps from point 3 to 4 on the curve. Similarly, when the forcing frequency $\omega$ is slowly decreased, the amplitude of vibration follows the curve along the points $5, 4, 6, 7, 2, 1$ and makes a jump from point 6 to 7. This behavior is known as the *jump phenomenon*. It is evident that there exist two amplitudes of vibration for a given forcing frequency, as shown in the shaded regions of the curves of Fig. 11.5. The shaded region can be thought of as unstable in some sense.

## 11.5.2   Ritz Averaging Method
In the Ritz averaging method, an approximate solution of the problem is found by satisfying the governing nonlinear equation in the average. To see how the method works, let the nonlinear differential equation be represented as

$$E(x) = 0 \qquad\qquad (11.32)$$

An approximate solution of Eq. (11.32) can be assumed:

$$\underset{\sim}{x}(t) = a_1\phi_1(t) + a_2\phi_2(t) + \cdots + a_n\phi_n(t) \tag{11.33}$$

where $\phi_1(t), \phi_2(t), \ldots, \phi_n(t)$ are prescribed functions of time and $a_1, a_2, \ldots, a_n$ are weighting factors to be determined. The weighting factors are found by substituting Eq. (11.33) into Eq. (11.32), multiplying the differential equation by each of the functions $\phi_i(t)$, and then integrating the product over a period of the motion. This leads to

$$\int_0^T E(\underset{\sim}{x})\phi_i(t)\, dt = 0, \qquad i = 1, 2, \ldots, n \tag{11.34}$$

Equation (11.34) represents a system of $n$ algebraic equations that can be solved simultaneously to find the values of $a_1, a_2, \ldots, a_n$.

## EXAMPLE 11.1

Find the solution of the equation

$$E(x) = \ddot{x} + \alpha^2 x + \beta x^3 = 0 \tag{E.1}$$

by the Ritz averaging method.

*Solution.* As a first approximation, let the solution of Eq. (E.1) be taken as

$$x(t) = a_1\phi_1(t) = a_1\cos\alpha_1 t \tag{E.2}$$

The application of Eq. (11.34) for a cycle ($\tau = 2\pi$) gives

$$\int_0^{2\pi}\left[ -\alpha_1^2 a_1\cos\alpha_1 t + \alpha^2 a_1\cos\alpha_1 t + \beta a_1^3\cos^3\alpha_1 t \right]\cos\alpha_1 t \cdot dt = 0 \tag{E.3}$$

Since

$$\int_0^{2\pi}\cos^2\alpha_1 t \cdot dt = \frac{\pi}{\alpha_1}$$

and

$$\int_0^{2\pi}\cos^4\alpha_1 t \cdot dt = \frac{3\pi}{4\alpha_1}$$

Eq. (E.3) becomes

$$-\alpha_1^2 a_1\left(\frac{\pi}{\alpha_1}\right) + \alpha^2 a_1\left(\frac{\pi}{\alpha_1}\right) + \beta a_1^3\left(\frac{3\pi}{4\alpha_1}\right) = 0 \tag{E.4}$$

Equation (E.4) gives

$$\alpha_1^2 = \alpha^2 + \frac{3}{4}\beta a_1^2$$

or

$$a_1 = 2\left(\frac{\alpha_1^2 - \alpha^2}{3\beta}\right)^{1/2} \tag{E.5}$$

Thus the solution of Eq. (E.1) can be written as

$$x(t) = 2\left(\frac{\alpha_1^2 - \alpha^2}{3\beta}\right)^{1/2} \cos \alpha_1 t \tag{E.6}$$

As a better approximation, a two-term solution satisfying the condition of symmetry can be taken:

$$x(t) = a_1\phi_1(t) + a_2\phi_2(t) \tag{E.7}$$

where $\phi_1(t) = \cos \alpha_1 t$ and $\phi_2(t) = \cos 3\alpha_1 t$. The application of Eq. (11.34) with the solution of Eq. (E.7) leads to two simultaneous cubic equations which must be numerically solved for $a_1$ and $a_2$.

Other numerical methods, such as the perturbation techniques, the equivalent linearization scheme, and the harmonic balance procedure, are also available for solving nonlinear vibration problems [11.8–11.10]. Specific solutions found using these techniques include the free vibration response of single degree of freedom oscillators [11.11, 11.12], two degree of freedom systems [11.13], and elastic beams [11.14, 11.15], and the transient response of forced systems [11.16, 11.17].

## 11.6   GRAPHICAL METHODS FOR NONLINEAR SYSTEMS

### 11.6.1   Phase Plane Representation

Graphical methods can be used to obtain qualitative information about the behavior of the nonlinear system and also to integrate the equations of motion. We shall first consider a basic concept known as the *phase plane*. For a single degree of freedom system, two parameters are needed to describe the state of motion completely. These parameters are usually taken as the displacement and velocity of the system. When the parameters are used as coordinate axes, the resulting graphical representation of the motion is called the *phase plane representation*. Thus each point in the phase plane represents a possible state of the system. As time changes, the state of the system changes. A typical or representative point in the phase plane (such as the point representing the state of the system at time $t = 0$) moves and traces a curve known as the *trajectory*. The trajectory shows how the solution of the system varies with time.

### EXAMPLE 11.2

Find the trajectories of a simple harmonic oscillator.

*Solution.* The equation of motion of an undamped linear system is given by

$$\ddot{x} + \omega_n^2 x = 0 \tag{E.1}$$

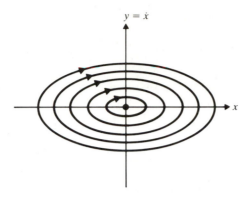

**Figure 11.6**

By setting $y = \dot{x}$, Eq. (E.1) can be written as

$$\frac{dy}{dt} = -\omega_n^2 x$$

$$\frac{dx}{dt} = y \tag{E.2}$$

from which we can obtain

$$\frac{dy}{dx} = -\frac{\omega_n^2 x}{y} \tag{E.3}$$

Integration of Eq. (E.3) leads to

$$y^2 + \omega_n^2 x^2 = c^2 \tag{E.4}$$

where $c$ is a constant. The value of $c$ is determined by the initial conditions of the system. Equation (E.4) shows that the trajectories of the system in the phase plane ($x$-$y$ plane) are a family of ellipses, as shown in Fig. 11.6. It can be observed that the point ($x = 0$, $y = 0$) is surrounded by closed trajectories. Such a point is called a *center*.

---

To see some of the characteristics of trajectories, consider a single degree of freedom nonlinear oscillatory system whose governing equation is of the form

$$\ddot{x} + f(x, \dot{x}) = 0 \tag{11.35}$$

By defining

$$\frac{dx}{dt} = \dot{x} = y \tag{11.36}$$

and

$$\frac{dy}{dt} = \dot{y} = -f(x, y) \tag{11.37}$$

we obtain

$$\frac{dy}{dx} = \frac{(dy/dt)}{(dx/dt)} = -\frac{f(x, y)}{y} = \phi(x, y), \text{ say.} \qquad (11.38)$$

Thus there is a unique slope of the trajectory at every point $(x, y)$ in the phase plane, provided that $\phi(x, y)$ is not indeterminate. If $y = 0$ and $f \neq 0$ (that is, if the point lies on the $x$ axis), the slope of the trajectory is infinite. This means that all trajectories must cross the $x$ axis at right angles. If $y = 0$ and $f = 0$, the point is called a *singular point*, and the slope is indeterminate at such points. A singular point corresponds to a state of equilibrium of the system—the velocity $y = \dot{x}$ and the force $\ddot{x} = -f$ are zero at a singular point. Further investigation is necessary to establish whether the equilibrium represented by a singular point is stable or unstable.

### 11.6.2   Phase Velocity
The velocity $\vec{v}$ with which a representative point moves along a trajectory is called the *phase velocity*. The components of phase velocity parallel to the $x$ and $y$ axes are

$$v_x = \dot{x}, \qquad v_y = \dot{y} \qquad (11.39)$$

and the magnitude of $\vec{v}$ is given by

$$|\vec{v}| = \sqrt{v_x^2 + v_y^2} = \sqrt{\left(\frac{dx}{dt}\right)^2 + \left(\frac{dy}{dt}\right)^2} \qquad (11.40)$$

We can note that if the system has a periodic motion, its trajectory in the phase plane is a closed curve. This follows from the fact that the representative point, having started its motion along a closed trajectory at an arbitrary point $(x, y)$, will return to the same point after one period. The time required to go around the closed trajectory (the period of oscillation of the system) is finite because the phase velocity is bounded away from zero at all points of the trajectory.

### 11.6.3   Method of Constructing Trajectories
We shall now consider a method known as the *method of isoclines* for constructing the trajectories of a dynamical system with one degree of freedom. By writing the equations of motion of the system as

$$\frac{dx}{dt} = f_1(x, y)$$

$$\frac{dy}{dt} = f_2(x, y) \qquad (11.41)$$

where $f_1$ and $f_2$ are nonlinear functions of $x$ and $y = \dot{x}$, the equation for the integral curves can be obtained as

$$\frac{dy}{dx} = \frac{f_2(x, y)}{f_1(x, y)} = \phi(x, y), \text{ say.} \qquad (11.42)$$

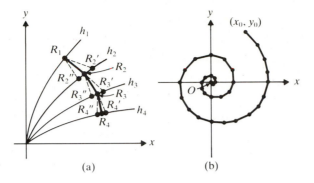

**Figure 11.7** Method of isoclines.

The curve

$$\phi(x, y) = c \tag{11.43}$$

for a fixed value of $c$ is called an *isocline*. An isocline can be defined as the locus of points at which the trajectories passing through them have the constant slope $c$. In the method of isoclines we fix the slope $(dy)/(dx)$ by giving it a definite number $c_1$ and solve Eq. (11.43) for the trajectory. The curve $\phi(x, y) - c_1 = 0$ thus represents an isocline in the phase plane. We plot several isoclines by giving different values $c_1, c_2, \dots$ to the slope $\phi = (dy)/(dx)$. Let $h_1, h_2, \dots$ denote these isoclines in Fig. 11.7(a). Suppose that we are interested in constructing the trajectory passing through the point $R_1$ on the isocline $h_1$. We draw two straight line segments from $R_1$: one with a slope $c_1$, meeting $h_2$ at $R_2'$, and the other with a slope $c_2$, meeting $h_2$ at $R_2''$. The middle point between $R_2'$ and $R_2''$ lying on $h_2$ is denoted as $R_2$. Starting at $R_2$, this construction is repeated, and the point $R_3$ is determined on $h_3$. This procedure is continued until the polygonal trajectory with sides $R_1R_2, R_2R_3, R_3R_4, \dots$ is taken as an approximation to the actual trajectory passing through the point $R_1$. Obviously, the larger the number of isoclines, the better is the approximation obtained by this graphical method. A typical final trajectory looks like that in Fig. 11.7(b).

## EXAMPLE 11.3

Construct the trajectories of a simple harmonic oscillator by the method of isoclines.

**Solution.** The differential equation defining the trajectories of a simple harmonic oscillator is given by Eq. (E.3) of Example 11.2. Hence the family of isoclines is given by

$$c = -\frac{\omega_n^2 x}{y} \qquad \text{or} \qquad y = \frac{-\omega_n^2}{c} x \tag{E.1}$$

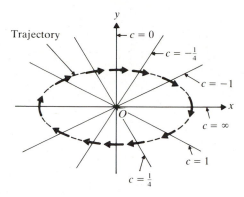

**Figure 11.8** Isoclines of a simple harmonic oscillator.

This equation represents a family of straight lines passing through the origin, with $c$ representing the slope of the trajectories on each isocline. The isoclines given by Eq. (E.1) are shown in Fig. 11.8. Once the isoclines are known, the trajectory can be plotted as indicated above.

### 11.6.4 Obtaining Time Solution from Phase Plane Trajectories

The trajectory plotted in the phase plane is a plot of $\dot{x}$ as a function of $x$, and time ($t$) does not appear explicitly in the plot. For the qualitative analysis of the system, the trajectories are enough, but in some cases we may need the variation of $x$ with time $t$. In such cases, it is possible to obtain the time solution $x(t)$ from the phase plane diagram, although the original differential equation cannot be solved for $x$ and $\dot{x}$ as functions of time. The method of obtaining a time solution is essentially a step-by-step procedure; several different schemes may be used for this purpose. In this section, we shall present a method based on the relation $\dot{x} = (\Delta x)/(\Delta t)$.

For small increments of displacement and time ($\Delta x$ and $\Delta t$), the average velocity can be taken as $\dot{x}_{\mathrm{av}} = (\Delta x)/(\Delta t)$, so that

$$\Delta t = \frac{\Delta x}{\dot{x}_{\mathrm{av}}} \tag{11.44}$$

In the phase plane trajectory shown in Fig. 11.9, the incremental time needed for the representative point to traverse the incremental displacement $\Delta x_{AB}$ is shown as $\Delta t_{AB}$. If $\dot{x}_{AB}$ denotes the average velocity during $\Delta t_{AB}$, we have $\Delta t_{AB} = \Delta x_{AB}/\dot{x}_{AB}$. Similarly, $\Delta t_{BC} = \Delta x_{BC}/\dot{x}_{BC}$, etc. Once $\Delta t_{AB}, \Delta t_{BC}, \ldots$ are known, the time solution $x(t)$ can be plotted easily, as shown in Fig. 11.9(b). It is evident that for good accuracy, the incremental displacements $\Delta x_{AB}, \Delta x_{BC}, \ldots$ must be chosen small enough that the corresponding incremental changes in $\dot{x}$ and $t$ are reasonably small. Note that $\Delta x$ need not be constant; it can be changed depending on the nature of the trajectories.

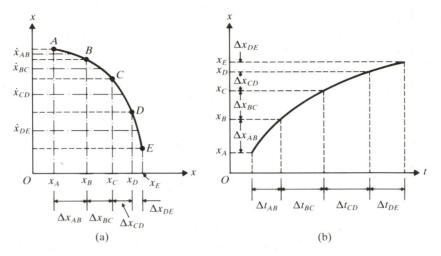

**Figure 11.9** Obtaining time solution from a phase plane plot.

## 11.7 STABILITY OF EQUILIBRIUM STATES IN NONLINEAR SYSTEMS

Consider a single degree of freedom nonlinear vibratory system described by two first order differential equations, Eq. (11.41). For this system, the slope of the trajectories in the phase plane is given by

$$\frac{dy}{dx} = \frac{\dot{y}}{\dot{x}} = \frac{f_2(x, y)}{f_1(x, y)} \tag{11.45}$$

Let $(x_0, y_0)$ be a singular point or an equilibrium point so that $(dy)/(dx)$ has the form $0/0$:

$$f_1(x_0, y_0) = f_2(x_0, y_0) = 0 \tag{11.46}$$

A study of Eqs. (11.41) in the neighborhood of the singular point provides us with answers as to the stability of equilibrium. We first note that there is no loss of generality if we assume that the singular point is located at the origin $(0,0)$. This is because the slope $(dy)/(dx)$ of the trajectories does not vary with a translation of the coordinate axes $x$ and $y$ to $x'$ and $y'$:

$$x' = x - x_0$$
$$y' = y - y_0 \tag{11.47}$$
$$\frac{dy}{dx} = \frac{dy'}{dx'}$$

Thus we assume that $(x = 0, y = 0)$ is a singular point, so that

$$f_1(0,0) = f_2(0,0) = 0 \tag{11.48}$$

If $f_1$ and $f_2$ are expanded in terms of Taylor's series about the singular point $(0,0)$,

we obtain

$$\dot{x} = f_1(x, y) = a_{11}x + a_{12}y + \text{higher order terms}$$

$$\dot{y} = f_2(x, y) = a_{21}x + a_{22}y + \text{higher order terms} \qquad (11.49)$$

where

$$a_{11} = \left.\frac{\partial f_1}{\partial x}\right|_{(0,0)}, \quad a_{12} = \left.\frac{\partial f_1}{\partial y}\right|_{(0,0)}, \quad a_{21} = \left.\frac{\partial f_2}{\partial x}\right|_{(0,0)}, \quad a_{22} = \left.\frac{\partial f_2}{\partial y}\right|_{(0,0)} \qquad (11.50)$$

In the neighborhood of $(0,0)$, $x$ and $y$ are small; $f_1$ and $f_2$ can be approximated by linear terms only, so that Eqs. (11.49) can be written as

$$\begin{Bmatrix} \dot{x} \\ \dot{y} \end{Bmatrix} = \begin{bmatrix} a_{11} & a_{12} \\ a_{21} & a_{22} \end{bmatrix} \begin{Bmatrix} x \\ y \end{Bmatrix} \qquad (11.51)$$

Let $\lambda_1$ and $\lambda_2$ be the eigenvalues and $\begin{Bmatrix} x_1 \\ y_1 \end{Bmatrix}$ and $\begin{Bmatrix} x_2 \\ y_2 \end{Bmatrix}$ be the corresponding

eigenvectors of the matrix $\begin{bmatrix} a_{11} & a_{12} \\ a_{21} & a_{22} \end{bmatrix}$. Then the transformation

$$\begin{Bmatrix} x \\ y \end{Bmatrix} = \begin{bmatrix} x_1 & x_2 \\ y_1 & y_2 \end{bmatrix} \begin{Bmatrix} \alpha \\ \beta \end{Bmatrix} \equiv [T] \begin{Bmatrix} \alpha \\ \beta \end{Bmatrix} \qquad (11.52)$$

where $[T]$ is the modal matrix and $\alpha$ and $\beta$ are the generalized coordinates, will uncouple the equations (11.51):

$$\begin{Bmatrix} \dot{\alpha} \\ \dot{\beta} \end{Bmatrix} = \begin{bmatrix} \lambda_1 & 0 \\ 0 & \lambda_2 \end{bmatrix} \begin{Bmatrix} \alpha \\ \beta \end{Bmatrix} \qquad (11.53)$$

This leads to

$$\dot{\alpha} = \lambda_1 \alpha$$

$$\dot{\beta} = \lambda_2 \beta \qquad (11.54)$$

The solution of Eqs. (11.54) can be expressed as

$$\alpha(t) = e^{\lambda_1 t}$$

$$\beta(t) = e^{\lambda_2 t} \qquad (11.55)$$

Hence Eq. (11.52) becomes

$$x(t) = x_1\alpha + x_2\beta = x_1 e^{\lambda_1 t} + x_2 e^{\lambda_2 t}$$

$$y(t) = y_1\alpha + y_2\beta = y_1 e^{\lambda_1 t} + y_2 e^{\lambda_2 t}$$

It can be seen that the stability of the singular point depends on the eigenvalues $\lambda_1$ and $\lambda_2$, which are given by the characteristic equation

$$\begin{vmatrix} a_{11} - \lambda & a_{12} \\ a_{21} & a_{22} - \lambda \end{vmatrix} = 0$$

that is,

$$\lambda_1, \lambda_2 = \frac{1}{2}\left( p \pm \sqrt{p^2 - 4q} \right) \qquad (11.56)$$

where

$$p = a_{11} + a_{22}$$

and

$$q = a_{11}a_{22} - a_{12}a_{21}$$

Thus we find that:

if $(p^2 - 4q) < 0$, the motion is oscillatory;
if $(p^2 - 4q) > 0$, the motion is aperiodic;
if $p > 0$, the system is unstable;
if $p < 0$, the system is stable.

## 11.8 NUMERICAL METHODS FOR NONLINEAR SYSTEMS

Most of the numerical methods described in the earlier chapters can be used for finding the response of nonlinear systems. The Runge-Kutta method described in Sec. 10.4 is directly applicable for nonlinear systems and is illustrated in Sec. 11.27.1. The central difference, Houbolt, Wilson, and Newmark methods considered in Chapter 10 can also be used for solving nonlinear multidegree of freedom vibration problems with slight modification. Let a multidegree of freedom system be governed by the equation

$$[m]\ddot{\vec{x}}(t) + [c]\dot{\vec{x}}(t) + \vec{P}(\vec{x}(t)) = \vec{F}(t) \tag{11.57}$$

where the internal set of forces opposing the displacements, $\vec{P}$, are assumed to be nonlinear functions of $\vec{x}$. For the linear case, $\vec{P} = [k]\vec{x}$. In order to find the displacement vector $\vec{x}$ that satisfies the nonlinear equilibrium in Eq. (11.57), it is necessary to perform an equilibrium iteration sequence in each time step. In implicit methods (Houbolt, Wilson, and Newmark methods), the equilibrium conditions are considered at the same time for which solution is sought. If the solution is known for time $t_i$ and we wish to find the solution for time $t_{i+1}$, then the following equilibrium equations are considered

$$[m]\ddot{\vec{x}}_{i+1} + [c]\dot{\vec{x}}_{i+1} + \vec{P}_{i+1} = \vec{F}_{i+1} \tag{11.58}$$

where $\vec{F}_{i+1} = \vec{F}(t = t_{i+1})$ and $\vec{P}_{i+1}$ is computed as

$$\vec{P}_{i+1} = \vec{P}_i + [k_i]\Delta\vec{x} = \vec{P}_i + [k_i](\vec{x}_{i+1} - \vec{x}_i) \tag{11.59}$$

where $[k_i]$ is the linearized or tangent stiffness matrix computed at time $t_i$. Substitution of Eq. (11.59) in Eq. (11.58) gives

$$[m]\ddot{\vec{x}}_{i+1} + [c]\dot{\vec{x}}_{i+1} + [k_i]\vec{x}_{i+1} = \hat{\vec{F}}_{i+1} = \vec{F}_{i+1} - \vec{P}_i + [k_i]\vec{x}_i \tag{11.60}$$

Since the right hand side of Eq. (11.60) is completely known, it can be solved for $\vec{x}_{i+1}$ using any of the implicit methods directly. The $\vec{x}_{i+1}$ found is only an approximate vector due to the linearization process used in Eq. (11.59). To improve the accuracy of the solution and to avoid the development of numerical instabilities, an iterative process has to be used within the current time step [11.40].

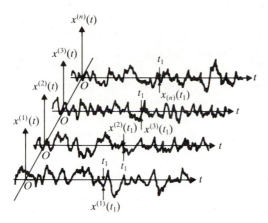

**Figure 11.10**   Ensemble of a random process ( $x^{(i)}(t)$ is the $i$th sample function of the ensemble).

## 11.9  INTRODUCTION TO RANDOM VIBRATION

If vibrational response characteristics such as displacement, acceleration, and stress are known precisely as functions of time, the vibration is known as *deterministic vibration*. This implies a deterministic system (or structure) and a deterministic loading (or excitation); deterministic vibration exists only if there is perfect control over all the variables that influence the structural properties and the loading. In practice, however, there are many processes and phenomena whose parameters cannot be precisely predicted. Such processes are therefore called *random processes* [11.18, 11.19]. An example of a random process is pressure fluctuation at a particular point on the surface of an aircraft flying in air. If several records of these pressure fluctuations are taken under the same flight speed, altitude, and load factor, they might look as indicated in Fig. 11.10. The records are not identical even though the measurements are taken under seemingly identical conditions. Similarly, a building subjected to ground acceleration due to an earthquake, a water tank under wind loading, and a car running on a rough road represent random processes. An elementary treatment of random vibration is presented in Secs. 11.10–11.20.

## 11.10  RANDOM VARIABLES AND RANDOM PROCESSES

Most phenomena in real life are nondeterministic. For example, the tensile strength of steel and the dimensions of a machined part are nondeterministic. If many samples of steel are tested, their tensile strengths will not be the same—they will fluctate about its mean or average value. Any quantity, like the tensile strength of steel, whose magnitude cannot be precisely predicted, is known as a *random variable* or a *probabilistic quantity*. If experiments are conducted to find the value of a random variable $x$, each experiment will give a number that is not a function of any parameter. For example, if 20 samples of steel are tested, their tensile strengths might be $x^{(1)} = 28.4$, $x^{(2)} = 30.2$, $x^{(3)} = 26.9, \ldots, x^{(20)} = 29.8$ kg/mm². Each of these outcomes is called a *sample point*. If $n$ experiments are conducted, all the $n$ possible outcomes of the random variable constitute what is known as the *sample space* of the random variable.

There are other types of probabilistic phenomena for which the outcome of an experiment is a function of some parameter such as time or spatial coordinate. Quantities such as the pressure fluctuations shown in Fig. 11.10 are called *random processes*. Each outcome of an experiment, in the case of a random process, is called a *sample function*. If $n$ experiments are conducted, all the $n$ possible outcomes of a random process constitute what is known as the *ensemble* of the process [11.20]. Notice that if the parameter $t$ is fixed at a particular value $t_1$, $x(t_1)$ is a random variable whose sample points are given by $x^{(1)}(t_1)$, $x^{(2)}(t_1), \ldots, x^{(n)}(t_1)$.

## 11.11 PROBABILITY DISTRIBUTION

Consider the random time function shown in Fig. 11.11. To find the probability of realizing the value of $x(t)$ smaller than some specified value, say $x_1$, we can draw a horizontal line at $x_1$ and sum the time intervals $t_i$ during which $x(t)$ is less than $x_1$:

$$P(x_1) = \text{Prob}\big[x(t) < x_1\big] = \lim_{t \to \infty} \frac{1}{t} \sum_i \Delta t_i \qquad (11.61)$$

If $x(t)$ denotes a physical quantity, the magnitude of $x(t)$ will always be a finite number, so $\text{Prob}[x(t) < -\infty] = P(-\infty) = 0$ (impossible event), and $\text{Prob}[x(t) < \infty] = P(\infty) = 1$ (certain event). The typical variation of $P(x)$ with $x$ is shown in Fig. 11.12(a). The function $P(x)$ is called the *probability distribution function* of $x$. The derivative of $P(x)$ with respect to $x$ is known as the *probability density function* and is denoted as $p(x)$. Thus

$$p(x) = \frac{dP(x)}{dx} = \lim_{\Delta x \to 0} \frac{P(x + \Delta x) - P(x)}{\Delta x} \qquad (11.62)$$

The quantity $P(x + \Delta x) - P(x)$ denotes the probability of realizing $x(t)$ between

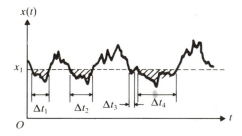

**Figure 11.11**

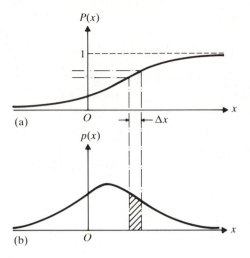

**Figure 11.12**

the values $x$ and $x + \Delta x$. Since $p(x)$ is the derivative of $P(x)$, we have

$$P(x) = \int_{-\infty}^{x} p(x')\,dx' \tag{11.63}$$

As $P(\infty) = 1$, Eq. (11.63) gives

$$P(\infty) = \int_{-\infty}^{\infty} p(x')\,dx' = 1 \tag{11.64}$$

which means that the total area under the curve of $p(x)$ is unity.

## 11.12 MEAN VALUE AND STANDARD DEVIATION

If $f(x)$ denotes a function of the random variable $x$, the expected value of $f(x)$ is defined as

$$E[f(x)] = \int_{-\infty}^{\infty} f(x)p(x)\,dx = \overline{f(x)} \tag{11.65}$$

If $f(x) = x$, Eq. (11.65) gives the expected value, also known as the *mean value*, of $x$:

$$E[x] = \bar{x} = \int_{-\infty}^{\infty} xp(x)\,dx \tag{11.66}$$

Similarly, if $f(x) = x^2$, we get the mean square value of $x$:

$$E[x^2] = \overline{x^2} = \int_{-\infty}^{\infty} x^2 p(x)\,dx \tag{11.67}$$

The variance of $x$, denoted as $\sigma^2(x)$, is defined as the mean square value of $x$ about the mean,

$$\sigma^2(x) = E\big[(x-\bar{x})^2\big] = \int_{-\infty}^{\infty} (x-\bar{x})^2 p(x)\,dx = \overline{(x^2)} - (\bar{x})^2 \qquad (11.68)$$

The positive square root of the variance, $\sigma(x)$, is called the *standard deviation* of $x$. When the mean value is zero, $\sigma(x) = \sqrt{\overline{x^2}}$, and the standard deviation is equal to the root mean square (rms) value.

## 11.13 JOINT PROBABILITY DISTRIBUTION OF SEVERAL RANDOM VARIABLES

When two or more random variables are being considered simultaneously, their joint behavior is determined by a *joint probability distribution function*. For example, while testing the tensile strength of steel specimens, we can obtain the values of yield strength and ultimate strength in each experiment. If we are interested in knowing the relation between these two random variables, we must know the joint probability density function of yield strength and ultimate strength. The probability distributions of single random variables are called *univariate distributions*; the distributions that involve two random variables are called *bivariate distributions*. In general, if a distribution involves more than one random variable, it is called a *multivariate distribution*.

The bivariate density function of the random variables $x_1$ and $x_2$ is defined as

$$p(x_1, x_2) = \text{Prob}[\,x_1 < x_1' < x_1 + dx_1,\, x_2 < x_2' < x_2 + dx_2\,] \qquad (11.69)$$

that is, the probability of realizing the value of the first random variable between $x_1$ and $x_1 + dx_1$ and the value of the second random variable between $x_2$ and $x_2 + dx_2$. The joint probability density function has the property that

$$\int_{-\infty}^{\infty} \int_{-\infty}^{\infty} p(x_1, x_2)\,dx_1\,dx_2 = 1 \qquad (11.70)$$

The joint distribution function of $x_1$ and $x_2$ is

$$P(x_1, x_2) = \text{Prob}[\,x_1' < x_1,\, x_2' < x_2\,]$$
$$= \int_{-\infty}^{x_1} \int_{-\infty}^{x_2} p(x_1', x_2')\,dx_1'\,dx_2' \qquad (11.71)$$

## 11.14 CORRELATION FUNCTIONS OF A RANDOM PROCESS

If $t_1, t_2, \ldots$ are fixed values of $t$, we use the abbreviations $x_1, x_2, \ldots$ to denote the values of $x(t)$ at $t_1, t_2, \ldots$, respectively. By forming the products of the random variables $x_1, x_2, \ldots$ (values of $x(t)$ at different times) and averaging the products over the set of all possibilities, we obtain a sequence of functions:

$$K(t_1, t_2) = E\big[x(t_1)x(t_2)\big] = E[x_1 x_2]$$
$$K(t_1, t_2, t_3) = E\big[x(t_1)x(t_2)x(t_3)\big] = E[x_1 x_2 x_3] \qquad (11.72)$$

and so on. These functions describe the statistical connection between the values of $x(t)$ at different times $t_1, t_2, \ldots$ and are called *correlation functions* [11.21, 11.22].

**Autocorrelation Function.** The mathematical expectation of $x_1 x_2$—the correlation function $K(t_1, t_2)$—is also known as the *autocorrelation function*, designated as $R(t_1, t_2)$. Thus

$$R(t_1, t_2) = E[x_1 x_2] \tag{11.73}$$

If the joint probability density function of $x_1$ and $x_2$ is known to be $p(x_1, x_2)$, the autocorrelation function can be expressed as

$$R(t_1, t_2) = \int_{-\infty}^{\infty} \int_{-\infty}^{\infty} x_1 x_2 \, p(x_1, x_2) \, dx_1 \, dx_2 \tag{11.74}$$

Experimentally, we can find $R(t_1, t_2)$ by taking the product of $x^{(i)}(t_1)$ and $x^{(i)}(t_2)$ in the $i$th sample function and averaging over the ensemble:

$$R(t_1, t_2) = \frac{1}{n} \sum_{i=1}^{n} x^{(i)}(t_1) x^{(i)}(t_2) \tag{11.75}$$

where $n$ denotes the number of sample functions in the ensemble.

**Covariance Function.** Analogous to the variance of a random process as defined in Eq. (11.68), the covariance of a random process is defined as the average of the product of the deviations from the means of $x(t)$ at two instants. Thus the covariance function of $x(t)$ can be expressed as

$$\text{Covariance of } x(t) = E\big[(x_1 - E[x_1])(x_2 - E[x_2])\big]$$

$$= \int_{-\infty}^{\infty} \int_{-\infty}^{\infty} (x_1 - E[x_1])(x_2 - E[x_2]) \, p(x_1, x_2) \, dx_1 \, dx_2$$

$$= E[x_1 x_2] - E[x_1] E[x_2] \tag{11.76}$$

It can be seen that the covariance is identical to the autocorrelation function when the mean values of $x_1$ and $x_2$ are zero.

## 11.15 STATIONARY RANDOM PROCESS

A stationary random process is one for which the probability distributions remain invariant under a shift of the time scale; the family of probability density functions applicable now also applies 5 hours from now or 500 hours from now. Thus the probability density function $p(x_1)$ becomes a universal density function $p(x)$, independent of time. Similarly, the joint density function $p(x_1, x_2)$, to be invariant under a shift of the time scale, becomes a function of $\tau = t_2 - t_1$, but not a function of $t_1$ or $t_2$ individually. Thus $p(x_1, x_2)$ can be written as $p(t, t + \tau)$. The expected value of a stationary random process $x(t)$ can be written as

$$E[x(t_1)] = E[x(t_1 + t)] \quad \text{for any } t \tag{11.77}$$

and the autocorrelation function as

$$R(t_1, t_2) = E[x_1 x_2] = E[x(t)x(t+\tau)] = R(\tau) \quad \text{for any } t \qquad (11.78)$$

where $\tau = t_2 - t_1$. If we set $\tau = 0$, $R(0)$ becomes the mean square value of $x(t)$, $E[x^2]$. We shall use subscripts to $R$ to identify the random process when more than one random process is involved. For example, we shall use $R_x(\tau)$ and $R_y(\tau)$ to denote the autocorrelation functions of the random processes $x(t)$ and $y(t)$, respectively.

**Ergodic Process.** An ergodic process is a stationary random process for which we can obtain all the probability information from a single sample function and assume that it is applicable to the entire ensemble. If $x^{(i)}(t)$ represents a typical sample function of duration $T$, the averages can be computed by averaging with respect to time along $x^{(i)}(t)$. Such averages are called *temporal averages*. By using the notation $\langle x(t) \rangle$ to represent the temporal average of $x(t)$ (the time average of $x$), we can write

$$E[x] = \langle x(t) \rangle = \lim_{T \to \infty} \frac{1}{T} \int_{-T/2}^{T/2} x^{(i)}(t) \, dt \qquad (11.79)$$

where $x^{(i)}(t)$ has been assumed to be defined from $t = -T/2$ to $t = T/2$ with $T \to \infty$ ($T$ very large). Similarly,

$$E[x^2] = \langle x^2(t) \rangle = \lim_{T \to \infty} \frac{1}{T} \int_{-T/2}^{T/2} [x^{(i)}(t)]^2 \, dt \qquad (11.80)$$

and

$$R(\tau) = \langle x(t)x(t+\tau) \rangle = \lim_{T \to \infty} \frac{1}{T} \int_{-T/2}^{T/2} x^{(i)}(t) x^{(i)}(t+\tau) \, dt \qquad (11.81)$$

## 11.16  GAUSSIAN RANDOM PROCESS

The most commonly used distribution for modeling physical random processes is called the *Gaussian* or *normal random process*. The Gaussian process has a number of remarkable properties that permit the computation of the random vibration characteristics in a simple manner. The probability density function of a Gaussian process $x(t)$ is given by

$$p(x) = \frac{1}{\sqrt{2\pi}\,\sigma_x} e^{-\frac{1}{2}\left(\frac{x - \bar{x}}{\sigma_x}\right)^2} \qquad (11.82)$$

where $\bar{x}$ and $\sigma_x$ denote the mean value and standard deviation of $x$. The mean ($\bar{x}$) and standard deviation ($\sigma_x$) of $x(t)$ vary with $t$ for a nonstationary process but are constants (independent of $t$) for a stationary process. A very important property of the Gaussian process is that the forms of its probability distributions are invariant with respect to linear operations. This means that if the excitation of a linear system

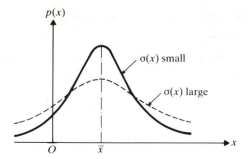

**Figure 11.13**

is a Gaussian process, the response is generally a different random process, but still a normal one. The only changes are that the magnitudes of the mean and standard deviations of the response are different from those of the excitation.

The graph of a Gaussian probability density function is a bell-shaped curve, symmetric about the mean value; its spread is governed by the value of the standard deviation, as shown in Fig. 11.13. By defining a standard normal variable $z$ as

$$z = \frac{x - \bar{x}}{\sigma_x} \tag{11.83}$$

Eq. (11.82) can be expressed as

$$p(x) = \frac{1}{\sqrt{2\pi}} e^{-\frac{1}{2}z^2} \tag{11.84}$$

The probability of $x(t)$ lying in the interval $-c\sigma$ and $+c\sigma$ where $c$ is any positive number can be found, assuming $\bar{x} = 0$:

$$\text{Prob}\left[-c\sigma \le x(t) \le c\sigma\right] = \int_{-c\sigma}^{c\sigma} \frac{1}{\sqrt{2\pi}\,\sigma} e^{-\frac{1}{2}\frac{x^2}{\sigma^2}}\,dx \tag{11.85}$$

The probability of $x(t)$ lying outside the range $\mp c\sigma$ is one minus the value given by Eq. (11.85). This can also be expressed as

$$\text{Prob}\left[|x(t)| > c\sigma\right] = \frac{2}{\sqrt{2\pi}\,\sigma} \int_{c\sigma}^{\infty} e^{-\frac{1}{2}\frac{x^2}{\sigma^2}}\,dx \tag{11.86}$$

The integrals in Eqs. (11.85) and (11.86) have been evaluated numerically and tabulated [11.20]; some typical values are indicated in the following table.

| Value of $c$ | 1 | 2 | 3 | 4 |
|---|---|---|---|---|
| Prob$[-c\sigma \le x(t) \le c\sigma]$ | 0.6827 | 0.9545 | 0.9973 | 0.999937 |
| Prob$[|x(t)| > c\sigma]$ | 0.3173 | 0.0455 | 0.0027 | 0.000063 |

## **11.17** FOURIER TRANSFORMS IN RANDOM VIBRATION ANALYSIS

We have seen in Chap. 1 that any periodic function $x(t)$ can be expressed in the form of a complex Fourier series (see Prob. 1.32):

$$x(t) = \sum_{n=-\infty}^{\infty} c_n e^{in\omega t} \tag{11.87}$$

where the complex coefficients $c_n$ are given by

$$c_n = \frac{1}{\tau} \int_{-\tau/2}^{\tau/2} x(t) e^{-in\omega t} \, dt \tag{11.88}$$

Equation (11.87) shows that the function $x(t)$ of period $\tau$ can be expressed as a sum of an infinite number of harmonics. The harmonics have amplitudes given by Eq. (11.88) and frequencies which are multiples of the fundamental frequency $\omega = 2\pi/\tau$. The difference between any two consecutive frequencies is $\Delta\omega = 2\pi/\tau = \omega$. Thus the larger the period $\tau$, the denser the frequency spectrum becomes. Note that for real functions $x(t)$, the Fourier coefficients satisfy the relationship

$$c_n = c_{-n}^* \tag{11.89}$$

where $c_{-n}^*$ is the complex conjugate of $c_{-n}$. The mean square value of $x(t)$ can be determined (see Problem 11.32):

$$\overline{x^2(t)} = \frac{1}{\tau} \int_{-\tau/2}^{\tau/2} x^2(t) \, dt = \sum_{-\infty}^{\infty} |c_n|^2 \tag{11.90}$$

A nonperiodic function can be treated as a periodic function having an infinite period ($\tau \to \infty$). In this case, the frequency spectrum becomes continuous, the fundamental frequency becomes infinitesimal, and the Fourier series becomes the Fourier integral. Since the fundamental frequency $\omega$ is very small, we can denote it as $\Delta\omega$ and combine Eqs. (11.87) and (11.88) to obtain

$$x(t) = \sum_{-\infty}^{\infty} \frac{\Delta\omega}{2\pi} \int_{-\tau/2}^{\tau/2} x(t) e^{-in\Delta\omega t} \, dt \, e^{in\Delta\omega t} \tag{11.91}$$

If we let $\Delta\omega \to d\omega$, the summation in Eq. (11.91) must include a continuous sequence of frequencies so that the summation becomes an integral and $n\Delta\omega$ becomes simply $\omega$. This results in

$$x(t) = \int_{-\infty}^{\infty} \left[ \frac{d\omega}{2\pi} \int_{-\infty}^{\infty} x(t) e^{-i\omega t} \, dt \, e^{i\omega t} \right] \tag{11.92}$$

This may be written as two equations corresponding to Eqs. (11.87) and (11.88):

$$x(t) = \frac{1}{2\pi} \int_{-\infty}^{\infty} A(i\omega) e^{i\omega t} \, d\omega \tag{11.93}$$

with

$$A(i\omega) = \int_{-\infty}^{\infty} x(t) e^{-i\omega t} \, dt \tag{11.94}$$

Equations (11.87) and (11.88) are known as a *discrete Fourier transform pair* and Eqs. (11.93) and (11.94) are known as an *integral Fourier transform pair* [11.23, 11.24]. If cycles per second are used as the units of the frequency, we set $\omega = 2\pi f$ in Eqs. (11.93) and (11.94) and obtain

$$x(t) = \int_{-\infty}^{\infty} A(if) e^{i2\pi ft} \, df \tag{11.95}$$

with

$$A(if) = \int_{-\infty}^{\infty} x(t) e^{-i2\pi ft} \, dt \tag{11.96}$$

The quantity $A(if)$ is usually complex. From the property of complex numbers, we have

$$A(if) A^*(if) = |A(if)|^2 \tag{11.97}$$

where $A^*(if)$ is the complex conjugate of $A(if)$. We can write

$$
\begin{aligned}
\int_{-\infty}^{\infty} x^2(t) \, dt = \int_{-\infty}^{\infty} x(t) x(t) \, dt &= \int_{-\infty}^{\infty} x(t) \left[ \int_{-\infty}^{\infty} A(if) e^{i2\pi ft} \, df \right] dt \\
&= \int_{-\infty}^{\infty} A(if) \left[ \int_{-\infty}^{\infty} x(t) e^{i2\pi ft} \, dt \right] df \\
&= \int_{-\infty}^{\infty} A(if) A^*(if) \, df \\
&= \int_{-\infty}^{\infty} |A(if)|^2 \, df
\end{aligned}
\tag{11.98}
$$

If we consider only positive frequencies, Eq. (11.98) can be rewritten as

$$\int_{-\infty}^{\infty} x^2(t) \, dt = 2 \int_{0}^{\infty} |A(if)|^2 \, df \tag{11.99}$$

because $|A(if)|^2$ is an even function of $f$. Thus the mean square value of a function $x(t)$ can be expressed as

$$\overline{x^2(t)} = \frac{1}{\tau} \int_{-\tau/2}^{\tau/2} x^2(t) \, dt = \frac{2}{\tau} \int_{0}^{\infty} |A(if)|^2 \, df \tag{11.100}$$

## 11.18 SPECTRAL DENSITY FUNCTION OF A RANDOM PROCESS

In practical applications, it is important to know the concentration of harmonics along the frequency axis. A function known as the *spectral density function* is used as a measure of the concentration of harmonics [11.25]. For a discrete spectrum $c_n$ of a function $x(t)$, the discrete spectral density function $S_x(nf)$ is defined as

$$S_x(nf) = \frac{2 c_n c_n^*}{\Delta f} \tag{11.101}$$

For a continuous spectrum $A(if)$ of a function $x(t)$, the continuous spectral density function, $S_x(f)$, is defined as

$$S_x(f) = \lim_{\tau \to \infty} \left\{ \frac{2}{\tau} |A(if)|^2 \right\} \tag{11.102}$$

The mean square value of $x(t)$ can be expressed in terms of the spectral density function as

$$\overline{x^2(t)} = \sum_{n=1}^{\infty} S_x(nf) \Delta f \qquad \text{for a discrete spectrum}$$

$$= \int_0^{\infty} S_x(f) \, df \qquad \text{for a continuous spectrum} \tag{11.103}$$

The autocorrelation function defined earlier is the Fourier transform of the spectral density function such that [11.32]

$$R_x(\tau) = \int_{-\infty}^{\infty} S_x(\omega) e^{i\omega\tau} \, d\omega \tag{11.104}$$

and

$$S_x(\omega) = \frac{1}{2\pi} \int_{-\infty}^{\infty} R_x(\tau) e^{-i\omega\tau} \, d\tau \tag{11.105}$$

When $\tau = 0$, we have from Eq. (11.104),

$$R_x(0) = \int_{-\infty}^{\infty} S_x(\omega) \, d\omega = E[x^2] \tag{11.106}$$

If $x(t)$ denotes the displacement, $R(0)$ represents the average energy. From Eq. (11.106), it is clear that $S_x(\omega)$ represents the energy density associated with the frequency $\omega$. Thus $S_x(\omega)$ indicates the spectral distribution of energy in a system. Equations (11.104) and (11.105) are called the Wiener-Khintchine relations. From Eqs. (11.104) and (11.105), the power spectral density and autocorrelation functions can be seen to be even functions of $\omega$ and $\tau$, respectively, so that

$$S_x(-\omega) = S_x(\omega)$$
$$R_x(-\tau) = R_x(\tau) \tag{11.107}$$

**EXAMPLE 11.4**

Find the autocorrelation function and spectral density functions for the rectangular pulse shown in Fig. 11.14(a).

*Solution.* The pulse is given by

$$x(t) = \begin{cases} x_0 & \text{for } |t| \leq t_0 \\ 0 & \text{for } |t| > t_0 \end{cases} \tag{E.1}$$

The autocorrelation function of $x$ is given by

$$R(\tau) = \lim_{T \to \infty} \frac{1}{T} \int_{-T/2}^{T/2} x(t) x(t + \tau) \, dt \tag{E.2}$$

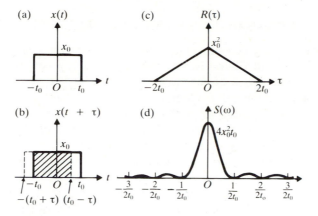

**Figure 11.14**

Since the pulse has a finite duration of $2t_0$, we need not let $T \to \infty$; instead, we select $T = 2t_0$ and rewrite Eq. (E.2) as

$$R(\tau) = \frac{1}{2t_0} \int_{-t_0}^{t_0 - \tau} x_0^2 \, dt \tag{E.3}$$

since the range of integration, for $0 < \tau < 2t_0$, is given by the shaded area shown in Fig. 11.14(b). Equation (E.3) gives

$$R(\tau) = \frac{x_0^2}{2t_0} \left[ t_0 - \tau - (-t_0) \right] = x_0^2 \left( 1 - \frac{\tau}{2t_0} \right) \tag{E.4}$$

For $\tau > 2t_0$, $R(\tau) = 0$. Further, since $R(\tau)$ is an even function of $\tau$, we have $R(\tau) = R(-\tau)$. Thus the graph of $R(\tau)$ looks as shown in Fig. 11.14(c). The spectral density function is given by (see Prob. 11.33):

$$S(f) = 4 \int_0^\infty R(\tau) \cos 2\pi f \tau \, d\tau \tag{E.5}$$

Since

$$R(\tau) = \begin{cases} x_0^2 \left( 1 - \dfrac{\tau}{2t_0} \right) & \text{for } |\tau| \le 2t_0 \\ 0 & \text{for } |\tau| > 2t_0 \end{cases} \tag{E.6}$$

Eq. (E.5) gives

$$S(f) = 4x_0^2 \int_0^{2t_0} \left( 1 - \frac{\tau}{2t_0} \right) \cos 2\pi f \tau \, d\tau$$

$$= 4x_0^2 t_0 \left( \frac{\sin 2\pi f t_0}{2\pi f t_0} \right)^2 \tag{E.7}$$

Equation (E.7) is shown graphically in Fig. 11.14(d).

## 11.19  WIDE-BAND AND NARROW-BAND RANDOM PROCESSES

A wide-band process is a stationary random process whose spectral density function $S(\omega)$ has significant values over a range or band of frequencies which is approximately the same order of magnitude as the center frequency of the band. An example of a wide-band random process is shown in Fig. 11.15. The pressure fluctuations on the surface of a rocket due to acoustically transmitted jet noise or due to supersonic boundary layer turbulence are examples of physical processes that are typically wide-band. A narrow-band random process is a stationary process whose spectral density function $S(\omega)$ has significant values only in a range or band of frequencies whose width is small compared to the magnitude of the center frequency of the process. Figure 11.16 shows the sample function and the corresponding spectral density and autocorrelation functions of a narrow-band process.

A random process whose power spectral density is constant over a frequency range is called *white noise*, by analogy with the white light which spans the visible spectrum more or less uniformly. It is called *ideal white noise* if the band of frequencies $\omega_2 - \omega_1$ is infinitely wide. Ideal white noise is a physically unrealizable concept, since the mean square value of such a random process would be infinite because the area under the spectrum would be infinite. It is called *band-limited white noise* if the band of frequencies has finite cut-off frequencies $\omega_1$ and $\omega_2$ [11.26]. The mean square value of a band-limited white noise is given by the total area under the

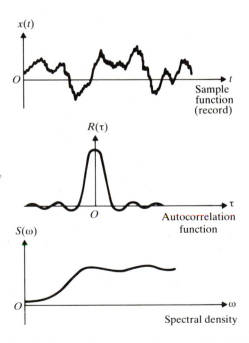

**Figure 11.15**  Wide-band stationary random process.

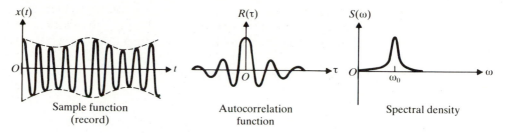

**Figure 11.16**   Narrow-band stationary random process.

spectrum, namely, $2S_0(\omega_2 - \omega_1)$, where $S_0$ denotes the constant value of the spectral density function.

---

### EXAMPLE 11.5

The spectral density of a random signal is given by $S(f) = 8 \times 10^{-5}$ m²/cps between 10 and 1000 cps and zero outside this frequency range. Find the standard deviation and the root mean square value of the signal if its mean value is 0.05 m.

***Solution.***   The mean square value of the signal is given by

$$\overline{x^2} = \int_0^\infty S(f)\, df = \int_{10}^{1000} 8 \times 10^{-5}\, df = 0.0792 \text{ m}^2$$

$$\sqrt{\overline{x^2}} = \text{rms value} = 0.2814 \text{ m}$$

The mean square value is

$$E[x^2] = \overline{x^2} = (\bar{x})^2 + \sigma_x^2$$

and hence

$$\sigma_x^2 = \overline{x^2} - (\bar{x})^2 = 0.0792 - (0.05)^2 = 0.0767 \text{ m}^2$$

$$\sigma_x = 0.2769 \text{ m}$$

---

## 11.20   RESPONSE TO RANDOM EXCITATION

The response of a linear single degree of freedom system subjected to a harmonic excitation $F_0 e^{i\omega t}$ has been found to be

$$x(t) = F_0 H(i\omega) e^{i\omega t} \tag{11.108}$$

where

$$H(i\omega) = \frac{1}{-\omega^2 m + i\omega c + k} \tag{11.109}$$

is called the *frequency response function*. If the system is subjected to a general forcing function $F(t)$, the function $F(t)$ can be written as a continuous sum of harmonic components by its Fourier integral as

$$F(t) = \int_{-\infty}^{\infty} A(if)e^{i2\pi ft}\,df \qquad (11.110)$$

where

$$A(if) = \int_{-\infty}^{\infty} F(t)e^{-i2\pi ft}\,dt \qquad (11.111)$$

Since the system is linear, the solution for $x(t)$ reduces to finding the solution to each harmonic term, $A(if)e^{i2\pi ft}\,df$, and then integrating the result over all the frequencies $f$:

$$x(t) = \int_{-\infty}^{\infty} A(if)H(if)e^{i2\pi ft}\,df \qquad (11.112)$$

where

$$H(if) = \frac{1}{-m(2\pi f)^2 + i2\pi fc + k} \qquad (11.113)$$

The mean square value of the response is given by

$$\overline{x^2} = \lim_{\tau \to \infty} \frac{1}{\tau} \int_{-\tau/2}^{\tau/2} x^2(t)\,dt \qquad (11.114)$$

Equations (11.112) and (11.114) give

$$\overline{x^2} = \lim_{\tau \to \infty} \frac{2}{\tau} \int_{0}^{\infty} |A(if)|^2 |H(if)|^2\,df$$

$$= \int_{0}^{\infty} \left\{ \lim_{\tau \to \infty} \frac{2}{\tau} |A(if)|^2 \right\} |H(if)|^2\,df$$

$$= \int_{0}^{\infty} S_F(f)|H(if)|^2\,df \qquad (11.115)$$

A comparison of Eqs. (11.115) and (11.103) shows that

$$S_x(f) = S_F(f)|H(if)|^2 \qquad (11.116)$$

The random vibration response of single and multidegree of freedom systems is presented in Refs. [11.28–11.30]. The random vibration analysis of road vehicles is given in [11.27, 11.31].

---

**EXAMPLE 11.6**

Find the expected value of the response of a single degree of freedom system subjected to a force whose spectral density is a white noise, $S_F(\omega) = S_0$.

**Solution.** The expected value of the response is given by Eq. (11.115):

$$\overline{x^2} = S_0 \int_0^\infty \frac{df}{\left[ -m(2\pi f)^2 + i2\pi fc + k \right]^2}$$

$$= S_0 \int_0^\infty \frac{df}{k^2 \left[ 1 - \dfrac{(2\pi f)^2}{(2\pi f_n)^2} + i\dfrac{2\pi f 2\zeta}{2\pi f_n} \right]^2}$$

$$= \frac{S_0}{k^2} \int_0^\infty \frac{df}{\left[ 1 - \left( \dfrac{f}{f_n} \right)^2 + i2\zeta \left( \dfrac{f}{f_n} \right) \right]^2}$$

$$= \frac{S_0 f_n}{k^2} \int_0^\infty \frac{d\xi}{(1 - \xi^2 + i2\zeta\xi)^2}$$

where $\xi = f/f_n$. Thus

$$\overline{x^2} = \frac{S_0 f_n}{k^2} \int_0^\infty \frac{d\xi}{(1 - \xi^2)^2 + (2\zeta\xi)^2}$$

Since the value of the integral in the above equation is equal to $\pi/4\zeta$ (see Prob. 11.20 and Ref. [11.32]), we have

$$\overline{x^2} = \frac{S_0 f_n}{k^2} \left( \frac{\pi}{4\zeta} \right) = \frac{S_0}{k^2} \frac{\omega_n}{2\pi} \frac{\pi}{4\zeta} = \frac{S_0}{k^2} \frac{1}{4} \frac{k}{c} = \frac{S_0}{4kc}.$$

## 11.21   BASIC CONCEPT OF THE FINITE ELEMENT METHOD

The increasing complexity of machine and structural systems and the sophistication of digital computers have been instrumental in the development of the finite element procedure [11.33]. The finite element method is a numerical procedure in which a complex structure is considered as an assemblage of a number of smaller elements, where each element is a continuous structural member called a *finite element* (see Fig. 1.1). The elements are assumed to be interconnected at certain points known as *nodes* or *joints*.* Since it is very difficult to find the exact solution (such as the displacements) of the original structure under the specified loads, a convenient approximate solution is assumed in each finite element. By requiring that the displacements be compatible and the forces balance at the joints, the entire structure is compelled to behave as a single entity. The application of the finite element method to simple vibration problems is presented in Secs. 11.22–11.26. Although the

---

*The term *node* is commonly used in the finite element literature. However, the term *joint* is used in this work to avoid confusion with the term *node* used with a different meaning in the earlier chapters.

method can be applied to more complex problems involving two- and three-dimensional finite elements, only the use of one-dimensional elements is considered in the numerical treatment.

## 11.22 EQUATIONS OF MOTION OF A FINITE ELEMENT

In the finite element method, the displacement within an element is expressed in terms of the displacements at the corners or joints of the element. For example, consider the plate shown in Fig. 11.17. The plate is divided into triangular finite elements by means of fictitious cuts, and the elements are assumed to be connected to each other only at the joints. The transverse displacement within an element is assumed to be $w(x, y, t)$. The values of $w$, $(\partial w)/(\partial x)$, and $(\partial w)/(\partial y)$ at joints 1, 2, and 3, namely $w(x_1, y_1, t)$, $(\partial w)/(\partial x)(x_1, y_1, t)$, $(\partial w)/(\partial y)(x_1, y_1, t),\ldots,(\partial w)/(\partial y)(x_3, y_3, t)$, are treated as unknowns and are denoted as $w_1(t), w_2(t), w_3(t),\ldots, w_9(t)$. The displacement $w(x, y, t)$ can be expressed in terms of the unknown joint displacements $w_i(t)$ in the form

$$w(x, y, t) = \sum_{i=1}^{n} N_i(x, y)w_i(t) \tag{11.117}$$

where $N_i(x, y)$ is called the *shape function* corresponding to the joint displacement $w_i(t)$ and $n$ is the number of unknown joint displacements ($n = 9$ in Fig. 11.17). If a distributed load $f(x, y, t)$ acts on the element, it can be converted into equivalent joint forces $f_i(t)$ ($i = 1, 2,\ldots, 9$). If concentrated forces act at the joints, they can also be added to the appropriate joint force $f_i(t)$. We shall now derive the equations of

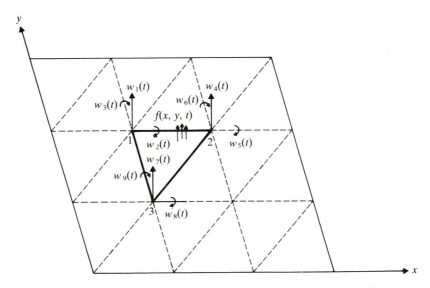

**Figure 11.17**

motion for determining the joint displacements $w_i(t)$ under the prescribed joint forces $f_i(t)$. By using Eq. (11.117), the kinetic energy $T$ and the strain energy $V$ of the element can be expressed as

$$T = \frac{1}{2} \dot{\vec{W}}^T [m] \dot{\vec{W}} \tag{11.118}$$

$$V = \frac{1}{2} \vec{W}^T [k] \vec{W} \tag{11.119}$$

where

$$\vec{W} = \begin{Bmatrix} w_1(t) \\ w_2(t) \\ \vdots \\ w_n(t) \end{Bmatrix}, \qquad \dot{\vec{W}} = \begin{Bmatrix} \dot{w}_1(t) \\ \dot{w}(t) \\ \vdots \\ \dot{w}_n(t) \end{Bmatrix} = \begin{Bmatrix} dw_1/dt \\ dw_2/dt \\ \vdots \\ dw_n/dt \end{Bmatrix}$$

and $[m]$ and $[k]$ are the mass and stiffness matrices of the element. By substituting Eqs. (11.118) and (11.119) into Lagrange's equations, Eq. (6.38), the equations of motion of the finite element can be obtained as

$$[m]\ddot{\vec{W}} + [k]\vec{W} = \vec{f} \tag{11.120}$$

where $\vec{f}$ is the vector of joint forces and $\ddot{\vec{W}}$ is the vector of joint accelerations given by

$$\ddot{\vec{W}} = \begin{Bmatrix} \ddot{w}_1 \\ \ddot{w}_2 \\ \vdots \\ \ddot{w}_n \end{Bmatrix} = \begin{Bmatrix} d^2 w_1/dt^2 \\ d^2 w_2/dt^2 \\ \vdots \\ d^2 w_n/dt^2 \end{Bmatrix}$$

Note that the shape of the finite elements and the number of unknown joint displacements may differ for different applications. Although the equations of motion of a single element, Eq. (11.120), are not useful directly (as our interest lies in the dynamic response of the assemblage of elements), the mass matrix $[m]$, the stiffness matrix $[k]$, and the joint force vector $\vec{f}$ of individual elements are necessary for the final solution. We shall derive the element mass and stiffness matrices and the joint force vectors for some one-dimensional elements in the next section.

## 11.23 FINITE ELEMENT MASS MATRIX, STIFFNESS MATRIX, AND FORCE VECTOR

### 11.23.1 Bar Element

Consider the uniform bar element shown in Fig. 11.18. For this one-dimensional element, the joints are merely the two end points. When the element is subjected to axial loads $f_1(t)$ and $f_2(t)$, the axial displacement within the element can be

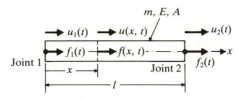

**Figure 11.18**

expressed as

$$u(x,t) = N_1(x)u_1(t) + N_2(x)u_2(t) \tag{11.121}$$

where $u_1(t)$ and $u_2(t)$ are the joint displacements and $N_1(x)$ and $N_2(x)$ are the shape functions. Since $u(0,t) = u_1(t)$ and $u(l,t) = u_2(t)$, the shape functions must satisfy the boundary conditions

$$N_1(0) = 1, \qquad N_1(l) = 0 \tag{11.122}$$

$$N_2(0) = 0, \qquad N_2(l) = 1 \tag{11.123}$$

where $l$ is the length of the bar element. Since two values are known for each of the functions $N_1(x)$ and $N_2(x)$, we assume $N_i(x)$ to be a polynomial with two constants (that is, a linear equation):

$$N_i(x) = a_i + b_i x, \qquad i = 1,2 \tag{11.124}$$

By using the boundary conditions (11.122) and (11.123), the constants $a_i$ and $b_i$ in Eq. (11.124) can be determined to obtain

$$N_1(x) = 1 - \frac{x}{l}, \qquad N_2(x) = \frac{x}{l} \tag{11.125}$$

and hence, from Eq. (11.121),

$$u(x,t) = \left(1 - \frac{x}{l}\right)u_1(t) + \frac{x}{l}u_2(t) \tag{11.126}$$

The kinetic energy of the bar element can be expressed as

$$T(t) = \frac{1}{2}\int_0^l m\left\{\frac{\partial u(x,t)}{\partial t}\right\}^2 dx = \frac{1}{2}\int_0^l m\left\{\left(1 - \frac{x}{l}\right)\frac{du_1(t)}{dt} + \left(\frac{x}{l}\right)\frac{du_2(t)}{dt}\right\}^2 dx$$

$$= \frac{1}{2}\frac{ml}{3}\left(\dot{u}_1^2 + \dot{u}_1\dot{u}_2 + \dot{u}_2^2\right) \tag{11.127}$$

where

$$\dot{u}_1 = \frac{du_1(t)}{dt}, \qquad \dot{u}_2 = \frac{du_2(t)}{dt}$$

and $m$ is the mass per unit length.

By expressing Eq. (11.127) in matrix form,

$$T(t) = \frac{1}{2}\dot{u}(t)^T[m]\dot{u}(t) \tag{11.128}$$

where

$$\dot{\vec{u}}(t) = \left\{ \begin{array}{c} \dot{u}_1(t) \\ \dot{u}_2(t) \end{array} \right\}$$

and the superscript $T$ indicates the transpose, the mass matrix $[m]$ can be identified as

$$[m] = \frac{ml}{6} \begin{bmatrix} 2 & 1 \\ 1 & 2 \end{bmatrix} \tag{11.129}$$

The strain energy of the element can be written as

$$V(t) = \frac{1}{2} \int_0^l EA \left\{ \frac{\partial u(x,t)}{\partial x} \right\}^2 dx = \frac{1}{2} \int_0^l EA \left\{ -\frac{1}{l} u_1(t) + \frac{1}{l} u_2(t) \right\}^2 dx$$

$$= \frac{1}{2} \frac{EA}{l} \left( u_1^2 - 2u_1 u_2 + u_2^2 \right) \tag{11.130}$$

where $u_1 = u_1(t)$, $u_2 = u_2(t)$, $E$ is Young's modulus, and $A$ is the cross-sectional area of the element. By expressing Eq. (11.130) in matrix form as

$$V(t) = \frac{1}{2} \vec{u}(t)^T [k] \vec{u}(t) \tag{11.131}$$

where

$$\vec{u}(t) = \left\{ \begin{array}{c} u_1(t) \\ u_2(t) \end{array} \right\} \qquad \text{and} \qquad \vec{u}(t)^T = \{ u_1(t) \quad u_2(t) \}$$

the stiffness matrix $[k]$ can be identified as

$$[k] = \frac{EA}{l} \begin{bmatrix} 1 & -1 \\ -1 & 1 \end{bmatrix} \tag{11.132}$$

The force vector

$$\vec{f} = \left\{ \begin{array}{c} f_1(t) \\ f_2(t) \end{array} \right\}$$

can be derived from the virtual work expression. If the bar is subjected to the distributed force $f(x,t)$, the virtual work $\delta W$ can be expressed as

$$\delta W(t) = \int_0^l f(x,t) \delta u(x,t) \, dx = \int_0^l f(x,t) \left\{ \left( 1 - \frac{x}{l} \right) \delta u_1(t) + \left( \frac{x}{l} \right) \delta u_2(t) \right\} dx$$

$$= \left( \int_0^l f(x,t) \left( 1 - \frac{x}{l} \right) dx \right) \delta u_1(t) + \left( \int_0^l f(x,t) \left( \frac{x}{l} \right) dx \right) \delta u_2(t) \tag{11.133}$$

By expressing Eq. (11.133) in matrix form as

$$\delta W(t) = \delta \vec{u}(t)^T \vec{f}(t) \equiv f_1(t) \delta u_1(t) + f_2(t) \delta u_2(t) \tag{11.134}$$

the joint forces can be identified as

$$f_1(t) = \int_0^l f(x, t)\left(1 - \frac{x}{l}\right) dx \tag{11.135}$$

$$f_2(t) = \int_0^l f(x, t)\left(\frac{x}{l}\right) dx \tag{11.136}$$

### 11.23.2    Torsion Element

Consider a uniform torsion element with the $x$ axis taken along the centroidal axis, as shown in Fig. 11.19. Let $I_p$ denote the polar moment of inertia about the centroidal axis and $GJ$ represent the torsional stiffness ($J = I_p$ for a circular cross section). The torsional displacement (rotation) within the element can be expressed in terms of the joint rotations $\theta_1(t)$ and $\theta_2(t)$ as

$$\theta(x, t) = N_1(x)\theta_1(t) + N_2(x)\theta_2(t) \tag{11.137}$$

where $N_1(x)$ and $N_2(x)$ are the same as in Eq. (11.125). The kinetic energy, the strain energy, and the virtual work for pure torsion are given by

$$T(t) = \frac{1}{2} \int_0^l \rho I_p \left\{ \frac{\partial \theta(x, t)}{\partial t} \right\}^2 dx \tag{11.138}$$

$$V(t) = \frac{1}{2} \int_0^l GJ \left\{ \frac{\partial \theta(x, t)}{\partial x} \right\}^2 dx \tag{11.139}$$

$$\delta W(t) = \int_0^l f(x, t)\delta\theta(x, t)\, dx \tag{11.140}$$

where $\rho$ is the mass density and $f(x, t)$ is the distributed torque per unit length. Using the procedures employed in Sec. 11.23.1, we can derive the element mass and stiffness matrices and the force vector:

$$[m] = \frac{\rho I_p l}{6} \begin{bmatrix} 2 & 1 \\ 1 & 2 \end{bmatrix} \tag{11.141}$$

$$[k] = \frac{GJ}{l} \begin{bmatrix} 1 & -1 \\ -1 & 1 \end{bmatrix} \tag{11.142}$$

$$\vec{f} = \begin{Bmatrix} f_1(t) \\ f_2(t) \end{Bmatrix} = \begin{Bmatrix} \int_0^l f(x, t)\left(1 - \frac{x}{l}\right) dx \\ \int_0^l f(x, t)\left(\frac{x}{l}\right) dx \end{Bmatrix} \tag{11.143}$$

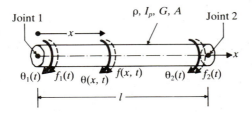

Joint 1       $\rho, I_p, G, A$      Joint 2

$\theta_1(t)$   $f_1(t)$   $\theta(x, t)$    $f(x, t)$    $\theta_2(t)$   $f_2(t)$

$l$

**Figure 11.19**

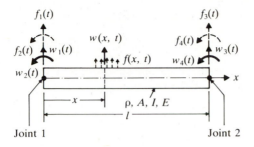

**Figure 11.20**

### 11.23.3  Beam Element

We now consider a beam element according to the Euler-Bernoulli theory.* Figure 11.20 shows a uniform beam element subjected to the transverse force distribution $f(x, t)$. In this case, the joints undergo both translational and rotational displacements, so the unknown joint displacements are labeled as $w_1(t)$, $w_2(t)$, $w_3(t)$, and $w_4(t)$. Thus there will be linear joint forces $f_1(t)$ and $f_3(t)$ corresponding to the linear joint displacements $w_1(t)$ and $w_3(t)$ and rotational joint forces (bending moments) $f_2(t)$ and $f_4(t)$ corresponding to the rotational joint displacements $w_2(t)$ and $w_4(t)$, respectively. We express the transverse displacement within the element as

$$w(x, t) = \sum_{i=1}^{4} N_i(x)w_i(t) \tag{11.144}$$

where $N_i(x)$ are the shape functions. Since

$$w(0, t) = w_1(t), \quad \frac{\partial w(0, t)}{\partial x} = w_2(t), \quad w(l, t) = w_3(t), \quad \frac{\partial w(l, t)}{\partial x} = w_4(t)$$

$$\tag{11.145}$$

$N_i(x)$ must satisfy the following boundary conditions:

$$N_1(0) = 1, \quad N_1(l) = 0, \quad \frac{dN_1}{dx}(0) = 0, \quad \frac{dN_1}{dx}(l) = 0 \tag{11.146}$$

$$N_2(0) = 0, \quad N_2(l) = 0, \quad \frac{dN_2}{dx}(0) = 1, \quad \frac{dN_2}{dx}(l) = 0 \tag{11.147}$$

$$N_3(0) = 0, \quad N_3(l) = 1, \quad \frac{dN_3}{dx}(0) = 0, \quad \frac{dN_3}{dx}(l) = 0 \tag{11.148}$$

$$N_4(0) = 0, \quad N_4(l) = 0, \quad \frac{dN_4}{dx}(0) = 0, \quad \frac{dN_4}{dx}(l) = 1 \tag{11.149}$$

Since four conditions are known for each $N_i(x)$, we assume a polynomial involving

---

*The beam element, according to the Timoshenko theory, was considered in Refs. [11.34, 11.37].

four constants (that is, a cubic equation) for $N_i(x)$:

$$N_i(x) = a_i + b_i x + c_i x^2 + d_i x^3, \qquad i = 1, 2, 3, 4 \qquad (11.150)$$

Substituting the four sets of boundary conditions of Eqs. (11.146) to (11.149) into Eq. (11.150) we obtain the shape functions:

$$N_1(x) = 1 - 3\left(\frac{x}{l}\right)^2 + 2\left(\frac{x}{l}\right)^3 \qquad (11.151)$$

$$N_2(x) = x - 2l\left(\frac{x}{l}\right)^2 + l\left(\frac{x}{l}\right)^3 \qquad (11.152)$$

$$N_3(x) = 3\left(\frac{x}{l}\right)^2 - 2\left(\frac{x}{l}\right)^3 \qquad (11.153)$$

$$N_4(x) = -l\left(\frac{x}{l}\right)^2 + l\left(\frac{x}{l}\right)^3 \qquad (11.154)$$

Thus the transverse displacement within the element becomes

$$w(x, t) = \left(1 - 3\frac{x^2}{l^2} + 2\frac{x^3}{l^3}\right) w_1(t) + \left(\frac{x}{l} - 2\frac{x^2}{l^2} + \frac{x^3}{l^3}\right) l w_2(t)$$

$$+ \left(3\frac{x^2}{l^2} - 2\frac{x^3}{l^3}\right) w_3(t) + \left(-\frac{x^2}{l^2} + \frac{x^3}{l^3}\right) l w_4(t) \qquad (11.155)$$

The kinetic energy, bending strain energy, and virtual work of the element can be expressed as

$$T(t) = \frac{1}{2} \int_0^l m \left\{ \frac{\partial w(x, t)}{\partial t} \right\}^2 dx \equiv \frac{1}{2} \dot{\vec{w}}(t)^T [m] \dot{\vec{w}}(t) \qquad (11.156)$$

$$V(t) = \frac{1}{2} \int_0^l EI \left\{ \frac{\partial^2 w(x, t)}{\partial x^2} \right\}^2 dx \equiv \frac{1}{2} \vec{w}(t)^T [k] \vec{w}(t) \qquad (11.157)$$

$$\delta W(t) = \int_0^l f(x, t) \, \delta w(x, t) \, dx \equiv \delta \vec{w}(t)^T \vec{f}(t) \qquad (11.158)$$

where $m$ is the mass of the beam per unit length, $E$ is Young's modulus, $I$ is the moment of inertia of the cross section, and

$$\vec{w}(t) = \begin{Bmatrix} w_1(t) \\ w_2(t) \\ w_3(t) \\ w_4(t) \end{Bmatrix}, \quad \dot{\vec{w}}(t) = \begin{Bmatrix} dw_1/dt \\ dw_2/dt \\ dw_3/dt \\ dw_4/dt \end{Bmatrix}, \quad \delta \vec{w}(t) = \begin{Bmatrix} \delta w_1(t) \\ \delta w_2(t) \\ \delta w_3(t) \\ \delta w_4(t) \end{Bmatrix}, \quad \vec{f}(t) = \begin{Bmatrix} f_1(t) \\ f_2(t) \\ f_3(t) \\ f_4(t) \end{Bmatrix}$$

By substituting Eq. (11.155) into Eqs. (11.156) to (11.158) and carrying out the necessary integrations, we obtain

$$[m] = \frac{ml}{420} \begin{bmatrix} 156 & 22l & 54 & -13l \\ 22l & 4l^2 & 13l & -3l^2 \\ 54 & 13l & 156 & -22l \\ -13l & -3l^2 & -22l & 4l^2 \end{bmatrix} \tag{11.159}$$

$$[k] = \frac{EI}{l^3} \begin{bmatrix} 12 & 6l & -12 & 6l \\ 6l & 4l^2 & -6l & 2l^2 \\ -12 & -6l & 12 & -6l \\ 6l & 2l^2 & -6l & 4l^2 \end{bmatrix} \tag{11.160}$$

$$f_i(t) = \int_0^l f(x,t) N_i(x)\, dx, \qquad i = 1,2,3,4 \tag{11.161}$$

## 11.24 TRANSFORMATION OF FINITE ELEMENT MATRICES AND VECTORS

As stated earlier, the finite element method considers the given dynamical system as an assemblage of elements. The joint displacements of an individual element are selected in a convenient direction, depending on the nature of the element. For example, for the bar element shown in Fig. 11.18, the joint displacements $u_1(t)$ and $u_2(t)$ are chosen along the axial direction of the element. However, other bar elements can have different orientations in an assemblage, as shown in Fig. 11.21. Here $x$ denotes the axial direction of an individual element and is called a *local coordinate axis*. If we use $u_1(t)$ and $u_2(t)$ to denote the joint displacements of different bar elements, there will be one joint displacement at joint 1, three at joint 2, two at joint 3, and 2 at joint 4. However, the displacements of joints can be specified more conveniently using reference or global coordinate axes $X$ and $Y$. Then the

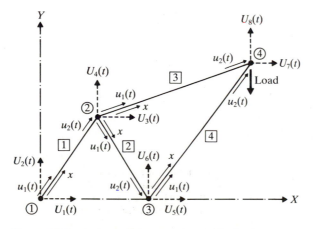

**Figure 11.21**   A dynamical system (truss) idealized as an assemblage of four bar elements.

displacement components of joints parallel to the $X$ and $Y$ axes can be used as the joint displacements in the global coordinate system. These are shown as $U_i(t)$, $i = 1, 2, \ldots, 8$ in Fig. 11.21. The joint displacements in the local and the global coordinate system for a typical bar element $e$ are shown in Fig. 11.22. The two sets of joint displacements are related as follows:

$$u_1(t) = U_{2i-1}(t)\cos\theta + U_{2i}(t)\sin\theta$$

$$u_2(t) = U_{2j-1}(t)\cos\theta + U_{2j}(t)\sin\theta \tag{11.162}$$

which can be rewritten as

$$\vec{u}(t) = [\lambda]\vec{U}(t) \tag{11.163}$$

where $[\lambda]$ is the coordinate transformation matrix given by

$$[\lambda] = \begin{bmatrix} \cos\theta & \sin\theta & 0 & 0 \\ 0 & 0 & \cos\theta & \sin\theta \end{bmatrix} \tag{11.164}$$

and $\vec{u}(t)$ and $\vec{U}(t)$ are the vectors of joint displacements in the local and the global coordinate system, respectively, and are given by

$$\vec{u}(t) = \begin{Bmatrix} u_1(t) \\ u_2(t) \end{Bmatrix}, \qquad \vec{U}(t) = \begin{Bmatrix} U_{2i-1}(t) \\ U_{2i}(t) \\ U_{2j-1}(t) \\ U_{2j}(t) \end{Bmatrix}$$

It is useful to express the mass matrix, stiffness matrix, and joint force vector of an element in terms of the global coordinate system while finding the dynamical response of the complete system. Since the kinetic and strain energies of the element

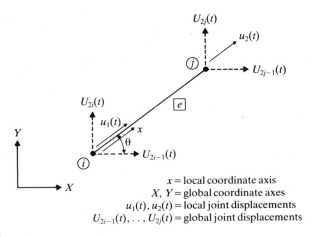

$x$ = local coordinate axis
$X, Y$ = global coordinate axes
$u_1(t), u_2(t)$ = local joint displacements
$U_{2i-1}(t), \ldots, U_{2j}(t)$ = global joint displacements

**Figure 11.22**   Local and global joint displacements of element $e$.

must be independent of the coordinate system, we have

$$T(t) = \frac{1}{2}\dot{u}(t)^T[m]\dot{u}(t) = \frac{1}{2}\dot{\vec{U}}(t)^T[\overline{m}]\dot{\vec{U}}(t) \qquad (11.165)$$

$$V(t) = \frac{1}{2}\vec{u}(t)^T[k]\vec{u}(t) = \frac{1}{2}\vec{U}(t)^T[\overline{k}]\vec{U}(t) \qquad (11.166)$$

where $[\overline{m}]$ and $[\overline{k}]$ denote the element mass and stiffness matrices, respectively, in the global coordinate system and $\dot{\vec{U}}(t)$ is the vector of joint velocities in the global coordinate system, related to $\dot{\vec{u}}(t)$ as in Eq. (11.163):

$$\dot{\vec{u}}(t) = [\lambda]\dot{\vec{U}}(t) \qquad (11.167)$$

By inserting Eqs. (11.163) and (11.167) into (11.165) and (11.166), we obtain

$$T(t) = \frac{1}{2}\dot{\vec{u}}(t)^T[\lambda]^T[m][\lambda]\dot{\vec{u}}(t) \equiv \frac{1}{2}\dot{\vec{U}}(t)^T[\overline{m}]\dot{\vec{U}}(t) \qquad (11.168)$$

$$V(t) = \frac{1}{2}\vec{u}(t)^T[\lambda]^T[k][\lambda]\vec{u}(t) \equiv \frac{1}{2}\vec{U}(t)^T[\overline{k}]\vec{U}(t) \qquad (11.169)$$

Equations (11.168) and (11.169) yield

$$[\overline{m}] = [\lambda]^T[m][\lambda] \qquad (11.170)$$

$$[\overline{k}] = [\lambda]^T[k][\lambda] \qquad (11.171)$$

Similarly, by equating the virtual work in the two coordinate systems,

$$\delta W(t) = \delta\vec{u}(t)^T\vec{f}(t) = \delta\vec{U}(t)^T\vec{\overline{f}}(t) \qquad (11.172)$$

we find the vector of element joint forces in the global coordinate system $\vec{\overline{f}}(t)$:

$$\vec{\overline{f}}(t) = [\lambda]^T\vec{f}(t) \qquad (11.173)$$

Equations (11.170), (11.171), and (11.173) can be used to obtain the equations of motion of a single finite element in the global coordinate system:

$$[\overline{m}]\ddot{\vec{U}}(t) + [\overline{k}]\vec{U}(t) = \vec{\overline{f}}(t) \qquad (11.174)$$

Although this equation is not of much use, since our interest lies in the equations of motion of an assemblage of elements, the matrices $[\overline{m}]$ and $[\overline{k}]$ and the vector $\vec{\overline{f}}$ are useful in deriving the equations of motion of the complete system, as indicated in the following section.

## 11.25 EQUATIONS OF MOTION OF THE COMPLETE SYSTEM OF FINITE ELEMENTS

Since the complete structure is considered an assemblage of several finite elements, we shall now extend the equations of motion obtained for single finite elements in the global system to the complete structure. We shall denote the joint displacements of the complete structure in the global coordinate system as $U_1(t), U_2(t), \ldots, U_M(t)$

or, equivalently, as a column vector:

$$\vec{U}(t) = \begin{Bmatrix} U_1(t) \\ U_2(t) \\ \vdots \\ U_M(t) \end{Bmatrix}$$

For convenience, we shall denote the quantities pertaining to an element $e$ in the assemblage by the superscript $e$. Since the joint displacements of any element $e$ can be identified in the vector of joint displacements of the complete structure, the vectors $\vec{U}^{(e)}(t)$ and $\vec{U}(t)$ are related:

$$\vec{U}^{(e)}(t) = [A^{(e)}]\vec{U}(t) \tag{11.175}$$

where $[A^{(e)}]$ is a rectangular matrix composed of zeros and ones. For example, for element 1 in Fig. 11.21, Eq. (11.175) becomes

$$\vec{U}^{(1)}(t) \equiv \begin{Bmatrix} U_1(t) \\ U_2(t) \\ U_3(t) \\ U_4(t) \end{Bmatrix} = \begin{bmatrix} 1 & 0 & 0 & 0 & 0 & 0 & 0 & 0 \\ 0 & 1 & 0 & 0 & 0 & 0 & 0 & 0 \\ 0 & 0 & 1 & 0 & 0 & 0 & 0 & 0 \\ 0 & 0 & 0 & 1 & 0 & 0 & 0 & 0 \end{bmatrix} \begin{Bmatrix} U_1(t) \\ U_2(t) \\ \vdots \\ U_8(t) \end{Bmatrix} \tag{11.176}$$

The kinetic energy of the complete structure can be obtained by adding the kinetic energies of individual elements:

$$T = \sum_{e=1}^{E} \frac{1}{2} \dot{\vec{U}}^{(e)^T}[\overline{m}]\dot{\vec{U}}^{(e)} \tag{11.177}$$

where $E$ denotes the number of finite elements in the assemblage. By differentiating Eq. (11.175), the relation between the velocity vectors can be derived:

$$\dot{\vec{U}}^{(e)}(t) = [A^{(e)}]\dot{\vec{U}}(t) \tag{11.178}$$

Substitution of Eq. (11.178) into (11.177) leads to

$$T = \frac{1}{2} \sum_{e=1}^{E} \dot{\vec{U}}^T[A^{(e)}]^T[\overline{m}^{(e)}][A^{(e)}]\dot{\vec{U}} \tag{11.179}$$

The kinetic energy of the complete structure can also be expressed in terms of joint velocities of the complete structure $\dot{\vec{U}}$:

$$T = \frac{1}{2}\dot{\vec{U}}^T[M]\dot{\vec{U}} \tag{11.180}$$

where $[M]$ is called the mass matrix of the complete structure. A comparison of Eqs.

(11.179) and (11.180) gives the relation

$$[\underset{\sim}{M}] = \sum_{e=1}^{E} [A^{(e)}]^T [\bar{m}^{(e)}][A^{(e)}] \tag{11.181}$$

Similarly, by considering strain energy, the stiffness matrix of the complete structure, $[\underset{\sim}{K}]$, can be expressed as

$$[\underset{\sim}{K}] = \sum_{e=1}^{E} [A^{(e)}]^T [\bar{k}^{(e)}][A^{(e)}] \tag{11.182}$$

Finally the consideration of virtual work yields the vector of joint forces of the complete structure, $\vec{F}$:

$$\vec{F} = \sum_{e=1}^{E} [A^{(e)}]^T \vec{f}^{(e)} \tag{11.183}$$

Once the mass and stiffness matrices and the force vector are known, Lagrange's equations of motion for the complete structure can be expressed as

$$[\underset{\sim}{M}]\ddot{\vec{U}} + [\underset{\sim}{K}]\vec{U} = \vec{F} \tag{11.184}$$

Note that the joint force vector $\vec{F}$ in Eq. (11.184) was generated by considering only the distributed loads acting on the various elements. If there is any concentrated load acting along the joint displacement $U_i(t)$, it must be added to the $i$th component of $\vec{F}$.

## 11.26 INCORPORATION OF BOUNDARY CONDITIONS IN FINITE ELEMENT ANALYSIS

In the preceding derivation, no joint was assumed to be fixed. Thus the complete structure is capable of undergoing rigid body motion under the joint forces. This means that $[\underset{\sim}{K}]$ is a singular matrix (see Sec. 6.11). In most cases, the structure is supported such that the displacements are zero at a number of joints, to avoid rigid body motion of the structure. A simple way of incorporating the zero displacement conditions is to eliminate the corresponding rows and columns from the matrices $[\underset{\sim}{M}]$ and $[\underset{\sim}{K}]$ and the vector $\vec{F}$. The final equations of motion of the restrained structure can be expressed as

$$\underset{N \times N}{[M]} \ \underset{N \times 1}{\ddot{\vec{U}}} + \underset{N \times N}{[K]} \ \underset{N \times 1}{\vec{U}} = \underset{N \times 1}{\vec{F}} \tag{11.185}$$

where $N$ denotes the number of free joint displacements of the structure.

Note the following points concerning finite element analysis:

1. The approach used in the above presentation is called the *displacement method* of finite element analysis because it is the displacements of an element that are directly approximated. Other methods, such as the force method, the mixed method, and hybrid methods, are also available [11.38].

2. The stiffness matrix, mass matrix, and force vector for other finite elements, including two-dimensional and three-dimensional elements, can be derived in a similar manner, provided the shape functions are known [11.33].

3. In the Rayleigh-Ritz method discussed in Sec. 8.8, the displacement of the continuous system is approximated by a sum of assumed functions, where each function denotes a deflection shape of the entire structure. In the finite element method, an approximation using shape functions (similar to the assumed functions) is also used for a finite element instead of the entire structure. Thus the finite element procedure can also be considered a Rayleigh-Ritz method.

4. Error analysis of the finite element method can also be conducted [11.39].

### EXAMPLE 11.7

Find the equations of motion of a uniform circular shaft subjected to a uniformly distributed torque $f(x, t) = f(t)$. The shaft is fixed at $x = 0$ and free at $x = L$, as shown in Fig. 11.23.

**Solution.** For simplicity, we divide the shaft into two elements of equal length $l = L/2$. Since the problem is one-dimensional, the global and local coordinate directions are the same. The mass and stiffness matrices are given by Eqs. (11.141) and (11.142):

$$[\overline{m}^{(e)}] = [m^{(e)}] = \frac{\rho I_p L}{12} \begin{bmatrix} 2 & 1 \\ 1 & 2 \end{bmatrix}, \qquad e = 1, 2$$

$$[\overline{k}^{(e)}] = [k^{(e)}] = \frac{2GJ}{L} \begin{bmatrix} 1 & -1 \\ -1 & 1 \end{bmatrix}, \qquad e = 1, 2$$

The relation between the rotational displacements of the elements and the complete structure leads to (according to Eq. (11.175))

$$[A^{(1)}] = \begin{bmatrix} 1 & 0 & 0 \\ 0 & 1 & 0 \end{bmatrix}, \qquad [A^{(2)}] = \begin{bmatrix} 0 & 1 & 0 \\ 0 & 0 & 1 \end{bmatrix}$$

The mass and stiffness matrices of the structure can be obtained from Eqs. (11.181)

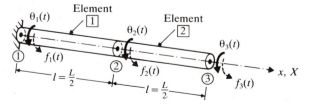

**Figure 11.23**

and (11.182):

$$[\underset{\sim}{M}] = \sum_{e=1}^{2} [A^{(e)}]^T [\overline{m}^{(e)}][A^{(e)}] = \frac{\rho I_p L}{12} \begin{bmatrix} 2 & 1 & 0 \\ 1 & 4 & 1 \\ 0 & 1 & 2 \end{bmatrix}$$

$$[\underset{\sim}{K}] = \sum_{e=1}^{2} [A^{(e)}]^T [\overline{k}^{(e)}][A^{(e)}] = \frac{2GJ}{L} \begin{bmatrix} 1 & -1 & 0 \\ -1 & 2 & -1 \\ 0 & -1 & 1 \end{bmatrix}$$

The components of the load vector can be computed from Eq. (11.143):

$$f_1^{(e)} = \int_0^l f(x,t)\left(1 - \frac{x}{l}\right) dx = f(t) \int_0^l \left(1 - \frac{x}{l}\right) dx = \frac{1}{2} f(t) l$$

$$f_2^{(e)} = \int_0^l f(x,t) \frac{x}{l} dx = f(t) \int_0^l \frac{x}{l} dx = \frac{1}{2} f(t) l$$

so that

$$\vec{f}^{(e)} = \vec{f}^{(e)} = \frac{1}{2} f(t) l \begin{Bmatrix} 1 \\ 1 \end{Bmatrix}, \qquad e = 1,2$$

The load vector of the complete structure is given by

$$\vec{F} = \sum_{e=1}^{2} [A^{(e)}]^T \vec{f}^{(e)} = \frac{1}{2} f(t) l \begin{Bmatrix} 1 \\ 2 \\ 1 \end{Bmatrix}$$

Thus the equations of motion of the complete shaft are given by

$$\frac{\rho I_p L}{12} \begin{bmatrix} 2 & 1 & 0 \\ 1 & 4 & 1 \\ 0 & 1 & 2 \end{bmatrix} \ddot{\vec{\theta}}(t) + \frac{2GJ}{L} \begin{bmatrix} 1 & -1 & 0 \\ -1 & 2 & -1 \\ 0 & -1 & 1 \end{bmatrix} \vec{\theta}(t) = \frac{1}{2} f(t) l \begin{Bmatrix} 1 \\ 2 \\ 1 \end{Bmatrix}$$

where

$$\vec{\theta}(t) = \begin{Bmatrix} \theta_1(t) \\ \theta_2(t) \\ \theta_3(t) \end{Bmatrix}$$

Since the end $x = 0$ is fixed, the condition $\theta_1(t) = 0$ can be incorporated by deleting the first row and first column of $[\underset{\sim}{M}]$ and $[\underset{\sim}{K}]$ and the first row of $\vec{F}$. The final equations of motion become

$$\frac{\rho I_p L}{12} \begin{bmatrix} 4 & 1 \\ 1 & 2 \end{bmatrix} \ddot{\vec{\theta}}(t) + \frac{2GJ}{L} \begin{bmatrix} 2 & -1 \\ -1 & 1 \end{bmatrix} \vec{\theta}(t) = \frac{1}{2} f(t) l \begin{Bmatrix} 2 \\ 1 \end{Bmatrix}$$

where

$$\vec{\theta}(t) = \begin{Bmatrix} \theta_2(t) \\ \theta_3(t) \end{Bmatrix}$$

## EXAMPLE 11.8

Find the natural frequencies of the simply supported beam shown in Fig. 11.24 using one finite element.

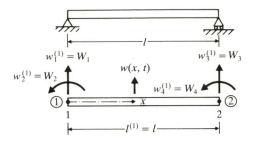

**Figure 11.24**

*Solution.* Since the beam is idealized using only one element, the element joint displacements are the same in both local and global systems. The stiffness and mass matrices of the beam are given by

$$[\underset{\sim}{K}] = [K^{(1)}] = \frac{EI}{l^3} \begin{bmatrix} 12 & 6l & -12 & 6l \\ 6l & 4l^2 & -6l & 2l^2 \\ -12 & -6l & 12 & -6l \\ 6l & 2l^2 & -6l & 4l^2 \end{bmatrix} \tag{E.1}$$

$$[\underset{\sim}{M}] = [M^{(1)}] = \frac{\rho Al}{420} \begin{bmatrix} 156 & 22l & 54 & -13l \\ 22l & 4l^2 & 13l & -3l^2 \\ 54 & 13l & 156 & -22l \\ -13l & -3l^2 & -22l & 4l^2 \end{bmatrix} \tag{E.2}$$

and the vector of joint displacements by

$$\vec{W} = \begin{Bmatrix} W_1 \\ W_2 \\ W_3 \\ W_4 \end{Bmatrix} \equiv \begin{Bmatrix} w_1^{(1)} \\ w_2^{(1)} \\ w_3^{(1)} \\ w_4^{(1)} \end{Bmatrix} \tag{E.3}$$

The boundary conditions corresponding to the simply supported ends ($W_1 = 0$ and $W_3 = 0$) can be incorporated* by deleting the rows and columns corresponding to $W_1$ and $W_3$ in Eqs. (E.1) and (E.2). This leads to the overall matrices

$$[K] = \frac{2EI}{l} \begin{bmatrix} 2 & 1 \\ 1 & 2 \end{bmatrix} \tag{E.4}$$

$$[M] = \frac{\rho Al^3}{420} \begin{bmatrix} 4 & -3 \\ -3 & 4 \end{bmatrix} \tag{E.5}$$

and the eigenvalue problem can be written as

$$\left[ \frac{2EI}{l} \begin{bmatrix} 2 & 1 \\ 1 & 2 \end{bmatrix} - \frac{\rho Al^3 \omega^2}{420} \begin{bmatrix} 4 & -3 \\ -3 & 4 \end{bmatrix} \right] \begin{Bmatrix} W_2 \\ W_4 \end{Bmatrix} = \begin{Bmatrix} 0 \\ 0 \end{Bmatrix} \tag{E.6}$$

---

*The bending moment cannot be set equal to zero at the simply supported ends explicitly, since there is no degree of freedom (joint displacement) involving the second derivative of the displacement $w$.

By multiplying throughout by $l/(2EI)$, Eq. (E.6) can be expressed as

$$\begin{bmatrix} 2-4\lambda & 1+3\lambda \\ 1+3\lambda & 2-4\lambda \end{bmatrix} \begin{Bmatrix} W_2 \\ W_4 \end{Bmatrix} = \begin{Bmatrix} 0 \\ 0 \end{Bmatrix} \tag{E.7}$$

where

$$\lambda = \frac{\rho A l^4 \omega^2}{840 EI} \tag{E.8}$$

By setting the determinant of the coefficient matrix in Eq. (E.7) equal to zero, we obtain the frequency equation

$$\begin{vmatrix} 2-4\lambda & 1+3\lambda \\ 1+3\lambda & 2-4\lambda \end{vmatrix} = (2-4\lambda)^2 - (1+3\lambda)^2 = 0 \tag{E.9}$$

The roots of Eq. (E.9) give the natural frequencies of the beam as

$$\lambda_1 = \frac{1}{7} \qquad \text{or} \qquad \omega_1 = \left( \frac{120 EI}{\rho A l^4} \right)^{1/2} \tag{E.10}$$

$$\lambda_2 = 3 \qquad \text{or} \qquad \omega_2 = \left( \frac{2520 EI}{\rho A l^4} \right)^{1/2} \tag{E.11}$$

## 11.27 COMPUTER PROGRAMS

### 11.27.1   Solution of Nonlinear Vibration Problems

A Fortran program is given for numerically solving the following free nonlinear vibration problem, using the fourth order Runge-Kutta method:

$$m\ddot{x} + c\dot{x} + kx + k^*x^3 = 0 \tag{E.1}$$

with $m = 0.01$, $c = 0.1$, $k = 2.0$, $k^* = 0.5$, and the initial conditions $x(0) = 7.5$ and $\dot{x}(0) = 0$. Equation (E.1) is written as a system of two first order differential equations:

$$\dot{x}_1 = f_1 = x_2$$

$$\dot{x}_2 = f_2 = -\frac{c}{m}x_2 - \frac{k}{m}x_1 - \frac{k^*}{m}x_1^3 \tag{E.2}$$

with the initial conditions $x_1(0) = 7.5$ and $x_2(0) = 0$. The time step $\Delta t$ is taken as 0.0025, the number of steps $NSTEP$ as 400, and the number of equations $N$ as 2. The values of $m$, $c$, $k$, and $k^*$ are denoted as YM, YC, YK and YKS and are transferred to subroutine FUN through a common statement. Subroutine FUN provides the values of $f_1 = F(1)$ and $f_2 = F(2)$ at specified values of array $X = \begin{Bmatrix} x_1 \\ x_2 \end{Bmatrix}$ and $T = $ time $t$. The listing of the program and the results follow.

```
C ===============================================================
C
C PROGRAM 22
C MAIN PROGRAM FOR SOLVING A NONLINEAR VIBRATION PROBLEM USING THE
C SUBROUTINE RK4
C
C ===============================================================
C DIMENSIONS: TIME(NSTEP),X(NSTEP,N),XX(N),F(N),XI(N),XJ(N),XK(N),XL(N),
C             UU(N)
C             XX(1),...,XX(N) CONTAIN INITIAL VALUES
C FOLLOWING 9 LINES CONTAIN PROBLEM-DEPENDENT DATA
      DIMENSION TIME(400),X(400,2),XX(2),F(2),XI(2),XJ(2),XK(2),XL(2),
     2 UU(2)
      COMMON /BLOCK1/ YM,YC,YK,YKS
      XX(1)=7.5
      XX(2)=0.0
      N=2
      NSTEP=400
      DT=0.0025
      DATA YM,YC,YK,YKS/0.01,0.1,2.0,0.5/
C END OF PROBLEM-DEPENDENT DATA
      T=0.0
      PRINT 10
10    FORMAT (//,40H SOLUTION OF NONLINEAR VIBRATION PROBLEM,/,
     2   35H BY FOURTH ORDER RUNGE-KUTTA METHOD)
      PRINT 20,YM,YC,YK,YKS
20    FORMAT (//,6H DATA:,/,6H YM =,E15.6,/,6H YC  =,E15.6,/,
     2   6H YK  =,E15.6,/,6H YKS =,E15.6,/)
      PRINT 60
      DO 50 I=1,NSTEP
      CALL RK4 (T,DT,N,XX,F,XI,XJ,XK,XL,UU)
      TIME(I)=T
      DO 30 J=1,N
30    X(I,J)=XX(J)
      PRINT 40, I, TIME(I),(X(I,J),J=1,N)
40    FORMAT (2X,I5,F12.6,2E15.6)
50    CONTINUE
60    FORMAT (/,9H RESULTS:,//,5X,2H I,4X,8H TIME(I),5X,7H X(I,1),
     2   5X,7H X(I,2),/)
      STOP
      END
C ===============================================================
C
C SUBROUTINE FUN FOR USE IN SUBROUTINE RK4
C THIS SUBROUTINE IS PROBLEM-DEPENDENT
C
C ===============================================================
      SUBROUTINE FUN (X,F,N,T)
      DIMENSION X(N),F(N)
      COMMON /BLOCK1/ YM,YC,YK,YKS
      F(1)=X(2)
      F(2)=-(YC/YM)*X(2)-(YK/YM)*X(1)-(YKS/YM)*(X(1)**3)
      RETURN
      END
```

SOLUTION OF NONLINEAR VIBRATION PROBLEM
BY FOURTH ORDER RUNGE-KUTTA METHOD

DATA:
YM  =   0.100000E-01
YC  =   0.100000E+00
YK  =   0.200000E+01
YKS =   0.500000E+00

RESULTS:

| I | TIME(I) | X(I,1) | X(I,2) |
|---|---|---|---|
| 1 | 0.002500 | 0.743030E+01 | -0.552857E+02 |
| 2 | 0.005000 | 0.722711E+01 | -0.106350E+03 |
| 3 | 0.007500 | 0.690398E+01 | -0.150944E+03 |
| 4 | 0.010000 | 0.647900E+01 | -0.187662E+03 |
| 5 | 0.012500 | 0.597268E+01 | -0.216014E+03 |
| 6 | 0.015000 | 0.540567E+01 | -0.236317E+03 |
| 7 | 0.017500 | 0.479708E+01 | -0.249463E+03 |
| 8 | 0.020000 | 0.416331E+01 | -0.256674E+03 |
| 9 | 0.022500 | 0.351755E+01 | -0.259270E+03 |
| 10 | 0.025000 | 0.286976E+01 | -0.258502E+03 |
| ⋮ | | | |
| 395 | 0.987511 | 0.906302E-01 | -0.951725E+00 |
| 396 | 0.990011 | 0.882246E-01 | -0.972474E+00 |
| 397 | 0.992511 | 0.857693E-01 | -0.991504E+00 |
| 398 | 0.995011 | 0.832685E-01 | -0.100883E+01 |
| 399 | 0.997511 | 0.807265E-01 | -0.102448E+01 |
| 400 | 1.000011 | 0.781475E-01 | -0.103847E+01 |

### 11.27.2  Solution of Beam Vibration Problems Using the Finite Element Method

A Fortran program is given for the eigenvalue analysis of the stepped beam shown in Fig. 11.25, using the finite element method. Although the program is written for the beam of Fig. 11.25, it can be easily generalized to find the natural frequencies and mode shapes of any beam with a specified number of steps (see Problem 11.38).

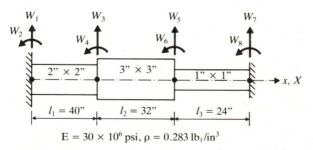

$E = 30 \times 10^6$ psi, $\rho = 0.283$ lb$_f$/in$^3$

**Figure 11.25**

The input data includes XL(I) = length of element $I$, XI(I) = moment of inertia of element $I$, A(I) = area of cross section of element $I$, BJ(I, J) = global degree of freedom number corresponding to the local $J$th degree of freedom of element $I$, E = Young's modulus, and RHO = mass density of the material.

The listing and the output of the program are given below.

```
C =============================================================
C
C PROGRAM 23
C PROGRAM FOR FINITE ELEMENT VIBRATION ANALYSIS OF STEPPED BEAM
C
C =============================================================
C DIMENSIONS: XL(NE),XI(NE),A(NE),XMAS(NE),BJ(NE,4),XM(4,4),XK(4,4),
C             AI(4,8),AIT(8,4),XKA(4,8),XMA(4,8),AKA(8,8),AMA(8,8),
C             BIGM(N,N),BIGK(N,N),BM(ND,ND),BK(ND,ND)
C             U(ND,ND),UI(ND,ND),UTI(ND,ND),BMU(ND,ND),UMU(ND,ND),
C             XF(ND,ND),EV(ND,ND)
C
C NE = NUMBER OF ELEMENTS
C N  = TOTAL NUMBER OF DEGREES OF FREEDOM
C ND = TOTAL NUMBER OF DEGREES OF FREEDOM AFTER DELETING ZERO D.O.F.
C
      DIMENSION XL(3),XI(3),A(3),XMAS(3),BJ(3,4),XM(4,4),XK(4,4),
     2    AI(4,8),AIT(8,4),XKA(4,8),XMA(4,8),AKA(8,8),AMA(8,8),BIGM(8,8),
     3    BIGK(8,8),BM(4,4),BK(4,4)
      DIMENSION U(4,4),UI(4,4),UTI(4,4),BMU(4,4),UMU(4,4),XF(4,4),
     2    EV(4,4)
      INTEGER BJ
      DATA XL/40.0,32.0,24.0/
      DATA XI/1.333333,6.75,0.083333/
      DATA A/4.0,9.0,1.0/
      DATA BJ/1,3,5,2,4,6,3,5,7,4,6,8/
      E=30.0E+06
      RHO=0.283/386.4
      DO 10 I=1,3
10    XMAS(I)=A(I)*RHO
      DO 20 I=1,8
      DO 20 J=1,8
      BIGM(I,J)=0.0
20    BIGK(I,J)=0.0
      DO 100 II=1,3
      DO 30 I=1,4
      DO 30 J=1,8
30    AI(I,J)=0.0
      I1=BJ(II,1)
      I2=BJ(II,2)
      I3=BJ(II,3)
      I4=BJ(II,4)
      AI(1,I1)=1.0
      AI(2,I2)=1.0
      AI(3,I3)=1.0
      AI(4,I4)=1.0
```

```
            XM(1,1)=156.0
            XM(1,2)=22.0*XL(II)
            XM(1,3)=54.0
            XM(1,4)=-13.0*XL(II)
            XM(2,2)=4.0*(XL(II)**2)
            XM(2,3)=13.0*XL(II)
            XM(2,4)=-3.0*(XL(II)**2)
            XM(3,3)=156.0
            XM(3,4)=-22.0*XL(II)
            XM(4,4)=4.0*(XL(II)**2)
            XK(1,1)=12.0
            XK(1,2)=6.0*XL(II)
            XK(1,3)=-12.0
            XK(1,4)=6.0*XL(II)
            XK(2,2)=4.0*(XL(II)**2)
            XK(2,3)=-6.0*XL(II)
            XK(2,4)=2.0*(XL(II)**2)
            XK(3,3)=12.0
            XK(3,4)=-6.0*XL(II)
            XK(4,4)=4.0*(XL(II)**2)
            DO 40 I=1,4
            DO 40 J=1,4
            XM(J,I)=XM(I,J)
       40   XK(J,I)=XK(I,J)
            DO 50 I=1,4
            DO 50 J=1,4
            XM(I,J)=(XMAS(II)*XL(II)/420.0)*XM(I,J)
       50   XK(I,J)=(E*XI(II)/(XL(II)**3))*XK(I,J)
            DO 60 I=1,8
            DO 60 J=1,4
       60   AIT(I,J)=AI(J,I)
            CALL MATMUL (XKA,XK,AI,4,4,8)
            CALL MATMUL (XMA,XM,AI,4,4,8)
            CALL MATMUL (AKA,AIT,XKA,8,4,8)
            CALL MATMUL (AMA,AIT,XMA,8,4,8)
            DO 70 I=1,8
            DO 70 J=1,8
            BIGM(I,J)=BIGM(I,J)+AMA(I,J)
       70   BIGK(I,J)=BIGK(I,J)+AKA(I,J)
      100   CONTINUE
C APPLICATION OF BOUNDARY CONDITIONS
C ROWS AND COLUMNS CORRESPONDING TO ZERO DISPLACEMENTS ARE DELETED
            DO 110 I=1,4
            DO 110 J=1,4
            BM(I,J)=BIGM(I+2,J+2)
      110   BK(I,J)=BIGK(I+2,J+2)
C DECOMPOSING THE MATRIX [BK] INTO TRIANGULAR MATRICES
            ND=4
            CALL DECOMP (BK,U,ND)
C FINDING THE INVERSE OF THE UPPER TRIANGULAR MATRIX [U]
            DO 120 I=1,ND
            DO 120 J=1,ND
      120   UI(I,J)=0.0
            DO 130 I=1,ND
```

```
130  UI(I,I)=1.0/U(I,I)
     DO 150 J=1,ND
     DO 150 II=1,ND
     I=ND-II+1
     IF (I .GE. J) GO TO 150
     IP=I+1
     SUM=0.0
     DO 140 K=IP,J
140  SUM=SUM+U(I,K)*UI(K,J)
     UI(I,J)=-SUM/U(I,I)
150  CONTINUE
     DO 160 I=1,ND
     DO 160 J=1,ND
160  UTI(I,J)=UI(J,I)
     CALL MATMUL (BMU,BM,UI,ND,ND,ND)
     CALL MATMUL (UMU,UTI,BMU,ND,ND,ND)
     CALL JACOBI (UMU,ND,EV,1.0E-05,200)
     CALL MATMUL (XF,UI,EV,ND,ND,ND)
     DO 170 I=1,ND
170  UMU(I,I)=SQRT(1.0/UMU(I,I))
     PRINT 180,(UMU(I,I),I=1,ND)
180  FORMAT (//,40H NATURAL FREQUENCIES OF THE STEPPED BEAM,//,
    2    4(1X,E15.6))
     PRINT 190
190  FORMAT (//,12H MODE SHAPES)
     DO 200 J=1,ND
200  PRINT 210, J, (XF(I,J),I=1,ND)
210  FORMAT (/,I4,5X,4(1X,E15.6))
     STOP
     END
```

NATURAL FREQUENCIES OF THE STEPPED BEAM

    0.160083E+03     0.617460E+03     0.225198E+04     0.712653E+04

MODE SHAPES

    1      0.103333E-01     0.189147E-03     0.141626E-01     0.445258E-04

    2     -0.376593E-02     0.202976E-03     0.471097E-02     0.259495E-03

    3      0.167902E-03    -0.181687E-03     0.135672E-02     0.207580E-03

    4      0.182648E-03     0.607247E-04     0.373775E-03     0.163869E-03

## REFERENCES

**11.1.** C. Hayashi, *Nonlinear Oscillations in Physical Systems*, McGraw-Hill, New York, 1964.

**11.2.** A. A. Andronow and C. E. Chaikin, *Theory of Oscillations* (English language edition), Princeton University Press, Princeton, N.J., 1949.

**11.3.** N. V. Butenin, *Elements of the Theory of Nonlinear Oscillations*, Blaisdell Publishing Co., New York, 1965.

**11.4.** J. N. J. Cunningham, *Introduction to Nonlinear Analysis*, McGraw-Hill, New York, 1958.

**11.5.** J. J. Stoker, *Nonlinear Vibrations in Mechanical and Electrical Systems*, Interscience Publishers, New York, 1950.

**11.6.** N. Minorsky, *Nonlinear Oscillations*, D. Van Nostrand Co., Princeton, N.J., 1962.

**11.7.** D. G. Ashwell and A. P. Chauhan, "A study of $\frac{1}{2}$-subharmonic oscillations by the method of harmonic balance," *Journal of Sound and Vibration*, Vol. 27, 1973, pp. 313–324.

**11.8.** R. E. Mickens, "Perturbation solution of a highly non-linear oscillation equation," *Journal of Sound and Vibration*, Vol. 68, 1980, pp. 153–155.

**11.9.** B. V. Dasarathy and P. Srinivasan, "Study of a class of non-linear systems reducible to equivalent linear systems," *Journal of Sound and Vibration*, Vol. 7, 1968, pp. 27–30.

**11.10.** G. L. Anderson, "A modified perturbation method for treating non-linear oscillation problems," *Journal of Sound and Vibration*, Vol. 38, 1975, pp. 451–464.

**11.11.** B. L. Ly, "A note on the free vibration of a non-linear oscillator," *Journal of Sound and Vibration*, Vol. 68, 1980, pp. 307–309.

**11.12.** V. A. Bapat and P. Srinivasan, "Free vibrations of non-linear cubic spring mass systems in the presence of Coulomb damping," *Journal of Sound and Vibration*, Vol. 11, 1970, pp. 121–137.

**11.13.** H. R. Srirangarajan, P. Srinivasan, and B. V. Dasarathy, "Analysis of two degrees of freedom systems through weighted mean square linearization approach," *Journal of Sound and Vibration*, Vol. 36, 1974, pp. 119–131.

**11.14.** S. R. Woodall, "On the large amplitude oscillations of a thin elastic beam," *International Journal of Nonlinear Mechanics*, Vol. 1, 1966, pp. 217–238.

**11.15.** D. A. Evenson, "Nonlinear vibrations of beams with various boundary conditions," *AIAA Journal*, Vol. 6, 1968, pp. 370–372.

**11.16.** M. E. Beshai and M. A. Dokainish, "The transient response of a forced non-linear system," *Journal of Sound and Vibration*, Vol. 41, 1975, pp. 53–62.

**11.17.** V. A. Bapat and P. Srinivasan, "Response of undamped non-linear spring mass systems subjected to constant force excitation," *Journal of Sound and Vibration*, Vol. 9, 1969, Part I: pp. 53–58 and Part II: pp. 438–446.

**11.18.** D. E. Newland, *An Introduction to Random Vibrations and Spectral Analysis*, Longman, London, 1975.

**11.19.** J. D. Robson, *An Introduction to Random Vibration*, Edinburgh University Press, Edinburgh, 1963.

**11.20.** A. Papoulis, *Probability, Random Variables and Stochastic Processes*, McGraw-Hill, New York, 1965.

**11.21.** J. S. Bendat and A. G. Piersol, *Engineering Applications of Correlation and Spectral Analysis*, John Wiley, New York, 1980.

**11.22.** F. Kandianis, "Notes on the properties of correlation functions in linear system analysis," *Journal of Sound and Vibration*, Vol. 36, 1974, pp. 207–223.

**11.23.** M. H. Richardson, "Fundamentals of the discrete Fourier transform," *Sound and Vibration*, Vol. 12, March 1978, pp. 40–46.

**11.24.** E. O. Brigham, *The Fast Fourier Transform*, Prentice-Hall, Englewood Cliffs, N.J., 1974.

**11.25.** J. B. Roberts and R. E. D. Bishop, "A simple illustration of spectral density analysis," *Journal of Sound and Vibration*, Vol. 2, 1965, pp. 37–41.

**11.26.** J. B. Roberts, "The response of a simple oscillator to band-limited white noise," *Journal of Sound and Vibration*, Vol. 3, 1966, pp. 115–126.

**11.27.** S. Kaufman, W. Lapinski, and R. C. McCaa, "Response of a single-degree-of-freedom isolator to a random disturbance," *Journal of the Acoustical Society of America*, Vol. 33, 1961, pp. 1108–1112.

**11.28.** J. K. Hammond, "On the response of single and multidegree of freedom systems to non-stationary random excitations," *Journal of Sound and Vibration*, Vol. 7, 1968, pp. 393–416.

**11.29.** T. K. Caughey and H. J. Stumpf, "Transient response of a dynamic system under random excitation," *Journal of Applied Mechanics*, Vol. 28, 1961, pp. 563–566.

**11.30.** S. H. Crandall, G. R. Khabbaz, and J. E. Manning, "Random vibration of an oscillator with nonlinear damping," *Journal of the Acoustical Society of America*, Vol. 36, 1964, pp. 1330–1334.

**11.31.** C. J. Chisholm, "Random vibration techniques applied to motor vehicle structures," *Journal of Sound and Vibration*, Vol. 4, 1966, pp. 129–135.

**11.32.** S. H. Crandall and W. D. Mark, *Random Vibration in Mechanical Systems*, Academic Press, New York, 1963.

**11.33.** S. S. Rao, *The Finite Element Method in Engineering*, Pergamon Press, Oxford, 1982.

**11.34.** R. Davis, R. D. Henshell, and G. B. Warburton, "A Timoshenko beam element," *Journal of Sound and Vibration*, Vol. 22, 1972, pp. 475–487.

**11.35.** D. L. Thomas, J. M. Wilson, and R. R. Wilson, "Timoshenko beam finite elements," *Journal of Sound and Vibration*, Vol. 31, 1973, pp. 315–330.

**11.36.** J. Thomas and B. A. H. Abbas, "Finite element model for dynamic analysis of Timoshenko beams," *Journal of Sound and Vibration*, Vol. 41, 1975, pp. 291–299.

**11.37.** R. S. Gupta and S. S. Rao, "Finite element eigenvalue analysis of tapered and twisted Timoshenko beams," *Journal of Sound and Vibration*, Vol. 56, 1978, pp. 187–200.

**11.38.** H. Alaylioglu and R. Ali, "Analysis of an automotive structure using hybrid stress finite elements," *Computers and Structures*, Vol. 8, 1978, pp. 237–242.

**11.39.** I. Fried, "Accuracy of finite element eigenproblems," *Journal of Sound and Vibration*, Vol. 18, 1971, pp. 289–295.

**11.40.** D. R. J. Owen, "Implicit finite element methods for the dynamic transient analysis of solids with particular reference to nonlinear situations," pp. 123–152 in *Advanced Structural Dynamics*, J. Donéa (ed.), Applied Science Publishers, London, 1980.

## REVIEW QUESTIONS

**11.1.** How do you recognize a nonlinear vibration problem?

**11.2.** What are the various sources of nonlinearity in a vibration problem?

**11.3.** What is the source of nonlinearity in Duffing's equation?

**11.4.** How is the frequency of the solution of Duffing's equation affected by the nature of the spring?

**11.5.** What are subharmonic oscillations?

**11.6.** Explain the jump phenomenon.

**11.7.** What principle is used in the Ritz averaging method?

**11.8.** Define these terms: phase plane, trajectory, singular point, phase velocity.

**11.9.** What is the method of isoclines?

**11.10.** Define probability density function and probability distribution function.

**11.11.** What is a bivariate distribution function?

**11.12.** What is autocorrelation function?

**11.13.** Explain the difference between a stationary process and a nonstationary process.

**11.14.** Define an ergodic process.

**11.15.** What is a Gaussian random process? Why is it frequently used in vibration analysis?

**11.16.** What is a discrete Fourier transform pair?

**11.17.** Define the following terms: spectral density function, white noise, band-limited white noise, wide-band process, narrow-band process.

**11.18.** What is the basic idea behind the finite element method?

**11.19.** What is a shape function?

**11.20.** What is the role of transformation matrices in the finite element method?

**11.21.** How are fixed boundary conditions incorporated in finite element equations?

## PROBLEMS

The problem assignments are organized as follows:

| Problems | Section covered | Topic covered |
|---|---|---|
| 11.1–11.3 | 11.3, 11.4 | Exact solution of nonlinear problems |
| 11.4–11.6 | 11.6 | Graphical methods |
| 11.7 | 11.5 | Approximate analytical methods |
| 11.8 | 11.8 | Numerical methods |
| 11.9 | 11.2 | Introduction to nonlinear vibration |
| 11.10 | 11.7 | Stability analysis |
| 11.11–11.12 | 11.12 | Mean and standard deviations |
| 11.13, 11.15 | 11.14 | Autocorrelation function |
| 11.14 | 11.17 | Fourier transform |
| 11.16 | 11.18 | Spectral density function |
| 11.17 | 11.12 | Mean and standard deviations |
| 11.18–11.19 | 11.14, 11.18 | Spectral density function |
| 11.20 | 11.20 | Random vibration response |
| 11.21 | 11.18 | Spectral density function |
| 11.22 | 11.20 | Random vibration response |
| 11.23–11.30 | 11.23–11.26 | Finite element analysis |

**11.1.** Two springs, having different stiffnesses $k_1$ and $k_2$ with $k_2 > k_1$, are placed on either side of a mass $m$, as shown in Fig. 11.26. When the mass is in its equilibrium position, no spring is in contact with the mass. However, when the mass is displaced from its equilibrium position, only one spring will be compressed. If the mass is given an initial velocity $\dot{x}_0$ at $t = 0$, determine (a) the maximum deflection and (b) the period of vibration of the mass.

**11.2.** Find the equation of motion of the mass shown in Fig. 11.27. Draw the spring force versus $x$ diagram.

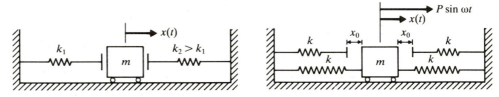

**Figure 11.26**                              **Figure 11.27**

**11.3.** Find the natural time period of oscillation of the pendulum shown in Fig. 11.1(a) when it oscillates between the limits $\theta = -\frac{\pi}{2}$ and $\theta = \frac{\pi}{2}$, using Eqs. (11.1) and (11.9).

**11.4.** The equation of motion of a single degree of freedom system is given by

$$2\ddot{x} + 0.8\dot{x} + 1.6x = 0$$

with initial conditions $x(0) = -1$ and $\dot{x}(0) = 2$. (a) Plot the graph $x(t)$ versus $t$ for $0 \le t \le 10$. (b) Plot a trajectory in the phase plane.

**11.5.** Find the equilibrium position and plot the trajectories in the neighborhood of the equilibrium position corresponding to the following equation:

$$\ddot{x} + 0.1(x^2 - 1)\dot{x} + x = 0$$

**11.6.** Obtain the phase trajectories for a system governed by the equation

$$\ddot{x} + 0.4\dot{x} + 0.8x = 0$$

with the initial conditions $x(0) = 2$ and $\dot{x}(0) = 1$ using the method of isoclines.

**11.7.** A string of length $2l_0$ is under a tension $P$ and carries a mass $m$ at the middle. Derive the equation of motion for large transverse displacements of the mass $m$. Assume the area of cross section as $A$ and Young's modulus as $E$ for the string.

**11.8.** Find the maximum steady-state displacement of the rocket described in Problem 4.27 using the Runge-Kutta method.

**11.9.** Consider the nonlinear differential equation $m\ddot{x} + k_1 x + k_2 x^3 = 0$. Show that if $f_1(t)$ and $f_2(t)$ are solutions of this equation, their superposition $f_1(t) + f_2(t)$ is not a solution.

**11.10.** Determine the eigenvalues and eigenvectors of the equations

$$\dot{x} = 5x - y$$
$$\dot{y} = x + 2y$$

**11.11.** The probability density function of a random variable $x$ is given by

$$p(x) = \begin{cases} 0 & \text{for } x < 0 \\ 0.5 & \text{for } 0 \le x \le 2 \\ 0 & \text{for } x > 2 \end{cases}$$

Determine $E[x]$, $E[x^2]$, and $\sigma_x$.

**11.12.** If $x$ and $y$ are statistically independent, then $E[xy] = E[x]E[y]$. That is, the expected value of the product $xy$ is equal to the product of the separate mean values. If $z = x + y$, where $x$ and $y$ are statistically independent, show that $E[z^2] = E[x^2] + E[y^2] + 2E[x]E[y]$.

**11.13.** Find the autocorrelation functions of the periodic functions shown in Fig. 11.28.

**11.14.** Find the complex form of the Fourier series for the wave shown in Fig. 11.28(b).

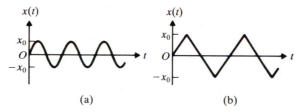

(a)                                                (b)

**Figure 11.28**

**11.15.** Compute the autocorrelation function of a periodic square wave with zero mean value and compare this result with that of a sinusoidal wave of the same period. Assume the amplitudes to be the same for both the waves.

**11.16.** A random signal has a spectral density of $S(f) = 0.0001$ m²/cycle per second between 10 Hz and 1000 Hz. Outside this range, the spectral density is zero. Find the standard deviation and the root mean square value of the signal by assuming its mean value to be 0.05 m.

**11.17.** Find the temporal mean value and the mean square value of the function $x(t) = x_0 \sin(\pi t/2)$.

**11.18.** The autocorrelation function of a Gaussian random process representing the unevenness of a road surface is given by

$$R_x(\tau) = \sigma_x^2 e^{-\alpha|v\tau|} \cos \beta v \tau$$

where $\sigma_x^2$ is the variance of the random process and $v$ is the velocity of the vehicle.

The values of $\sigma_x$, $\alpha$, and $\beta$ for different types of road are as follows:

| Type of road | $\sigma_x$ | $\alpha$ | $\beta$ |
|---|---|---|---|
| Asphalt | 1.1 | 0.2 | 0.4 |
| Paved | 1.6 | 0.3 | 0.6 |
| Gravel | 1.8 | 0.5 | 0.9 |

Compute the spectral density of the road surface for the different types of road.

**11.19.** Compute the autocorrelation function corresponding to the ideal white noise spectral density.

**11.20.** Show that

$$\int_0^\infty \frac{d\xi}{\left(1-\xi^2\right)^2+\left(2\zeta\xi\right)^2} = \frac{\pi}{4\zeta} \qquad \text{for } \zeta \ll 1.$$

**11.21.** Starting from Eqs. (11.104) and (11.105), derive the relations

$$R(\tau) = \int_0^\infty S(f)\cos 2\pi f\tau \cdot df$$

$$S(f) = 4\int_0^\infty R(\tau)\cos 2\pi f\tau \cdot d\tau$$

**11.22.** The motion of a lifting surface about the steady flight path due to atmospheric turbulence can be represented by the equation

$$\ddot{x}(t)+2\zeta\omega_n\dot{x}(t)+\omega_n^2 x(t) = \frac{1}{m}F(t)$$

where $\omega_n$ is the natural frequency, $m$ is the mass, and $\zeta$ is the damping coefficient of the system. The forcing function $F(t)$ denotes the random lift due to the air turbulence and its spectral density is given by

$$S_F(\omega) = \frac{S_T(\omega)}{\left(1+\frac{\pi\omega c}{v}\right)}$$

where $c$ is the chord length and $v$ is the forward velocity of the lifting surface and $S_T(\omega)$ is the spectral density of the upward velocity of air due to turbulence, given by

$$S_T(\omega) = A^2 \frac{1+\left(\frac{L\omega}{v}\right)^2}{\left\{1+\left(\frac{L\omega}{v}\right)^2\right\}^2}$$

where $A$ is a constant and $L$ is the scale of turbulence (constant). Find the mean square value of the response $x(t)$ of the lifting surface.

**11.23.** Find the undamped natural frequencies of longitudinal vibration of the stepped bar shown in Fig. 11.29 with the following data:

$$l_1 = l_2 = l_3 = 0.2 \text{ m}, \ A_1 = 2A_2 = 4A_3 = 0.4\times10^{-3} \text{ m}^2,$$

$$E = 2.1\times10^{11} \text{ N/m}^2, \text{ and } \rho = 7.8\times10^3 \text{ kg/m}^3.$$

**11.24.** Find the natural frequencies of a cantilever beam of length $l$, cross-sectional area $A$, moment of inertia $I$, Young's modulus $E$, and density $\rho$, using one finite element.

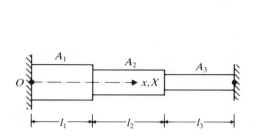

**Figure 11.29**

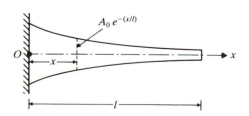

**Figure 11.30**

**11.25.** Find the natural frequencies of vibration of the beam shown in Fig. 11.30, using two beam elements. Also find the load vector if a uniformly distributed transverse load $p$ is applied to element 1.

**11.26.** Find the natural frequencies of a beam of length $l$, which is pin connected at $x = 0$ and fixed at $x = l$, using one beam element.

**11.27.** Find the global stiffness matrix of each of the four elements of the truss shown in Fig. 11.21 using the following data:
Nodal coordinates: $(X_1, Y_1) = (0,0)$, $(X_2, Y_2) = (50, 100)$ in., $(X_3, Y_3) = (100, 0)$ in., $(X_4, Y_4) = (200, 150)$ in.
Cross-sectional areas: $A_1 = A_2 = A_3 = A_4 = 2$ in.$^2$.
Young's modulus of all members: $30 \times 10^6$ lb$_f$/in.$^2$.

**11.28.** Using the result of Problem 11.27, find the assembled stiffness matrix of the truss and formulate the equilibrium equations if the vertical downward load applied at node 4 is 1000 lb$_f$.

**11.29.** Find the natural frequencies of torsional vibration of the stepped shaft shown in Fig. 11.31. Assume that $\rho_1 = \rho_2 = \rho$, $G_1 = G_2 = G$, $I_{p_1} = 2I_{p_2} = 2I_p$, $J_1 = 2J_2 = 2J$, and $l_1 = l_2 = l$.

**11.30.** Find the natural frequencies of vibration of the uniform shaft shown in Fig. 11.23, using two finite elements of equal length.

**11.31.** Derive the stiffness matrix of the bar element in longitudinal vibration whose cross-sectional area varies as $A(x) = A_0 e^{-(x/l)}$, where $A_0$ is the area at the root (see Fig. 11.32).

**11.32.** Derive Eq. (11.90) for a periodic function.

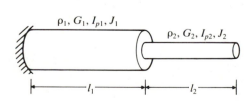

**Figure 11.31**

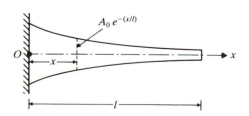

**Figure 11.32**

**11.33.** Derive Eq. (E.5) of Example 11.4.

**11.34.** Solve the equation of motion $\ddot{x} + 0.5\dot{x} + x + 0.4x^3 = 0.8\cos 0.4t$ using the Runge-Kutta method with $\Delta t = 0.05$, $t_{max} = 5.0$, and $x_0 = \dot{x}_0 = 0$. Plot the variation of $x$ with $t$.

**11.35.** Find the time variation of the angular displacement of a simple pendulum (i.e., the solution of Eq. (11.2)) for $g/l = 0.5$, using the initial conditions $\theta_0 = 45°$ and $\dot{\theta}_0 = 0$. Use the Runge-Kutta method.

**11.36.** Write a computer program for finding the period of vibration corresponding to Eq. (11.11). Use a suitable numerical integration procedure. Using this program, find the solution of Problem 11.35.

**11.37.** Write a computer program for finding the assembled stiffness matrix of a general planar truss.

**11.38.** Generalize the computer program of Sec. 11.27.2 to make it applicable to the solution of any stepped beam having a specified number of steps.

**11.39.** Find the natural frequencies and mode shapes of the beam shown in Fig. 11.25 with $l_1 = l_2 = l_3 = 10$ in. and a uniform cross section of 1 in. × 1 in. throughout the length, using the computer program of Sec. 11.27.2. Compare your results with those given in Chap. 8. (Hint: Only the data XL, XI, and A need to be changed.)

# Matrices

Arthur Cayley (1821–1895) was a British mathematician and professor of mathematics at Cambridge University. His greatest work, produced with James Joseph Sylvester, was the development of the theory of invariants, which played a crucial role in the theory of relativity. He made many important contributions to $n$-dimensional geometry and invented and developed the theory of matrices.

Courtesy The Granger Collection

## A.1 DEFINITIONS

**Matrix.** A matrix is a rectangular array of numbers. An array having $m$ rows and $n$ columns enclosed in brackets is called an $m$-by-$n$ matrix. If $[A]$ is an $m \times n$ matrix, it is denoted as

$$[A] = [a_{ij}] = \begin{bmatrix} a_{11} & a_{12} & \cdots & a_{1n} \\ a_{21} & a_{22} & \cdots & a_{2n} \\ \vdots & & & \\ a_{m1} & a_{m2} & \cdots & a_{mn} \end{bmatrix} \tag{A.1}$$

where the numbers $a_{ij}$ are called the *elements* of the matrix. The first subscript $i$ denotes the row and the second subscript $j$ specifies the column in which the element $a_{ij}$ appears.

**Square Matrix.** When the number of rows ($m$) is equal to the number of columns ($n$), the matrix is called a *square matrix of order n*.

**Column Matrix.** A matrix consisting of only one column—that is, an $m \times 1$ matrix—is called a *column matrix* or more commonly a *column vector*. Thus if $\vec{a}$ is a column vector having $m$ elements, it can be represented as

$$\vec{a} = \begin{Bmatrix} a_1 \\ a_2 \\ \vdots \\ a_m \end{Bmatrix} \tag{A.2}$$

**Row Matrix.** A matrix consisting of only one row—that is, a $1 \times n$ matrix—is called a *row matrix* or a *row vector*. If $\lfloor b \rfloor$ is a row vector, it can be denoted as

$$\lfloor b \rfloor = [b_1 \quad b_2 \quad \cdots \quad b_n] \tag{A.3}$$

**Diagonal Matrix.** A square matrix in which all the elements are zero except those on the principal diagonal is called a *diagonal matrix*. For example, if $[A]$ is a diagonal matrix of order $n$, it is given by

$$[A] = \begin{bmatrix} a_{11} & 0 & 0 & \cdots & 0 \\ 0 & a_{22} & 0 & \cdots & 0 \\ 0 & 0 & a_{33} & \cdots & 0 \\ \vdots & & & & \\ 0 & 0 & 0 & \cdots & a_{nn} \end{bmatrix} \tag{A.4}$$

**Identity Matrix.** If all the elements of a diagonal matrix have a value 1, then the matrix is called an *identity matrix* or *unit matrix* and is usually denoted as $[I]$.

**Zero Matrix.** If all the elements of a matrix are zero, it is called a *zero* or *null matrix* and is denoted as $[0]$. If $[0]$ is of order $2 \times 4$, it is given by

$$[0] = \begin{bmatrix} 0 & 0 & 0 & 0 \\ 0 & 0 & 0 & 0 \end{bmatrix} \tag{A.5}$$

**Symmetric Matrix.** A square matrix for which the upper right half can be obtained by flipping the matrix about the diagonal is called a *symmetric matrix*. This means that if $[A]$ is a symmetric matrix, we have $a_{ji} = a_{ij}$. For example,

$$[A] = \begin{bmatrix} 4 & -1 & -3 \\ -1 & 0 & 7 \\ -3 & 7 & 5 \end{bmatrix} \tag{A.6}$$

is a symmetric matrix of order 3.

**Transpose of a Matrix.** The transpose of an $m \times n$ matrix $[A]$ is the $n \times m$ matrix obtained by interchanging the rows and columns of $[A]$ and is denoted as $[A]^T$. Thus if

$$[A] = \begin{bmatrix} 2 & 4 & 5 \\ 3 & 1 & 8 \end{bmatrix} \tag{A.7}$$

then $[A]^T$ is given by

$$[A]^T = \begin{bmatrix} 2 & 3 \\ 4 & 1 \\ 5 & 8 \end{bmatrix} \tag{A.8}$$

Note that the transpose of a column matrix (vector) is a row matrix (vector), and vice versa.

**Trace.** The sum of the main diagonal elements of a square matrix $[A] = [a_{ij}]$ is called the *trace* of $[A]$ and is given by

$$\text{Trace}[A] = a_{11} + a_{22} + \cdots + a_{nn} \tag{A.9}$$

**Determinant.** If $[A]$ denotes a square matrix of order $n$, then the determinant of $[A]$ is denoted as $|[A]|$. Thus

$$|[A]| = \begin{vmatrix} a_{11} & a_{12} & \cdots & a_{1n} \\ a_{21} & a_{22} & \cdots & a_{2n} \\ \vdots & & & \\ a_{n1} & a_{n2} & \cdots & a_{nn} \end{vmatrix} \tag{A.10}$$

The value of a determinant can be found by obtaining the minors and cofactors of the determinant.

The *minor* of the element $a_{ij}$ of the determinant $|[A]|$ of order $n$ is a determinant of order $(n-1)$ obtained by deleting the row $i$ and the column $j$ of the original determinant. The minor of $a_{ij}$ is denoted as $M_{ij}$.

The *cofactor* of the element $a_{ij}$ of the determinant $|[A]|$ of order $n$ is the minor of the element $a_{ij}$, with either a plus or a minus sign attached; it is defined as

$$\text{cofactor of } a_{ij} = \beta_{ij} = (-1)^{i+j} M_{ij} \tag{A.11}$$

where $M_{ij}$ is the minor of $a_{ij}$. For example, the cofactor of the element $a_{32}$ of

$$\det[A] = \begin{vmatrix} a_{11} & a_{12} & a_{13} \\ a_{21} & a_{22} & a_{23} \\ a_{31} & a_{32} & a_{33} \end{vmatrix} \tag{A.12}$$

is given by

$$\beta_{32} = (-1)^5 M_{32} = -\begin{vmatrix} a_{11} & a_{13} \\ a_{21} & a_{23} \end{vmatrix} \tag{A.13}$$

The value of a second order determinant $\|[A]\|$ is defined as

$$\det[A] = \begin{vmatrix} a_{11} & a_{12} \\ a_{21} & a_{22} \end{vmatrix} = a_{11}a_{22} - a_{12}a_{21} \tag{A.14}$$

The value of an $n$th order determinant $\|[A]\|$ is defined as

$$\det[A] = \sum_{j=1}^{n} a_{ij}\beta_{ij} \text{ for any specific row } i$$

or

$$\det[A] = \sum_{i=1}^{n} a_{ij}\beta_{ij} \text{ for any specific column } j \tag{A.15}$$

For example, if

$$\det[A] = \|[A]\| = \begin{vmatrix} 2 & 2 & 3 \\ 4 & 5 & 6 \\ 7 & 8 & 9 \end{vmatrix} \tag{A.16}$$

then, by selecting the first column for expansion, we obtain

$$\det[A] = 2\begin{vmatrix} 5 & 6 \\ 8 & 9 \end{vmatrix} - 4\begin{vmatrix} 2 & 3 \\ 8 & 9 \end{vmatrix} + 7\begin{vmatrix} 2 & 3 \\ 5 & 6 \end{vmatrix}$$
$$= 2(45-48) - 4(18-24) + 7(12-15) = -3 \tag{A.17}$$

**Properties of Determinants**

1. The value of a determinant is not affected if rows (or columns) are written as columns (or rows) in the same order.

2. If all the elements of a row (or a column) are zero, the value of the determinant is zero.

3. If any two rows (or two columns) are interchanged, the value of the determinant is multiplied by $-1$.

4. If all the elements of one row (or one column) are multiplied by the same constant $a$, the value of the new determinant is $a$ times the value of the original determinant.

5. If the corresponding elements of two rows (or two columns) of a determinant are proportional, the value of the determinant is zero. For example,

$$\det[A] = \begin{vmatrix} 4 & 7 & -8 \\ 2 & 5 & -4 \\ -1 & 3 & 2 \end{vmatrix} = 0 \tag{A.18}$$

**Adjoint Matrix.** The adjoint matrix of a square matrix $[A] = [a_{ij}]$ is defined as the matrix obtained by replacing each element $a_{ij}$ by its cofactor $\beta_{ij}$ and then transposing. Thus

$$\text{Adjoint}[A] = \begin{bmatrix} \beta_{11} & \beta_{12} & \cdots & \beta_{1n} \\ \beta_{21} & \beta_{22} & \cdots & \beta_{2n} \\ \vdots & & & \\ \beta_{n1} & \beta_{n2} & \cdots & \beta_{nn} \end{bmatrix}^T = \begin{bmatrix} \beta_{11} & \beta_{21} & \cdots & \beta_{n1} \\ \beta_{12} & \beta_{22} & \cdots & \beta_{n2} \\ \vdots & & & \\ \beta_{1n} & \beta_{2n} & \cdots & \beta_{nn} \end{bmatrix} \quad (A.19)$$

**Inverse Matrix.** The inverse of a square matrix $[A]$ is written as $[A]^{-1}$ and is defined by the following relationship:

$$[A]^{-1}[A] = [A][A]^{-1} = [I] \quad (A.20)$$

where $[A]^{-1}[A]$, for example, denotes the product of the matrix $[A]^{-1}$ and $[A]$. The inverse matrix of $[A]$ can be determined (see Ref. [A.1]):

$$[A]^{-1} = \frac{\text{adjoint}[A]}{\det[A]} \quad (A.21)$$

when $\det[A]$ is not equal to zero. For example, if

$$[A] = \begin{bmatrix} 2 & 2 & 3 \\ 4 & 5 & 6 \\ 7 & 8 & 9 \end{bmatrix} \quad (A.22)$$

its determinant has a value $\det[A] = -3$. The cofactor of $a_{11}$ is

$$\beta_{11} = (-1)^2 \begin{vmatrix} 5 & 6 \\ 8 & 9 \end{vmatrix} = -3 \quad (A.23)$$

In a similar manner, we can find the other cofactors and determine

$$[A]^{-1} = \frac{\text{adjoint}[A]}{\det[A]} = \frac{1}{-3} \begin{bmatrix} -3 & 6 & -3 \\ 6 & -3 & 0 \\ -3 & -2 & 2 \end{bmatrix} = \begin{bmatrix} 1 & -2 & 1 \\ -2 & 1 & 0 \\ 1 & 2/3 & -2/3 \end{bmatrix} \quad (A.24)$$

**Singular Matrix.** A square matrix is said to be singular if its determinant is zero.

## A.2 BASIC MATRIX OPERATIONS

**Equality of Matrices.** Two matrices $[A]$ and $[B]$, having the same order, are equal if and only if $a_{ij} = b_{ij}$ for every $i$ and $j$.

**Addition and Subtraction of Matrices.** The sum of the two matrices $[A]$ and $[B]$, having the same order, is given by the sum of the corresponding elements. Thus if $[C] = [A] + [B] = [B] + [A]$, we have $c_{ij} = a_{ij} + b_{ij}$ for every $i$ and $j$. Similarly, the difference between two matrices $[A]$ and $[B]$ of the same order, $[D]$, is given by $[D] = [A] - [B]$ with $d_{ij} = a_{ij} - b_{ij}$ for every $i$ and $j$.

**Multiplication of Matrices.** The product of two matrices $[A]$ and $[B]$ is defined only if they are conformable—that is, if the number of columns of $[A]$ is equal to the number of rows of $[B]$. If $[A]$ is of order $m \times n$ and $[B]$ is of order $n \times p$, then the product $[C] = [A][B]$ is of order $m \times p$ and is defined by $[C] = [c_{ij}]$, with

$$c_{ij} = \sum_{k=1}^{n} a_{ik} b_{kj} \tag{A.25}$$

This means that $c_{ij}$ is the quantity obtained by multiplying the $i$th row of $[A]$ and the $j$th column of $[B]$ and summing these products. For example, if

$$[A] = \begin{bmatrix} 2 & 3 & 4 \\ 1 & -5 & 6 \end{bmatrix} \quad \text{and} \quad [B] = \begin{bmatrix} 8 & 0 \\ 2 & 7 \\ -1 & 4 \end{bmatrix} \tag{A.26}$$

then

$$\begin{aligned}
[C] = [A][B] &= \begin{bmatrix} 2 & 3 & 4 \\ 1 & -5 & 6 \end{bmatrix} \begin{bmatrix} 8 & 0 \\ 2 & 7 \\ -1 & 4 \end{bmatrix} \\
&= \begin{bmatrix} 2 \times 8 + 3 \times 2 + 4 \times (-1) & 2 \times 0 + 3 \times 7 + 4 \times 4 \\ 1 \times 8 + (-5) \times 2 + 6 \times (-1) & 1 \times 0 + (-5) \times 7 + 6 \times 4 \end{bmatrix} \\
&= \begin{bmatrix} 18 & 37 \\ -8 & -11 \end{bmatrix} \tag{A.27}
\end{aligned}$$

If the matrices are conformable, the matrix multiplication process is associative:

$$([A][B])[C] = [A]([B][C]) \tag{A.28}$$

and is distributive:

$$([A] + [B])[C] = [A][C] + [B][C] \tag{A.29}$$

Note that $[A][B]$ is the premultiplication of $[B]$ by $[A]$ or the postmultiplication of $[A]$ by $[B]$. Also, the product $[A][B]$ is not necessarily equal to $[B][A]$.

The transpose of a matrix product is given by the product of the transposes of the separate matrices in reverse order. Thus, if $[C] = [A][B]$,

$$[C]^T = ([A][B])^T = [B]^T [A]^T \tag{A.30}$$

The inverse of a matrix product is given by the product of the inverses of the separate matrices in reverse order. Thus if $[C] = [A][B]$,

$$[C]^{-1} = ([A][B])^{-1} = [B]^{-1}[A]^{-1} \tag{A.31}$$

## REFERENCES

**A.1.** S. Barnett, *Matrix Methods for Engineers and Scientists*, McGraw-Hill, New York, 1982.

# Laplace Transform Pairs

Pierre Simon Laplace (1749–1827) was a French mathematician, remembered for his fundamental contributions to probability theory, mathematical physics, and celestial mechanics; the name Laplace occurs in both mechanical and electrical engineering. Much use is made of Laplace transforms in vibrations and applied mechanics, and the Laplace equation is applied extensively in the study of electric and magnetic fields. Courtesy of Brown Brothers.

Courtesy Brown Brothers

| Laplace domain $$\bar{f}(s)=\int_0^\infty f(t)e^{-st}\,dt$$ | Time domain $f(t) = \mathcal{L}^{-1}\{F(s)\}$ $x(t)$ |
|---|---|
| 1. $c_1\bar{f}(s)+c_2\bar{g}(s)$ | $c_1 f(t)+c_2 g(t)$ |
| 2. $\bar{f}\left(\dfrac{s}{a}\right)$ | $f(a\cdot t)a$ |
| 3. $\bar{f}(s)\bar{g}(s)$ | $\displaystyle\int_0^t f(t-\tau)g(\tau)\,d\tau = \int_0^t f(\tau)g(t-\tau)\,d\tau$ |
| 4. $s^n\bar{f}(s)-\displaystyle\sum_{j=1}^n s^{n-j}\dfrac{d^{j-1}f}{dt^{j-1}}(0)$ | $\dfrac{d^n f}{dt^n}(t)$ |
| 5. $\dfrac{1}{s^n}\bar{f}(s)$ | $\underbrace{\displaystyle\int_0^t\cdots\int_0^t f(\tau)\,d\tau\cdots d\tau}_{n}$ |
| 6. $\bar{f}(s+a)$ | $e^{-at}f(t)$ |
| 7. $\dfrac{a}{s(s+a)}$ | $1-e^{-at}$ |
| 8. $\dfrac{s+a}{s^2}$ | $1+at$ |
| 9. $\dfrac{a^2}{s^2(s+a)}$ | $at-(1-e^{-at})$ |
| 10. $\dfrac{s+b}{s(s+a)}$ | $\dfrac{b}{a}\left\{1-\left(1-\dfrac{a}{b}\right)e^{-at}\right\}$ |
| 11. $\dfrac{a}{s^2+a^2}$ | $\sin at$ |
| 12. $\dfrac{s}{s^2+a^2}$ | $\cos at$ |
| 13. $\dfrac{a^2}{s(s^2+a^2)}$ | $1-\cos at$ |
| 14.* $\dfrac{1}{s^2+2\zeta\omega_n s+\omega_n^2}$ | $\dfrac{1}{\omega_d}e^{-\zeta\omega_n t}\sin\omega_d t$ |
| 15.* $\dfrac{s}{s^2+2\zeta\omega_n s+\omega_n^2}$ | $-\dfrac{\omega_n}{\omega_d}e^{-\zeta\omega_n t}\sin(\omega_d t-\phi_1)$ |
| 16.* $\dfrac{s+2\zeta\omega_n}{s^2+2\zeta\omega_n s+\omega_n^2}$ | $\dfrac{\omega_n}{\omega_d}e^{-\zeta\omega_n t}\sin(\omega_d t+\phi_1)$ |
| 17.* $\dfrac{\omega_n^2}{s(s^2+2\zeta\omega_n s+\omega_n^2)}$ | $1-\dfrac{\omega_n}{\omega_d}e^{-\zeta\omega_n t}\sin(\omega_d t+\phi_1)$ |
| 18.* $\dfrac{s+\zeta\omega_n}{s(s^2+2\zeta\omega_n s+\omega_n^2)}$ | $e^{-\zeta\omega_n t}\sin(\omega_d t+\phi_1)$ |

$\omega_n^2=\dfrac{k}{m}$

$\zeta=\dfrac{c}{c_c}$

$c_c=2m\omega_n$

19.   1

Unit impulse at $t = 0$

20.   $\dfrac{e^{-as}}{s}$

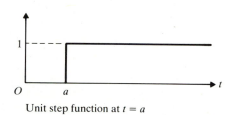

Unit step function at $t = a$

---

$^*\omega_d = \omega_n\sqrt{1 - \zeta^2}\,;\quad \zeta < 1$

$\phi_1 = cos^{-1}\zeta;\quad \zeta < 1$

# Units

Heinrich Rudolf Hertz (1857–1894), a German physicist and a professor of physics at the Polytechnic Institute in Karlsruhe and later at the University of Bonn, gained fame through his experiments on radio waves. His investigations in the field of elasticity formed a relatively small part of his achievements, but are of vital importance to engineers. His work on the analysis of elastic bodies in contact is referred to as ''Hertzian stresses,'' and is very important in the design of ball and roller bearings. The unit of frequency of periodic phenomena, measured in cycles per second, is named Hertz in SI units.

Courtesy The Granger Collection

The English system of units is now being replaced by the International System of units (SI). The SI system is the modernized version of the metric system of units. Its name in French is Système International; hence the abbreviation *SI*. The SI system has seven basic units. All other units can be derived from these seven [C.1–C.2]. The three basic units of concern in the study of vibrations are meter for length, kilogram for mass, and second for time.

The consistent set of units for various quantities is given in Table C.1. The common prefixes for multiples and submultiples of SI units are given in Table C.2. In the SI system, the combined units must be abbreviated with care. For example, a torque of 4 kg × 2 m must be stated as 8 kg m or 8 kg·m with either a space or a dot between kg and m. It should not be written as kgm. Another example is 8 m × 5 s = 40 m s or 40 m·s or 40 meter-seconds. If it is written as 40 ms, it means 40 milliseconds.

### Conversion of Units

To convert the units of any given quantity from one system to another, we use the equivalence of units given on the inside back cover. The following examples illustrate the procedure.

---

### EXAMPLE C.1

*Mass moment of inertia:*

$$\begin{pmatrix} \text{mass moment of} \\ \text{inertia in SI units} \end{pmatrix} = \begin{pmatrix} \text{mass moment of inertia} \\ \text{in English units} \end{pmatrix} \times \begin{pmatrix} \text{multiplication} \\ \text{factor} \end{pmatrix}$$

$$\begin{aligned}
\left( \text{kg} \cdot \text{m}^2 \right) \equiv \left( \text{N} \cdot \text{m} \cdot \text{s}^2 \right) &= \left( \frac{\text{N}}{\text{lb}_\text{f}} \cdot \text{lb}_\text{f} \right) \left( \frac{\text{m}}{\text{in.}} \cdot \text{in.} \right) \left( \text{sec}^2 \right) \\
&= \left( \text{N per 1 lb}_\text{f} \right) \left( \text{m per 1 in.} \right) \left( \text{lb}_\text{f}\text{-in.-sec}^2 \right) \\
&= (4.448222)(0.0254) \left( \text{lb}_\text{f}\text{-in.-sec}^2 \right) \\
&= 0.1129848 \left( \text{lb}_\text{f}\text{-in.-sec}^2 \right)
\end{aligned}$$

---

### EXAMPLE C.2

*Stress:*

$$(\text{Stress in SI units}) = (\text{stress in English units}) \times \begin{pmatrix} \text{multiplication} \\ \text{factor} \end{pmatrix}$$

$$\begin{aligned}
(\text{Pa}) \equiv \left( \text{N/m}^2 \right) &= \left( \frac{\text{N}}{\text{lb}_\text{f}} \cdot \text{lb}_\text{f} \right) \frac{1}{\left( \dfrac{\text{m}}{\text{in.}} \cdot \text{in.} \right)^2} = \frac{\text{N}}{\text{lb}_\text{f}} \frac{1}{\left( \dfrac{\text{m}}{\text{in.}} \right)^2} \left( \text{lb}_\text{f}/\text{in.}^2 \right) \\
&= \frac{\left( \text{N per 1 lb}_\text{f} \right)}{\left( \text{m per 1 in.} \right)^2} \left( \text{lb}_\text{f}/\text{in.}^2 \right) \\
&= \frac{(4.448222)}{(0.0254)^2} \left( \text{lb}_\text{f}/\text{in.}^2 \right) \\
&= 6894.757 \left( \text{lb}_\text{f}/\text{in.}^2 \right)
\end{aligned}$$

---

**TABLE C.1   Consistent set of units**

| Quantity | SI Unit | | English Unit* (symbol) |
|---|---|---|---|
| | Unit | Symbol | |
| Mass | kilogram | kg | $lb_f\text{-}sec^2/in.$ |
| Length | meter | m | in. |
| Time | second | s | sec |
| Force | Newton $= kg \cdot m/s^2$ | N | $lb_f$ |
| Stress or pressure or elastic modulus | Pascal $= N/m^2$ | Pa | $lb_f/in^2 = psi$ |
| Moment of a force | Newton meter $= kg \cdot m^2/s^2$ | $N \cdot m$ | $in.\text{-}lb_f$ |
| Work or energy | Joule $= N \cdot m$ | J | $in.\text{-}lb_f$ |
| Power | Watt $= J/s$ | W | $in.\text{-}lb_f/sec$ |
| Frequency | Hertz $= 1/s$ | Hz | Hz |
| Mass density | kilogram per cubic meter | $kg/m^3$ | $lb_f\text{-}sec^2/in^4$ |
| Velocity or speed | meter per second | m/s | in./sec |
| Acceleration | meter per second squared | $m/s^2$ | $in./sec^2$ |
| Angular displacement or plane angle | radian | rad | rad |
| Angular velocity | radian per second | rad/s | rad/sec |
| Angular acceleration | radian per second squared | $rad/s^2$ | $rad/sec^2$ |
| Area moment of inertia | $(meter)^4$ | $m^4$ | $in^4$ |
| Mass moment of inertia | kilogram $\cdot$ meter squared | $kg \cdot m^2$ | $in.\text{-}lb_f\text{-}sec^2$ |
| Stiffness coefficient: | | | |
|   Rectilinear (translational) | Newton per meter | N/m | $lb_f/in.$ |
|   Rotational (torsional) | Newton meter per radian | $N \cdot m/rad$ | $lb_f\text{-}in./rad$ |
| Damping coefficient: | | | |
|   Rectilinear (translational) | Newton second per meter | $N \cdot s/m$ | $lb_f\text{-}sec/in.$ |
|   Rotational (torsional) | Newton meter second per radian | $N \cdot m \cdot s/rad$ | $lb_f\text{-}in.\text{-}sec/rad$ |
| Mass: | | | |
|   Rectilinear (translational) | kilogram | kg | $lb_f\text{-}sec^2/in.$ |
|   Rotational (torsional) | kilogram meter squared | $kg \cdot m^2$ | $lb_f\text{-}sec^2\text{-}in.$ |

*ft (foot) can be interchanged with in. (inch).*

**TABLE C.2 Prefixes for Multiples and Submultiples of SI Units**

| Multiple | Prefix | Symbol | Submultiple | Prefix | Symbol |
|---|---|---|---|---|---|
| 10 | deka | da | $10^{-1}$ | deci | d |
| $10^2$ | hecto | h | $10^{-2}$ | centi | c |
| $10^3$ | kilo | k | $10^{-3}$ | milli | m |
| $10^6$ | mega | M | $10^{-6}$ | micro | $\mu$ |
| $10^9$ | giga | G | $10^{-9}$ | nano | n |
| $10^{12}$ | tera | T | $10^{-12}$ | pico | p |

# REFERENCES

**C.1.** E. A. Mechtly, "The International System of Units" (Second Revision), NASA SP-7012, 1973.

**C.2.** C. Wandmacher, "Metric Units in Engineering—Going SI," Industrial Press, New York, 1978.

# Answers to
# Selected
# Problems

# Chapter 1

**1.2.** $k_{eq} = k_t + k_1 l^2 + k_2 l^2$   **1.4.** 247.2 MN/m

**1.6.** $k_{eq} = 3k\cos^2\alpha$, $c_{eq} = 3c\cos^2\alpha$   **1.9.** $d = 49.35$ mm

**1.12.** $m_{eq} = \dfrac{W}{g} + \dfrac{2J_0}{r^2}$   **1.14.** $J_{eq} = J_1 + J_2\left(\dfrac{n_1}{n_2}\right)^2$

**1.16.** $c_t = \dfrac{\pi\mu R^4}{2l}$   **1.18.** $A = 5.3852$, $\theta = 21.8014°$

**1.21.** (i) $A = 0.02022$ m, $\alpha = -81.4692°$   (ii) $A_1 = 0.003$ m, $A_2 = 0.02$ m

**1.24.** $x_2(t) = 6.1966\sin(\omega t + 83.7938°)$   **1.26.** No

**1.29.** $\tau = 0.1$ s, $\dot{x}_{max} = 3.1416$ m/s, $\ddot{x}_{max} = 197.393$ m/s$^2$

**1.31.** 62.4642 Hz   **1.34.** $x(t) = \dfrac{A}{2} - \dfrac{4A}{\pi^2}\displaystyle\sum_{n=1,3,\ldots}^{\infty}\dfrac{1}{n^2}\cos n\omega t$

**1.39.** $x(t) = \dfrac{A}{2} + \dfrac{A}{\pi}\displaystyle\sum_{n=1}^{\infty}\dfrac{1}{n}\sin n\omega t$

**1.41.** $a_0 = 36.0$, $a_1 = -10.4243$, $a_2 = -0.8534$, $a_3 = -2.7014$, $b_1 = -2.0830$, $b_2 = -5.7249$, $b_3 = 1.8017$

**1.45.** $a_0 = 13.0833$, $a_1 = -10.2959$, $a_2 = 0.0223$, $a_3 = 0.5933$, $a_4 = 0.4999$, $a_5 = 0.5719$, $a_6 = 0.1666$, $b_1 = 25.8611$, $b_2 = 1.8438$, $b_3 = 3.9101$, $b_4 = 2.0208$, $b_5 = -0.8263$, $b_6 = 0.4167$

# Chapter 2

**2.2.** 0.1715 s   **2.4.** 0.0993 s

**2.6.** $\left[k_1 k_2\left(\dfrac{l_1}{l_2}\right)^2 \middle/ \left(m\left\{k_2 + k_1\left(\dfrac{l_1}{l_2}\right)^2\right\}\right)\right]^{1/2}$   **2.8.** $(k/m)^{1/2}$

**2.10.** $k = 52.6381$ N/m, $m = 0.3333$ kg   **2.12.** $\sqrt{\dfrac{k}{2m}}$

**2.14.** $\tau_n = 0.2204$ s, $A = 0.07015$ m   **2.19.** $\tau_n = \dfrac{\pi\sqrt{M}}{h}\left(\dfrac{2h^2 + a^2}{2k}\right)^{1/2}$

**2.21.** 162.5743 rad/s   **2.24.** 454.7935 N/m

**2.26.** 1.4185 s   **2.29.** 0.0043 m   **2.32.** 20.0734 rad/s

**2.36.** 2.0258 s   **2.40.** $\omega_n = \sqrt{\dfrac{2k}{m}}$

**2.43.** $\omega_n = \sqrt{\dfrac{3\left(k_t + k_1 a^2 + k_2 l^2\right)}{ml^2}}$

**2.46.** $x_{max} = \dfrac{1}{\omega_n}e^{-\omega_n t}(x_0\omega_n + \dot{x}_0)$ where $t = \dot{x}_0/\{\omega_n(\dot{x}_0 + \omega_n x_0)\}$

**2.48.** 344.1686 N-s/m   **2.50.** 26.6775 %   **2.53.** 0.09541°

**2.55.** 4.6 s   **2.58.** (i) 8.514 m/s   (ii) 0.78 s   (iii) 0.4446 m

**2.61.** Coulomb; $\omega_n = 14.1421$ rad/s

**2.63.** 5.8 mm     **2.65.** $\beta = 0.03032$, $c_{eq} = 0.04288$ N-s/m, $\Delta W = 0.019050$ N-m
**2.68.** $\beta = 0.3220$     **2.70.** $x(t = 0.1) = 0.11198$ m, $\dot{x}(t = 0.1) = -2.289$ m/s,
$\ddot{x}(t = 0.1) = -12.7520$ m/s$^2$

# Chapter 3

**3.1.** (i) $\delta = 0.0125$ m   (ii) $\delta_{st} = 0.015$ m   (iii) $X = 0.0185$ m
**3.4.** 2.3749 kg     **3.7.** $\zeta = 0.151382$

**3.12.** (iii) $\dfrac{c\omega kX}{(k^2 + \omega^2 c^2)}[-k\sin\omega t + c\omega\cos\omega t]$

**3.15.** $1/\sqrt{1 - \zeta^2}$
**3.18.** $y(t) = 0.0001739[\cos 7.0711t + \cos 25t]$ m, where $y(t) = x(t) - x_g(t)$
**3.20.** 57.2268 mm     **3.22.** 0.10598 m     **3.26.** 2.587 mm     **3.32.** 0.0006984 m
**3.34.** 0.1     **3.36.** 0.1511°     **3.39.** $b = 0.028685$, $\gamma = 2.254964$
**3.42.** for viscous damping     **3.44.** (i) 20.4053 N-m   (ii) 81.6212 N-m
**3.47.** 12.5% at $r^* = 3$     **3.48.** 5.34% at $r^* = 3.0$
**3.51.** 13.99% at $r^* = 0.6$     **3.56.** $m = 0.01396$ kg, $k = 3527.1762$ N/m
**3.58.** 35.2635 Hz     **3.61.** (i) 1.4222 Hz   (ii) 3.0630 Hz
**3.65.** with $N = 20$, $x(t = 20\Delta t) = 0.01048$, $\dot{x}(t = 20\Delta t) = 2.1295$, $\ddot{x}(t = 20\Delta t) = 0.4372$

# Chapter 4

**4.1.** 4.0556 mm

**4.5.** $x(t) = \dfrac{4F_0}{\pi k}\displaystyle\sum_{n=1,3,5,\ldots}\dfrac{1}{n}\dfrac{1}{\sqrt{(1 - n^2 r^2)^2 + (2\zeta nr)^2}}\sin(n\omega t - \phi_n)$, where $r = \dfrac{\omega}{\omega_n}$ and $\phi_n$

$= \tan^{-1}\left(\dfrac{2\zeta nr}{1 - n^2 r^2}\right)$

**4.11.** $x(t) = 0.016 + 0.0034613\cos 41.888t + 0.0106529\sin 41.888t - 0.0011542\cos 83.776t + 0.0008386\sin 83.776t - 0.0002157\cos 125.664t - 0.0001567\sin 125.664t + \cdots$ m

**4.16.** $x(t) = A\sin(10t + \phi) + 0.00015708 - 0.0001441e^{-3t}$ m, ($A$ and $\phi$ are determined from initial conditions)

**4.20.** $\tan\omega_n\sqrt{1 - \zeta^2}\cdot t_m = \dfrac{\sqrt{1 - \zeta^2}}{\zeta}$

**4.24.** $z_{max} = -2.9090$
**4.27.** (1) $(2000 - 10t)\ddot{x} + c\dot{x} + kx = 20000$   (2) 0.002667 m

**4.30.** For $t > t_0$: $z_{max} = -\dfrac{2\dot{x}_0}{\omega_n t_0}\sin\dfrac{\omega_n t_0}{2}$

**4.32.** 37.6975 MN

**4.35.** $x(t) = \dfrac{F_0}{m\omega_n^2}[(1 - \cos\omega_n t) - (1 - \cos\omega_n\{t - t_0\})]$ for $t \geq t_0$

**4.42.** $x(20) = 0.01867$ (Idealization 1, Fig. 4.18), 0.01745 (Idealization 1, Fig. 4.19), 0.01867 (Idealization 2, Fig. 4.20), 0.01808 (Idealization 3, Fig. 4.21)

# Chapter 5

**5.1.**  $\omega_1 = 3.6603$ rad/s, $\omega_2 = 13.6603$ rad/s     **5.3.**  $\omega_1 = \sqrt{k/m}$, $\omega_2 = \sqrt{2k/m}$

**5.5.**  $\omega_1 = 7.3892$ rad/s, $\omega_2 = 58.2701$ rad/s     **5.7.**  $\omega_1 = 0.7654\sqrt{g/l}$, $\omega_2 = 1.8478\sqrt{g/l}$

**5.10.**  $\omega_1 = 12.8817$ rad/s, $\omega_2 = 30.5624$ rad/s

**5.13.**  $x_1(t) = 0.1046 \sin 40.4225t + 0.2719 \sin 58.0175t$,
$x_2(t) = 0.1429 \sin 40.4225t - 0.09952 \sin 58.0175t$

**5.15.**  $\omega_1 = 3.7495\sqrt{\dfrac{EI}{mh^3}}$, $\omega_2 = 9.0524\sqrt{\dfrac{EI}{mh^3}}$

**5.18.**  $\omega_1 = 12.7564$ rad/s, $\omega_2 = 60.1176$ rad/s

**5.20.**  $\omega_1 = 0.5176\sqrt{k_t/J_0}$, $\omega_2 = 1.9319\sqrt{k_t/J_0}$

**5.23.**  $\omega_1 = 0.38197\sqrt{k_t/J_0}$, $\omega_2 = 2.61803\sqrt{k_t/J_0}$

**5.28.**  $\omega_1 = 0.5904\sqrt{k/m}$, $\omega_2 = 1.4668\sqrt{k/m}$

**5.31.**  16.7934 m/s (bounce), 23.8916 m/s (pitch)

**5.37.**  $x_1(t) = 0.009773 \sin 4\pi t$ (meters), $x_2(t) = 0.016148 \sin 4\pi t$ (meters)

**5.39.**  $x_2(t) = \left( \dfrac{1}{60} - \dfrac{1}{40} \cos 10t + \dfrac{1}{120} \cos 10\sqrt{3}\, t \right) u(t)$

**5.42.**  $\omega_1 = 0$, $\omega_2 = 19.3649$ rad/s

**5.48.**  (b) $x_1(t) = 1.44722 \cos 5.86319t + 0.55279 \cos 15.35t$,
$x_2(t) = 2.34165 \cos 5.86319t - 0.34164 \cos 15.35t$

# Chapter 6

**6.2.**  $[k] = \begin{bmatrix} (k_1 + k_2) & -k_2 & 0 \\ -k_2 & (k_2 + k_3) & -k_3 \\ 0 & -k_3 & (k_3 + k_4) \end{bmatrix}$

**6.4.**  $[a] = \dfrac{l^3}{EI} \begin{bmatrix} 9/64 & 1/6 & 13/192 \\ 1/6 & 1/3 & 1/6 \\ 13/192 & 1/6 & 9/64 \end{bmatrix}$

**6.9.**  $2k$     **6.11.**  $x^2 + y^2 + z^2 - l^2 = 0$

**6.14.**  $m_1 \ddot{x}_1 + x_1(k_1 + k_2) - k_2 x_2 = 0$
$m_2 \ddot{x}_2 - k_2 x_1 + x_2(k_2 + k_3) - k_3 x_3 = 0$
$m_3 \ddot{x}_3 - k_3 x_2 + x_3(k_3 + k_4) = 0$

**6.17.**  $\omega_1 = 0.44504\sqrt{k/m}$, $\omega_2 = 1.2471\sqrt{k/m}$, $\omega_3 = 1.8025\sqrt{k/m}$

**6.20.**  $\omega_1 = 0.533399\sqrt{k/m}$, $\omega_2 = 1.122733\sqrt{k/m}$, $\omega_3 = 1.669817\sqrt{k/m}$

**6.23.**  $\lambda_1 = 2.21398$, $\lambda_2 = 4.16929$, $\lambda_3 = 10.6168$

**6.26.** $\omega_1 = 0.644798\sqrt{g/l}$, $\omega_2 = 1.514698\sqrt{g/l}$, $\omega_3 = 2.507977\sqrt{g/l}$

**6.29.** $\omega_1 = 0.562587\sqrt{\dfrac{P}{ml}}$, $\omega_2 = 0.915797\sqrt{\dfrac{P}{ml}}$, $\omega_3 = 1.584767\sqrt{\dfrac{P}{ml}}$

**6.32.** $[X] = \dfrac{1}{2}\begin{bmatrix} 1 & 1 & 0 \\ -1 & 1 & \sqrt{2/3} \\ 1 & 1 & \sqrt{8/3} \end{bmatrix}$

**6.35.** $\omega_1 = 0$, $\omega_2 = 0.752158\sqrt{k/m}$, $\omega_3 = 1.329508\sqrt{k/m}$

**6.38.** (a) $\omega_1 = 0.44497\sqrt{k_t/J_0}$, $\omega_2 = 1.24700\sqrt{k_t/J_0}$, $\omega_3 = 1.80194\sqrt{k_t/J_0}$

(b) $\vec{\theta}(t) = \begin{Bmatrix} -0.0000025 \\ 0.0005190 \\ -0.0505115 \end{Bmatrix} \cos 100t$ radians

**6.46.** $x^3 - 17x^2 + 77x - 98 = 0$

**6.49.** with $\Delta t = 0.1$, $x_1(t = 20\,\Delta t) = 0.2499$, $x_2(t = 20\,\Delta t) = -0.2554$, $x_3(t = 20\,\Delta t) = 0.2742$

# Chapter 7

**7.1.** (a) $\omega_1 \simeq 2.6917\sqrt{\dfrac{EI}{ml^3}}$ (b) $\omega_1 \simeq 2.7994\sqrt{\dfrac{EI}{ml^3}}$

**7.3.** $4.5\sqrt{\dfrac{EI}{ml^3}}$ **7.5.** $1.6357\sqrt{\dfrac{EI}{Ml^3}}$

**7.7.** $0.4082\sqrt{k/m}$ **7.9.** $1.0954\sqrt{\dfrac{T}{lm}}$ **7.11.** $0.5498\sqrt{k/m}$

**7.13.** $\omega_1 = 0$, $\omega_2 \simeq 6.2220$ rad/s, $\omega_3 \simeq 25.7156$ rad/s

**7.16.** $\lambda_1 = 0.5$, $\lambda_2 = 2.0$, where $\lambda_i = \dfrac{\omega_i^2 mh^3}{24EI}$

**7.18.** $\omega_1 = 0.3104$, $\omega_2 = 0.4472$, $\omega_3 = 0.6869$ where $\omega_i = 1/\sqrt{\lambda_i}$

**7.24.** $\omega_1 = 0.2583$, $\omega_2 = 3.0$, $\omega_3 = 7.7417$

**7.30.** $[D] = \begin{bmatrix} 0.7500 & 0.5303 & 0.3062 & 0.4330 \\ 0.5303 & 1.3750 & 0.7939 & 1.1227 \\ 0.3062 & 0.7939 & 1.1250 & 1.5910 \\ 0.4330 & 1.1227 & 1.5910 & 5.2500 \end{bmatrix}$

**7.33.** $\omega_1 = 0.2430$, $\omega_2 = 0.5728$, $\omega_3 = 7.1842$

**7.36.** $\omega_1 = 0.4450$, $\omega_2 = 1.2470$, $\omega_3 = 1.8019$

**7.38.** $\omega_1 = 5.8694$, $\omega_2 = 85.5832$, $\omega_3 = 293.5470$

# Chapter 8

**8.1.** 28.2843 m/s **8.3.** (a) 12000 Hz (b) $\omega_1$ increased by 9.54%

**8.6.** 245.0448 N

**8.9.** $\quad w(x, t) = \dfrac{8al}{\pi^3 c} \displaystyle\sum_{n=1,3,5,\ldots} \dfrac{(-1)^{(n-1)/2}}{n^3} \sin\dfrac{n\pi x}{l} \sin\dfrac{n\pi ct}{l}$

**8.14.** $\quad \omega_n = 50n$ Hz      **8.16.**    $M = 1.1397$ m

**8.19.** $\quad \tan\dfrac{\omega l}{c} + \dfrac{\omega l}{c}\left(\dfrac{GJ}{lk_t}\right) = 0$

**8.22.** $\quad \omega_n = \dfrac{(2n+1)\pi}{2}\sqrt{\dfrac{G}{\rho l^2}} \; ; \; n = 0,1,2,\ldots$

**8.25.**    15908 rad/s      **8.28.**    $\cos\beta l\cosh\beta l = 1$      **8.30.**    $\tan\beta l - \tanh\beta l = 0$

**8.32.**    202.3278 N-m      **8.35.**    $\cos\beta l\cosh\beta l = 1$, and $\tan\beta l - \tanh\beta l = 0$

**8.45.** $\quad \omega_{mn}^2 = \dfrac{\gamma_n P}{\rho}$, where $J_m(\gamma_n R) = 0$; $m = 0,1,2,\ldots$; $n = 1,2,\ldots$

**8.48.** $\quad w(x, y, t) = \dfrac{\dot{w}_0}{\omega_{12}} \sin\dfrac{\pi x}{a} \sin\dfrac{2\pi y}{b} \sin\omega_{12}t$

**8.51.**    $4.4721\sqrt{\dfrac{EI}{\rho Al^4}}$      **8.53.**    $7.7460\sqrt{\dfrac{EI_0}{\rho A_0 l^4}}$      **8.56.**    $2.4146\sqrt{\dfrac{EA_0}{m_0 l^2}}$

**8.61.** $\quad \omega_1 = 2.4062\sqrt{\dfrac{EA_0}{m_0 l^2}}$ , $\omega_2 = 5.5299\sqrt{\dfrac{EA_0}{m_0 l^2}}$

**8.64.**    4.73004, 7.85320, 10.9956, 14.1372, 17.2788

# Chapter 9

**9.1.**    3.2977 oz; $\alpha = 143.9038°$ CW

**9.3.**    $\vec{B}_L = 4.5949 \big/ 102.1849°$ oz, $\vec{B}_R = 3.3292 \big/ 144.3574°$ oz

**9.5.**    1.6762 oz; $\alpha = 75.6261°$ CW      **9.7.**    $d_D = 1.7060$ cm, $\theta_D = -55.0720°$

**9.9.**    662.88 rpm      **9.12.**    left plane: $m/3$, down; right plane: $m/3$, up

**9.14.**    (a) $\omega < $ rpm    (b) $\omega > 276.7803$ rpm      **9.16.**    2.385 mm

**9.18.**    add a mass of 15.9527 kg to instrument

**9.20.**    $m_2 = 10.2084$ kg, $k_2 = 1.7912$ MN/m

**9.22.**    $k_2 = 0.7896$ MN/m; $\Omega_1 = 683.4$ cpm, $\Omega_2 = 1404.6$ cpm

**9.25.**    $0.9764 \le \left(\dfrac{\omega}{\omega_2}\right) \le 1.05125$

**9.27.**    $m_2 = 102.09$ kg, $k_2 = 2.5185$ MN/m, $X_2 = 0.1959$ mm

# Chapter 10

**10.2.** $\quad \left.\dfrac{d^4x}{dt^4}\right|_i = \dfrac{x_i - 4x_{i-1} + 6x_{i-2} - 4x_{i-3} + x_{i-4}}{(\Delta t)^4}$

**10.4.**    $x(t = 5) = -1$ with $\Delta t = 1$ and $-0.9733$ with $\Delta t = 0.5$

**10.6.**    $x_{10} = -0.0843078$, $x_{15} = 0.00849639$

**10.9.** $x(t = 0.1) = 0.131173$, $x(t = 0.4) = -0.0215287$, $x(t = 0.8) = -0.0676142$

**10.14.** With $\Delta t = 0.07854$, $x_1 = x$ and $x_2 = \dot{x}$, $x_1(t = 0.2356) = 0.100111$,
$x_2(t = 0.2356) = 0.401132$, $x_1(t = 1.5708) = 1.040726$, $x_2 = (t = 1.5708) = -0.378066$

**10.20.**

| $t$ | $x_1$ | $x_2$ |
|------|---------|---------|
| 0.25 | 0.07813 | 1.1860 |
| 1.25 | 2.3360 | -0.2832 |
| 3.25 | -0.6363 | 2.3370 |

**10.22.** $\omega_1 = 3.06147\sqrt{\dfrac{E}{\rho l^2}}$ , $\omega_2 = 5.65685\sqrt{\dfrac{E}{\rho l^2}}$ , $\omega_3 = 7.39103\sqrt{\dfrac{E}{\rho l^2}}$

**10.25.** $\omega_1 = 17.9274\sqrt{\dfrac{EI}{\rho A l^4}}$ , $\omega_2 = 39.1918\sqrt{\dfrac{EI}{\rho A l^4}}$ , $\omega_3 = 57.1193\sqrt{\dfrac{EI}{\rho A l^4}}$

**10.28.** With $\Delta t = 0.24216267$,

| $t$ | $x_1$ | $x_2$ |
|--------|---------|--------|
| 0.2422 | 0.0000 | 0.1466 |
| 2.4216 | 0.8691 | 1.9430 |
| 4.1168 | -0.1154 | 0.6498 |

**10.35.** With $\Delta t = 0.24216267$,

| $t$ | $x_1$ | $x_2$ |
|--------|---------|--------|
| 0.2422 | 0.01776 | 0.1335 |
| 2.4216 | 0.7330 | 1.8020 |
| 4.1168 | 0.1059 | 0.8573 |

# Chapter 11

**11.1.** (a) $x_1 = \sqrt{\dfrac{m}{k_1}}\,\dot{x}_0$ (b) $\tau_n = \pi\left(\sqrt{\dfrac{m}{k_1}} + \sqrt{\dfrac{m}{k_2}}\right)$

**11.3.** $\tau = 4\sqrt{\dfrac{l}{g}}\displaystyle\int_0^{\pi/2}\dfrac{d\phi}{\sqrt{1 - k^2\sin^2\phi}}$ , where $k = \sin(\theta_0/2)$

**11.5.** equilibrium position is $(0,0)$ **11.7.** $m\ddot{x} + \dfrac{2P}{l_0}x + \dfrac{AE}{l_0^3}x^3 = 0$

**11.10.** $\lambda_1 = 2.382$, $\lambda_2 = 4.618$ **11.11.** $E[x] = 1.0$, $E[x^2] = 1.3333$, $\sigma_x^2 = 0.3333$

**11.16.** RMS value $= 0.3146$ m, $\sigma = 0.3106$ m

**11.18.** For gravel surface: $S_X(\omega) =$

$$0.51566v\left[\dfrac{\omega^2 + 1.06v^2}{\left\{0.25v^2 + (\omega - 0.9v)^2\right\}\left\{0.25v^2 + (\omega + 0.9v)^2\right\}}\right]$$

**11.24.** $\omega_1 = 3.5327 \sqrt{\dfrac{EI}{\rho A l^4}}$ , $\omega_2 = 34.8069 \sqrt{\dfrac{EI}{\rho A l^4}}$

**11.26.** $\omega_1 = 20.4939 \sqrt{\dfrac{EI}{\rho A l^4}}$

**11.29.** $\omega_1 = 1.9839 \sqrt{\dfrac{GJ}{\rho I_p l^2}}$ , $\omega_2 = 5.1584 \sqrt{\dfrac{GJ}{\rho I_p l^2}}$

**11.31.** $[k] = \dfrac{EA_0}{l}(0.6321) \begin{bmatrix} 1 & -1 \\ -1 & 1 \end{bmatrix}$

**11.34.**

| $t$ | $x$ | $\dot{x}$ |
|-----|-----|-----|
| 0.1 | 0.00393 | 0.07788 |
| 1.0 | 0.30945 | 0.51051 |
| 5.0 | −0.24925 | −0.48372 |

**11.36.** $\tau = 8.7627392$

**11.39.** $\omega_1 = 1654.5$ rad/s, $\omega_2 = 3984.2$ rad/s, $\omega_3 = 10807.4$ rad/s, $\omega_4 = 17060.5$ rad/s

# Name Index

# Subject Index

# Spring Constants

**1.** $N$ springs in series

$$k = \frac{1}{\dfrac{1}{k_1} + \dfrac{1}{k_2} + \cdots + \dfrac{1}{k_N}}$$

**2.** $N$ axial springs in parallel (end bars do not tilt)

$$k = k_1 + k_2 + \cdots + k_N$$

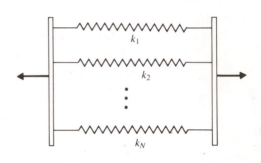

**3.** Uniform bar under axial load

$$k = \frac{EA}{l}$$

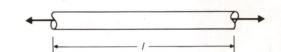

**4.** Tapered bar under axial load

$$k = \frac{\pi E D_1 D_2}{4l}$$

$D_1, D_2$ = end diameters of the bar

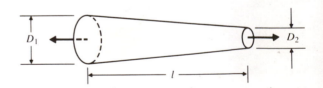

**5.** Helical spring under axial load

$$k = \frac{Gd^4}{64 n R^3}$$

$n$ = number of active turns
$d$ = wire diameter
$l \gg R$

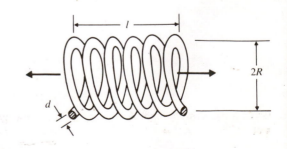